T0240691

# Fundamentals of Modern Electric Circuit Analysis and Filter Synthesis

Afshin Izadian

# Fundamentals of Modern Electric Circuit Analysis and Filter Synthesis

A Transfer Function Approach

Second Edition

 Springer

Afshin Izadian
Associate Professor
Purdue University
Indianapolis, IN, USA

ISBN 978-3-031-21910-8     ISBN 978-3-031-21908-5   (eBook)
https://doi.org/10.1007/978-3-031-21908-5

This Springer imprint is published by the registered company Springer Nature Switzerland AG
The registered company address is: Gewerbestrasse 11, 6330 Cham, Switzerland

*To*
*Pardis, Anahita, and Atousa*

# Preface

In the second edition, the content of each chapter has been enhanced with numerous solved problems. Some topics have been added to cover most of the material taught in electric circuits for undergraduate students. The end-of-chapter problems have been advanced to provide a guided study for students. However, each chapter's depth of solved problems lays out the foundation expected for most undergraduate-level circuit requirements. The focus on filters and their transfer functions has advanced to understand the subjects better.

The book can cover two semesters of electric circuits. The first course may include Chaps. 1, 2, 3, 4, 5, and 6, emphasizing electric circuits' foundations. The second part, Chaps. 7, 8, 9, 10, 11, and 12, covers the analysis of electric circuits in the frequency domain with the transfer function approach.

Chapter 1 covers electric circuits' basics by introducing some measurement units and their conversions. Circuit elements and their schematics are introduced, and several examples of electric circuits are provided. Students familiarize themselves with the units used in the book and understand the importance and application of electric circuits.

Chapter 2 covers the circuit elements, OHM's law, and the voltage-current relations in each electric circuit element. Resistors, inductors, and capacitors can be connected in various ways. Series, parallel, delta, and star and their equivalent models are covered for each circuit element. However, analysis of RL, RC, and RLC is introduced in Chapters 4 and 5. Several switching elements are introduced with their detailed operations. Mechanical and electrical switches and their applications are discussed. Dependent and independent sources are introduced. Linear time-invariant circuits and superposition are introduced through several examples.

Chapter 3 covers fundamental waveforms, their analysis, and mathematical expressions. Sinusoidal waveform, phasor, and rectangular-polar coordinations are introduced. The application of phasors in circuit analysis is introduced in Chapter 5.

Chapter 4 covers the concept of the order of a circuit. The first-order circuits and second-order circuits are introduced as an introduction to time-based circuit analysis. Several examples are introduced to help students classify the circuits and predict their behavior. Natural response and forced response are discussed. Students also learn to obtain the circuit parameters by analyzing the time-based responses.

Chapter 5 covers the basics of sinusoidal AC circuit analysis in LTI. This starts with a coverage of complex math calculations. The foundations of phase shift, lead, and lag circuits are discussed. The idea of reactance is introduced, and several circuits are analyzed. Accordingly, the power factor and resonance are introduced through several examples. The power calculations in AC circuits are introduced and covered in-depth. This section also includes the Thevenin-Norton equivalent for AC circuits. The maximum power transfer for the AC circuits is also introduced.

Chapter 6 covers the fundamentals of mutual inductance and their equivalent circuits. The concept of induced voltage through mutual inductance is introduced. Accordingly, transformers are introduced, and their power, voltage, and current are analyzed.

Chapter 7 covers the fundamentals of the Laplace transformation and its application in circuit analysis. Basic mathematical analysis of the Laplace transformations and operations are also introduced for signals and waveforms. They are identified as the input and output of the circuits to prepare for the transfer function analysis. The frequency domain analysis of simple circuits with various inputs is introduced.

Chapter 8 covers the transfer functions, their definition, various forms of system combination, and feedback systems. The transfer function is obtained for single-loop systems and systems with multiple inputs and multiple outputs. The initial and final value theorems are introduced, and the systems' steady-state responses are analyzed accordingly. First-order system transfer functions second order systems are analyzed in the frequency domain. The frequency response of the systems is analyzed, and their Bode diagrams are obtained. This chapter also covers the state space representation of the systems.

Chapter 9 covers the design and analysis of passive filters. Various filters are introduced and designed in the form of hardware (circuits) and software (computer code). Filter transfer functions are analyzed, and their basic operations are improved to design better filters. A free-style software design that matches the filter circuit transfer functions is introduced. RLC resonant circuit is used for this purpose to form filters. Butterworth filters are designed to improve their operation. Higher-order filters are also designed to obtain the desired performance.

Chapter 10 covers the principles of operational amplifiers. Several design principles are covered, and simple circuits are introduced to help students understand the approach to analyzing the Opamp circuits. Opamps are also utilized to solve differential equations. This chapter introduces the design of analog computers and PID controllers.

Chapter 11 covers the active filters in various forms. Simple circuits are designed with the approach to control the feedback impedance or the input impedance. Filter hardware is analyzed, and the transfer functions are obtained. Non-resonant-based filters are introduced, and their circuits are analyzed. Multi FeedBack (MFB) circuits are introduced, and methods to obtain various filters are discussed.

Chapter 12 covers two-port networks. Several representation methods in impedance, admittance, and transmission matrices are introduced. These matrices are often obtained through several methods explained in numerous examples.

The new edition of this textbook is intended to be useful for all engineering and other majors. Thanks to Ms. Han Shi, who tirelessly helped prepare the figures. The content of this book is constantly in improvement. Your feedback enriches the topics and sheds light on shortcomings. Therefore, please forward your suggestions to improve the book to my institutional email address or Springer.

Sincerely

Indianapolis, IN, USA                                                      Afshin Izadian

# Contents

# Chapter 1
# Introduction to Electric Circuits

## Introduction

A flashlight has batteries to provide power to a light bulb, wires to bring the power
from batteries to the bulb, and a switch to control the on-off action. The operation of
this flashlight can be modeled through an electric circuit with schematics allocated to
each of the circuit components. Of course, it has mechanical components to hold the
batteries, wire insulator, switch, and bulb in a container to hold the system structure
in place. The wires and the switch have electric insulators to protect the flow of
electric current in the wires. Nevertheless, these components have no electrical
importance when it comes to the distribution of the power and modeling of the
circuit. In an electric system circuit, parts and components that conduct the electric
current and mechanical actions that show the current flow are shown.

In some cases, currents that leak through insulators might be interesting and can
also be modeled. However, in a simple form, Fig. 1.1 shows a flashlight with its
internal components and equivalent circuit. The batteries are modeled by their circuit
schematic, conductors' wires, mechanical switch that allows current to flow or cut at
the desired time, and the bulb modeled as a resistor.

Another example of electric circuits is in the modeling of electromechanical
devices. An electric motor has rotating mechanical parts and some electric circuits
that have to build an electromagnetic field. Coils are wrapped around a magnetic
material core and receive electric current to build magnetic fields. This is a pure
electric circuit. However, the magnetic field enters the rotor circuit and pulls it to
rotate. This is an electromagnetic behavior but can be modeled as an electric circuit.
The rotor speeds up and generates a voltage interacting with the main source. This
mutual reaction is an electromagnetic phenomenon but is modeled in electric
circuits. The motor's electric circuit requires specific circuit elements whose values
change as the motor takes a mechanical load and its speed varies. The circuit that can
connect electric, electromagnetic, and mechanical properties is called *an equivalent
circuit*. Figure 1.2 shows a wind turbine system utilizing fluid or hydraulic systems

© The Author(s), under exclusive license to Springer Nature Switzerland AG 2023
A. Izadian, *Fundamentals of Modern Electric Circuit Analysis and Filter Synthesis*,
https://doi.org/10.1007/978-3-031-21908-5_1

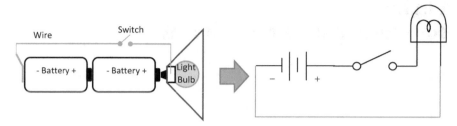

**Fig. 1.1** A flashlight and its equivalent circuit. The battery, wire, switch, and bulb are shown in circuit schematics. There is no need to model the nonelectric parts of the circuit unless they participate in an active role in the circuit operation

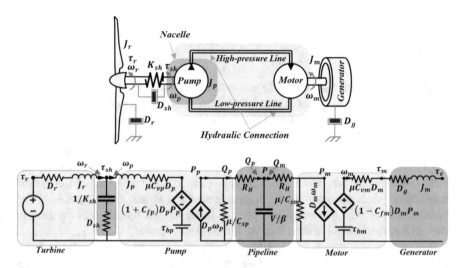

**Fig. 1.2** Equivalent electric circuit of a wind turbine with the hydraulic drivetrain. Each part of the hydraulic (fluid power) circuit has its electric equivalent

to transfer power from the turbine to the generator. The system is purely mechanical but can also be presented as an electric circuit. The equivalent circuit of each part is also shown in Fig. 1.2.

Electric power distributions are a great example of electric circuits and their application. Electric power is generated from various sources that can be modeled as electric circuits and brought to the users through distribution lines, and the voltages may require stepping up or down. Electric circuits and their theories can provide models for electrical components such as flashlights, power distribution systems, electric motors, generators, transformers, integrated circuits, transistors, cell phones, computers, and many more (Fig. 1.3).

Solar cells generate electric power by converting the sunray photons into electric current. However, this physical phenomenon can also be modeled as an electric circuit. Solar cells can be modeled in several ways, each showing specific characteristics observed when the cell is utilized in a specific application. The equivalent

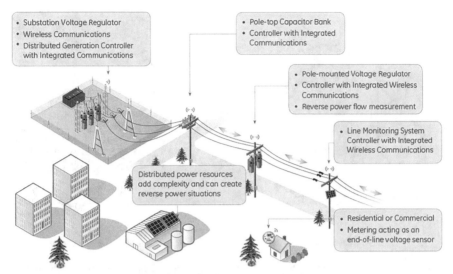

GE's Portfolio of Volt-Var Control Solutions

**Fig. 1.3** Electric power distribution GE concept. (Image Courtesy of GE © 2022)

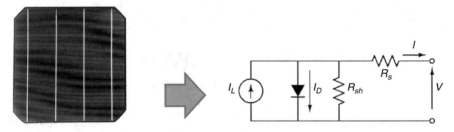

**Fig. 1.4** A single solar cell and electric circuit equivalent of the cell. A single diode with series and parallel resistance demonstrates various operations of the cell

electric circuits are utilized in a way to facilitate further analysis of the systems that are powered by solar cells. They convert the energy of the photons in sunrays into electric currents. A solar cell and its electric equivalent used in power systems and control systems are shown in Fig. 1.4.

## Electric Circuit Topologies

Consider a complicated system with many components and a complex network of wires to connect them. These components and their connecting wires might fit into a 3D space, meaning that there are wires that come out of the surface to connect components that themselves might be crossing other elements. Figure 1.5 shows a

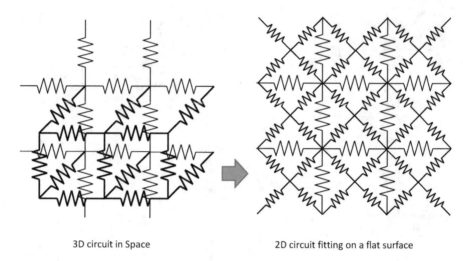

3D circuit in Space          2D circuit fitting on a flat surface

**Fig. 1.5** A 3D circuit and an acceptable flat version of the circuit. Resistors and wire connectors do not pass over each other

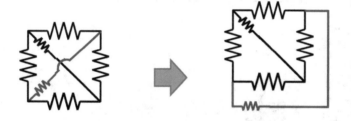

Unacceptable: crossing of elements       Acceptable: spread out version of the same circuit

**Fig. 1.6** A 2D circuit with two elements or the wires crossing should be presented in a topology with no wires/elements crossing. A separate path on the surface can be found to prevent crossing

3D circuit. However, the type of circuit studied in this book must fit only in a 2D plane, and no wire or element should pass any other wire or element.

An unacceptable example of a circuit with components and wires crossing each other and its acceptable version is shown in Fig. 1.6.

## Hinged Circuits

If an electric circuit is such that it splits into two halves such that the connecting point does not have any current passing in any direction toward the two parts, the circuit is called electrically hinged. This makes the two parts of the circuit essentially

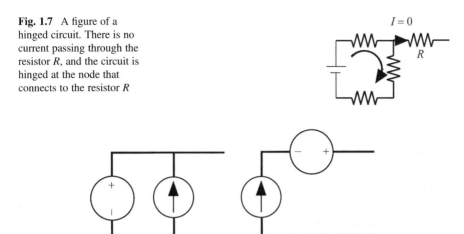

**Fig. 1.7** A figure of a hinged circuit. There is no current passing through the resistor $R$, and the circuit is hinged at the node that connects to the resistor $R$

**Fig. 1.8** The current source in the left circuit is redundant, and the circuit is hinged. The voltage source in the right circuit is redundant, and the circuit is hinged. These circuits are discussed in more detail in the later chapters

independent unless there is a magnetic coupling or other forms of couplings that influence their interdependency (Fig. 1.7).

Hinged circuits may also occur due to voltage and current source configurations in the circuit. This is when a voltage source feeds a current source in series, and this redundancy may cause a hinged circuit (Fig. 1.8).

## *Measurement Units*

Throughout the world, quantities are measured in various units. For instance, the speed in the USA is measured in miles per hour or MPH, but in European countries, it is measured in kilometers per hour KPH. Generally, there are "imperial units" of measurement or BU and "International Standard of Units" or SI. The quantities in these measurement units refer to different values. In this book, the SI unit is used to measure the quantities as follows:

- Length (**L**) measured in meters (**m**)
- Mass (**M**) measured in kilograms (**kg**)
- Time (**T**) measured in seconds (**s**)
- Electric current (**I**) measured in amperes (**A**)
- Electric charge (**q**) measured in Coulombs (**C**)

The quantities to be measured might vary from very small to very large. The scales are useful for presenting these quantities. A list of these numbers is as follows:

## Scales and Units

- 1 Zetta $= 10^{21} = 1e21$
- 1 Exa $= 10^{18} = 1e18$
- 1 Peta $= 10^{15} = 1e15$
- 1 Tera $= 10^{12} = 1e12$
- 1 Giga $= 10^{9} = 1e9$
- 1 Mega $= 10^{6} = 1e6$
- 1 Kilo $= 10^{3} = 1e3$
- 1 Hecto $= 10^{2} = 1e2$
- 1 Deca $= 10 = 1e1$
- 1 Deci $= \frac{1}{10^{1}} = 1e-1$
- 1 Centi $= \frac{1}{10^{2}} = 1e-2$
- 1 Mili $= \frac{1}{10^{3}} = 1e-3$
- 1 Micro $= \frac{1}{10^{6}} = 1e-6$
- 1 Nano $= \frac{1}{10^{9}} = 1e-9$
- 1 Pico $= \frac{1}{10^{12}} = 1e-12$
- 1 Femto $= \frac{1}{10^{15}} = 1e-15$
- 1 Atto $= \frac{1}{10^{18}} = 1e-18$
- 1 Zepto $= \frac{1}{10^{21}} = 1e-21$

**Example 1.1 The current passing through a wire is $1.2e-3\ A$. Show the current in mA.**
*Solution.* The current is $1.2e-3$, and $1e-3$ equals 1 mA. Therefore, the current $1.2e-3\ A = 1.2$ mA.

**Example 1.2 The capacitance of a capacitor is 21 nF. Show the capacitance in PF.**
*Solution.* Each nF equals 1000 PF. Therefore, $21 \times 1000$ PF $= 21{,}000$ PF.

**Example 1.3 The power generation of a power plant is 31 GW. Show the power generation in MW.**
*Solution.* Each GW equals 1000 MW. Therefore, 31 GW becomes $31 \times 1000$ MW or $31{,}000$ MW.

## *Most Common Electric Circuit Symbols*

| Schematic | Circuit element | Symbol | Unit of measurement |
|---|---|---|---|
| | Resistor | $R$ | Ohm, $\Omega$ |
| | Variable resistor | $R$ | Ohm, $\Omega$ |
| | Capacitor | $C$ | Farad, F |
| | Variable capacitor | $C$ | Farad, F |
| | Inductor | $L$ | Henry, H |
| | Variable inductor | $L$ | Henry, H |
| | Ground, Earth | – | |
| | Single-cell battery | $E$ | Volt, V |

(continued)

| Schematic | Circuit element | Symbol | Unit of measurement |
|---|---|---|---|
| | Multi-cell battery | $E$ | Volt, V |
| | Voltage source | $V$ | Volt, V |
| | Current source | $I$ | Ampere, A |
| | Transformers or mutual inductance | $M$ | Henry, H |

## Problems

1.1. Find examples of electric circuits, and explain how an electric circuit can help understand them better.

1.2. Find examples of mechanical/fluid power systems, and explain how an equivalent electric circuit can help understand them better.

1.3. Find examples of electromechanical systems, and explain how an equivalent electric circuit can help understand them better.

1.4. Find examples of chemical systems, and explain how an equivalent electric circuit can help understand them better.

1.5. The distance between the two points is 100 kilometers. How much is the distance in meters? In centimeters? and millimeters?

1.6. How many watts are in 1.5 megawatts of power?

1.7. How many kilowatts is in 1.5 Megawatts of power?

1.8. How many Farads is in 12 picofarad of capacitance?

1.9. How many millifarads are in 150 nanofarads of capacitance?

1.10. A battery is designed to provide 0.4 kilovolts of voltage. How many volts is this voltage?

1.11. A transformer is designed to provide a full load current of 15.6 kiloamps of current. How many amps is this current?

1.12. A cellphone battery is rated at 5200 mili-amp-hours (mAh). How many amp-hours is this battery rating? How long this battery lasts 15 mA current is drawn from it.

# Chapter 2
# Circuit Components, Voltage, and Current Laws

## Introduction

Electric circuit analysis is the collection of methods and tools to determine the voltages and currents and the power consumption and generation in electric circuits and components. The electric elements' relations depend on the circuit's elements and their configuration or topology. That is how the circuit elements are connected.

This chapter introduces the circuit elements, and the laws determining the relation of voltages and currents in various elements are studied.

## Definition of Voltage

Consider different electric charges on two points of an object. The voltage is defined as the charge difference between these two points. Considering potential energy at these two points, any difference in the potential energy applies forces on electrons to be displaced. The difference in these potential energies is known as voltage, and the displacement of electrons results in electric current flow. For instance, chemical reactions can generate different potentials on the anode and cathode of a battery. The difference in these potential energies can be 1.3 or 1.5 V, which leads to 1.3 or 1.5 V batteries (Fig. 2.1).

Definition: Electric potential or voltage, which is measured in units of volts, is precisely defined as

$$1 \text{ Volt} = \frac{1 \text{ Joul of potential energy}}{1 \text{ Coulomb of electric charge}}$$

A. Izadian, *Fundamentals of Modern Electric Circuit Analysis and Filter Synthesis*, https://doi.org/10.1007/978-3-031-21908-5_2

**Fig. 2.1** The potentials $V_1$ and $V_2$ are shown on an object. In this example, the difference between these potential energies results in a voltage equivalent to 1.5 V

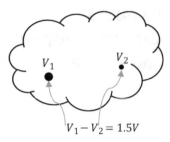

This means that 1 V of electric potential energy is generated if it requires the equivalent of 1 J of work to move 1 coulomb of electrons away from an oppositely charged place. One coulomb of electron equals $6.24 \times 10^{18}$ electrons.

## Definition of Current

The difference in potential energy between two points in a circuit creates a voltage across the points. This means that potential energy exists that can force the electrons to move. This movement is from the higher potential to the lower potential points. Providing a path through a conductor allows for electrons to pass. The current is the number of electrons passing through a cross-section of the conductor per unit of time, 1 s. The current of 1-ampere $i = 1$ A equals the charge of $q = 1$ C coulomb passing a conductor in $t = 1$ s.

$$i = \frac{dq}{dt} (A)$$

The current density $J$ is defined as the amount of current $i$ per cross-section area of the conductor $A$ and is measured in A/m$^2$ as

$$J = \frac{i}{A} \left(\frac{A}{m^2}\right)$$

## Resistor

A resistor is an element of electric circuits that limits current flow. A resistor, shown by the symbol $R$, limits the current in several ways. Consider a wire made with a material that has specific resistance $\rho$ against electric current. The physical dimensions of this wire, such as length $l$ and cross-section area $A$, influence the number of passing electrons and, consequently, its resistance. If the length of the wire is long, the resistance increases. If the cross-section area is large, the resistance is less as it provides more room for the electrons to pass. Therefore, the resistance of that piece

**Fig. 2.2** Pieces of
conductors with length *l* and
cross-section *A*

of material which is measured in ohms $\Omega$ (also shown as $R(\Omega)$) can be calculated as
(Fig. 2.2)

$$R = \rho \frac{l}{A} \quad (\Omega)$$

The units of these elements are listed as

$$R(\Omega), \rho(\Omega.m), l(m), A(m^2)$$

**Example 2.1 A piece of material with specific resistance $\rho = 3000e - 6\ \Omega.$ m has a length of $l = 15$ cm and a cross-section of $A = 1$ mm². Find the resistance of the material.**
*Solution.*

$$R = \frac{\rho l}{A} = 3000e - 6 \frac{15 \times 1e - 2(cm \rightarrow m)}{1 \times 1e - 6(mm^2 \rightarrow m^2)} = 450\ \Omega$$

**Example 2.2 A piece of material is formed into a round cross-section wire. How much does the resistance change if the radius is cut in half and the length is increased by 50%?**
*Solution.*

The length is increased by 50% means $l_2 = 1.5l_1$. The area is $A = \pi r^2$. This
results in $\frac{A_1}{A_2} = \frac{\pi r_1^2}{\pi r_2^2}$ since $r_2 = \frac{1}{2}r_1$, then $\frac{A_1}{A_2} = 4$ or $A_2 = \frac{1}{4}A_1$. The material of the
conductor has not changed; hence, $\rho_1 = \rho_2$.

Therefore,

$$\frac{R_1}{R_2} = \frac{\rho_1}{\rho_2} \frac{l_1}{l_2} \frac{A_2}{A_1} = \frac{l_1}{1.5l_1} \frac{\frac{1}{4}A_1}{A_1} = \frac{1}{4 \times 1.5} = \frac{1}{6}$$

$$R_2 = 6R_1$$

This means that the resistance is six times higher.

## Conductors, Insulators, and Semiconductors

Specific resistance value depends on the type of material. A low specific resistance value makes materials conductors, and high specific resistance makes other materials insulators. Examples of conductors are aluminum ($\rho = 2.8e - 8$ Ω. m), brass ($\rho = 6e - 8$ Ω. m), and iron ($\rho = 9.8e - 8$ Ω. m). Examples of insulators are amber ($\rho = 5e14$ Ω. m), rubber ($\rho = 1e16$ Ω. m), and glass ($\rho = 1e11$ Ω. m).

There is a third class of material, "semiconductors," in which their specific resistance is not fixed; rather, it is influenced by external stimulators such as electric charge accumulation and light excitation.

## Effect of Temperature on Resistance

In *conductors*, the resistance value is increased linearly by increasing the temperature. As the material's temperature changes, its specific resistance changes, which changes the material's resistance. Each material has a changing slope in specific resistance known as temperature coefficient $\alpha$. Consider the resistance of material at 0 °C, known as $R_0$. The amount of resistance at temperature $t$ or $R_t$ is

$$R_t = R_0(1 + \alpha t)$$

The amount of resistance due to a temperature change from $t_0 \rightarrow t$ can be found as

$$R_t = R_{t_0}(1 + \alpha(t - t_0))$$

In *insulators* and *electrolytes*, the resistance value decreases linearly by increasing the temperature, and these materials show a negative value temperature coefficient.

If the temperature is decreased, the resistance value will decrease in conductors and increase in insulators and electrolytes. At very low temperatures, the resistance is saturated to a very low value, at which conductors become superconductors.

A list of materials with their specific resistances and temperature coefficients is provided in the following table. This includes conductors and insulators.

| Material | $\rho$ (Ω. m) | $\alpha/$° C |
|---|---|---|
| Silver | $1.59 \times 10^{-8}$ | 0.0038 |
| Copper | $1.68–1.72 \times 10^{-8}$ | 0.00386 |
| Aluminum | $2.65 \times 10^{-8}$ | 0.00429 |
| Tungsten | $5.6 \times 10^{-8}$ | 0.0045 |
| Iron | $9.71 \times 10^{-8}$ | 0.00651 |
| Platinum | $10.6 \times 10^{-8}$ | 0.003927 |
| Manganin | $48.2 \times 10^{-8}$ | 0.000002 |

(continued)

| Material | $\rho$ ($\Omega.$ m) | $\alpha$/ ° C |
|----------|----------------------|--------------|
| Mercury | $98 \times 10^{-8}$ | 0.0009 |
| Constantan | $49 \times 10^{-8}$ | |
| Carbon | $3000-5000 \times 10^{-5}$ | -0.0005 |
| Germanium | $1-500 \times 10^{-3}$ | -0.05 |
| Silicon | 0.1-60 | -0.07 |
| Glass | $1-10000 \times 10^{9}$ | |
| Amber | $5 \times 10^{14}$ | |
| Quartz | $7.5 \times 10^{17}$ | |
| Rubber | $1 \times 10^{16}$ | |

**Example 2.3 The resistance of a silver resistor is measured to be 100 $\Omega$ at 20 ° C. Find the resistance of the resistor at 80 ° C.**
*Solution.* The silver has a temperature coefficient $\alpha = 0.0038$, according to the table. The resistance at a different temperature is obtained as follows:

$$R_t = R_0(1 + \alpha t)$$

$$R_t = 100(1 + 0.0038(80 - 20)) = 122.28 \ \Omega$$

## Conductance

The ability of a material to pass electric current is known as conductance, shown by the symbol $G$ and measured by various units such as siemens, MHO, or $\Omega^{-1}$. There is a reciprocal relationship between resistance and conductance, and they relate as follows:

$$G = \frac{1}{R} \ (\Omega^{-1})$$

## Series Connection of Circuit Elements

Elements are connected in series if the entire current exiting from one element enters another connected to the same node. The connection is still in series if more than one element is connected to the same node if the simplification of all elements constitutes two groups of elements connected in tandem (Fig. 2.3).

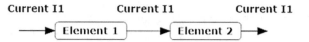

**Fig. 2.3** Two circuit elements in series connection. The same current passes through the elements

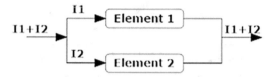

**Fig. 2.4** Two circuit elements are in parallel connection, and the current is shared between the elements but shares the same voltage across their terminals

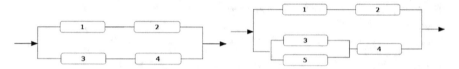

**Fig. 2.5** A combination of series and parallel connections. In the left circuit, elements 1 and 2 are in series, 3 and 4 in series, and the entire 1–2 and 3–4 in parallel. It may read (1 + 2)||(3 + 4). In the circuit to the right, elements 1 and 2 are in series, and the parallel combination of 3 and 5 is in series with 4. The 1–2 is parallel to the series of 4 and 3 and 5 combined. It may read (1 + 2)||((3||5) + 4)

## Parallel Connection of Circuit Elements

Elements are connected in parallel if they share similar nodes at both ends. Elements in parallel connections share the same voltage across themselves. The parallel connection provides a path to share the current (Fig. 2.4).

## Mixed Connection of Circuit Elements

Series and parallel connections may be combined. In this case, special attention should be paid to identifying the elements connected in tandem and those with similar ending nodes. This may be a group of elements being in parallel connected in series to another group (Figs. 2.5 and 2.6).

There might be a connection that does not fit into series or parallel forms. These forms can be known as the Y connection (star connection) on the left and Δ connection (delta connection) on the right circuit. The solution to these circuits is the Kirchhoff voltage law (KVL) or Kirchhoff current law (KCL).

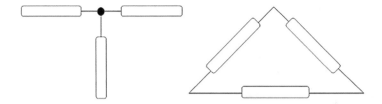

**Fig. 2.6** These elements are not series or parallel, forming a star and delta connection

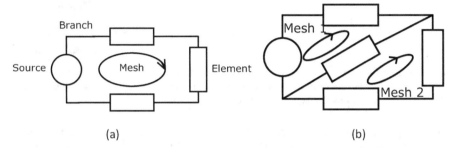

**Fig. 2.7** A mesh is formed by connecting circuit elements in a closed circuit that forms a loop. (**a**) A single-loop circuit (mesh) and (**b**) a multi-loop circuit (mesh 1 and mesh 2). Note that the circuit is planar, and elements do not cross each other to connect nodes. Elements might be shared in two or more loops, but each element is considered once in each loop

## *Mesh*

Consider an electric circuit with a path to carry current from the source or elements to other elements. These paths are also called branches. A closed path that starts and ends at the same element and contains no branches inside is called a mesh (Fig. 2.7).

## *Node*

A node is a circuit point connecting at least two elements. The elements connected to the node take a share of the current entering the node but have the same voltage (potential) at the shared node (Fig. 2.8).

## Ohm's Law

Consider an object with total resistance or $R$ ($\Omega$) (read $R$ ohms). This object, when connected to a voltage source at voltage $V$ (Volts) (e.g., battery) between its ports, passes a certain amount of current $I$ (A). The current passing through the element is

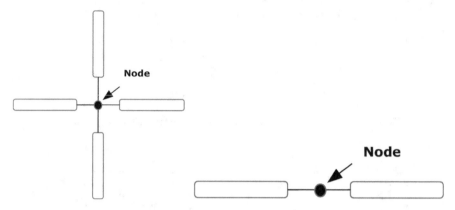

**Fig. 2.8** Node is formed when at least two elements are connected. The voltages of the nodes are important as they determine how much current passes through each element. The more potential node differences, the more current passes through the elements

directly proportional to the applied voltage. The voltage and current increase ratio are always constant, which equals the element's resistance. The Ohm's law equation is expressed as

$$\frac{V}{I} = R = \text{constant}$$

This equation also shows that the voltage drop across a resistor equals the resistance times of the current passing through the resistor. It should be noted that the positive polarity of voltage drop is always at the terminal that receives the current. If the current direction changes, the voltage drop's polarity also changes.

**Example 2.4 The current through a 10 Ω resistor is 1.5 A. Find the voltage drop measured across the resistor. What is the voltage if the voltmeter probes are reversely connected to the terminals of the resistor?**
*Solution.* $V = RI$ with $R = 10$ and $I = 1.5$ leads to the voltage as follows:

$$V = 10 \times 1.5 = 15 \text{ V}$$

Suppose the probes are reverse connected (meaning that the positive probe is connected to the negative polarity and the negative probe is connected to the positive polarity). In that case, the voltmeter measures the voltage drop from the negative terminal with respect to the positive terminal. It can also be interpreted as a reverse direction of the current. Therefore, the voltage drop is measured as

$$V = 10 \times (-1.5) = -15 \text{ V}$$

**Example 2.5 A 1 kΩ resistor shows a voltage drop of 2 V. Find the current passing the resistor.**
*Solution.*

$$V = RI \rightarrow I = \frac{V}{R} = \frac{2}{1e3} = 2 \text{ mA}$$

**Example 2.6 A resistor is required to drain 10 A current from a 200 V node to the ground. Find a suitable resistor value.**
*Solution.*

$$V = RI \rightarrow R = \frac{V}{I} = \frac{200}{10} = 20 \ \Omega$$

## Kirchhoff Voltage Law (KVL)

Consider a circuit with multiple loops and nodes. KVL theory indicates zero summation of voltage drops across elements in a loop, and this involves all elements and is true in all meshes that either stand alone or share some branches with their adjacent loops. In the case multiple meshes exist in a circuit, the summation of voltage drops in all individual meshes is independent and must be zero.

**Example 2.7 Consider the circuit shown in Fig. 2.9. Write the KVL and find the loop's current.**

*Solution.* The KVL suggests considering a direction for the current and following the flow of current, writing the voltage drop of elements around the loop while paying attention to the polarity of the voltage drop across the element. The algebraic summation of these voltages in a closed loop must be zero. Now, the loop has a current $I$ with the direction shown. Starting from any element, let us start with the negative terminal of the source, and following the direction of the current, the voltage drops across elements are as follows:

- Across the voltage source: $-V$ (negative, because the current enters the negative terminal of the source)
- Across the element 1: $+V_1$ (positive, because the current entering any passive element generates *a* positive (+) polarity voltage drop)

(continued)

**Example 2.7** (continued)

- Across the element 2: $+V_2$
- Across the element 3: $+V_3$

    Adding these voltages in KVL results in the following:

$$-V + V_1 + V_2 + V_3 = 0$$

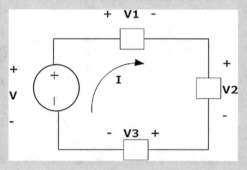

**Fig. 2.9** The circuit of Example 2.7

**Example 2.8 Consider the circuit of Fig. 2.10 with a voltage source $V_{dc}$ (battery) and a resistor load $R$ connected in series. The source forces a current in the circuit through the resistor. The resistor prevents the current from passing; according to Ohm's law, the current $I$ flows in the circuit. The source and the resistor form a loop that, according to KVL, can be analyzed to calculate the current $I$. Start with any arbitrary point in the loop, and calculate the voltage drop of elements. Write the KVL in this loop.**

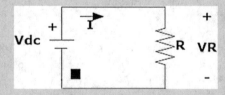

**Fig. 2.10** In this example, starting from the (■) sign and rotating clockwise, the first element is the voltage source. Following the current direction, it is observed that the voltage of the source is measured as $-V_{dc}$. This negative sign shows that the voltage is measured from the negative terminal with respect to its positive terminal, hence a negative value. Following the current direction in the loop, the voltage drop across the resistor is $+V_R$. The current should be calculated

(continued)

**Example 2.8** (continued)

Therefore, combining all elements in this KVL results in

$$\sum \Delta V = 0$$

$$- V_{dc} + V_R = 0$$

According to Ohm's law, the actual value of VR can be obtained as $V_R = RI$. Replacing the KVL equation results in

$$- V_{dc} + RI = 0$$

Solving for $I$ yields

$$RI = V_{dc}$$

$$I = \frac{V_{dc}}{R}$$

**Note 2.1** The voltage drop across passive elements always shows positive polarity at the entry terminal. Passive elements are $R$, $L$, and $C$; therefore, the resistor's voltage drop is $+V_R$.

## Kirchhoff Current Law (KCL)

Consider a node in a circuit that connects two or more elements (node A in Fig. 2.11, for instance). Some of these elements feed the current into the node, and some drain the current out of the node. KCL theory indicates that all currents entering or leaving

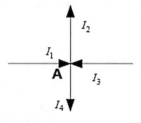

**Fig. 2.11** Node $A$ connects four elements. Element 1 directs the current $I_1$ into the node, carrying a negative KCL value. Element 2 directs the current $I_2$ out of the node and carries a positive value in KCL. Element 3 directs the current $I_3$ into the node; hence, it carries a negative value in KCL, and current $I_4$ directs the current out of the node; hence, it carries a positive value in KCL

a node must summation must be zero. In node $n$, the balance of currents can be written as

$$\sum I_n = 0$$

**Note 2.2** Currents may enter or leave a node. Consider all currents entering a node as a negative value and all currents leaving a node as a positive value.

**Note 2.3** Current sources force the current in or out of the node at a fixed value.

**Note 2.4** If passive elements are connected to a node, they always drain the current out of the node.

**Note 2.5** These rules apply to elements connected to each node regardless of all other considerations at the other nodes. For instance, the current direction at both sides of a resistor is always inward to the terminals.

**Example 2.9 Consider the circuit of Fig. 2.11. Write KCL for node $A$. KCL indicates that the summation of all currents entering and leaving a node must be zero. Therefore,**

$$\sum I_A = 0$$

Since $I_1$ and $I_4$ are entering the node, they will be written with a negative sign, and since $I_2$ and $I_3$ are leaving the node, they are considered positive values. Therefore,
KCL:

$$-I_1 + I_2 + I_3 - I_4 = 0$$

**Example 2.10 In the circuit shown in Fig. 2.12, find the voltages $V_1$ and $V_2$ once by using KCL and once by using KVL.**

**Solution 1 (Using KCL)**
The circuit has two nodes which are labeled 1 and 2. Each node has a voltage $V_1$ and $V_2$. These voltages determine the current directions. However, as these nodes are connected to either current source or passive elements, their voltage needs to be determined.

Since there are two nodes, two KCLs need to be written.

(continued)

**Example 2.10** (continued)

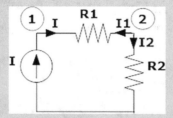

**Fig. 2.12** The circuit of Example 2.10. A current source forces the current into the node ①, and node ② is connected to two passive elements

KCL ①: The current source forces the current $I$ in the node, showing a negative value. The resistor $R_1$ is a passive element that drains the current out of node ①. Hence, the KCL in this node is written as

$$-I + I_1 = 0$$

The current $I_1$ can be obtained from Ohm's law for resistors. The current is the voltage drop across the resistor divided by the resistance. As the current direction is from $V_1$ to $V_2$, $I_1 = \frac{V_1 - V_2}{R_1}$.

The KCL can now be completed as

$$-I + \frac{V_1 - V_2}{R_1} = 0$$

KCL ②: This node is connected to two resistors, $R_1$ and $R_2$. Hence, the currents $I_1$ and $I_2$ exit the node. The KCL becomes

$$I_1 + I_2 = 0$$

The currents can be obtained considering their direction and voltages of the nodes as follows:

- $I_1 = \frac{V_2 - V_1}{R_1}$ because the current is leaving node ②; $V_2$ is considered larger than $V_1$.
- $I_2 = \frac{V_2}{R_2}$ because the voltage drop across the resistor $R_2$ is $(V_2 - 0)$.

Replacing the values, the KCL ② can be written as

$$\frac{V_2 - V_1}{R_1} + \frac{V_2}{R_2} = 0$$

(continued)

**Example 2.10** (continued)

Considering KCL equations, there is a set of two equations with two unknowns, $V_1$ and $V_2$. Solving for $V_1$ and $V_2$ yields the following:

Simplifying the equations as follows:

$$\begin{cases} -I + \dfrac{V_1 - V_2}{R_1} = 0 \\ \dfrac{V_2 - V_1}{R_1} + \dfrac{V_2}{R_2} = 0 \end{cases}$$

$$\begin{cases} \dfrac{V_1}{R_1} - \dfrac{V_2}{R_1} = I \\ \dfrac{-V_1}{R_1} + V_2 \left( \dfrac{1}{R_1} + \dfrac{1}{R_2} \right) = 0 \end{cases}$$

Adding two equations eliminates $V_1$ and results in

$$\frac{V_2}{R_2} = I$$

or

$$V_2 = R_2 I$$

The voltage $V_1$ becomes

$$V_1 = (R_1 + R_2)I$$

**Solution 2 (Using KVL)**

This problem can be solved using KVL. The existence of the current source imposes the same current $I$ through $R_1$ and $R_2$ as a known value. Therefore, the purpose of KVL is already served. The voltage drop across each element can be obtained from Ohm's law.

Therefore, knowing the current $I$,

$$V_2 = R_2 I$$

Since $V_1$ is measured from node ① to the ground, the total resistance from this node to the ground must be considered in voltage calculations. Hence, $V_1$ becomes

$$V_1 = (R_1 + R_2)I$$

**Fig. 2.13** Resistors are
connected in series. The
equivalent becomes a
summation of all resistors

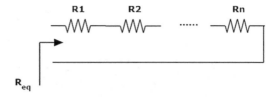

## The Equivalent of Resistors in Series

The equivalent resistance of several resistors in a series is the summation of those elements. The equivalent resistance is the amount of one resistor that can be replaced with the complex that is being considered to follow the same Ohm's law as the individual elements. The equivalent of n resistors in series can be obtained as follows (Fig. 2.13):

$$R_{eq} = \sum_{i=1}^{n} R_i$$

**Example 2.11** Two resistors of $R_1 = 1\ k\Omega$ and $R_2 = 5\ k\Omega$ are connected in series to a 110 V source (as shown in Fig. 2.14). Find the current passing through the circuit and the voltage drop across each resistor.

*Solution.* The equivalent resistor is $R_{eq} = R_1 + R_2 = 1\ k + 5\ k = 6\ k\Omega$. The current passing through the circuit becomes

$$I = \frac{V}{R_{eq}} = \frac{110}{6\ k} = 18.3\ mA$$

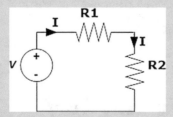

**Fig. 2.14** The circuit of Example 2.11

(continued)

**Example 2.11** (continued)

The current passes each resistor, resulting in a voltage drop across each element proportional to its resistance. Therefore,

$$V_1 = IR_1 = 18.3e - 3 \times 1e3 = 18.3 \text{ V}$$
$$V_2 = IR_2 = 18.3e - 3 \times 5e3 = 91.7 \text{ V}$$

## The Equivalent of Resistors in Parallel

*Resistors connected in parallel share the current* proportional to their resistance values. Consider the following circuit with $n$ resistors in parallel. The voltage across the circuit is $V$. Therefore, the current of the resistor $k$ is $I_k = \frac{V}{R_k}$ for $k = 1, \ldots, n$. The current drawn from the source can be calculated using KCL as follows (Fig. 2.15):

$$I = \sum_{i=1}^{n} I_i$$

$$I = \sum_{i=1}^{n} \left( \frac{V}{R_i} \right) = \left( \sum_{i=1}^{n} \frac{1}{R_i} \right) V = \frac{1}{R_{eq}} V$$

$$R_{eq} = \frac{1}{\sum_{i=1}^{n} \frac{1}{R_i}}$$

$$I_k = \frac{\frac{1}{R_k}}{\sum_{i=1}^{n} \frac{1}{R_i}}$$

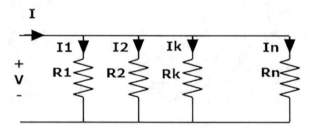

**Fig. 2.15** The connection of several resistors in parallel. They all share the same voltage across the complex, but the current is shared inversely proportional to their resistance. Higher resistance takes a lower current

**Example 2.12** For a circuit connecting $R_1$ and $R_2$ in parallel (circuit of Fig. 2.16), find the equivalent resistance from ports a and b.

*Solution.*

$$R_{\text{eq}} = \cfrac{1}{\frac{1}{R_1} + \frac{1}{R_2}}$$

This can be simplified as follows:

$$R_{\text{eq}} = \frac{R_1 R_2}{R_1 + R_2}$$

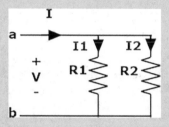

**Fig. 2.16** The circuit of Example 2.12

**Example 2.13** In the circuit of Fig. 2.16, the value of $R_1 = 10\ \Omega$ and $R_2 = 15\ \Omega$ are connected to a 50 V source. Find the equivalent resistance of the circuit, the current drawn from the source, and the current in each resistor.

*Solution.* The equivalent resistance is the parallel of $R_1$ and $R_2$ as

$$R_{\text{eq}} = R_1 \| R_2 = \frac{R_1 R_2}{R_1 + R_2}$$

$$R_{\text{eq}} = 10 \| 15 = \frac{10 \times 15}{10 + 15} = 6\ \Omega$$

Therefore, the current $I$ drawn from the source is

$$I = \frac{50}{6} = 8.33\ \text{A}$$

This current is shared between the resistors as follows

(continued)

**Example 2.13** (continued)

$$I_1 = \frac{V}{R_1} = \frac{50}{10} = 5 \text{ A}$$

$$I_2 = \frac{V}{R_1} = \frac{50}{15} = 3.33 \text{ A}$$

**Example 2.14 Simplify circuits A and B (Fig. 2.17), and find the equivalent resistance with the following resistance values $R_1 = 5\ \Omega$, $R_2 = 15\ \Omega$, $R_3 = 20\ \Omega$, $R_4 = 10\ \Omega$, and $R_5 = 30\ \Omega$.**

*Solution.*
*Circuit A.* The same current passes through $R_1$ and $R_2$; therefore, they are connected in series. $R_3$ and $R_4$ are connected in series. However, the equivalent of series $R_1 + R_2$ is connected in parallel to the series connection of $R_3 + R_4$. That reads as $(R_1 + R_2)\|(R_3 + R_4)$. The equivalent resistance of circuit A becomes

$$R_{eq} = (R_1 + R_2)\big\|(R_3 + R_4) = \frac{(R_1 + R_2)(R_3 + R_4)}{(R_1 + R_2) + (R_3 + R_4)}$$

Considering the numerical values of the resistors, the equivalent resistance becomes

$$R_{eq} = (5 + 15)\big\|(20 + 10) = \frac{20 \times 30}{20 + 30} = \frac{600}{50} = 12\ \Omega$$

*Circuit B.* In this circuit, $R_1$ and $R_2$ are in series and parallel to the other branch of $R_3\|R_5$ in series to $R_4$. Therefore,

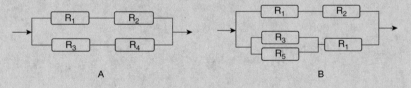

A                                                        B

**Fig. 2.17** The figure of Example 2.14

(continued)

**Example 2.14** (continued)

$$R_{eq} = (R_1 + R_2) \left\| \left( \left( R_3 \middle\| R_5 \right) + R_4 \right) \right.$$

$$R_{eq} = \frac{(R_1 + R_2) \left( \left( R_3 \middle\| R_5 \right) + R_4 \right)}{(R_1 + R_2) + \left( \left( R_3 \middle\| R_5 \right) + R_4 \right)}$$

Considering

$$R_3 \middle\| R_5 = \frac{R_3 R_5}{R_3 + R_5}$$

$$R_{eq} = \frac{(R_1 + R_2) \left( \left( \frac{R_3 R_5}{R_3 + R_5} \right) + R_4 \right)}{(R_1 + R_2) + \left( \left( \frac{R_3 R_5}{R_3 + R_5} \right) + R_4 \right)}$$

Considering the numerical values of the resistors, the equivalent resistance becomes

$$R_{eq} = (5 + 15) \left\| \left( \left( 20 \middle\| 30 \right) + 10 \right) = 20 \right\| \left( \frac{20 \times 30}{20 + 30} + 10 \right)$$

$$= 20 \middle\| (12 + 10) = \frac{20 \times 22}{20 + 22} = 10.476 \ \Omega$$

## Delta (Δ) and Star (Y) Connection

Figure 2.18 shows the connection of resistors such that they form a Δ or Y connection. The equivalent of the resistance in circuits with this type of connection requires a transformation from Δ to Y or vice versa. The Δ and Y connection of resistors can be converted together, as shown in Fig. 2.19.

**Fig. 2.18** Δ (left) and Y (right) connection of resistors. These connections can also be known as Π (pi) left and T (tee) right

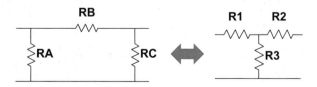

**Fig. 2.19** Overlay
connection of delta and
Y. These circuits are
equivalent to each other.
Using the $\Delta - Y$ conversion
formula, the values of the
desired configuration can be
obtained

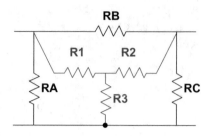

## $Y \rightarrow \Delta$ *Conversion*

This is when the values in a $Y$ circuit are given as $R_1$, $R_2$, $R_3$ and the values of the $\Delta$
equivalent circuit as $R_A$, $R_B$, $R_C$ is needed.

$$R_A = \frac{R_1 R_2 + R_2 R_3 + R_1 R_3}{R_2}$$

$$R_B = \frac{R_1 R_2 + R_2 R_3 + R_1 R_3}{R_3}$$

$$R_C = \frac{R_1 R_2 + R_2 R_3 + R_1 R_3}{R_1}$$

## $\Delta \rightarrow Y$ *Conversion*

This is when the values in a $\Delta$ circuit are given as $R_A$, $R_B$, $R_C$ and the values of the
$Y$ equivalent circuit as $R_1$, $R_2$, $R_3$ is needed.

$$R_1 = \frac{R_A R_B}{R_A + R_B + R_C}$$

$$R_2 = \frac{R_B R_C}{R_A + R_B + R_C}$$

$$R_3 = \frac{R_A R_C}{R_A + R_B + R_C}$$

**Example 2.15 Find the equivalent resistance of the circuit shown
in Fig. 2.20.**
*Solution.* The resistors of $R_A = 10\ \Omega$, $R_B = 40\ \Omega$, and $R_C = 30\ \Omega$ form a $\Delta$
connection. The Y equivalent can be found as follows:

(continued)

**Example 2.15** (continued)

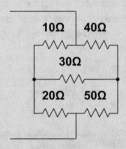

**Fig. 2.20** Circuit of Example 2.15

$$R_1 = \frac{R_A R_B}{R_A + R_B + R_C} = \frac{10 \times 40}{10 + 40 + 30} = 5\ \Omega$$

$$R_2 = \frac{R_B R_C}{R_A + R_B + R_C} = \frac{40 \times 30}{10 + 40 + 30} = 15\ \Omega$$

$$R_3 = \frac{R_A R_C}{R_A + R_B + R_C} = \frac{10 \times 30}{10 + 40 + 30} = 3.75\ \Omega$$

The circuit is converted to

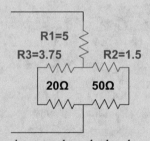

Now, the equivalent resistor can be calculated as

$$R_{eq} = R_1 + ((R_3 + 20)\ |\ |(R_2 + 50))$$

$$R_{eq} = 5 + ((3.75 + 20)\ |\ |(1.5 + 50))$$

$$R_{eq} = 5 + \frac{23.75 \times 51.5}{23.75 + 51.5} = 21.25\ \Omega$$

## Power and Energy in Resistors

Resistors are passive elements and cannot store electric energy. However, they can consume power and generate heat. The power loss through a resistor $P$ is measured in watts (W) and is directly proportional to the resistance and the square of the current as follows:

$$P = RI^2 = \frac{V^2}{R}$$

**Example 2.16 A 100 Ω resistor passes a current of 2.2 A. Find the voltage drop across the resistor and the amount of power loss in the resistor.**
*Solution.*

$$V = RI \rightarrow V = 100 \times 2.2 = 220 \text{ V}$$

The power of the resistor is

$$P = RI^2 = 100 \times 2.2^2 = 484 \text{ W}$$

**Example 2.17 An electric heater operates at a voltage of 110 V. If it takes 15 A to operate, find its resistance and power rating.**
*Solution.*

$$R = \frac{V}{I} = \frac{110}{15} = 7.33 \text{ A}$$

$$P = VI = 110 \times 15 = 1650 \text{ W}$$

**Example 2.18 Find a proper resistor (resistance and power rating) that can pass a 1.5 A current at a voltage drop of 100 V.**
*Solution.* The resistance can be

$$R = \frac{V}{I} = \frac{100}{1.5} = 66.6 \text{ Ω}$$

(continued)

**Example 2.18** (continued)

However, passing this current through the resistor generates power loss. The resistor has to be sized properly to be able to dissipate the heat, and the power loss is

$$P = VI = 100 \times 1.5 = 15 \text{ W}$$

Therefore, a 15-W resistor is needed.

## Definition of a Short Circuit

Part of an electric circuit can be called a "short circuit" if the total resistance connecting two points of that section becomes very small and ideally zero. For instance, if an electric switch is used to turn an electric bulb on and off, when the switch is closed, it fully conducts the current without any resistance (or small resistance), and in a sense, it shorts that part of the circuit. Short circuits happen for many reasons; some are intentional, e.g., a switch, and some are unwanted, e.g., a fault in the circuit. When two circuit points are shorted together, they are forced to become equipotential, which may cause a current to flow into the short circuit section. Most of the existing circuits' short circuits are due to a fault, and a significant current passes the short circuit part. The short circuit analysis finds the amount of current passing through the short circuit segment. Figure 2.21 shows the schematic of a short circuit.

## What Is an Inductor?

Consider a straight line of wire. When a current passes through this wire, it builds a magnetic field around the wire. However, the magnetic field can be increased if more wires are grouped to add their fields. One way to increase this field is to wrap the wire to form a cylindrical shape. The wire can be wrapped around a toroidal core to

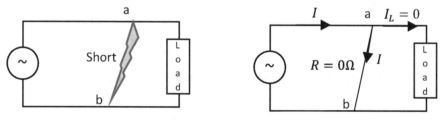

**Fig. 2.21** Creation of a short circuit between points *a* and *b*

**Fig. 2.22** The voltage drop across the terminals of an inductor depends on the time variation of the current passing through the inductor. Once the current variations are zero, the inductor shows a zero-volt drop, equivalent to a short circuit

form a toroidal inductor, or it can be circular but on a flat surface to form a winding (Fig. 2.22). Any of these shapes form an inductor, where its inductance directly depends on the square of the number of turns ($N^2$) as follows:

$$L \propto N^2$$

**Example 2.19 An inductor has an inductance of $L = 1$ mH. If the number of turns in this inductor is increased by 40%, what would be the new inductance value?**
*Solution.* The inductance value is directly proportional to the $N^2$, and for the new condition, $N_{new} = 1.4\, N$. Therefore,

$$\frac{L_{new}}{L} = \left(\frac{N_{new}}{N}\right)^2$$

$$\frac{L_{new}}{L} = \left(\frac{1.4N}{N}\right)^2 = 1.4^2 = 1.96$$

The inductance is increased by 196%.

## *Inductor's Voltage and Current Relation*

The voltage drop across an inductor is directly proportional to the current variation over time. The slope of this dependency is the inductance $L$ measured in henrys ($H$). The Ohm's law for an inductor $L$ is expressed as

$$v = L\frac{di}{dt}$$

Considering an initial current $I_0$ in the inductor, the instantaneous current becomes

$$i = \frac{1}{L} \int v \, dt + I_0$$

It can be interpreted that the time variation of inductor current induces a voltage across the element. Therefore, if there is no current variation across the inductor, the voltage generated across the inductor falls to zero. A zero voltage-induced value indicates a short circuit.

**Example 2.20 A 1 mH inductor experiences a 10 A current change in 2 ms. Find the voltage induced at the terminals of the inductor.**
*Solution.* The voltage induced at the terminal of an inductor is

$$v = L \frac{di}{dt} = L \frac{\Delta i}{\Delta t}$$

Therefore,

$$v = L \frac{\Delta i}{\Delta t} = 1e - 3 \frac{10}{2e - 3} = 5 \text{ V}$$

**Example 2.21 A 1 mH inductor experiences a 10 A current change in 2 μs. Find the voltage induced at the terminals of the inductor.**
*Solution.* The voltage induced at the terminal of an inductor is

$$v = L \frac{di}{dt} = L \frac{\Delta i}{\Delta t}$$

Therefore,

$$v = L \frac{\Delta i}{\Delta t} = 1e - 3 \frac{10}{2e - 6} = 5000 \text{ V}$$

## *Energy and Power of an Inductor*

Inductors store energy in the space around the coil wires. The amount of stored energy $W$ (J) (joules) in an inductor $L$ (H) when a current $I$ (A) is passing through can be calculated by

$$W = \frac{1}{2}LI^2 \ (\text{J})$$

The power of an inductor is the capability of discharging the stored energy over time. That is

$$P = \frac{W}{t} \ (\text{Watts})$$

**Example 2.22 Find the energy stored in a 1 H inductor that passes 10 A current.**
*Solution.* The amount of stored energy is

$$W = \frac{1}{2}LI^2 = \frac{1}{2} \times 1 \times 10^2 = 50 \ (\text{J})$$

**Example 2.23 An inductor is used as an energy storage unit. Find the inductance to feed 100 W of energy in 2 min at the current discharge of 5 A.**
*Solution.* The inductor must store $W = P \times t$ joules of energy. To be able to discharge 100 Watts in $2 \times 60 = 120$ s, it needs

$$W = 100 \times 120 = 12 \ \text{kJ}$$

The amount of inductance needed is

$$W = \frac{1}{2}LI^2$$

$$L = \frac{2W}{I^2} = \frac{2 \times 12,000}{5^2} = 960 \ \text{H}$$

The inductor in this example is large because the discharge rate is small. At higher discharge rates, the amount of inductance can be lower.

## *The Equivalent of Inductors in Series*

Equivalent inductances in series (Fig. 2.23) are the summation of the inductances. As the inductors share the same current, KVL determines that the summation of all voltages must add up to the source value. This leads to

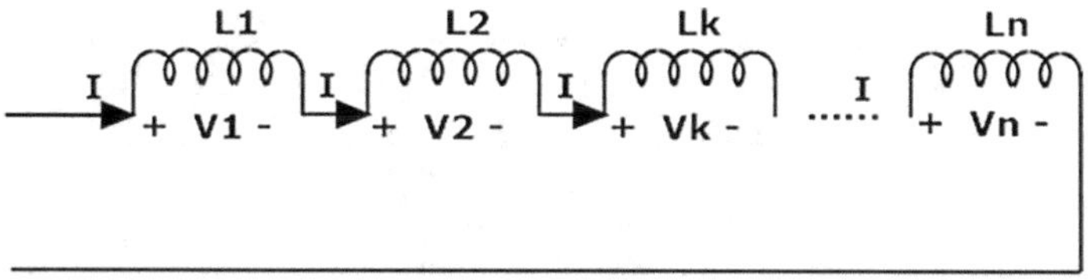

**Fig. 2.23** The connection of inductors in series

$$V = \sum_{k=1}^{n} V_k = \sum_{k=1}^{n} \left( L_k \frac{dI}{dt} \right).$$

Series connection results in the same current through all inductors.

$$V = \left( \sum_{k=1}^{n} L_k \right) \frac{dI}{dt} \triangleq L_{eq} \frac{dI}{dt}$$

$$L_{eq} = \sum_{k=1}^{n} L_k$$

## *The Equivalent of Inductors in Parallel*

Inductors connected in parallel share the current proportional to the voltage integral and the inductance value. Consider the following circuit with $n$ inductors in parallel. The voltage across the circuit is $V$. Therefore, the current of each inductor is $I_k = \frac{1}{L_k} \int V$ for $k = 1,\ldots,n$. The current drawn from the source can be calculated using KCL as follows:

$$I = \sum_{k=1}^{n} I_k$$

$$I = \sum_{k=1}^{n} \left( \frac{1}{L_k} \int V \right) = \left( \sum_{k=1}^{n} \frac{1}{L_k} \right) \int V$$

$$L_{eq} = \frac{1}{\sum_{k=1}^{n} \frac{1}{L_k}}$$

**Example 2.24** For a parallel circuit connecting L1 and L2, find the equivalent inductance from ports $a$ and $b$ (Fig. 2.24).

*Solution.*

$$L_{eq} = \frac{1}{\frac{1}{L_1} + \frac{1}{L_2}}$$

This can be simplified as follows:

$$L_{eq} = \frac{L_1 L_2}{L_1 + L_2}$$

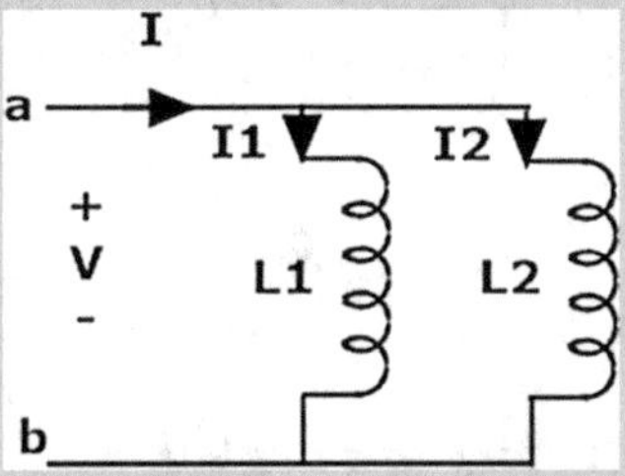

**Fig. 2.24** Circuit of Example 2.24

**Example 2.25** Find the equivalent inductance and current of each inductor in the circuit of Fig. 2.24 when the voltage $v = 2 \sin 10t$ and $L_1 = 15$ H, $L_2 = 20$ H.

*Solution.* The two inductors are connected in parallel. Therefore, the equivalent inductance is

$$L_{eq} = 15 \| 20 = \frac{15 \times 20}{15 + 20} = 8.57 \text{ H}$$

The current equivalent inductance is

$$v = L_{eq} \frac{di}{dt}$$

Therefore, the current becomes

(continued)

**Example 2.25** (continued)

$$i = \frac{1}{L_{eq}} \int v dt = \frac{1}{8.57} \int 2 \sin 10t dt$$

$$i = \frac{1}{8.57} \frac{-2}{10} \cos 10t = -0.023 \cos 10t \text{ A}$$

$$i_1 = \frac{1}{L_1} \int 2 \sin 10t \, dt = \frac{1}{15} \frac{-2}{10} \cos 10t = -0.013 \cos 10t \text{ A}$$

$$i_2 = \frac{1}{L_2} \int 2 \sin 10t \, dt = \frac{1}{20} \frac{-2}{10} \cos 10t = -0.01 \cos 10t \text{ A}$$

## What Is a Capacitor?

Consider two conductive plates facing each other, and form an overlap area $A$ (m$^2$) in a close distance $d$ (m). The shape of the blades and the shape of the distance are not important as long as they maintain the same area and a constant distance. When the effective area between the plates is filled with a dielectric material with permittivity $\epsilon$, the collection of the plates and the dielectric forms a capacitor in which its capacitance $C$ is measured in Farads $F$ as follows:

$$C = \epsilon \frac{A}{d}$$

The permittivity used in this formula consists of two parts as $\epsilon = \epsilon_0 \epsilon_r$, where the $\epsilon_0 = 8.85 \, 1e - 12 \frac{F}{m}$.

**Example 2.26 What is the capacitance of a capacitor with a relative dielectric constant of 40 and plates at a distance of 1 mm with an area of $A = 200$ mm$^2$?**
*Solution.*

$$C = \epsilon \frac{A}{d} = 40 \times 8.85 \times 1e - 12 \frac{200 \times 1e - 6}{1e - 3} = 70.8 \text{ pF}$$

## *Capacitor's Voltage and Current Relation*

The amount of voltage drop across a capacitor is directly proportional to the integral of the current passing through the capacitor, and the slope of this dependency is the inverse of capacitance $\frac{1}{C}$. Considering an initial voltage $V_0$ in the capacitor, the instantaneous voltage becomes

$$v = \frac{1}{C} \int di + V_0$$

Therefore, the current passing through the inductor becomes

$$i = C\frac{dv}{dt}$$

It can be interpreted that the voltage across a capacitor depends on the integral of the current over time. In another word, a capacitor's current depends on the voltage's instantaneous time variation. The current flows through a capacitor only when its terminal voltages change. If the voltage has no variation over time, like a DC source, the current of the capacitor reaches zero after it is fully charged. The fully charged capacitor under DC shows open-circuit behavior.

**Example 2.27 A 1 μF capacitor experiences a current change with a slope of 10 A/s. Find the voltage drop across the terminals of the capacitor.**
*Solution.* The current is increasing with a slope of 10 A/s resulting in a linear equation of $i(t) = 10t$. Therefore,

$$v = \frac{1}{C} \int idt$$

$$v = \frac{1}{1\mu} \int 10tdt = 1e6\frac{10}{2}t^2 = 5e6t^2\,\text{V}$$

**Example 2.28 A 5 μF capacitor is connected to a voltage of $v(t) = 110\sqrt{2}\sin 377t$. Find the current passing through the capacitor.**
*Solution.* The current of the capacitor is obtained by

$$i = C\frac{dv}{dt} = 5e - 6\frac{d}{dt}\left(110\sqrt{2}\sin 377t\right)$$
$$= 5e - 6 \times 110\sqrt{2} \times 377\cos 377t = 0.293\cos 377t$$

## Energy and Power of a Capacitor

Capacitors store energy from electric charges on the plates interfaced by a dielectric material. The amount of stored energy $W$ (J) depends on the applied voltage $V$ and the capacitance, expressed as follows:

$$W = \frac{1}{2}CV^2 \text{ (J)}$$

---

**Example 2.29 Find the amount of energy stored in a 1 mF capacitor charged at 480 V.**
*Solution.* The stored energy is

$$W = \frac{1}{2}CV^2 = \frac{1}{2}1e-3 \times 480^2 = 115.2 \text{ (J)}$$

---

**Example 2.30 An ultra-capacitor is used to store energy in an electric vehicle. The amount of energy needed is 1 MJ, delivered at a 400 V system. Find the size of the capacitor needed.**
*Solution.* The amount of energy and the operating voltage is given. Therefore,

$$1e6 = \frac{1}{2}C\,400^2$$

$$C = \frac{2 \times 1e6}{400^2} = 12.5 \text{ F}$$

---

**Example 2.31 A capacitor is needed to smooth out the output voltage of a 500 W power supply when rectifying the 60 Hz waveforms. Find the amount of capacitor needed at full load in a half-wave rectifier when operating at 12 V.**
*Solution.* The amount of time that is needed to deliver the power is half-cycle $\frac{1}{2} \times \frac{1}{60} = \frac{1}{120}$ s. The amount of energy needed is

(continued)

**Example 2.31** (continued)

$$W = P \times t$$

$$W = 500 \times \frac{1}{120} = 4.16 \text{ (J)}$$

Therefore,

$$C = \frac{2W}{V^2} = \frac{2 \times 4.16}{12^2} = 0.0577 \text{ F}$$

Or

$$C = 57.77 \text{ mF}$$

**Note 2.6** A half-wave rectifier generates a voltage waveform that only selects the positive peaks of a sinusoidal waveform. Figure 2.25 shows the waveform and the duration that the capacitor needs to feed the load (Fig. 2.26).

**Fig. 2.25** The current passing a capacitor depends on the time variation of voltage across its terminals. Once the capacitor is fully charged, the current reaches zero, becoming an open circuit

**Fig. 2.26** The figure of a half-wave rectifier

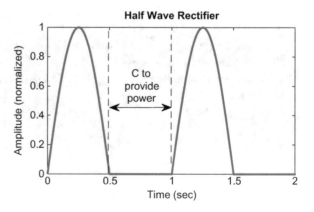

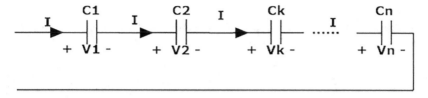

**Fig. 2.27** The connection of capacitors in series

## *The Equivalent of Capacitors in Series*

The series connection of capacitors suggests similar current passing through each capacitor. The voltages around a loop are added to hold KVL (Fig. 2.27).

Therefore, the voltage $V$ across the series network can be obtained by

$$V = \sum_{k=1}^{n} V_k$$

The voltage of the capacitor $k$ can be obtained from Ohm's law as

$$V_k = \frac{1}{C_k} \int I \text{ for } k = 1, \dots, n$$

Sharing the same current, the KVL can be rewritten as

$$V = \sum_{k=1}^{n} \left( \frac{1}{C_k} \int I \right) = \left( \sum_{k=1}^{n} \frac{1}{C_k} \right) \int I \triangleq \frac{1}{C_{eq}} \int I$$

Therefore,

$$\frac{1}{C_{eq}} = \left( \sum_{k=1}^{n} \frac{1}{C_k} \right)$$

Or

$$C_{eq} = \frac{1}{\sum_{k=1}^{n} \frac{1}{C_k}}$$

**Example 2.32 Find the equivalent capacitance of the series connection of two capacitors, $C_1$ and $C_2$, as shown in Fig. 2.28.**

*Solution.*

$$C_{eq} = \frac{1}{\frac{1}{C_1} + \frac{1}{C_2}} = \frac{C_1 C_2}{C_1 + C_2}$$

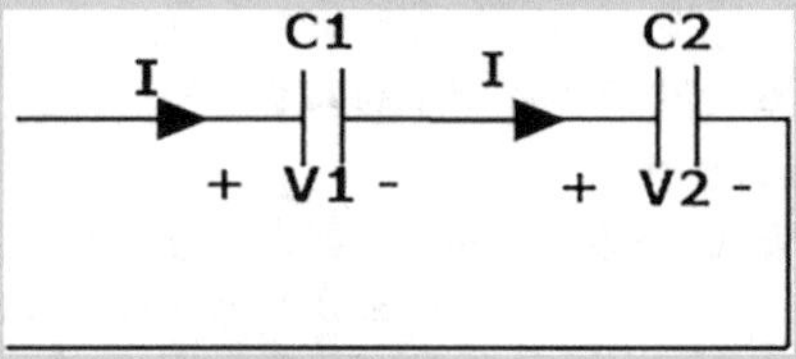

**Fig. 2.28** The circuit of Example 2.32

**Example 2.33 Find the equivalent capacitance of the series connection of a 20 mF and a 40 mF capacitor.**

*Solution.*

$$C_{eq} = \frac{20e - 3 \times 40e - 3}{20e - 3 + 40e - 3} = \frac{800}{60} e - 3 = 13.3 \text{ mF}$$

**Example 2.34 Find the equivalent of three equal 30 PF capacitors in series.**

*Solution.*

$$C_{eq} = \frac{1}{\frac{1}{30e - 12} + \frac{1}{30e - 12} + \frac{1}{30e - 12}} = \frac{1}{\frac{3}{30e - 12}} = \frac{30e - 12}{3} = 10e - 12 \text{ or } 10 \text{ PF}$$

## The Equivalent of Capacitors in Parallel

The equivalent of capacitors in parallel shares the same voltage across the complex. The voltage imposes current to pass through each capacitor $k$ as

**Fig. 2.29** The parallel connection of capacitors

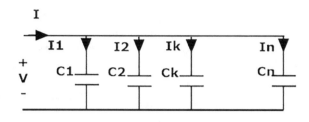

$$I_k = C_k \frac{dV}{dt} \text{ for } k = 1, \ldots, ns$$

The total current drawn from the source is shown in Fig. 2.29.

$$I = \sum_{k=1}^{n} I_k = \sum_{k=1}^{n} \left( C_k \frac{dV}{dt} \right)$$

As the voltage for all capacitors is the same, the KCL can be rewritten as

$$I = \left( \sum_{k=1}^{n} C_k \right) \frac{dV}{dt} \triangleq C_{eq} \frac{dV}{dt}$$

Parallel connection of capacitors results in equivalent capacitance as follows

$$C_{eq} = \sum_{k=1}^{n} C_k$$

**Example 2.35 Find the equivalent capacitance of the parallel connection of a 20 mF and a 40 mF capacitor.**
*Solution.*

$$C_{eq} = 20e - 3 + 40e - 3 = 60 \text{ mF}$$

**Example 2.36 Find the equivalent of three equal 30 PF capacitors in parallel.**
*Solution.*

$$C_{eq} = 30 \, P + 30 \, P + 30 \, P = 90 \text{ PF}$$

**Example 2.37 Three capacitors of 10 mF, 15 mF, and 30 mF are connected once in series and once in parallel to a 100 V voltage source. Calculate the total energy stored in the capacitors in each case.**
*Solution.*

*Series:* The capacitors in series connection provide an equivalent of

$$C_{eq} = \frac{1}{\frac{1}{10e-3} + \frac{1}{15e-3} + \frac{1}{30e-3}} = 5 \text{ mF}$$

The stored energy in this case is

$$W = \frac{1}{2} C_{eq} V^2$$

$$W = \frac{1}{2} 5e - 3 \times 100^2 = 25 \text{ J}$$

*Parallel:* The capacitors in parallel provide an equivalent of

$$C_{eq} = 10 + 15 + 30 \text{ mF} = 55 \text{ mF}$$

The stored energy in this case is

$$W = \frac{1}{2} C_{eq} V^2$$

$$W = \frac{1}{2} 50e - 3 \times 100^2 = 250 \text{ J}$$

**Example 2.38 Find the equivalent capacitor of the circuit shown in Fig. 2.30.**
*Solution.* The 10 mF and 25 mF are connected in parallel, and their equivalent is connected in series to the 30 mF capacitors. Therefore, this can be written as

$$C_{eq} = \frac{1}{\frac{1}{10m+25m} + \frac{1}{30m}} = 16.135 \text{ mF}$$

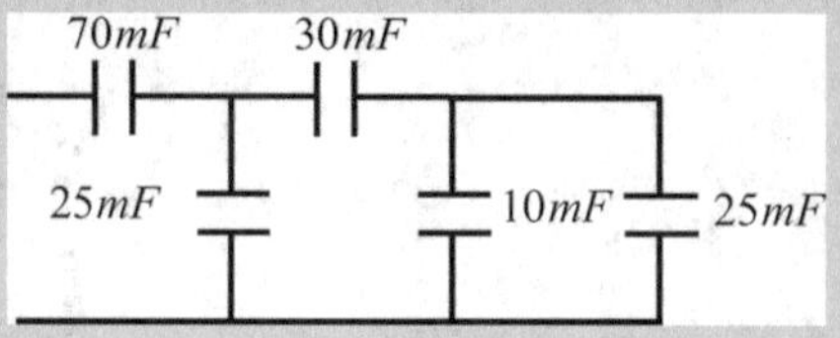

**Fig. 2.30** Circuit of Example 2.38

# Sources

Sources are generally divided into voltage source and current source. Each source can generate a voltage waveform across the terminals of the circuit, or it may force a specific current through the circuit. A voltage source tends to keep the voltage constant no matter how much current is fed to the terminals of the circuit. A current source tends to keep the current constant no matter how much voltage is required across the source. The value of the voltage in a voltage source or the value of current in a current source can be either a fixed number or may depend on other parameters of the circuit. For instance, a voltage source might be 12 V fixed, or it may depend on $\alpha$ times of a current that passes through a resistor, somewhere in the circuit, as $\alpha I$.

The sources with fixed values are called independent sources, and those whose values depend on other circuit parameters are called independent sources.

## *Independent Voltage Source*

A voltage source generates a voltage value regardless of the current drawn from the source. A voltage source keeps the voltage at the terminal by allowing the current to vary. $V(t)$ can have any waveforms discussed in this chapter. In reality, all sources have an internal resistance that causes a voltage drop as the current passes through. The voltage drop from the desired value is called voltage regulation. A schematic of an ideal independent voltage source is shown in Fig. 2.31.

A unit step voltage source keeps the voltage at a steady value and can ideally provide infinite current without a voltage drop. The voltage measured at the terminal of the voltage source (or its $V$–$I$ characteristics) is shown in Fig. 2.32. The figure demonstrates that the voltage amplitude is fixed at any given current. A sinusoidal waveform is shown in Fig. 2.33, and a bipolar step function is shown in Fig. 2.34.

**Fig. 2.31** Schematic of a voltage source with positive and negative terminals and amplitude $V$

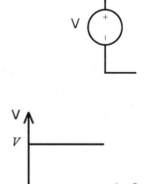

**Fig. 2.32** An independent voltage source and its $V$–$I$ characteristics. The voltage amplitude is constant at any drawn current

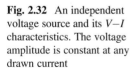

**Fig. 2.33** A voltage source
generates a sinusoidal
waveform

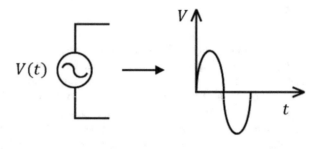

**Fig. 2.34** A step function
generator with bipolar
amplitude. It means the
voltage becomes positive
and negative

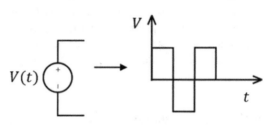

**Example 2.39 A 9 V battery ideally provides a fixed voltage equivalent to 9
volts. The waveform of this battery can be expressed as $V(t) = 9$. The
source ideally maintains this voltage under all load currents. However,
in reality, the terminal voltage drops as the load increases.**
The sinusoidal waveform of signal $V(t)$ has a standard form as follows:

$$V(t) = A \sin(\omega t + \phi)$$

wherein $A$ is the amplitude or the peak value of the waveform, $\omega$ is the angular
frequency in units of $\frac{rad}{sec}$, and $\phi$ is the initial phase measured in degrees or
radians. The amplitude $A$ can be measured in peak or root-mean-square
(RMS). In the sinusoidal waveform, the peak and RMS are related as follows:

$$A(\text{peak}) = \sqrt{2}\, A(\text{rms})$$

The angular frequency $\omega$ in $\left(\frac{rad}{sec}\right)$ and the frequency $f$ in (Hz) are related as
follows:

$$\omega\left(\frac{rad}{sec}\right) = 2\pi f \,(\text{Hz})$$

**Example 2.40 A household power outlet is 110 V RMS and has a frequency of 60 Hz with the initial phase of 0°. Find the waveform expression.**

*Solution.* The peak value of the waveform is $110 \times \sqrt{2} = 155.56$ V, the angular frequency $\omega = 2\pi f = 2\pi \times 60 = 120\pi \left(\frac{rad}{sec}\right)$. The waveform is

$$V(t) = 155.56 \sin(120\pi t)$$

## *Independent Current Source*

An independent current source generates a constant current value regardless of the voltage across the source. This source changes the voltage at the terminal to keep the current constant. $I(t)$ can force a desired amount of current through a circuit even if the voltage virtually reaches infinite. A schematic of an independent current source is shown in Fig. 2.35. The $V–I$ characteristics are shown in Fig. 2.36. As the figure shows, the amount of current is fixed at any voltage. In reality, the amount of current will drop because of the source's internal resistance.

**Note 2.7** A voltage source keeps the voltage of the nodes constant by source current to the branches connected to the terminals.

**Note 2.8** A current source keeps the current of a branch constant by varying the voltage at terminals.

**Note 2.9** A zero-volt (0 V) voltage source resembles a short circuit. Figure 2.37 shows an equivalent of a zero-volt voltage source.

**Fig. 2.35** Schematic of an ideal independent current source

**Fig. 2.36** $V–I$ characteristics of a current source. The current source can deliver the current at an arbitrary voltage level

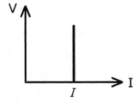

**Fig. 2.37** The equivalent of
a zero-volt voltage source

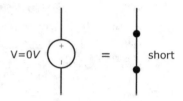

V=0V          =          short

**Fig. 2.38** The equivalent of
a zero-amp current source

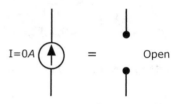

I=0A          =          Open

**Note 2.10** A zero-amp (0 A) current source resembles an open circuit. Figure 2.38
shows an equivalent of a zero-amp current source.

---

**Example 2.41 Find the currents $I, I_1$, and $I_2$ of this circuit, when $v(t) = 10$
$u(t)$ (Fig. 2.39).**

*Solution.* The voltage source generates a pulse at amplitude $V$. The voltage
source is connected in parallel to two resistors. Each of these resistors takes a
current proportional to its resistance. The current through $R_1$ is calculated to be
$i_1(t) = \frac{V(t)}{R_1}$ or $i_1(t) = \frac{10}{R_1}$, and the current of $R_2$ is $i_2(t) = \frac{V(t)}{R_2}$ or $i_2(t) = \frac{10}{R_2}$. The
total current taken from the source is the summation of $i_1(t)$ and $i_2(t)$. There-
fore, $i(t) = i_1(t) + i_2(t) = \frac{V(t)}{R_{eq}}$; $R_{eq} = R_1 \| R_2$. This reads $R_1$ in parallel to $R_2$.
The parallel of two resistors can be calculated as

$$\frac{1}{R_{eq}} = \frac{1}{R_1} + \frac{1}{R_2}$$

$R_{eq} = \frac{R_1 R_2}{R_1 + R_2}$ which results in a total current of $I = \frac{V(t)}{\frac{R_1 R_2}{R_1 + R_2}} = V(t) \frac{R_1 + R_2}{R_1 R_2} =$
$10 \frac{R_1 + R_2}{R_1 R_2}$

**Fig. 2.39** Circuit of
Example 2.41

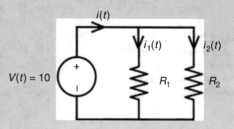

**Example 2.42 Find the voltages of $V_1$, $V_2$, and $V_3$, when $I_1(t) = 10$ (A) and $I_2(t) = -20$ (A) (Fig. 2.40).**

*Solution*. The circuit has two current sources of $I_1(t) = 10$ and $I_2(t) = -20$ (A), which operate at the same frequency. The circuit has three nodes at unknown voltages of $V_1$, $V_2$, and $V_3$. The voltages can be obtained using KCL written for each node.

KCL ①. The summation of entering current (negative sign) and exiting currents (positive sign) must be zero.

$$-I_1 + \frac{V_1 - V_2}{R_1} = 0$$

KCL ②

$$\frac{V_2 - V_1}{R_1} + \frac{V_2 - 0}{R_2} + \frac{V_2 - V_3}{R_3} = 0$$

KCL ③

$$-I_2 + \frac{V_3 - V_2}{R_3} = 0.$$

This results in three equations and three unknowns. Note that phasors replace the current source values.

Simplifying the equations as follows:

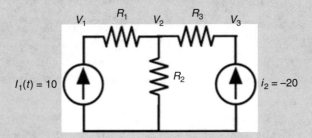

**Fig. 2.40** Circuit of Example 2.42

(continued)

**Example 2.42** (continued)

$$\begin{cases} \dfrac{V_1}{R_1} - \dfrac{V_2}{R_1} = I_1 \\[2mm] \dfrac{-V_1}{R_1} + V_2\left(\dfrac{1}{R_1} + \dfrac{1}{R_2} + \dfrac{1}{R_3}\right) + \dfrac{-V_3}{R_3} = 0 \\[2mm] \dfrac{-V_2}{R_3} + \dfrac{V_3}{R_3} = I_2 \end{cases}$$

Adding all three equations results in

$$\frac{V_2}{R_2} = I_1 + I_2$$

$$V_2 = R_2(I_1 + I_2)$$

$$V_2 = R_2(10 - 20) = -10R_2$$

Replacing the second equation and solving for $V_3$ result in

$$V_3 = R_3 I_2 + R_2(I_1 + I_2) = R_2 I_1 + (R_2 + R_3)I_2$$

$$V_3 = -10R_2 - 20R_3$$

Moreover, using the first equation with the value of $V_2$, $V_1$ becomes

$$V_1 = V_2 + R_1 I_1$$

$$V_1 = (R_1 + R_2)I_1 + R_2 I_2$$

$$V_1 = 10R_1 - 10R_2$$

**Example 2.43** The values of circuit Fig. 2.40 are $R_1 = 100\ \Omega$, $R_2 = 250\ \Omega$, and $R_3 = 50\ \Omega$. Find $V_1$, $V_2$, $V_3$ and the currents of the resistors $R_1$, $R_2$, $R_3$.

*Solution.* According to the solution provided earlier, the voltage of each node can be obtained as follows:

$$V_1 = 10R_1 - 10R_2$$

Substituting the numbers results in

(continued)

**Example 2.43** (continued)

$$V_1 = 10 \times 100 - 10 \times 250 = -1500 \text{ V}$$
$$V_2 = -10R_2 = -10 \times 250 = -2500 \text{ V}$$

And

$$V_3 = -10R_2 - 20R_3 = -10 \times 250 - 20 \times 50 = -3500 \text{ V}$$

The currents are forced by the current sources and can be calculated as follows:

The current passing through $R_1$ is forced by the source $I_1$ and equals

$$I_{R_1} = 10 \text{ A}$$

It can also be found as follows:

$$I_{R_1} = \frac{V_1 - V_2}{R_1} = \frac{-1500 - (-2500)}{100} = \frac{1000}{100} = 10 \text{ A}$$

The current through $R_3$ the resistor is forced by the current source of $-20$ A.

$$I_{R_3} = -20 \text{ A}$$

It can also be obtained as follows:

$$I_{R_3} = \frac{V_3 - V_2}{R_3} = \frac{-3500 - (-2500)}{50} = -\frac{1000}{50} = -20 \text{ A}$$

Accordingly, the current of $R_2$ is obtained through a KCL at this node or by dividing the voltage of the node $V_2$ by $R_2$ as follows:

$$I_{R_2} = \frac{-2500}{250} - 10 \text{ A}$$

Or

$$I_{R_2} = I_{R_1} + I_{R_3} = 10 - 20 = -10 \text{ A}$$

**Note 2.11** A mix of voltage and the current source might exist in a circuit. It should be noted that a voltage source keeps the voltage of the terminal nodes constant and a current source keeps the current of a circuit branch constant.

**Example 2.44 Find the voltage $V_2$ and the current $I_1$ when the DC source is $V = 100$ V, $I = 15$ A and $R_1 = 25$ Ω, $R_2 = 75$ Ω shown in Fig. 2.41.**
*Solution.* As the circuit demonstrates, the voltage source keeps the voltage of node ① at constant $V$ volts. Considering the voltage of node ② at $V_2$, the current $I_1$ can be found.

KCL ②

$$I_{R_1} + I_{R_2} - I = 0$$

**Note 2.12.** The current of passive elements should always drain the node, hence positive. The source's current, if forced to the node (in this circuit), is negative.

Therefore,

$$\frac{V_2 - V}{R_1} + \frac{V_2}{R_2} - I = 0$$

Solving for $V_2$ results in

$$V_2 \left( \frac{1}{R_1} + \frac{1}{R_2} \right) = \frac{V}{R_1} + I$$

$$V_2 = \left( \frac{R_1 R_2}{R_1 + R_2} \right) \left( \frac{V}{R_1} + I \right)$$

$$V_2 = \frac{25 \times 75}{25 + 75} \left( \frac{100}{25} + 15 \right) = 356.25 \text{ V}$$

Knowing the value of $V_2$, the current $I_1$ can be obtained as

$$I_1 = \frac{V - V_2}{R_1} = \frac{V}{R_1} - \frac{V_2}{R_1}$$

$$I_1 = \frac{356.25 - 100}{25} = -10.25 \text{ A}$$

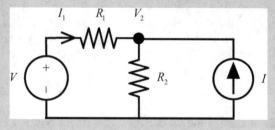

**Fig. 2.41** Circuit of Example 2.44

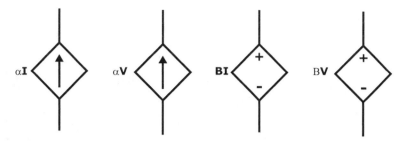

**Fig. 2.42** Dependent current and voltage sources. The dependency of the voltage or current generated from each source can be on a current or a voltage measured at any part of the circuit

## Dependent Sources

The value of dependent sources may change based on their dependency type to other circuit parameters. These dependencies may be the voltage of nodes or the current of branches in the circuit. Therefore, a dependent voltage source may be controlled by a node's voltage or a branch's current. A dependent current source may be controlled by a node's voltage or a branch's current (Fig. 2.42).

The fact that the source output voltage or current is dependent on circuit parameters makes no difference in their operation, such as KVL and KCL, as they are still voltage or current sources, but their values may depend on other circuit values. A dependent voltage source keeps the voltage of the terminal nodes constant to the value dictated by the control parameter, and a dependent current source still keeps the current of the branch constant by changing the voltage.

**Note 2.13** The only consideration is to correctly account for the dependency of the voltage or current generated by the circuit parameters.

**Note 2.14** Changes in the circuit or equivalent analysis of elements should be such that the dependency of voltage or current is not eliminated from the simplified circuit, i.e., keeping track of the circuit parameters.

**Note 2.15** A dependent voltage source or a dependent current source cannot be turned off unless the control parameter forces the source to be off.

**Example 2.45 In the circuit of Fig. 2.43, find $I_1$.**
*Solution.* The circuit has a current-controlled voltage source, and the control current is the current through resistor $R_1$. In the circuit simplification, this current, $I_2$, must be tracked to calculate the value of the voltage source. Therefore, although $R_2$ and $R_3$ are in parallel, it is not recommended to simplify these two elements.

(continued)

**Example 2.45** (continued)

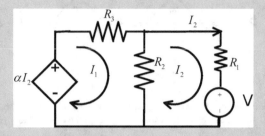

**Fig. 2.43** Circuit of Example 2.45

The circuit has two loops, I and II. KVL must be written for each of these loops (since there is no current source connected to any nodes around the loops).

KVL ①

$$-\alpha I_2 + V_{R_3} + V_{R_2} = 0$$

**Note 2.16**

The value of the dependent voltage source becomes negative because, following the suggested direction of the current in loop ①, the current enters the negative terminal of the voltage source.

The voltage drop across $R_3$ $V_{R_3} = I_1 R_3$ and $R_2$ $V_{R_2} = (I_1 - I_2)R_2$. The currents $I_1$ and $I_2$ are in opposite directions over the resistor $R_2$, and following the KVL in loop ①, the positive voltage drop suggests that the current value is $I_1 - I_2$.

Therefore, loop ① KVL results in

$$-\alpha I_2 + I_1 R_3 + (I_1 - I_2)R_2 = 0$$

KVL ②

$$V_{R_1} + V + V_{R_2} = 0$$

$V_{R_1} = I_2 R_1$ and $V_{R_2} = (I_2 - I_1)R_2$. In loop ②, the KVL follows the current $I_2$; hence, the positive voltage drop across the same resistor suggests $I_2 - I_1$.

Therefore, loop ② KVL results in

(continued)

**Example 2.45** (continued)

$$I_2R_1 + (I_2 - I_1)R_2 = -V$$

Simplifying and solving for $I_1$ and $I_2$ result in

$$\begin{cases} -\alpha I_2 + I_1 R_3 + (I_1 - I_2)R_2 = 0 \\ I_2 R_1 + (I_2 - I_1)R_2 = -V \end{cases}$$

Solving for $I_1$ and $I_2$ results in

$$\begin{cases} (R_3 + R_2)I_1 - (\alpha + R_2)I_2 = 0 \\ -R_2 I_1 + (R_1 + R_2)I_2 = -V \end{cases}$$

Solving for $I_1$ from the first equation $I_1 = \frac{(\alpha + R_2)}{(R_3 + R_2)} I_2$ and replacing it in the second equation yield

$$I_2 = \frac{V}{R_1 + R_2 - R_2 \frac{(\alpha + R_2)}{(R_3 + R_2)}}$$

The control parameter $\alpha$ is set to obtain specific characteristics from the circuit. A is mostly determined by transistors used in the circuit. MOSFETs and BJTs have different dependencies and generate dependable sources.

**Example 2.46** Find voltage $V_T$ and current $I_1$ in the circuit of Fig. 2.44. Evaluate the values when the circuit parameters are $V = 20$ V, $R_1 = 50\ \Omega$, $R_2 = 100\ \Omega$, and $B = 0.5$.

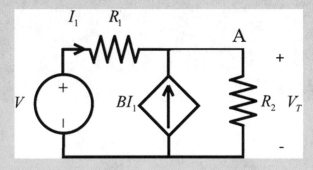

**Fig. 2.44** Circuit of Example 2.46

(continued)

**Example 2.46** (continued)

*Solution.* The output voltage $V_T$ is the voltage across the resistor $R_2$ at node A. Writing a KCL at this node relates the voltage of this node to other parameters in the circuit. The goal is to relate the voltage somehow $V_T$ to the source voltage, which is a known value.

Therefore, a KCL at node A involves three branches of resistor $R_1$, the voltage source $BI_1$ and the resistor $R_2$. The currents involved in the KCL are $-I_1$, $-BI_1$, and $+\frac{V_T}{R_2}$. Remember, the currents entering a node are considered negative, and the current leaving a node is considered positive. The KCL is

$$-I_1 - BI_1 + \frac{V_T}{R_2} = 0$$

$$-(1+B)I_1 + \frac{V_T}{R_2} = 0$$

The value of the source $I_1$ depends on the voltage of the source and the voltage at node A. Therefore,

$$I_1 = \frac{V - V_T}{R_1}$$

Replacing this value in the KCL equation results in

$$-(1+B)\frac{V - V_T}{R_1} + \frac{V_T}{R_2} = 0$$

$$-(1+B)\frac{V}{R_1} + V_T\left(\frac{1}{R_2} + \frac{(1+B)}{R_1}\right) = 0$$

Solving for $V_T$

$$V_T = \frac{(1+B)\frac{V}{R_1}}{\left(\frac{1}{R_2} + \frac{(1+B)}{R_1}\right)} = \frac{\frac{(1+B)}{R_1}}{\left(\frac{R_1 + (1+B)R_2}{R_1 R_2}\right)}V = \frac{(1+B)R_2}{R_1 + (1+B)R_2}\,V$$

The current $I_1$ can be obtained by replacing the value of $V_T$ into the equation as follows:

(continued)

**Example 2.46** (continued)

$$I_1 = \frac{V - V_T}{R_1} = \frac{V - \frac{(1+B)R_2}{R_1 + (1+B)R_2}V}{R_1} = \frac{1}{R_1 + (1+B)R_2}V$$

Considering the circuit values, the output voltage and the source current are

$$V_T = \frac{(1+B)R_2}{R_1 + (1+B)R_2}V = \frac{(1+0.5)100}{50 + (1+0.5)100}20 = 15\ \text{V}$$

$$I_1 = \frac{1}{R_1 + (1+B)R_2}V = \frac{1}{50 + (1+0.5)100}20 = \frac{20}{200} = 0.1\ \text{A}$$

## Switches

Some electric circuit elements such as resistors, inductors, and capacitors with their specific behavior have been introduced. As part of the electric circuit, the power sources should be connected to the circuit at the required time or disconnected at the right time. Therefore, some devices exist to start or stop the flow of electric current. Some examples of these devices are switches that operate mechanically. Some switches operate based on electric, optical, or magnetic stimulation. A mechanical toggle switch requires a mechanical action to start or stop the flow of the electric current. A group of these switches cut only one current line (on positive or negative wires), and other groups cut the current of both positive and negative lines simultaneously. This determines the number of *poles* a switch might have. Single-pole, double-pole, or even triple-pole switches have only one mechanical contact, two mechanical contacts, or three mechanical contacts close or open simultaneously. The schematic of these switches is shown in Fig. 2.45. The abbreviation "SP" stands for single pole, and consequently, "DP" stands for double pole, and "TP" stands for triple pole.

The crossbar that connects two poles or three poles shows that all these poles act at the same time. It is also important to note that the switches are identified based on their initial status, either "open" or "closed." This identifies the normal status of the switch to either cut or conduct the current. Once the switch is activated, it changes its initial status, e.g., a normally open switch is open at rest (normal) condition and

**SPST_OPEN  SPST_CLOSED   DPST_OPEN   DPST_CLOSED   TPST_CLOSED   TPST_OPEN**

**Fig. 2.45** Mechanical switches: SPST: single-pole single-throw, DPST double-pole single-throw, TPST triple-pole single-throw

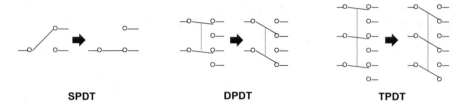

**SPDT**                          **DPDT**                          **TPDT**

**Fig. 2.46** Various double-throw switches. SPDT single-pole double-throw, DPDT double-pole double-throw, TPDT triple-pole double-throw. Each pole has two options for routing the current

becomes closed when excited. Consequently, a normally closed switch is closed in its initial mode of operation, and once activated, it opens and cuts the current. Figure 2.45 shows single-pole, double-pole, and triple-pole switches in open or closed status. In other word, these switches are single-throw switches identified as the "ST" part in their naming.

Consequently, SPST means a single-pole single-throw switch cuts the current only in one open or closed wire. DPST is a double-pole switch that cuts the current in two wires simultaneously and can be either open or closed. Similarly, TPST is a triple-pole single-throw switch.

If the action of a switch is closed or open, it provides a single-throw operation. As shown in Fig. 2.46, the switch can reroute the current to another circuit. These witches are double-throw switches. If an off position is added to the double-throw switch, it becomes a triple-throw switch.

**Example 2.47 Using an SPST switch, draw a circuit to turn a light on and off.**
*Solution.* The switch S1 in Fig. 2.47 is a single-pole single-throw switch with only on and off positions. In the left figure, the switch is in off mode, and the bulb is off. When the switch toggles, it turns on and allows the current to flow. The flow of current through the bulb turns it on. The bulb can only be controlled from one point in the circuit.

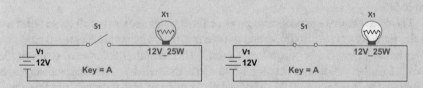

**Fig. 2.47** SPST switch to turn an electric bulb on and off from one point

**Example 2.48 Using SPDT switches, draw a circuit that can turn a bulb on and off from two different locations in the circuit.**
*Solution.* The circuit with two points of control is shown in Fig. 2.48. Switches S1 and S2 are connected so that their single-pole port is connected to the source or the bulb and their double-throw ports are connected directly, as shown. In this case, the circuit must be shown in an off position. As shown, two combinations of S1 and S2 can turn the bulb off. Turning switches will provide a path for the current to flow and the bulb to turn on. Figure 2.46 shows the case when switch S1 turns the bulb on. The position of switch S2 is connected to either of the throw ports, and switch S1 provides a selection to connect to either of these ports, hence control over the bulb at point S1. Figure 2.49 also shows the control at point S2 that can turn the bulb on and off.

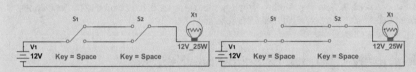

**Fig. 2.48** The off position of the bulb is controlled by two switches, S1 and S2

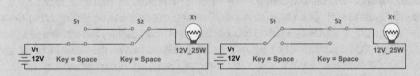

**Fig. 2.49** S1 controls the bulb's position in the left figure and S2 in the right figure

# Mechanical Relays and Contactors

The mechanical switching action can be originated from various sources. One is an operator that mechanically toggles the switches. The operator is simple but inaccurate in timing the on-off operations. For instance, the operator cannot switch the circuit on at time $t = 0.5$ s and turn it off at time $t = 1$ s. The mechanical contacts can move by the force of an electromagnet. The electromagnet receives current from another source and pulls the contacts to close or open the circuit. This is more accurate in timing the switch's on-off action. These types of switches are called "relays" or "contactors." Relays send commands at low current ratings, and contractors are used to starting and cutting higher currents. Depending on the type of relays, there might be several types of switches included in one unit. Figure 2.50 shows the electrically activated mechanical relays. The mechanical contacts may be single or multiple and have open or closed initial status. When the coil "K" is activated, the relay is turned on, and contacts change their positions.

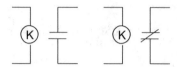

**Fig. 2.50**  Coil "K" is excited by an external source that changes the initial status of the contacts. A normally open relay is shown at the left and a normally closed one at the right

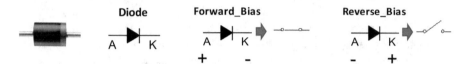

**Fig. 2.51**  Schematic of a diode with a gray bar in one end: "A" stands for the anode, and "K" stands for the cathode. When the anode is positive compared to the cathode, it is forward-biased and has the potential to flow the current. When the anode is negative compared to the cathode, it is reverse-biased and can potentially stop the current

## Electrically Operated or Solid-State Switches

As described earlier, the mechanical switches have latency in their response and cannot operate in fast switching. The metallic contacts of these switches have a limited lifetime and must be protected against the weather and operating conditions. However, fully enclosed switches can be designed and utilized, but switching speed still limits them. Electrically operated switches do not have mechanical parts and conduct current based on the specific characteristics of semiconductors. When a small voltage or current is excited, the switch starts to conduct. There are several types of switches that either operate autonomously or on-demand.

**Diode**  This is an autonomous switch similar to a check valve in hydraulic systems. When the anode "A" voltage exceeds the cathode "K," the switch starts to conduct electric current. In this case, the diode is
 called to be "forward biased." When the polarity of the applied voltage changes, i.e., the anode is connected to a negative voltage, and the cathode is connected to the positive voltage, the diode is "reverse-biased" and stops the current flow in the reverse direction. The diode has no control over the current flow except for the current's direction. For the current to stop, the diode must be reversed biased, and the direction of the current must reach zero from its forward flowing. The schematic of a diode is shown in Fig. 2.51. Diodes are designed at low and extra high currents and can operate in switching circuits at frequencies up to gigahertz. Cellphones and other processors use ultrafast circuit elements.

**Example 2.49 Determine if the diode in the circuit of Fig. 2.52 is on or off.**

*Solution.* The circuit shows that the positive voltage of the source is connected to the anode of the diode. The source sends the current in the same direction as the diode is forward connected. This lets the current pass through the diode, hence turning it on. *Therefore the circuit lets the current flow. The diode takes some voltage drop across itself to stay on.* That voltage is called the threshold voltage, or $V_T$. For most diodes, this voltage is about $0.1 - 0.7$ V. In this case, the current that passes through the circuit is

$$I = \frac{12 - 0.7}{100} = \frac{11.3}{100} = 113 \text{ mA}$$

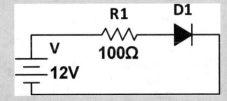

**Fig. 2.52** Circuit schematics of Example 2.49

**Example 2.50 Determine if the diode in the circuit of Fig. 2.53 is on or off.**

*Solution.* The positive port of the source is connected to the cathode of the diode, making it reverse-biased. This reverse-biased diode does not pass the current and is limited to a very small value in the microamp range. *The reverse-biased current is ideally zero.*

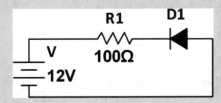

**Fig. 2.53** Circuit schematics of Example 2.50

**Example 2.51 Determine if the diode stays on or off in the circuit of Fig. 2.54.**

*Solution.* The amount of forward bias across the diode must be calculated to determine whether the diode is on or off. By looking at the circuit, the diode's direction and the source's polarity suggest a possible forward bias on the diode. However, is there enough voltage across it to keep it on?

If the diode is on, the voltage drop across the diode can be accurately calculated according to the currents on the $R_2$ and $R_3$. If the diode is on, the circuit has two loops. Considering this case, the currents of $I_1$, $I_2$ can be found as follows:

KVL Loop 1

$$-12 + 20I_1 + 100(I_1 - I_2) = 0$$

KVL Loop 2

In this loop, the voltage drop across the diode in forward bias is considered +0.6 V. The equations are solved for the amount of current in the diode branch to determine its actual voltage drop. The diode cannot stay on if the voltage drop does not reach 0.6 volts. Therefore

$$10I_2 + 0.6 + 100(I_2 - I_1) = 0$$

These equations are solved for $I_1$ and $I_2$.

$$120I_1 - 100I_2 = 12$$
$$-100I_1 + 110I_2 = -0.6$$

Multiplying the entire second equation by a factor of 1.2 results in

$$120I_1 - 100I_2 = 12$$
$$\rightarrow -120I_1 + 132I_2 = -0.72$$

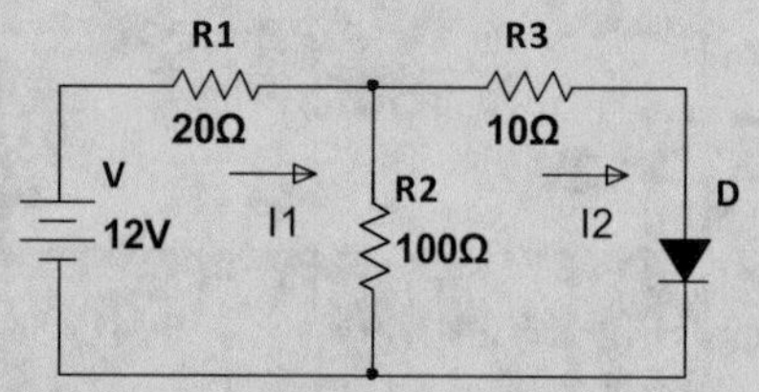

**Fig. 2.54** Circuit of Example 2.51

(continued)

**Example 2.51** (continued)

Adding these equations now can eliminate the $I_1$, as follows:

$$(120 - 120)I_1 + (-100 + 132)I_2 = 12 - 0.72 = 11.28$$

Therefore

$$I_2 = \frac{11.28}{32} = 0.3525 \text{ A}$$

And

$$I_1 = \frac{12 + 100 \times 0.3525}{120} = 0.39375 \text{ A}$$

The voltage drop across $R_2$ can be found as follows:

$$V_{R_2} = R_2 \times (I_1 - I_2)$$
$$V_{R_2} = 100 \times (0.39375 - 0.3525) = 4.125 \text{ V}$$

Knowing the voltage drop across the diode branch needs to be verified by knowing the current through the diode branch. If the equations result in a 0.6 V, then the earlier assumption is correct. Otherwise, it is not correct.

The voltage across the diode is

$$V_D = V_{R_2} - R_3 \times I_2$$
$$V_D = 4.125 - 10 \times 0.3525 = \mathbf{0.6}$$

The diode is *on*.


**Example 2.52 Repeat the previous example using KCL shown in Fig. 2.55.**
*Solution.* KCL in the node where $R_1$, $R_2$, and $R_3$ are connected, as shown in Fig. 2.55, is

$$I_{R_1} + I_{R_2} + I_{R_3} = 0$$

(continued)

**Example 2.52** (continued)

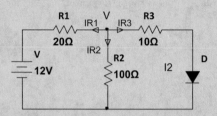

**Fig. 2.55** Circuit of Example 2.52

The direction of each current is considered exiting the node as all elements are passive elements (resistors).

Considering voltage $V$ at the KCL node and 0.6 V voltage drop across the diode to stay on, each current is calculated and replaced in the KCL equation as follows:

$$\frac{V-12}{20} + \frac{V}{100} + \frac{V-0.6}{10} = 0$$

Solve for $V$ as follows:

$$\frac{V}{20} - \frac{12}{20} + \frac{V}{100} + \frac{V}{10} - \frac{0.6}{10} = 0$$

$$0.05V - 0.6 + 0.01V + 0.1V + 0.1V - 0.06 = 0$$

$$0.16V = 0.66$$

$$V = 4.125 \text{ V}$$

This voltage is high enough to validate the earlier assumption that the diode stays on, provided that the voltage drop across $R_3$ does not exceed $4.125 - 0.6 = 3.525$ V. Therefore, the current passing through the diode is

$$I_{R_3} = \frac{3.525}{10} = 0.3525 \text{ A}$$

## Diode's Peak Inverse Voltage (PIV)

The diode is in reverse mode, as shown in the circuit of Example 2.50 opposes the current flow. In this case, the voltage across the diode is close to the voltage of the source. This is an inverse voltage. Like in Example 2.50, the voltage across the diode is $-12$ V as the source was 12 V and the diode was reversed across the source. The voltage drop across the resistor $R_1$ is zero as the current is zero.

If the source voltage increases, the inverse voltage across the diode can increase. As each diode can withstand a certain amount of voltage, the peak of inverse voltage or PIV is fixed for each diode. For instance, a diode with 40 PIV can withstand an inverse voltage of a maximum of 40 V. Any voltage above this value breaks the diode, and the current flows through the circuit. A broken diode becomes either open with no current passing or short with zero resistance. The PIV should be carefully considered in selecting diodes for any application. For instance, a diode with a PIV higher than 160 V is a suitable 110 V household application. The 110 V is an RMS or effective value and generates a $110 \times \sqrt{2} = 155.6$ V. As long as the peak voltage is less than the PIV limit of the diode, the diode is okay for the application, provided that it can pass the amount of current the application needs.

**Example 2.53 Find the bias of the diode and its PIV in the circuit shown in Fig. 2.56.**
*Solution.* The voltage at point $V$ determines the bias of the diode.

Assuming the diode is off, the voltage drop across the 150 $\Omega$ resistor determines PIV. Therefore, the current flow in the circuit from the source is

$$I = \frac{12}{50 + 150} = 0.06 \text{ A, or 60 mA}$$

The voltage across the 150 $\Omega$ resistor is calculated as

$$V_{150} = I \times 150 = 0.06 \times 150 = 9 \text{ V}$$

The voltage drop across the 150 $\Omega$ resistor is $+9$ V, and the connection of the diode to the resistor puts it in a reverse-biased position. Therefore, the earlier assumption is correct.

The PIV is the amount of voltage drop across the 150 $\Omega$ resistor or $-9$ V.

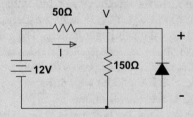

**Fig. 2.56** Circuit of Example 2.53

**Example 2.54 Find the PIV of the diode in the circuit shown in Fig. 2.57.**
*Solution*. The cathode of the diode is connected to the 200 V source, and its anode is connected to the 70 V source. Therefore, there is more voltage at the cathode than at the anode, which puts the diode in a reverse-biased position. Since the diode is reverse-biased, there is almost no current passing through the diode, and the current in the circuit is zero. The voltage drop across the 50 Ω resistor is zero. Therefore, the cathode is charged at +200 V, and the anode is charged at +70 V. Since the voltage across the diode is measured from anode to cathode, the voltage drop or PIV across the diode is

$$PIV = -200 + 70 = -130 \text{ V}$$

This can also be achieved by writing a KVL in the loop as follows: Starting from the negative port of the 200 V source

$$-200 + 50 \times 0 - V_d + 70 = 0$$

$$V_d = -130 \text{ V}$$

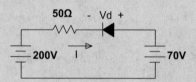

**Fig. 2.57** Circuit of Example 2.54

# Diode Current Carrying Capacity and Operating Frequency

Each electrical element is rated for specific power. For instance, a cellphone charger can provide enough power to charge the cellphone's battery, and a car charger has a higher power rating to charge a much larger battery. The amount of current passed through the circuit of a cellphone charger required diodes rated for less current, e.g., 1 A. However, the amount of current that passes through a car battery charger must meet the higher ratings, e.g., 35 A. Diodes with different current-carrying capacities must be selected for each of these applications.

Another consideration for diode operation is how fast it can switch from forward bias to reverse bias and vice versa. This is important for applications that need to operate at high frequency, for instance, in cellphone operations or ultrafast computing. If a diode is designed for low-frequency operation, its electrons and electronic structure cannot recover fast enough from forward bias to block the current in a full reverse bias. Therefore, the diode loses its functionality and starts to leak current in reverse bias. The higher the frequency, the more the current leakage. At some point, the diode does not operate and becomes obsolete.

**Thyristor** Diodes do not have an on-demand mechanism to turn on. They operate autonomously when they are forward-biased. To add a control mechanism to diodes, gate "G" was added. The new device is called a thyristor. To turn the device on, the gate must be excited, and the device must be forward-biased. There is no mechanism to turn the thyristor off on demand. However, like diodes, the thyristor can be turned off when it is reverse-biased, and the current reaches zero, called self-commutation. Some external circuits allow for such conditions to occur on demand. These circuits are called forced commutation circuits. Most of them work based on charging a capacitor with higher voltage and reverse connecting it to the thyristor to turn it off. Figure 2.58 shows the schematic of a thyristor. Similar to diodes, thyristors also conduct current only in one direction.

**GTO (gate turn-off)** A gate turn-off is the other type of switch that can turn high currents on demand. The same gate is used to turn on the device. When the gate of the GTO is positive with respect to the anode "A," the device is turned on. To turn it off, the gate "G" must be more positive than the cathode "K" for enough time. GOT directs the current only in one direction. Figure 2.59 shows the circuit schematic of GTO.

**Transistors** Figure 2.60 shows a transistor that has three ports. The gate "G" port is used to control the on-off action of the transistor. Once the gate is excited, the current flows from the drain "D" port to the source "S." The gate voltage must be positive with respect to the source port for the transistor to conduct. Otherwise, the transistor is turned off. To guarantee that the transistor is off, the voltage of the gate is either

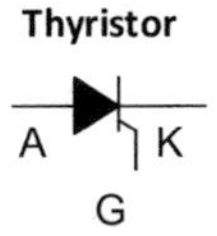

**Fig. 2.58** Schematic of a thyristor: "A" stands for the anode, "K" stands for the cathode, and "G" is the gate

Thyristor

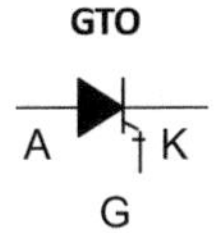

**Fig. 2.59** Schematic of a GTO: "A" stands for the anode, "K" stands for the cathode, and "G" is the gate

GTO

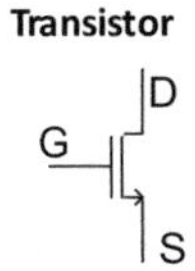

**Fig. 2.60** Schematic of a transistor: "G" is the gate, "D" is the drain, and "S" is the source of the transistor

Transistor

**Fig. 2.61** Schematic of a
triac. "A" stands for the
anode and "K" stands for the
cathode, and "G" is the gate

connected to the ground or the voltage of the gate has to become negative with
respect to the source. Transistors can operate at ultrahigh frequencies. Transistors
can also operate at very high voltages and conduct very high currents. Some
examples of these devices are inverters that generate AC voltage from DC sources.
Computer processors employ millions of transistors that work together to perform
digital calculations. Most of the transistors utilized in power circuits must have a
driver circuit to ensure proper voltage application at the gate of the transistor and
optimize its operation. Many transistors can be controlled by applying voltage or
current to the gate or base. Transistors, when turned on, can conduct current only in
one direction from their drain to their source.

**Triac**  Most of the switches studied so far (diodes, transistors, GTOs, and thyristors)
conduct the current only in one direction. They must be forward-biased, and proper
gate commands must be applied to all (except for the diode) to start conducting.
Triac is a device that can conduct the current in both directions, providing a suitable
gate command in either direction. Figure 2.61 shows the schematic of a triac.

## Linear Time-Invariant Circuits

The electric circuits have parameters that may include resistance of resistors, inductance or inductors, and capacitance of capacitors. The amount of resistance, inductance, and capacitance can be fixed or change over time. If the circuit parameters do
not change over time, the circuit is called time invariant; if the parameters change
over time, it is called time variant.

On another note, the circuit current and voltage relations over the elements may
follow the rules discussed earlier. That is, for instance, $R = \frac{v(t)}{i(t)}$. This is a linear
relationship between the voltage and current. However, there might be some circuits
in which the ratio of voltage and current is not linear. For instance, $v(t) = i^2(t) +
2i(t) + 1$ represents a nonlinear relationship between the voltage and current.
Inductors follow the $v(t) = L\frac{di}{dt}$, meaning that the ratio of the voltage and derivative
of current is fixed. A nonlinear inductor can change over time itself. For instance,
$v(t) = L(t)\frac{di}{dt}$ is a nonlinear circuit. The same is true for a capacitor; $i(t) = C(t)\frac{dv}{dt}$ is a
nonlinear capacitor.

**Example 2.55 In the circuit of Fig. 2.62, the voltage and current across the resistor is $v = i^2$. Find the power dissipated in $t = 1$ s.**
*Solution.* The power dissipation from a resistor is

$$P = v \times i$$

Replacing the values yields

$$P = i^2 \times i = i^3$$

The total amount of energy in $t = 1$ s becomes

$$W = \int P\, dt = \int_0^1 i^3\, dt = \frac{1}{4} i^4 \Big|_0^1 = \frac{1}{4}\ J$$

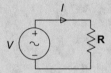

**Fig. 2.62** Circuit of Example 2.55

A linear circuit is called a circuit where the relationship of voltage and current in all components follows a linear rule, and the circuit topology does not change over time. An example of a varying topology circuit is a circuit in which its diode is turned on and off over time. Alternatively, a circuit with a switch inside is forced to open or close. These circuits are also called switching circuits.

This book analyses the linear time-invariant circuits or LTI circuits. Linear systems hold two important properties:

1. *Scalability:* meaning that if the source voltage is increased by a factor $\alpha$, the value of voltage and currents also increase by the same factor. The function $f$ with input $i$ and output $v$ is scalable if the following holds:

$$v = f(i)$$

$$\text{if} : i \rightarrow \alpha i \Longrightarrow v \rightarrow \alpha v$$

It must be proven as follows:

$$f(\alpha i) = \alpha f(i) = \alpha v$$

2. *Superposition:* meaning that the effect of multiple sources simultaneously equals the effect of summation of them individually. The function $f$ with two inputs of $i_1$ and $i_2$ and output of $v$ holds superposition if the following holds:

$$v_1 = f(i_1), v_2 = f(i_2)$$

If the input is a summation of $i_1 + i_2$, then the outputs in a linear circuit should add up as well. This can be interpreted as follows:

$$\text{if} : i = i_1 + i_2 \Longrightarrow v = v_1 + v_2$$

$$\text{Then} : i = i_1 + i_2 \rightarrow f(i_1 + i_2) = f(i_1) + f(i_2) = v_1 + v_2$$

**Example 2.56 Is the circuit with $v = 2i$ equation linear?**
*Solution.* Check for scalability. If $i \rightarrow \alpha i \Rightarrow v = (\alpha i)^2 = \alpha^2 i^2 = \alpha^2 v \neq \alpha v$. It shows that the circuit is not scalable.
   Check for the superposition.

$$\text{if} : i_1 \Longrightarrow v_1 \& i_2 \Longrightarrow v_2$$

$$\text{Then} : i = i_1 + i_2 \Longrightarrow v = 2(i_1 + i_2) = 2i_1 + 2i_2 = v_1 + v_2$$

Then the system holds superposition.
Since the circuit is scalable and holds superposition, it is a linear system.

**Example 2.57 Is the circuit with $v = i^2$ equation linear?**
*Solution.* Check for scalability. If $i \rightarrow \alpha i \Rightarrow v = 2(\alpha i) = \alpha 2i = \alpha(2i) = \alpha v$. It shows that the circuit is scalable.
   Check for the superposition.

$$\text{if } i_1 \Longrightarrow v_1 \& i_2 \Longrightarrow v_2$$

$$\text{Then} : i = i_1 + i_2 \Longrightarrow v = (i_1 + i_2)^2 = i_1^2 + i_2^2 + 2i_1 i_2 = v_1 + v_2 + 2i_1 i_2 \neq v_1 + v_2$$

Then the system does not hold superposition.
Since the circuit is not scalable and does not hold superposition, it is *not* a linear system.

## Superposition in Circuits

In linear circuits, the superposition holds and vice versa. This means that if there are multiple independent sources and the circuit is linear (two conditions hold, and the circuit does not change topology), the voltage of nodes and branches can be found due to individual sources and then added to determine the overall voltages and currents.

---

**Example 2.58 A repeat of Example 2.42. Find the voltages of $V_1$, $V_2$, and $V_3$, when $I_1(t) = 10$ (A) and $I_2(t) = -20$ (A) using superposition. $R_1 = 100\ \Omega$, $R_2 = 250\ \Omega$, and $R_3 = 50\ \Omega$ (Fig. 2.63).**

*Solution.* The circuit has two fixed independent sources and no switching device (if there was a switching device, it must have remained on or off the entire time). The circuit is linear, and then the superposition holds.

The source $I_1$ results in some voltages, and the source $I_2$ results in some other voltages. The voltages need to be identified and then added together.

**Effect of $I_1$: This means only $I_1$ is on, and the other sources must turn off, i.e., $I_2 = 0$. The second source must turn off; zero current means an open circuit. Therefore, the circuit becomes**

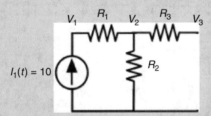

In this circuit, there is no current passing through the resistor $R_3$. Therefore, the voltage of the node $V_3$ equals the voltage of the node $V_2$. Two resistors of $R_1$ and $R_2$ are in series as the current exiting one enters the other.

The source forces the current $I_1 = 10$ A. Therefore, the voltage of the node $V_1 = 10(R_1 + R_2) = 10 \times (100 + 250) = 10 \times 350 = 3500$ V. The voltage of $V_2$ is the current passing through times, the resistance of $R_2$. $V_2 = 10\,R_2 = 10 \times 250 = 2500$ V. Hence, $V_3 = 2500$ V.

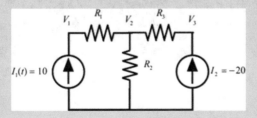

**Fig. 2.63** Circuit of Example 2.58

(continued)

**Example 2.58** (continued)

**Effect of $I_2$: This means only $I_2$ is on, and the other sources must turn off, i.e., $I_1 = 0$. The first source must turn off; zero current means an open circuit. Therefore, the circuit becomes**

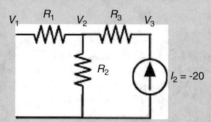

In this circuit, there is no current passing through the resistor $R_1$. Therefore, the voltage of the node $V_1$ equals the voltage of the node $V_2$. Two resistors of $R_3$ and $R_2$ are in series as the current exiting one enters the other.

The source forces the current $I_2 = -20$ A. Therefore, the voltage of the node $V_3 = -20(R_3 + R_2) = -20 \times (50 + 250) = -20 \times 300 = -6000$ V. The voltage of $V_2$ is the current passing through times, the resistance of $R_2$. $V_2 = -20R_2 = -20 \times 250 = -5000$ V. Hence, $V_1 = -5000$ V.

**Effect of both sources: (effect of source 1 + effect of source 2)**

$$\text{Overall}: V_1 = 3500 + (-5000) = 1500 \text{ V}$$

$$\text{Overall}: V_2 = 2500 + (-5000) = -2500 \text{ V}$$

$$\text{Overall}: V_3 = 2500 + (-6000) = -3500 \text{ V}$$

**Example 2.59 Redo Example 2.44 using superposition. Find the voltage $V_2$ and the current $I_1$ when the DC source is $V = 100$ V, $I = 15$ A, and $R_1 = 25\ \Omega$, $R_2 = 75\ \Omega$ shown in Fig. 2.64.**
*Solution.*

*Effect of the voltage source:* In this case, all other sources must turn off. This means that the current is off equal to 0 A, which resembles an open source.

(continued)

**Example 2.59** (continued)

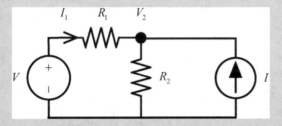

**Fig. 2.64** Circuit of Example 2.59

The circuit becomes as follows:

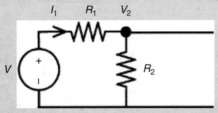

In this circuit, the resistors $R_1$ and $R_2$ are connected in series. Therefore, the current of the circuit and the voltage at the node $V_2$ can be found as follows:

$$I_1 = \frac{V}{R_1 + R_2} = \frac{100}{25 + 75} = 1 \text{ A}$$

$$V_2 = I \times R_2 = 1 \times 75 = 75 \text{ V}$$

*Effect of the current source:* In this case, the voltage source must turn off. A 0 V source resembles a short circuit shown as follows:

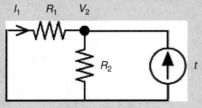

The current source feeds two parallel resistors. The direction of the measured current $I_1$ is in the opposite direction of the current source feed. Therefore, the current in $R_1$ is a negative value obtained as follows:

(continued)

**Example 2.59** (continued)

$$I_1 = -\frac{R_2}{R_1 + R_2} I = -\frac{75}{25 + 75} 15 = -11.25 \text{ A}$$

$$V_2 = I(R_1 \mid \mid R_2) = 15 \times (25 \mid \mid 75) = 15 \times \frac{25 \times 75}{25 + 75} = 15 \times 18.75 = 281.25 \text{ V}$$

*The overall effect of sources* is a summation of the effect of individual sources as follows:

$$\text{Overall}: I_1 = 1 + (-11.25) = -10.25 \text{ A}$$

$$\text{Overall}: V_2 = 75 + 281.25 = 356.25 \text{ V}$$

**Example 2.60 A copper wire has a cross-section of 1 mm². Select a length of the wire in which the voltage drop across the length of the wire does not exceed 15 V when a load of 25 A is fed.**
*Solution.* The specific resistance of copper is $9.71 \times 10^{-8}$ $\Omega$. m. The total resistance of the copper is

$$R = \frac{\rho L}{A}$$

$$R = 9.71 \times 10^{-8} \frac{L}{1 \times 1e - 6 \ (\text{mm}^2 \to \text{m}^2)} = 0.00971 \times L \ (\Omega)$$

The voltage drop across the wire is limited to 15 V. That is

$$\Delta V = RI = 0.00971 \times L \times 25 = 15$$

Solve for *L*:

$$L = \frac{15}{25 \times 0.00971} = 61.79 \ (\text{m})$$

The wire can run for 61.79 m and feed the current of 25 A.

**Example 2.61 In the wire of Example 2.60, if the resistance of both wires that connect to the positive and negative terminals of the battery are included, what would be the total length of the cable before the total voltage drop exceeds 15 V?**

*Solution.* Each of the wires has resistance and, therefore, a voltage drop. The 15 V drop limit is now applied to both wires, and each is allowed to drop 7.5 V. The length of the wire is cut in half, allowing the wire to connect to the positive and negative terminals.

Another way to look at this problem is to double the resistance, as both wires are concerned. This results in

$$2(\text{wire}) \times R = 2 \times 9.71 \times 10^{-8} \frac{L}{1 \times 1e - 6 \ (\text{mm}^2 \rightarrow \text{m}^2)} = 0.01942 \times L \ (\Omega)$$

The voltage drop across the wire is limited to 15 V. That is

$$\Delta V = R_{(2 - \text{wires})}I = 0.01942 \times L \times 25 = 15$$

Solve for $L$:

$$L = \frac{15}{25 \times 0.01943} = 30.88 \ (\text{mH})$$

**Example 2.62 Find the equivalent resistance in the circuit of Fig. 2.65.**
*Solution.*

$$R_{\text{eq}} = 50 + (150 \mid\mid (25 + 5)) + 15$$

$$R_{\text{eq}} = 50 + (150 \mid\mid (30)) + 15$$

$$R_{\text{eq}} = 50 + \frac{150 \times 30}{150 + 30} + 15$$

$$R_{\text{eq}} = 50 + 25 + 15$$

$$R_{\text{eq}} = 90 \ \Omega$$

**Fig. 2.65** Circuit of Example 2.62

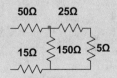

**Example 2.63 Find the equivalent resistance of the circuit shown in Fig. 2.66.**

*Solution.* The 150 Ω resistor forms a Y connection in combination with the 50 Ω and 25 Ω resistors. This can be converted to a Δ connection as follows:

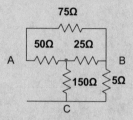

Δ is among the circuit's points A, B, and C. The equivalent resistor between these points can be calculated as follows:

$$R_{AC} = \frac{50 \times 150 + 50 \times 25 + 150 \times 25}{25} = 500 \ \Omega$$

$$R_{AB} = \frac{50 \times 150 + 50 \times 25 + 150 \times 25}{150} = 83.3 \ \Omega$$

$$R_{BC} = \frac{50 \times 150 + 50 \times 25 + 150 \times 25}{50} = 250 \ \Omega$$

The circuit with converted $Y$ to $\Delta$ equivalent is as follows:

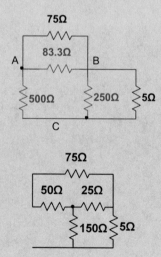

**Fig. 2.66**  Circuit of Example 2.63

(continued)

**Example 2.63** (continued)

The circuit now shows a couple of resistors in parallel, such as $75\Omega\|83.3\Omega$ and $250\Omega\|5\Omega$. Their equivalents are

$$75\Omega\|83.3\Omega = \frac{75 \times 83.3}{75 + 83.3} = 39.46\ \Omega$$

$$250\Omega\|5\Omega = \frac{250 \times 5}{250 + 5} = 4.9\ \Omega$$

The circuit becomes

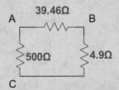

Considering ports A and C, the equivalent circuit becomes

$$R_{eq} = 500\|(39.46 + 4.9) = \frac{500 \times 44.36}{500 + 44.36} = 40.74\ \Omega$$

**Example 2.64 Find the voltages and currents in the circuit of Fig. 2.67.**
*Solution.* The circuit can be solved both through KVL and KCL. As a reminder, KCL provides the voltages of nodes, and KVL provides the currents of loops. The solution is provided using both approaches.
*Using KVL*
There are two loops in the circuit. Starting from the negative port of the 180 V source, the KVL is

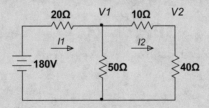

**Fig. 2.67** Circuit of Example 2.64

(continued)

**Example 2.64** (continued)

$$-180 + 20I_1 + 50(I_1 - I_2) = 0$$

In the second loop starting from the 40 $\Omega$ resistor, the KVL is

$$40I_2 + 50(I_2 - I_1) + 10I_2 = 0$$

Solving these two equations results in

$$(20 + 50)I_1 - 50I_2 = 180$$
$$-50I_1 + (40 + 50 + 10)I_2 = 0$$

Then,

$$70I_1 - 50I_2 = 180$$
$$-50I_1 + 100I_2 = 0$$

From the second equation, find $I_1$:

$$-50I_1 = -100I_2 \rightarrow I_1 = 2I_2$$

Replacing the first equation results in

$$70(2I_2) - 50I_2 = 180$$

Simplifying

$$140I_2 - 50I_2 = 180 \rightarrow 90I_2 = 180 \rightarrow I_2 = 2 \text{ A}$$
$$I_1 = 2I_2 = 2 \times 2 = 4 \text{ A}$$

The voltage of $V_1$ and $V_2$ can be found as follows:

$$V_1 = 50 \, (I_1 - I_2) = 50(4 - 2) = 100 \text{ V}$$
$$V_2 = 40I_2 = 40 \times 2 = 80 \text{ V}$$

*Using KCL*

(continued)

**Example 2.64**  (continued)

The voltage of nodes is 180, $V_1$ and $V_2$. Therefore, considering that all current is exiting the nodes through passive elements, the KCL for the nodes $V_1$ and $V_2$ is

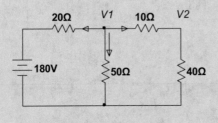

$$I_{20\Omega} + I_{50\Omega} + I_{10\Omega} = 0$$

Replacing all current with voltages of the nodes results in

$$\frac{V_1 - 180}{20} + \frac{V_1}{50} + \frac{V_1 - V_2}{10} = 0$$

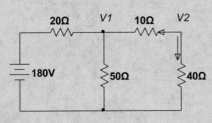

$$I_{10\Omega} + I_{40\Omega} = 0$$

Replacing all current with voltages of the nodes results in

$$\frac{V_2 - V_1}{10} + \frac{V_2}{40} = 0$$

Solve for $V_1$ and $V_2$.

$$V_1\left(\frac{1}{20} + \frac{1}{50} + \frac{1}{10}\right) - V_2\left(\frac{1}{10}\right) = \frac{180}{20}$$

(continued)

**Example 2.64** (continued)

$$V_1\left(-\frac{1}{10}\right) + V_2\left(\frac{1}{10}+\frac{1}{40}\right) = 0$$

Simplifying results in

$$0.17V_1 - 0.1V_2 = 9$$

$$-0.1V_1 + 0.125V_2 = 0$$

From the second equation, find $V_1$:

$$V_1 = \frac{0.125}{0.1} V_2 = 1.25V_2$$

Replacing the first equation, solve for $V_2$.

$$0.17(1.25\ V_2) - 0.1V_2 = 9$$

$$(0.2125 - 0.1)V_2 = 9$$

$$V_2 = \frac{9}{0.1125} = 80\ \text{V}$$

$$V_1 = 1.25\ V_2 = 1.25 \times 80 = 100\ \text{V}$$

From these voltages, the currents can be calculated as follows:

$$I_1 = \frac{180 - V_1}{20} = \frac{180 - 100}{20} = \frac{80}{20} = 4\ \text{A}$$

$$I_2 = \frac{V_1 - V_2}{10} = \frac{100 - 80}{10} = \frac{20}{10} = 2\ \text{A}$$

**Example 2.65 A space heater element is required to provide 1500 W of power to a blower. At the voltage of 110 V, find the resistance of the element.**
*Solution.* The power dissipation from the element is directly related to its resistance.

(continued)

**Example 2.65** (continued)

$$P = RI^2 = \frac{V^2}{R}$$

$$1500 = \frac{110^2}{R}$$

Therefore,

$$R = \frac{110^2}{1500} = 8.067 \ \Omega$$

**Example 2.66 Find the voltage drop across the heating element of Example 2.65.**
*Solution.* The voltage drop across a resistor needs the current and the resistance. Since the power is fixed at 1500 W, at the voltage of 110 V, the current can be calculated as

$$I = \frac{P}{V} = \frac{1500}{110} = 13.63 \ \text{A}$$

The amount of voltage drop across the resistor is

$$\Delta V = RI = 8.067 \times 13.63 = 107.77 \ \text{V}$$

**Example 2.67 Find the current $I_3$ in the circuit of Fig. 2.68.**
*Solution.* A KCL at the node is written by considering $I_1$ negative as it enters the node, $I_2$ negative as it enters the node, and $I_3$ positive as it exits the node. Arrows show the direction of these currents.

**Fig. 2.68** Circuit of Example 2.67

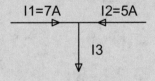

(continued)

**Example 2.67** (continued)

Therefore,

$$-I_1 - I_2 + I_3 = 0$$
$$-7 - 5 + I_3 = 0$$
$$I_3 = 12 \text{ A}$$

**Example 2.68 Find the voltage across the resistor $R$ as shown in the circuit of Fig. 2.69.**

*Solution.* KVL in the loop can be written starting from the negative port of the 9 V source as follows:

$$-9 - 5 + 1.5 - V = 0$$
$$V = 12.5 \text{ V}$$

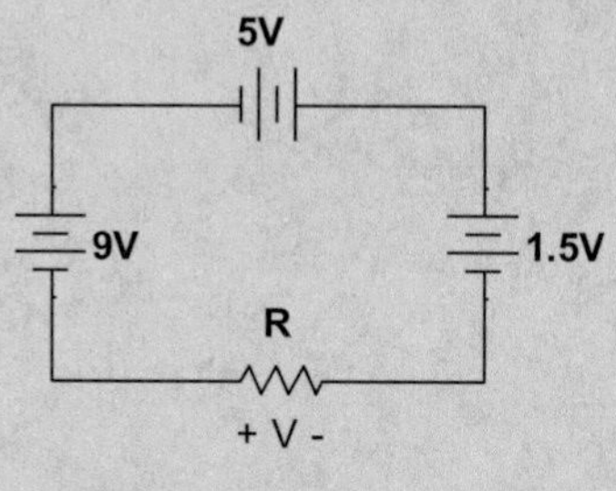

**Fig. 2.69** Circuit of Example 2.68

**Example 2.69 Find the voltage across all the circuit resistors shown in Fig. 2.70 using the voltage division technique.**

*Solution.*

The voltage of 1 kΩ is

$$V_{1k} = \frac{1k}{1k + 3k + 5k} \times 90 = 10 \text{ V}$$

(continued)

**Example 2.69** (continued)

**Fig. 2.70** Circuit of
Example 2.69

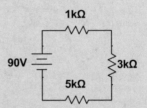

The voltage of 3 kΩ is

$$V_{3k} = \frac{3k}{1k + 3k + 5k} \times 90 = 30 \text{ V}$$

The voltage of 5 kΩ is

$$V_{5k} = \frac{5k}{1k + 3k + 5k} \times 90 = 50 \text{ V}$$

**Example 2.70 Find the current of each branch in the circuit of Fig. 2.71.**
*Solution.* The current of each branch can be found using the current division technique as follows:

$$I_{1k} = \frac{\frac{1}{1k}}{\frac{1}{1k} + \frac{1}{3k} + \frac{1}{5k}} 10 = 0.652 \times 10 = 6.52 \text{ A}$$

$$I_{3k} = \frac{\frac{1}{3k}}{\frac{1}{1k} + \frac{1}{3k} + \frac{1}{5k}} \times 10 = 0.217 \times 10 = 2.17 \text{ A}$$

$$I_{5k} = \frac{\frac{1}{5k}}{\frac{1}{1k} + \frac{1}{3k} + \frac{1}{5k}} \times 10 = 0.131 \times 10 = 1.31 \text{ A}$$

**Fig. 2.71** Circuit of
Example 2.70

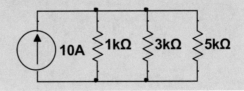

**Example 2.71 Find $V_1$, $V_2$ and $I$ in the circuit of Fig. 2.72.**

*Solution.* KCL generally provides a simpler solution to find the voltage of any node. The current source feeds the circuit at the node $V_1$ through the 1 kΩ resistor. Therefore, KCL at node $V_1$ results in

$$\frac{V_1 - 20}{10k} + \frac{V_1}{25k} - 2.5e - 3 = 0$$

$$0.0001\ V_1 - 0.002 + 0.00004\ V_1 - 0.025 = 0$$

$$0.00014 V_1 = 0.027$$

$$V_1 = \frac{0.027}{0.0014} = 19.285$$

The current through the 1 kΩ resistor is 2.5 mA, which drops a voltage of $1k \times 2.5 \times 1e - 3 = 2.5$ V.

Since the direction of the current is from $V_2$ to $V_1$. The voltage of $V_2 > V_1$. Therefore, the voltage of the node $V_2$ can be found as follows:

$$V_2 = V_1 + 2.5e - 3 \times 1e3 = 19.285 + 2.5 = 21.785 \text{ V}$$

The current $I$ can be found as

$$I = \frac{V_1}{25k} = \frac{19.285}{25k} = 0.00077 \text{ or } 0.77 \text{ mA}$$

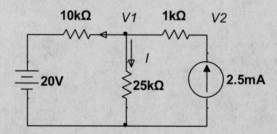

**Fig. 2.72** Circuit of Example 2.71

**Example 2.72 Find $V_1$, $V_2$ and $I$ in the circuit of Fig. 2.73.**
*Solution.* Considering that current drains out of the passive elements, KCL at node $V_1$ becomes

$$\frac{V_1 - 20}{10k} + \frac{V_1}{25k} + \frac{V_1 - V_2}{1k} = 0$$

$$0.0001V_1 - 0.002 + 0.00004V_1 + 0.001V_1 - 0.001V_2 = 0$$

$$0.00114V_1 - 0.001V_2 = 0.002$$

KCL at node $V_2$

$$-2.5m + \frac{V_2 - V_1}{1k} + \frac{V_2 - 20}{5k} = 0$$

$$-0.0025 + 0.001V_2 - 0.001V_1 + 0.0002V_2 - 0.004 = 0$$

$$-0.001V_1 + 0.0012V_2 = 0.0065$$

Solving two equations for $V_1$ and $V_2$ results in

$$\begin{cases} 0.00114V_1 - 0.001V_2 = 0.002 \\ -0.001V_1 + 0.0012V_2 = 0.0065 \end{cases}$$

Multiply the first equation by 1.2, and adding two equations eliminates $V_2$ as follows:

$$\begin{cases} 0.001368V_1 - 0.0012V_2 = 0.0024 \\ -0.001V_1 + 0.0012V_2 = 0.0065 \end{cases}$$

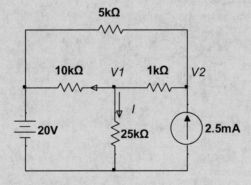

**Fig. 2.73** Circuit of Example 2.72

(continued)

**Example 2.72** (continued)

$$0.000368V_1 = 0.0089$$

$$V_1 = 24.184 \text{ V}$$

$$\rightarrow -0.001V_1 + 0.0012V_2 = 0.0065$$

$$-0.001 \times 24.184 + 0.0012V_2 = 0.0065$$

$$V_2 = \frac{0.0065 + 0.001 \times 24.184}{0.0012} = 25.57 \text{ V}$$

Therefore, the current $I$ becomes

$$I = \frac{V_1}{25k} = \frac{24.184}{25k} = 0.96 \text{ mA}$$

**Example 2.73 Find the power delivered or generated from each source in the circuit of Fig. 2.73.**

*Solution.* The power delivery from a source or to a source depends on the direction of the source's current. The 2.5 mA source feeds the 2.5 mA current to the circuit at the calculated voltage of $V_2 = 25.57$ V. Therefore, this source delivers a power of $P = 2.5 e - 3 \times 25.57 = 0.0639$ Watts. This is 63.9 mW of power.

The node of $V_1 = 24.181$ V, and the node $V_2 = 25.57$ V. Therefore, the direction of the current at the 20 V voltage source cannot be to the circuit. The summation of currents that flow through 5 k$\Omega$ and 10 k$\Omega$ feed this source.

The total current that feeds the source is

$$I_{20V} = \frac{V_2 - 20}{5k} + \frac{V_1 - 20}{10k}$$

$$I_{20V} = \frac{25.57 - 20}{5k} + \frac{24.184 - 20}{10k} = 0.001114 + 0.0004184 = 0.0015324$$

The power fed to the 20 V source is a negative value as follows:

$$P_{20V} = 0.0015324 \times 20 = 0.03064 \text{ Watts or } 30.64 \text{ mW.}$$

(continued)

**Example 2.73** (continued)

The current source provides 63.9 mW of power, and 30.64 mW of it is received by the 20 V source. The rest of the power is dissipated in the resistors as heat. That is

$$P_{\text{Loss}} = 63.9m - 30.64 = 33.26 \text{ mW}$$

**Example 2.74 Find the equivalent capacitance of the circuit in Fig. 2.74.**
*Solution.* The circuit shows a series of 1 μF and 2 μF in parallel with a series of 3 μF and 4 μF. Their equivalent is in series with the 5 μF. This results in
Series of 1 μF and 2 μF

$$C_{1,2} = \frac{1\mu \times 2\mu}{1\mu + 2\mu} = \frac{2}{3} \ \mu F$$

Series of 3 μF and 4 μF

$$C_{3,4} = \frac{3\mu \times 4\mu}{3\mu + 4\mu} = \frac{12}{7} \mu F$$

$$C_{1,2} \big\| C_{3,4} = C_{1,2} + C_{3,4} = \frac{2}{3} + \frac{12}{7} = 2.38 \ \mu F$$

Therefore, $C_{\text{eq}}$ is a series of 2.35 μF and 5 μF as follows

$$C_{\text{eq}} = \frac{2.35\mu \times 5\mu}{2.35\mu + 5\mu} = 1.612 \ \mu F$$

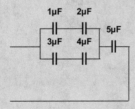

**Fig. 2.74** Circuit of Example 2.74

**Example 2.75 Find equivalent $Y$ capacitance of the $\Delta$ circuit shown in Fig. 2.75.**

*Solution.* $C_1$ shares a node at point $A$ with $C_{AB}$, $C_{AC}$; therefore,

$$\frac{1}{C_1} = \frac{\frac{1}{C_{AB}}\frac{1}{C_{AC}}}{\frac{1}{C_{AB}} + \frac{1}{C_{AC}} + \frac{1}{C_{BC}}}$$

$$\frac{1}{C_1} = \frac{\frac{1}{1} \times \frac{1}{2}}{\frac{1}{1} + \frac{1}{2} + \frac{1}{3}} \Rightarrow C_1 = 3.66\ \text{F}$$

$C_2$ shares a node at point $B$ with $C_{AB}$, $C_{BC}$; therefore,

$$\frac{1}{C_2} = \frac{\frac{1}{C_{AB}}\frac{1}{C_{BC}}}{\frac{1}{C_{AB}} + \frac{1}{C_{AC}} + \frac{1}{C_{BC}}}$$

$$\frac{1}{C_2} = \frac{\frac{1}{1} \times \frac{1}{3}}{\frac{1}{1} + \frac{1}{2} + \frac{1}{3}} \Rightarrow C_2 = 5.51\ \text{F}$$

$C_3$ shares a node at point $C$ with $C_{AC}$, $C_{BC}$; therefore,

$$\frac{1}{C_3} = \frac{\frac{1}{C_{BC}}\frac{1}{C_{AC}}}{\frac{1}{C_{AB}} + \frac{1}{C_{AC}} + \frac{1}{C_{BC}}}$$

$$\frac{1}{C_3} = \frac{\frac{1}{2} \times \frac{1}{3}}{\frac{1}{1} + \frac{1}{2} + \frac{1}{3}} \Rightarrow C_3 = 11.09\ \text{F}$$

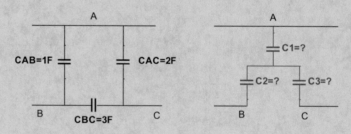

**Fig. 2.75** Circuit of Example 2.75

**Example 2.76 Find equivalent $Y$ inductance of the $\Delta$ circuit shown in Fig. 2.76.**

*Solution.*

$$L_{AB} = \frac{L_1L_2 + L_1L_3 + L_2L_3}{L_3} \rightarrow L_{AB} = \frac{1 \times 2 + 1 \times 3 + 2 \times 3}{3} = 3.66 \text{ H}$$

$$L_{AC} = \frac{L_1L_2 + L_1L_3 + L_2L_3}{L_2} \rightarrow L_{AC} = \frac{1 \times 2 + 1 \times 3 + 2 \times 3}{2} = 5.5 \text{ H}$$

$$L_{BC} = \frac{L_1L_2 + L_1L_3 + L_2L_3}{L_1} \rightarrow L_{BC} = \frac{1 \times 2 + 1 \times 3 + 2 \times 3}{1} = 11 \text{ H}$$

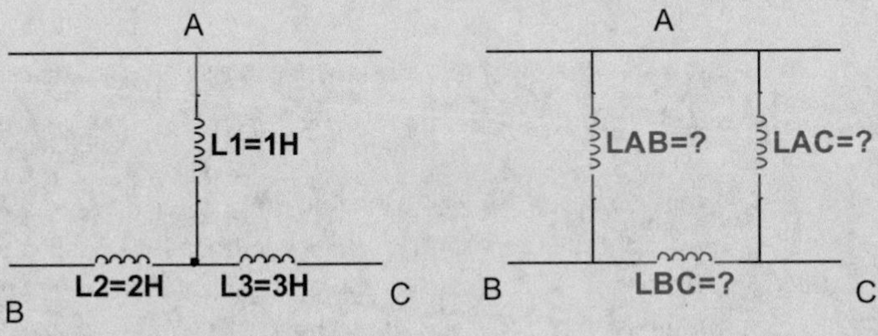

**Fig. 2.76** Circuit of Example 2.76

**Example 2.77 Find the amount of energy stored in the inductor shown in the circuit of Fig. 2.77.**

*Solution.* The DC source feeds the inductor until it is fully charged. A fully charged inductor acts as a short circuit. Therefore, the inductor in the circuit can be replaced by a short circuit, making two resistors of 100 $\Omega$ and 1 k$\Omega$ parallel. The current passing through the 1 k$\Omega$ resistor is the current of the inductor as well.

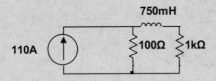

**Fig. 2.77** Circuit of Example 2.77

(continued)

**Example 2.77** (continued)

That can be found as follows:

$$I_{1k\Omega} = \frac{100}{100 + 1k} \times 110 = 10 \text{ A}$$

The amount of energy stored in the inductor is, therefore,

$$W = \frac{1}{2}LI^2 = \frac{1}{2}750e - 3 \times 10^2 = 37.5 \text{ Jouls}$$

**Example 2.78 Find the amount of energy stored in the circuit's capacitor in Fig. 2.78.**
*Solution.* The DC source starts to charge the capacitor. Once fully charged, the capacitor does not accept any more charges and becomes an open circuit. In this case, the current stops flowing through the resistor 500 $\Omega$. This results in a 0 V voltage drop on the resistor, making the voltage of the capacitor the same as the voltage across the 200 $\Omega$ resistor. This means $V_{200\Omega} = V_C$. The voltage division on the 100 $\Omega$ and 200 $\Omega$ resistor yields:

$$V_{200\Omega} = \frac{200}{100 + 200} \times 30 = 20 \text{ V}$$

The amount of energy stored in the capacitor is

$$W = \frac{1}{2}CV^2 = \frac{1}{2}100e - 3 \times 20^2 = 20 \text{ j}$$

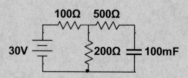

**Fig. 2.78** Circuit of Example 2.78

**Example 2.79 Find the amount of energy stored in the inductor and capacitor of the circuit shown in Fig. 2.79.**

*Solution.* The DC source charges the inductor and capacitor in two parallel paths. However, when fully charged, the inductor becomes a short circuit, and the capacitor becomes an open circuit. This changes the currents such that the 10 Ω resistor path has 0 A. Therefore, the voltage across the capacitor becomes the voltage measured across the inductor and 5 Ω resistor series. The current through the inductor branch depends on the resistances of 5 Ω and 20 Ω. The voltage drop across the fully charged inductor becomes zero. Therefore, the voltage of the capacitor is similar to the voltage drop across the 5 Ω resistor.

$$V_c = V_{5\Omega}$$

$$V_{5\Omega} = \frac{5}{20+5} \times 50 = 10 \text{ V}$$

$$I_L = \frac{50}{20+5} = 2 \text{ A}$$

Therefore, the amount of energy in the capacitor and inductor can be found as

$$W_C = \frac{1}{2}CV^2 = \frac{1}{2} \times 2 \times 10^2 = 100 \text{ J}$$

$$W_L = \frac{1}{2}LI^2 = \frac{1}{2}1 \times 2^2 = 2 \text{ J}$$

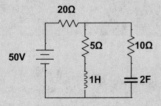

**Fig. 2.79** Circuit of Example 2.79

**Example 2.80** In vehicles whose windows are operated electrically, the driver door and each door are equipped with a switch that allows for two-point controls. However, the safety feature requires that the window be stopped if both controllers are commanded simultaneously and the window operates only if one of the control switches is pressed. To implement this logic, design a control circuit using two contact switches, one NC and one NO.

*Solution.* The logic requires that the switches not be on at the same time. That can be interpreted as the following logic:

$$\text{IF } SW_1 = 1 \& SW_2 = 1 \Rightarrow \text{motor : OFF}$$

$$\text{IF } SW_1 = 1 \& SW_2 = 0 \Rightarrow \text{motor : ON}$$

$$\text{IF } SW_1 = 0 \& SW_2 = 1 \Rightarrow \text{motor : ON}$$

$$\text{IF } SW_1 = 0 \& SW_2 = 0 \Rightarrow \text{motor : OFF}$$

The table that shows how these two switches can operate can be translated to the circuit shown in Fig. 2.80.

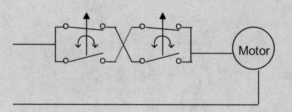

**Fig. 2.80**  The figure of Example 2.80

**Example 2.81 Design a circuit to turn three LED lights connected in series from two physically separated single-pole switches.**

*Solution.* Either switch SW1 and SW2 can turn all LEDs in series. The circuit is shown in Fig. 2.81.

(continued)

**Example 2.81** (continued)

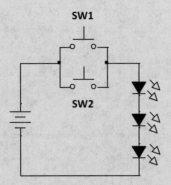

**Fig. 2.81** The figure of the circuit for Example 2.81

**Example 2.82 Using superposition, find the voltage and current of the 3 Ω resistor in the circuit of Fig. 2.82.**
*Solution.* There are two independent sources and one dependent source. The value of the dependent source is controlled by the current passing through the 9 Ω resistor, as shown in Fig. 2.82. The superposition holds in linear circuits, and the circuit has no time-variant parameters. The independent sources must turn on one by one and their effects recorded. Then the overall voltage *V* is the linear combination of the two sources.

*Effect of the current source.* The voltage source must turn off, i.e., 0 V, which means a short circuit. This shorts out the 5 Ω resistor. In this case, the 3 Ω and 9 Ω resistors are in parallel. This is shown as follows:

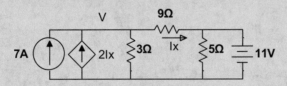

**Fig. 2.82** Circuit of Example 2.82

(continued)

**Example 2.82** (continued)

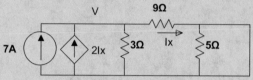

The KCL at node $V$ results in

$$-7 - 2I_X + \frac{V}{3} + \frac{V}{9} = 0$$

The current $I_X$ can be found as

$$I_X = \frac{V}{9}$$

Replacing this into the KCL equation results in

$$-7 - 2\left(\frac{V}{9}\right) + \frac{V}{3} + \frac{V}{9} = 0$$

Solve for $V$:

$$V\left(\frac{1}{3} - \frac{1}{9}\right) = 7$$

$$V = \frac{7 \times 27}{6} = 31.5 \text{ V} \rightarrow I_{3\Omega} = \frac{V}{3} = \frac{31.5}{3} = 10.5 \text{ A}$$

*Effect of the voltage source.* The current source must turn off, which means 0 A or an open circuit. The circuit is shown as follows:

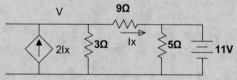

The KCL at the node $V$ is

$$-2I_X + \frac{V}{3} + \frac{V - 11}{9} = 0$$

The current $I_X$ can also be written as

(continued)

**Example 2.82** (continued)

$$I_X = \frac{V - 11}{9}$$

Replacing the KVL equation results in

$$-2\left(\frac{V-11}{9}\right) + \frac{V}{3} + \frac{V-11}{9} = 0$$

$$V\left(\frac{1}{3} - \frac{1}{9}\right) = -\frac{11}{9}$$

$$V\left(\frac{6}{27}\right) = -\frac{11}{9}$$

$$V = -\frac{11}{9} \times \frac{27}{6} = -5.5 \text{ V} \rightarrow I_{3\Omega} = -\frac{5.5}{3} = -1.83 \text{ A}$$

Therefore, the overall effect of both sources are

$$V = 31.5 - 5.5 = 26 \text{ V}$$

$$I_{3\Omega} = 10.5 - 1.83 = 8.67 \text{ A}$$

**Example 2.83 Determine if the following system is linear or not.**

$$\frac{dv}{dt} + 2v = 5$$

*Solution*. The system is not time variant, has fixed coefficients, and is a linear first-order differential equation. Therefore, it is LTI.

**Example 2.84 Determine if the following system is linear or not.**

$$\frac{d^2i}{dt^2} + 2\frac{di}{dt} + 5i = 0$$

*Solution.* The system has fixed coefficients and is a linear second-order differential equation. Therefore, it is LTI.

**Example 2.85 Determine if the following system is linear or not.**

$$\frac{di}{dt} + i^2 = 0$$

*Solution.* The system has fixed coefficients but is nonlinear with respect to $i$ because of the term $i^2$. Therefore, it is not LTI.

**Example 2.86 Determine if the following system is linear or not.**

$$f(x) = x^2 + 5$$

*Solution.*

$$\text{if } x \to f(x) = x^2 + 5$$
$$\text{if } \alpha x \to f(\alpha x) = (\alpha x)^2 + 5 = \alpha^2 x^2 + 5 \neq \alpha(x^2 + 5)$$
$$\therefore f(\alpha x) \neq \alpha f(x) \Rightarrow \text{not linear}$$

**Example 2.87 Determine if the following system is linear or not.**

$$f(x) = 11x$$

*Solution.*

$$\text{if } x \rightarrow f(x) = 11x$$
$$\text{if } \alpha x \rightarrow f(\alpha x) = 11(\alpha x) = \alpha 11x = \alpha f(x)$$

Also,

$$\text{if } x_1 \rightarrow f(x_1) \,\&\, \text{if } x_2 \rightarrow f(x_2) \Rightarrow \text{if } (x_1 + x_2) \rightarrow f(x_1) + f(x_2)$$

$$\text{if } x_1 \rightarrow 11x_1 \,\&\, \text{if } x_2 \rightarrow 11x_2 \Rightarrow \text{if } (x_1 + x_2) \rightarrow 11(x_1 + x_2) = 11x_1 + 11x_2$$
$$= f(x_1) + f(x_2)$$

Therefore, the system is linear and LTI.

---

**Example 2.88 Does superposition hold in the circuit of Fig. 2.83? Prove your answer.**
*Solution.* The circuit has a diode and two sources. If a superposition holds, the diode should not switch on or off and must stay in one position, i.e., the circuit topology must not change.

*Effect of* 20 *V source.* The 10 V source must turn off, i.e., 0 V and short circuit. If so, the voltage across the diode is forward-biased from the 20 V source, and the diode is on. It ideally takes no voltage to stay on. Therefore, the current flows through the diode is

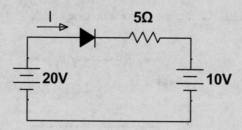

**Fig. 2.83** Circuit of Example 2.88

(continued)

**Example 2.88** (continued)

$$I = \frac{V}{R} = \frac{20}{5} = 4 \text{ A}$$

*Effect of* 10 *V source.* The 20 V source must turn off and short circuit. However, the 10 V source is reverse-biased across the diode. Therefore, no current passes, and the diode is turned off.

This triggers a change in the circuit topology, so superposition does not hold.

## Problems

2.1. Find the number of electrons that need to pass a section of a wire in a unit of time to create a current of 2.2 A.

2.2. Find the current density of a wire with a cross-section of 25 mm$^2$ that passes 1.2 A current.

2.3. Find the resistance of a 100 m wire made of copper, with a cross-section of 25 mm$^2$.

2.4. What is the resistance of a 100 g copper when it is shaped as a wire with a cross-section of 4 mm$^2$?

2.5. Find the equivalent resistance of the following circuit.

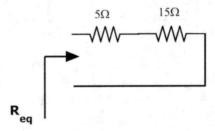

2.6. Find the equivalent resistance of the following circuit.

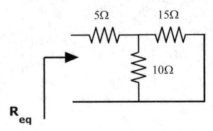

2.7. Find the equivalent resistance of the following circuit.

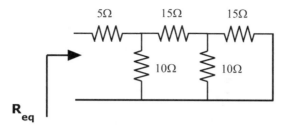

2.8. Find the equivalent resistance of the following circuit.

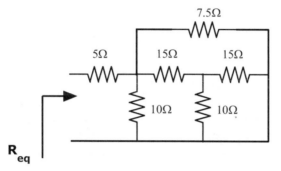

2.9. Find the current and voltage drop of all resistors in the following circuit.

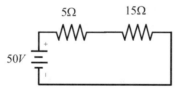

2.10. Find the current and voltage drop of all resistors in the following circuit. Find the power loss across the 15 Ω resistor.

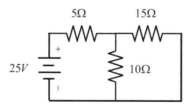

2.11. Find the current and voltage drop across 10 Ω resistors. Find the power loss across the 5 Ω resistor.

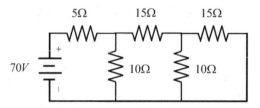

2.12. Find the current and voltage drop across a 7.5 Ω resistor. Find the power loss of 5 Ω and 7.5 Ω resistors.

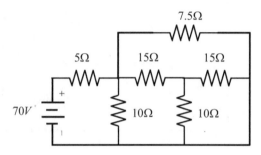

2.13. An inductor shows a voltage variation of 150 V when its current varies by 2.5 A in 2 ms. Find the inductance of the inductor. How much energy is stored in this inductor when connected to a 200 V DC source?

2.14. Find the equivalent inductance of the following circuit.

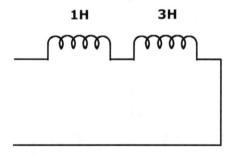

2.15.  Find the equivalent inductance of the following circuit.

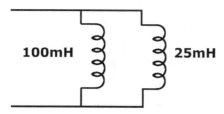

2.16.  Find the equivalent inductance of the following circuit.

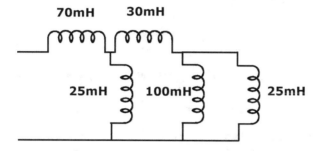

2.17.  Find the current in the following circuit when the voltage source varies 120 volts in 16 msec.

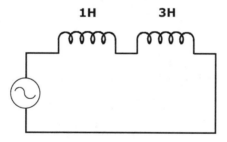

2.18.  Find the current in the following circuit when the source voltage changes 20 V in 10 msec.

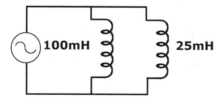

2.19. Find the voltage across inductors in the following circuit when the source current changes 5 A in 0.1 msec.

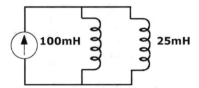

2.20. Find the source's current in the following circuit when the source voltage changes 120 V in 16 msec.

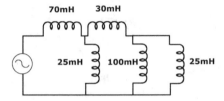

2.21. Find the energy stored in the circuits of problems 2.18–2.21 if the source currents and voltages do not change over time and remain constant.

2.22. Find the equivalent capacitance of the following circuit.

2.23. Find the equivalent capacitance of the following circuit.

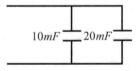

2.24. Find the equivalent capacitance of the following circuit.

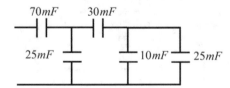

2.25. Find the current in the following circuit when the source voltage changes 120 V in 16 msec.

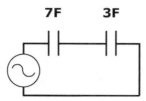

2.26. Find the current in the following circuit when the source voltage changes 20 V in 16 msec.

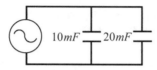

2.27. Find the source's current in the following circuit when the voltage source changes 120 V in 16 msec.

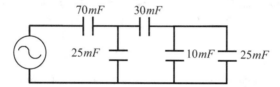

2.28. Find the voltage in the following circuit when the source current changes 5 A in 1 msec.

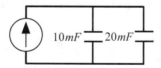

2.29. Calculate the energy stored in the circuits of problems 2.26–2.30 if the source currents and voltages do not change over time and remain constant.

2.30. A household power outlet is 220 V RMS and has a frequency of 50 Hz with the initial phase of 0°. Find the waveform expression.

2.31. Find the currents $I$, $I_1$, and $I_2$ of this circuit when $v(t) = 10 \, u(t)$ $R_1 = 20 \, \Omega$, $R_2 = 60 \, \Omega$.

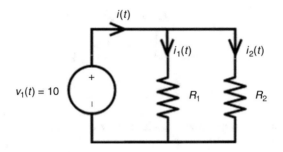

2.32. Find the voltages of $V_1$, $V_2$, and $V_3$, when $I_1(t) = 10$ (A) and $I_2(t) = -20$ (A), $R_1 = 10 \, \Omega$, $R_2 = 20 \, \Omega$, $R_3 = 30 \, \Omega$.

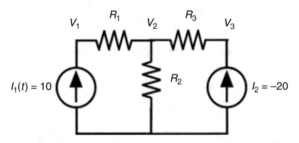

2.33. Find the voltage $V_2$ and the current $I_1$ when the DC source is $V = 250$ V, $I = 18$ A and $R_1 = 25 \, \Omega$, $R_2 = 75 \, \Omega$.

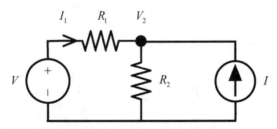

2.34. Find $I_1$, $R_1 = R_2 = 50 \, \Omega$, $R_3 = 150 \, \Omega$, $\alpha = 100$, $V = 45$ V.

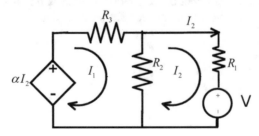

2.35. Determine the minimum resistance that is required to turn the diode off.

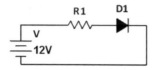

2.36. Determine the value of $R_2$ to turn the diode off.

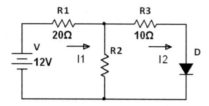

2.37. Find the bias of the diode and its PIV in the circuit.

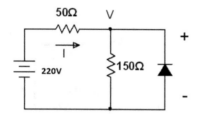

2.38. Is the circuit with $v = 2i + 5$ equation linear?

2.39. Find the voltages of $V_1$, $V_2$, and $V_3$, when $I_1(t) = 150$ (A) and $I_2(t) = -85$ (A) using superposition. $R_1 = 1$ kΩ, $R_2 = 2.5$ kΩ, and $R_3 = 15$ kΩ.

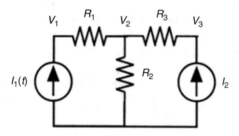

2.40. Find the voltages and currents.

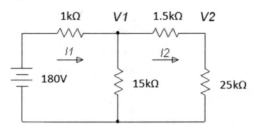

2.41. A DC load is rated at 4.5 kW. At the voltage of 400 V, find the element's resistance and the amount of current drawn from the source. Consider the efficiency of the load at 80 % .

2.42. Two DC loads of 10 kW and 15 kWare connected in parallel to a 450 V source with an internal resistance of 0.04 Ω. Find the current of each load, the current drawn from the source, and the voltage drop across the loads.

2.43. Find $V_1$, $V_2$,and $I$.

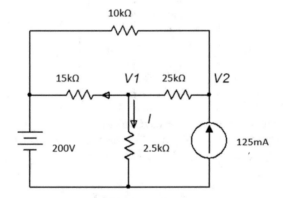

2.44. Find the amount of energy stored in the inductor.

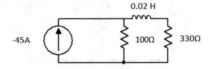

2.45. Find the amount of energy stored in the capacitor.

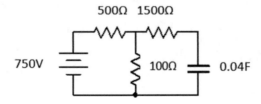

2.46. Find the amount of energy stored in the inductor and capacitor.

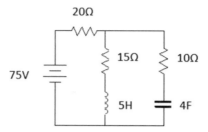

2.47. Using superposition, find the voltage and current of the 3 Ω resistor.

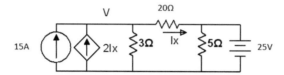

2.48. Find the R-value at which the superposition holds.

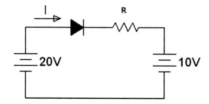

# Chapter 3
# Waveform and Source Analyses

## Introduction

Electric circuits consist of several components that form a particular topology. The drivers of the circuit can be voltage sources and current sources. They force the current to pass through the circuit by generating voltage drops across elements. A circuit performs particular tasks and has a separate output as well. The input sources to the circuit can generate various waveforms, which excite the circuit and cause different effects.

This chapter introduces the standard waveforms that might be seen from the sources or due to the excitation throughout the circuit. A critical aspect of waveform analysis is their mathematical expression, which allows the programming of the circuit in computer codes and obtaining their results without building them. Later in this chapter, the types of sources as independent and dependent are discussed, and their effects in the circuit are studied.

## Waveform Analysis

### Impulse Function f(t) = δ(t)

The impulse function, Dirac delta or chronicle impulse, $\delta(t)$, has value only at a single time, where the function's argument is zero. The impulse function is zero elsewhere. As the function is shown in Fig. 3.1, its mathematical expression is as follows:

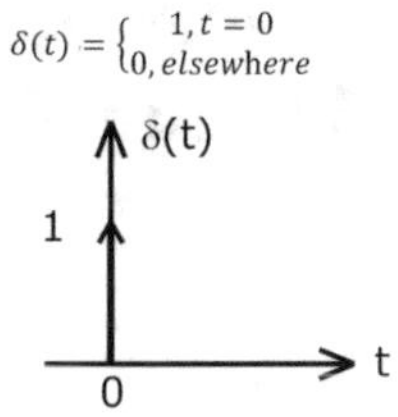

**Fig. 3.1** Impulse at time
$t = 0$

$$\delta(t) = \begin{cases} 1, t = 0 \\ 0, elsewhere \end{cases}$$

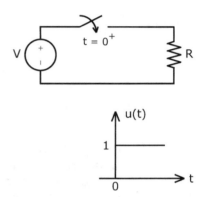

**Fig. 3.2** The voltage source
and the switch closing at
time $t = 0^+$ resemble a step
function applied to the
resistor

**Fig. 3.3** Step function of
amplitude 1 at time $t = 0^+$.
The time means that the
switching event occurs in a
moment after time $t = 0$

$$\delta(t) = \begin{cases} 1, & t=0 \\ 0, & \text{elsewhere} \end{cases}$$

**Note 3.1** The $\int_0^\infty f(t)\delta(t-a)dt$ has value only at $t - a = 0$ or $t = a$. The value
of the integral is, therefore, $f(a)$. For instance, $f(t) = 2e^{-t} + 1$ and $\delta(t - 1)$ yield
$\int_0^\infty (2e^{-t} + 1)\delta(t-1)dt = 2e^{-1} + 1 = 1.758$.

## *Unit Step Function* f(t) = u(t)

Consider an on-off switch that connects a source to a circuit and acts at a specific
time. When the switch is on, the source is connected to the circuit, and the current
flows, and when the switch is turned off, the source is disconnected, and the current
flow stops. To show this even mathematically, a unit step can be used. The switching
event changes the circuit topology, connecting or disconnecting a source to a circuit
and activating part of the circuit in its simplest form.

As Fig. 3.2 shows, the voltage at times less than zero (means before switching
action) is zero, and the voltage applied across the resistor jumps to a specific voltage
a moment after switching $t = 0^+$. The source amplitude determines the function's
amplitude, and the switching mechanism makes it a unit step.

As Fig. 3.3 shows the sketch of the unit step function in the time domain, this
function's amplitude is 1, starting at time $t = 0^+$.

The mathematical expression of the unit step function is identified as

$$u(t) = \begin{cases} 0, & t < 0 \\ 1, & t \geq 0 \end{cases}.$$

**Example 3.1** If the unit step's amplitude changes, it will be expressed as a coefficient. For instance, the current amplitude as a unit step function can be $i(t) = 22u(t)$. It means that the current started at time $t = 0^+$ with an amplitude of 22 A. The current was zero at $t < 0$ (Fig. 3.4).

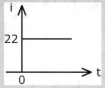

**Fig. 3.4** The current in the form of a step function with amplitude 22 A and starting at time $t = 0^+$

The step function can also start at any arbitrary time. A time shift in the function affects all terms with time element $t$. A shift to the right of the time axis by $a$ second means a delay or lagging effect in the function, which institutes a transform of all $t \rightarrow t - a$, and a shift to the left of the time axis by $a$ second means a leading effect in the function, which institutes a transform of all $t \rightarrow t + a$.

For instance, as Fig. 3.5 demonstrates, the shift of 7 s to the right transforms the function $V(t) = 3u(t)$ to $V(t) = 3u(t - 7)$, and a shift of the same time to the left makes the function transform to $V(t) = 3u(t + 7)$.

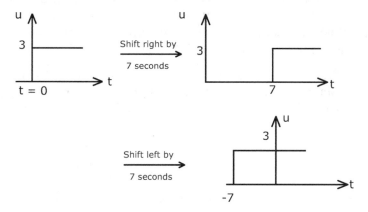

**Fig. 3.5** A shift of step function to the right (delay in switching) and shift left (early switching)

**Note 3.2** The start time of any function can be found by arguing the signal to zero and solving for time. For instance, the signal $V(t) = 3u(t - 7)$ has the time argument of $(t - 7)$. Making $t - 7 = 0$ and solving for the time yield $t = 7$ as the starting time of the signal. The same procedure can be used to find the start time of any function.

**Example 3.2 Sketch the following functions in time domain:**
**(a) $f(t) = 15u(t)$, (b) $f(t) = 15u(t - 10)$, and (c) $f(t) = 15u(3t - 18)$.**
*Solution*. All functions are unit steps with amplitude 15. The only difference is in their start time. Function $a$ has start time at $t = 0$; function $b$ has start time obtained from $t - 10 = 0$, $t = 10$ s. Function $c$ has start time where its argument is zero, or $3t - 18 = 0$, which yields $3t = 18$, or $t = 6$ s. Figure 3.6 shows the functions.

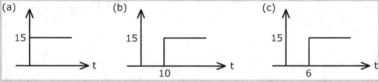

**Fig. 3.6** Waveforms of Example 3.1

**Note 3.3** A step function can eliminate a part of a function. Production of the function $f(t)$ by the unit step $u(t)$ guarantees that the waveform is forced to zero before the unit step starts. A shift in time, when required, must appear in both functions as $f(t - a)$ and $u(t - a)$. Figure 3.7 shows the results of an arbitrary function $f(t)$ (that exists for both time positive and negative) both before and after it is multiplied by the unit step function $u(t)$.

**Note 3.4** The actual value of a step function is its amplitude. When any function is multiplied by a step function, this only indicates that the function has a value when the step indicates. Therefore, in $\int_0^\infty f(t - a)u(t - a)dt$, the integral borders are set for $t \geq a$. The integral can be rewritten as $\int_a^\infty f(t - a)dt$.

**Note 3.5** When two-step functions, for instance, $f_1(t) = u(t - 1)$ and $f_2(t) = -u(t - 5)$, are being added together, the result is obtained by adding point-to-point values of both signals. Therefore, considering the functions, the summation

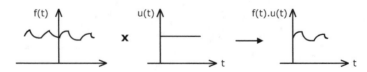

**Fig. 3.7** Product of a function $f(t)$ by a unit step function. The amplitude does not change; however, only the positive part of the waveform $f(t)$, where the unit step exits, is selected

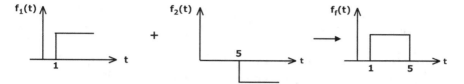

**Fig. 3.8** Summation of two unit steps and formation of a pulse. Functions $f_1(t)$ and $f_2(t)$ have opposite amplitudes but start at different times. Therefore, when adding them together, they cancel each other when they both exist, e.g., for $t > 5$, and for the time both have not started yet, e.g., $t < 1$

result has no value until $t = 1$ when the function $f_1$ starts. Within the window of $1 < t < 5$, only $f_1(t)$ has a value. Therefore, the summation has a function with amplitude 1 for $1 < t < 5$. When the function $f_2(t)$ starts at time $t = 5$, the result summation of both amplitudes becomes $+1 - 1 = 0$. Therefore, the amplitude of the summation becomes zero for any time $t > 5$. This yields a pulse function between $1 < t < 5$ with amplitude 1, as shown in Fig. 3.8.

**Example 3.3 Find the mathematical expression of the waveform shown in Fig. 3.9.**

*Solution.* The location (time) of changes that occur in waveform $f(t)$ are at times 0, 5, and 10. The amount of change (amplitude) at each of these locations is from @$t = 0$, $0 \rightarrow 3$ or $(3u(t))$; @ $t = 5$, $3 \rightarrow 7$ or $(4u(t - 5))$; and @ $t = 10$, $7 \rightarrow 15$ or $(8u(t - 10))$. Therefore, the mathematical expression is

$$f(t) = 3u(t) + 4u(t - 5) + 8u(t - 10)$$

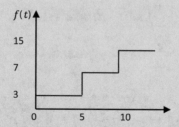

**Fig. 3.9** The waveform of Example 3.3

**Example 3.4 Find the mathematical expression of the waveform shown in Fig. 3.10.**

*Solution.* The location (time) of changes that occur in waveform $f(t)$ are at times 0, 3, 6, and 9. The amount of change (amplitude) at each of these locations is from @$t = 0$, $0 \rightarrow 2$ or $(2u(t))$; @ $t = 3$, $2 \rightarrow 4$ or $(2u(t - 3))$; @ $t = 6$, $4 \rightarrow 6$ or $(2u(t - 6))$; and @ $t = 9$, $6 \rightarrow 0$ or $(-6u(t - 9))$. Therefore, the mathematical expression is

$$f(t) = 2u(t) + 2u(t - 3) + 2u(t - 6) - 6u(t - 9)$$

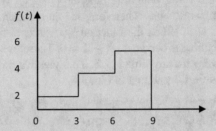

**Fig. 3.10** Waveform of Example 3.4

**Note 3.6** The derivative of a unit step function is a delta function, and the integral of a delta function is the unit step function at the same amplitude. The time shift of the signals is preserved.

$$\int k\delta(t - a) = ku(t - a)$$

$$\frac{d}{dt}(Mu(t - a)) = M\delta(t - a)$$

**Example 3.5 Find the derivative of the waveform shown in Fig. 3.11.**

*Solution.* The amount of change in the waveform amplitude is observed at times 0, 3, 6, and 9. The amount of change in the step function at each of these locations generates a delta function of the same amplitude. That means $\frac{d}{dt}f(t) = ($@$t = 0$, $0 \rightarrow 2 \equiv 2\delta(t)$; @$t = 3$, $2 \rightarrow 4 \equiv 2\delta(t - 3)$; @$t = 6$, $4 \rightarrow 6 \equiv 2\delta(t - 6)$; @$t = 9$, $6 \rightarrow 0 \equiv -6\delta(t - 9))$.

$$\frac{d}{dt}f(t) = 2\delta(t) + 2\delta(t - 3) + 2\delta(t - 6) - 6\delta(t - 9)$$

(continued)

**Example 3.5** (continued)

The derivative waveform can be shown in Fig. 3.12.

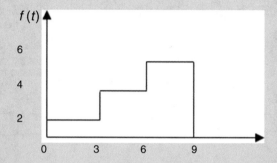

**Fig. 3.11** The waveform of Example 3.5

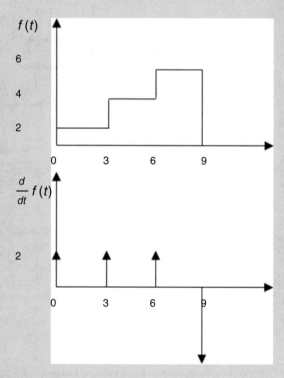

**Fig. 3.12** Derivative of the waveform

**Example 3.6 Find the derivative of the waveform shown in Fig. 3.13.**
*Solution.* The amount of change in the waveform amplitude is observed at times 3 and 10. The amount of change in the step function at each of these locations generates a delta function of the same amplitude. That means $\frac{d}{dt}f(t) = (@t = 3, 0 \rightarrow 11 \equiv 11\delta(t-3); @t = 10, 11 \rightarrow 0 \equiv -11\delta(t-10))$.

$$\frac{d}{dt}f(t) = 11\delta(t-3) - 11\delta(t-10).$$

This can be shown in Fig. 3.14.

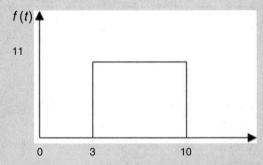

**Fig. 3.13**  Waveform of Example 3.6

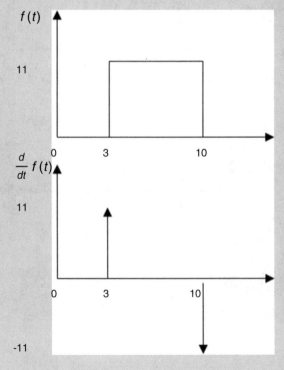

**Fig. 3.14**  Derivative of the waveform

**Example 3.7 Find the integral of the waveform shown in Fig. 3.15.**
*Solution.* The waveform shows four delta functions at times 0, 3, 6, and 9. This means that there are four changes added together to build the integral waveform. The integral will be

$$\int f(t) = \int 2\delta(t) + \int 2\delta(t-3) + \int 3\delta(t-6) + \int 2\delta(t-9)$$

$$\int f(t) = 2u(t) + 2u(t-3) + 2u(t-6) + 2u(t-9)$$

The waveform of the integral is shown in Fig. 3.16.

**Fig. 3.15** Waveform of
Example 3.7

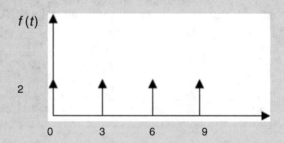

**Fig. 3.16** Integral of the
waveform Fig. 3.15

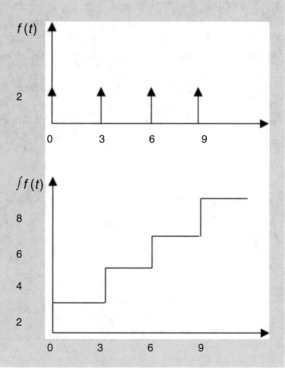

**Example 3.8 The current passing through a 20 Ω resistor is measured as $i(t) = 2.5u(t)$. Find the voltage drop across the resistor.**
*Solution.* The Ohm's law across a resistor is expressed as $v(t) = Ri(t)$. Therefore, the voltage drop across the resistor is

$$v(t) = 20 \times 2.5u(t) = 50u(t) \text{ V}$$

**Example 3.9 The current passing through a 1.5 H inductor is measured as $i(t) = 2.5u(t)$. Find the voltage drop across the inductor.**
*Solution.* The Ohm's law across an inductor is expressed as $v(t) = L\frac{di}{dt}$. Therefore, the voltage drop across the inductor is

$$v(t) = 1.5\frac{d}{dt}(2.5\ u(t)) = 1.5 \times 2.5\frac{d}{dt}u(t) = 3.75\delta(t)\ \text{V}$$

**Example 3.10 The voltage drop across a 1.5 F capacitor is measured as $v(t) = 2.5u(t)$. Find the current passing through the capacitor.**
*Solution.* The Ohm's law across a capacitor is expressed as $i(t) = c\frac{dv}{dt}$. Therefore, the current passing through the capacitor is

$$i(t) = 1.5\frac{d}{dt}(2.5u(t)) = 1.5 \times 2.5\frac{d}{dt}(u(t)) = 3.75\delta(t) \text{ A}$$

**Example 3.11 A pulse voltage of $v(t) = 11u(t-3) - 11u(t-5)$ is applied across a 10 F capacitor. Find the current passing through the capacitor.**
*Solution.* The current is $i(t) = c\frac{dv}{dt}$. Therefore,

$$i(t) = 10\frac{d}{dt}(11\ u(t-3) - 11u(t-5))$$
$$= 10 \times 11\ (\delta(t-3) - \delta(t-5)) = 110\ (\delta(t-3) - \delta(t-5))$$

Example 3.12 **The voltage drop recorded across a 10 mH inductor is** $v(t) = 5\delta(t-1) - 5\delta(t-7)$. **Find the current passing through the inductor.**

*Solution.* In an inductor, the voltage and current are related as follows:

$$v = L\frac{di}{dt}$$

The current becomes

$$i = \frac{1}{L}\int v\, dt$$

Therefore, the current is

$$i = \frac{1}{10e-3}\int 5\delta(t-1) - 5\delta(t-7) = 100(5u(t-1) - 5u(t-7))$$

$$= 500\,((t-1) - u(t-7))\ \text{A}$$

## *Ramp Function* f(t) = r(t)

If the value of the function linearly (proportionally) increases over time, the function is called a ramp. The slope of this line, $k$, determines the rate of increment. The function is written in mathematical form as

$$r(t) = ktu(t).$$

The ramp function can also be expressed as

$$r(t) = \begin{cases} kt, & t \geq 0 \\ 0, & t < 0 \end{cases}$$

The ramp function has zero value for time negative and increases by a $k$ every second. The slope is defined as the ratio of the value gained over the time it took to gain the value. Figure 3.17 demonstrates the ramp function and its slope calculation.

**Note 3.7** A shift in the ramp function shifts the waveform's starting time but keeps the slope intact.

**Fig. 3.17** A ramp function
with slope $k$. The slope is the
ratio of the gain in the
amplitude over the time it
takes $k = \frac{\text{Gain}}{\text{Time}}$

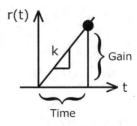

**Example 3.13 Find the mathematical expression of the waveform shown in Fig. 3.18.**
*Solution.* The slope of this ramp is the amount of gain (9) over the time it took for this gain (7). The slope is, therefore,

$$k = \frac{9}{7}$$

The ramp equation that starts from time $t = 0^+$ has a $u(t)$ to show the start time. The function is

$$f(t) = ktu(t)$$
$$f(t) = \frac{9}{7}tu(t)$$

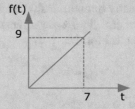

**Fig. 3.18** The ramp of Example 3.18

**Example 3.14 Find the mathematical expression of the waveform shown in Fig. 3.19.**
*Solution.* There is a shift of $t = 5$ s in the start time of the signal. The slope is the gain of 9 over the time of 2 s. The signals start at time 5, and it gains 9 units in 2 s at $t = 7$. The slope is, therefore,

(continued)

**Example 3.14** (continued)

$$k = \frac{9}{2}$$

Since the start time has a shift of +5 s, the signals become

$$f(t) = \frac{9}{2}(t-5)u(t-5)$$

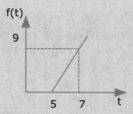

**Fig. 3.19** Waveform of Example 3.14

**Example 3.15 Considering functions $r_1$ and $r_2$ shown in Fig. 3.20, sketch the summation of the functions, and express the result in mathematical terms.**

*Solution.* Function $r_1(t) = 10tu(t)$ is shown in Fig. 3.20. The function reaches amplitude 10 in 1 s and 20 in 2 s. It reaches 50 in 5 s and continues to gain amplitude at the same rate of 10 units/s. Function $r_2(t)$ starts at time $t = 5$ s. It has a slope of $-10$, which means that each second, it drops by $-10$. Therefore, in 1 s after it starts (or $t = 5 + 1 = 6$ s), it reaches $-10$ and continues to drop $-10$ each second. The mathematical expression of this function is $r_2(t) = -10(t - 5)u(t - 5)$, with slope $-10$ and start time 5.

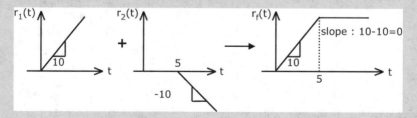

**Fig. 3.20** Ramp functions of Example 3.15. When ramp functions are added together, their slopes are algebraically added. It is essential to consider the slope changes when any functions experience a slope change. For instance, the points of $t = 0$ and $t = 5$ are needed to be observed for the slope changes because $r_1(t)$ and $r_2(t)$ start at these times

(continued)

**Example 3.15** (continued)

Now, to add $r_1(t)$ and $r_2(t)$ point by point, the result starts from $t = 0$. The slope is 10 in time $0 < t < 5$, and it shows a change of slope at $t = 5$ to $(+10-10 = 0)$. Since the effect of two signals has been considered and there is no change in them for $t > 5$, the result summation function continues with slope 0 onward.

**Example 3.16 Consider the previous example, and add a third function**
$r_3(t) = -10(t - 7)u(t - 7)$ **to the functions $r_1(t)$ and $r_2(t)$. Sketch**
**the summation result.**
*Solution.* Function $r_3(t)$ starts at $t = 7$, which means no change in the summation of $r_1(t)$ and $r_2(t)$ until $t = 7$. From $5 \leq t < 7$, the slope of $r_1(t) + r_2(t)$ is zero. At time $= 7$, $r_3(t)$ with slope $-10$ is added to the signals. The slope is therefore changed to $0 - 10 = -10$. The function of $r_1(t) + r_2(t) + r_3(t)$ is shown in Fig. 3.21.

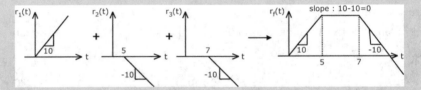

**Fig. 3.21** Ramp functions of Example 3.4

**Note 3.8** Adding ramp functions together changes their slope at the starting time of the functions.

**Example 3.17 Note that the function of $r_1(t) + r_2(t) + r_3(t)$ continues**
**with slope $-10$ even after it reaches zero at time $t > 12$. Suggest a fourth**
**function $r_4(t)$, to stop it from proceeding to negative values and make its**
**slope zero at $t = 12$ and onward.**
*Solution.* The functions $r_1(t) + r_2(t) + r_3(t)$ result in slope $-10$ at $t > 12$, and to reach slope zero, a function $r_4(t)$ with slope $+10$ must be added. However, the function must be started at the time $t = 12$. Therefore, if a shift of 12 s is applied, the function that needs to be added is obtained as $r_4(t) = +10(t - 12)$ $u(t - 12)$. The summation shown in Fig. 3.22.

(continued)

**Example 3.17** (continued)

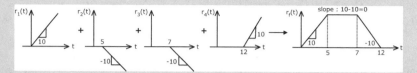

**Fig. 3.22** Ramp functions of Example 3.17

**Note 3.9** Note that the term of unit step function $u(t - 12)$ presented in the function $r_4(t) = 10(t - 12)u(t - 12)$ guarantees that the value of $r_4(t)$ remains zero for all $t < 12$.

**Example 3.18 Write the mathematical expression of the signal shown in Fig. 3.23.**
*Solution.* The function starts at time $t = 0$ with a slope of 3/8. This suggests the existence of a signal $r_1(t) = \frac{3}{8} tu(t)$. The function continues with this slope until time $t = 8$. At this point, the slope has reached zero. This requires a $-\frac{3}{8}$ slope change to the existing signal of $r_1(t)$ but starting at time $t = 8$. Therefore, $r_2(t) = -\frac{3}{8}(t - 8)u(t - 8)$, it must be added to the function $r_1(t)$. The summation of these two signals continues with slope zero until the time $t = 10$. At this point, the slope needs to reach $-6$. This requires a $0 - 6 = -6$ slope change introduced by a third function at $t = 10$ or $r_3(t) = -6(t - 10)u(t - 10)$. With slope $-6$, the amplitude of the resultant signal reaches zero in 0.5 s. The summation of these three signals leaves the slope $-6$ for time $10 \leq t < 10.5$.

After this time, or $t \geq 10.5$, the amplitude and slope should remain at 0. Therefore, another slope change of +6 must be added to the signals starting at $t = 10.5$. Therefore, the fourth signal is $r_4(t) = +6(t - 10.5)u(t - 10.5)$ (Fig. 3.24).

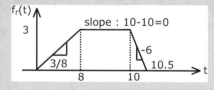

**Fig. 3.23** Ramp function of Example 3.18

(continued)

**Example 3.18** (continued)

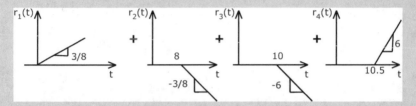

**Fig. 3.24** Split of ramp functions in Example 3.23

**Example 3.19 The waveform of Fig. 3.23 is the current of a 10 mH inductor. Find the voltage drop across the inductor.**

*Solution.* The current can be expressed as

$$i(t) = \frac{3}{8}tu(t) - \frac{3}{8}(t-8)u(t-8) - 6(t-10)u(t-10) + 6(t-10.5)u(t-10.5)$$

The voltage drop across an inductor is

$$v = L\frac{di}{dt}$$

$$v = 10e - 3\frac{d}{dt}$$

$$\times \left( \frac{3}{8}tu(t) - \frac{3}{8}(t-8)u(t-8) - 6(t-10)u(t-10) + 6(t-10.5)u(t-10.5) \right)$$

$$v = 10e - 3\left( \frac{3}{8}u(t) - \frac{3}{8}u(t-8) - 6u(t-10) + 6u(t-10.5) \right) \text{ V}$$

**Example 3.20 The current of a 10 mF capacitor is shown in Fig. 3.23. Find the voltage drop across the capacitor.**

*Solution.* In a capacitor,

$$i = C\frac{dv}{dt} \rightarrow v = \frac{1}{C}\int idt$$

Since

$$i(t) = \frac{3}{8}t(ut) - \frac{3}{8}(t-8)u(t-8) - 6(t-10)u(t-10) + 6(t-10.5)u(t-10.5)$$

Then:

$$v = \frac{1}{10e-3}\int\left(\frac{3}{8}tu(t) - \frac{3}{8}(t-8)u(t-8) - 6(t-10)u(t-10)\right.$$
$$\left. + 6(t-10.5)u(t-10.5)\right)$$

$$v = 100\left(\frac{3}{8}\times\frac{1}{2}\ t^2u(t) - \frac{3}{8}\times\frac{1}{2}(t-8)^2u(t-8) - 6\times\frac{1}{2}(t-10)^2u(t-10) + 6\right.$$
$$\left. \times\frac{1}{2}(t-10.5)^2u(t-10.5)\right)$$

**Example 3.21 Sketch the function $f(t) = 10tu(t-2)$.**

*Solution.* The function has a significant shift of 2 s determined from the $u(t)$ function as $u(t-2)$. However, the other time variables are not shifted by the same amount. A solution to this problem is to shift all the other times by the amount dictated by the unit step function. To do this, the variable $t$ is converted to $t-2+2$. This amount does not change. However, it now shows the right amount of time shift. Therefore,

$$f(t) = 10(t-2+2)u(t-2)$$

Separating the expressions and keeping the terms $t-2$ as one component results in

$$f(t) = 10(t-2)u(t-2) + 10\times 2\ u(t-2)$$

Now all the times in each part of $f(t)$ are shifted by the same amount.

(continued)

**Example 3.21** (continued)

$f(t) = 10(t - 2)u(t - 2)$ is the ramp function of slope 10 and the start time of 2, and $+10 \times 2\ u(t - 2)$ is a unit step function with amplitude 20 and start time 2.

Therefore, the sketch is (Fig. 3.25)

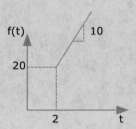

**Fig. 3.25**   Waveform of Example 3.21

**Example 3.22 Sketch the function $f(t) = 5(t - 10)u(t - 7)$.**
*Solution*. The argument of the unit step function and the time shift of the ramp are different. In this case, an operation is needed to make the ramp format $k(t - a)u(t - a)$. It is critical to have the same time shift for the ramp and its unit step function.

In this example, 10 is split to a 7 and a 3 as $f(t) = 5(t - 7 - 3)u(t - 7)$. Then, an expansion results in

$$f(t) = 5(t - 7)u(t - 7) + 5(-3)u(t - 7)$$

The function becomes a combination of a ramp $f_1(t) = 5(t - 7)u(t - 7)$ and a unit step function $f_2(t) = 5(-3)u(t - 7)$. The ramp has a slope of 5 and a start time of 7. The step function has an amplitude of $-15 = 5(-3)$ and a start time of 7.

The waveform is shown in Fig. 3.26.

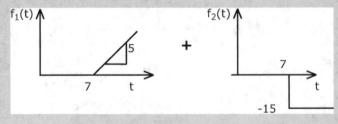

**Fig. 3.26**   The figure of function in Example 3.22

## *Power Function* $f(t) = \frac{At^n}{n!} u(t)$

For a given natural number $n$, the function has an amplitude $\frac{A}{n!}$ and a time factor of $t^n$. For $n = 1$, the function becomes a ramp $f(t) = Atu(t)$, shown in Fig. 3.27. For $n = 2$ the function becomes a parabolic that exists in positive time (the effect of $u(t)$), as $f(t) = \frac{At^2}{2!} u(t)$ shown in Fig. 3.28.

**Note 3.10** Derivative of a ramp function $r(t) = ktu(t)$ is a unit step function with amplitude $k$. Likewise, the integral of a step function $ku(t)$ is a ramp function $r(t) = ktu(t)$. Figure 3.28 shows the derivative and integral functions of the ramp and unit step (Fig. 3.29).

$$\frac{d}{dt} r(t) = \frac{d}{dt} ktu(t) = ku(t)$$

$$\int ku(t) dt = ktu(t)$$

**Note 3.11** Derivative of a step function $ku(t)$ is an impulse with amplitude $k\delta(t)$. Likewise, the integral of $k\delta(t)$ is $ku(t)$ (Fig. 3.30).

**Fig. 3.27** Power function becomes a ramp function when $n = 1$. Coefficient $A$ becomes the slope

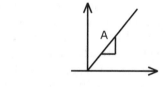

**Fig. 3.28** Power function becomes a parabolic function when $n = 2$

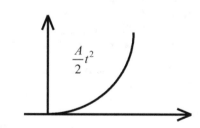

**Fig. 3.29** Derivative of a ramp function with slope $k$ becomes a unit step with amplitude $k$

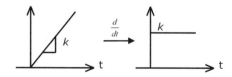

**Fig. 3.30** Derivative of a
unit step function with
amplitude $k$ becomes an
impulse with amplitude $k$

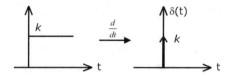

$$\frac{d}{dt}ku(t) = k\delta(t)$$

$$\int k\delta(t)dt = ku(t)$$

**Note 3.12**  The derivative of an impulse function $k\delta(t)$ is $k\dot{\delta}(t)$. Likewise, the integral $k\dot{\delta}(t)$ is $k\delta(t)$.

$$\frac{d}{dt}k\delta(t) = k\dot{\delta}(t)$$

$$\int k\dot{\delta}(t)dt = k\delta(t)$$

## *Exponential Function* $\mathbf{f(t) = Ae^{\alpha t}u(t)}$

This function shows an exponentially steady rise or decay. The decay factor $\alpha$ determines the rate of change, where a positive value shows a rise and a negative value shows a decay in the amplitude. The functions for $\alpha > 0$ and $\alpha < 0$ are shown in Fig. 3.31. The value of function $f(t) = Ae^{\alpha t}u(t)$ at $t = 0$ is $f(t = 0) = A$. The value of $1/\alpha$ shows the time constant of the signal.

The decay factor determines how fast the signal reaches zero. For instance, $\alpha = -1$ has a slower decay rate than $\alpha = -2$. Figure 3.32 shows the effect of the decay factor on the shape of the signal.

For $\alpha > 0$, the waveform exponentially increases. Figure 3.33 shows the waveforms $\alpha = +1, -1$ and $\alpha = +2, -2$.

**Note 3.13** The derivative of an exponential function repeats itself and generates coefficients as follows:

$$\frac{d}{dt}(Ae^{\alpha t}) = \alpha(Ae^{\alpha t})$$

$$\int Ae^{\alpha t}dt = A\frac{1}{\alpha}e^{\alpha t}$$

**Fig. 3.31** Derivative of an impulse with amplitude $k$ becomes a function with amplitudes split to $-k$ in negative time $t = 0^-$ and $+k$ in positive time $t = 0^+$

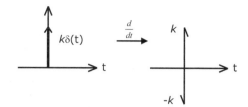

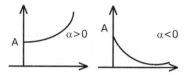

**Fig. 3.32** Exponential function with positive damping factor $\alpha > 0$. The amplitude increases with time. Exponential function with negative damping factor $\alpha < 0$. The amplitude decreases over time

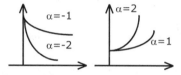

**Fig. 3.33** As the damping rate becomes more negative, the rate of amplitude decrease accelerates, and the function reaches zero faster. As the damping rate becomes more positive, the rate of amplitude increase accelerates, and the function reaches infinite faster

**Example 3.23 Do the math.**

(a) $\int 10e^{-3t}dt = \frac{10}{-3}e^{-3t}$

(b) $\frac{d}{dt}(5e^{-7t}) = 5(-7)e^{-7t}$

(c) $\int \left(\frac{7}{8} e^{\frac{3}{16}t}\right)dt = \frac{7}{8}\frac{1}{\frac{3}{16}}e^{\frac{3}{16}t} = \frac{7}{8}\frac{16}{3}e^{\frac{3}{16}t} = \frac{14}{3}e^{\frac{3}{16}t}$

## *Sinusoidal Function* **f(t) = A sin (ωt + φ)**

The sinusoidal function is periodical and repeats every $2\pi$ degrees. Maximum amplitude reaches $A$ at frequency $\omega$ rad/s. The frequency in Hertz is obtained by $f = \frac{\omega}{2\pi}$. The signal can lead or lag in crossing the origin by $\varphi$ degrees. Therefore, the angle $\varphi$ is called a phase shift.

**Fig. 3.34** A sinusoidal waveform with amplitude $A = 10$ (peak) and phase shift of angle $\varphi = -\frac{\pi}{6}$ or $-30°$ can be mathematically expressed as $A \sin(\omega t - 30)$

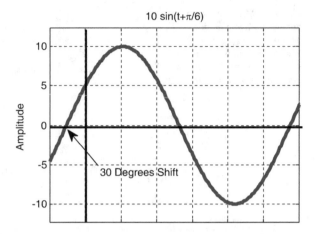

**Note 3.14**  A lead signal has a positive phase shift, and a lag signal has a negative phase shift (Fig. 3.34).

**Note 3.15**  The derivative of a sinusoidal function is as follows:

$$\frac{d}{dt} A \sin(\omega t + \varphi) = A\omega \cos(\omega t + \varphi)$$

$$\int A \sin(\omega t + \varphi) dt = -\frac{A}{\omega} \cos(\omega t + \varphi)$$

**Note 3.16**  A sinusoidal function presents the amplitude, phase shift, and frequency information. When a circuit is excited by a sinusoidal waveform, the frequency remains constant throughout the circuit. In every element of the circuit, whether current or voltage, the frequency is the frequency of the source. However, the amplitude and phase of the voltages and currents are influenced by the circuit topology and its element values. Therefore, when representing values of a circuit, a sinusoidal function can be presented by conveying their amplitudes and phase information. A phasor equivalent of a sinusoidal conveys this information and is presented as follows:

$$A \sin(\omega t + \varphi) \equiv A \angle \varphi$$

## *Polar to Cartesian (Rectangle) Conversion*

Phasor is a polar representation of the function. The polar coordinates can be converted to the rectangle as follows. Consider a function $f(t)$ as

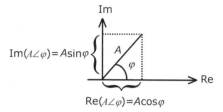

**Fig. 3.35** Polar representation of a sinusoidal waveform (amplitude and phase) can be converted by projecting the vector onto the real and imaginary axes to obtain the rectangular representation

$$f(t) = A \angle \varphi$$

Rectangle presentation of this phasor, shown in Fig. 3.35, is obtained by projecting the vector on the real and imaginary axes as follows:

$$A \angle \varphi = A \cos (\varphi) + jA \sin (\varphi) = \text{Re} (f(t)) + j\text{Im}(f(t)) = p + jq.$$

The real part of this function $p = \text{Re} (A \angle \varphi) = A \cos (\varphi)$, and the imaginary part of this function is $q = \text{Im} (A \angle \varphi) = A \sin (\varphi)$.

## Cartesian (Rectangle) to Polar Conversion

Any complex conjugate number $p + jq$ can be converted to polar coordinates as follows:

$$A \angle \varphi = p + jq = \sqrt{p^2 + q^2} \angle \tan^{-1} \frac{q}{p}$$

**Example 3.24 Find polar transform of the following numbers**

| | | |
|---|---|---|
| (a) | $2 + j2$ | Answer: $\sqrt{2^2 + 2^2} \angle \tan^{-1} \frac{2}{2} = 2\sqrt{2} \angle 45$ |
| (b) | $2 - j2$ | Answer: $\sqrt{2^2 + (-2)^2} \angle \tan^{-1} \frac{-2}{2} = 2\sqrt{2} \angle - 45$ |
| (c) | $1 + j0.866$ | Answer: $\sqrt{1^2 + 0.866^2} \angle \tan^{-1} \frac{0.866}{1} = 1.3228 \angle 40.91$ |
| (d) | $0 + j1$ | Answer: $\sqrt{0^2 + 1^2} \angle \tan^{-1} \frac{1}{0} = 1 \angle 90$ |
| (e) | $-j1.$ | Answer: $\sqrt{0^2 + (-1)^2} \angle \tan^{-1} \frac{-1}{0} = 1 \angle - 90$ |

**Example 3.25 Find the rectangle transform of the following numbers**

| (a) | $10 \angle 30$ | Answer: $10 \cos (30) + j10 \sin (30) = 8.66 + j5$ |
|-----|----------------|------------------------------------------------------|
| (b) | $10 \angle 0$  | Answer: $10 \cos (0) + j10 \sin (0) = 10 + j0$       |
| (c) | $10 \angle 90$ | Answer: $10 \cos (90) + j10 \sin (90) = 0 + j10$     |
| (d) | $10 \angle -90$ | Answer: $10 \cos (-90) + j10 \sin (-90) = 0 - j10$  |

**Example 3.26 Find the value for $j, j^2$, and $j^3$.**
*Solution.*   Considering   $j = \sqrt{-1}s$,   then   $j^2 = \sqrt{-1}\sqrt{-1} = -1$,   and
$j^3 = jj^2 = j(-1) = -j$.

**Example 3.27 The signal 10 sin (377t + 40) has a peak amplitude of 10 and angular frequency of 377 rad/s, with a phase angle of 40°. This can also be presented in phasor as 10 ∠ 40.**

**Note 3.17**   Sine and cosine functions can be converted to each other in mathematical operations. Following is the list of these conversions obtained from the unit circle, as shown in Fig. 3.36.

$$\cos (\omega t) = \sin (\omega t + 90)$$
$$- \cos (\omega t) = \sin (\omega t - 90)$$
$$\sin (\omega t) = \cos (\omega t - 90)$$
$$- \sin (\omega t) = \cos (\omega t + 90)$$

**Fig. 3.36**   Unit circle, sin, and cos axes are shown

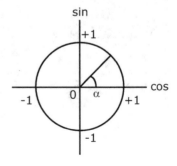

**Note 3.18** To add or subtract several sine and cosine functions, (1) all must have the same frequency, and (2) all must be converted to either sine or cosine. Otherwise, a phase shift of $\pm 90°$ will be lost in the calculations.

**Example 3.28 Show the following functions in sine form and phasor:**

$$f(t) = 10 \sin (377t + 30)$$

*Solution.* The function is already in sine form; therefore, the phasor presentation is $10 \angle 30$.

**Example 3.29 The function needs to be converted to a sine function.**

$$f(t) = 5 \cos (377t + 60)$$

*Solution.* The phase shift, according to Note 3.17, must be considered:

$$\cos (\omega t) = \sin (\omega t + 90)$$

Therefore, $f(t) = 5 \sin (377t + 60 + 90)$. The 90° phase shift is added to convert the cosine to sine. The phasor representation is $5 \angle (60 + 90)$ or $5 \angle 150$.

## Mathematical Operation of Polar and Complex Numbers

It is recommended to add complex numbers in Cartesian coordinates and multiply numbers in polar coordinates. Adding numbers in polar coordinates and multiplying numbers in Cartesian coordinates require longer mathematical operations.

### Adding Complex Numbers

Add the real parts together, and add imaginary parts together. It becomes

$$(a + jb) + (c + jd) = (a + c) + j(b + d)$$

**Product of Complex Numbers**

Expansion production results in

$$(a+jb)(c+jd) = (ac + jad + jbc + jjbd)$$

$jj = j^2 = -1$, which leads to

$$(ac + jad + jbc - bd)$$

Now, collecting real parts and imaginary parts separately results in

$$(ac - bd) + j(ad + jbc)$$

**Product of Polar Numbers**

Amplitudes are multiplied, and angles are summed to result in

$$(A\angle\varphi)(B\angle\theta) = (AB)\angle(\varphi + \theta)$$

**Division of Polar Numbers**

Amplitudes are divided, and angles are subtracted as

$$\frac{(A\angle\varphi)}{(B\angle\theta)} = \left(\frac{A}{B}\right)\angle(\varphi - \theta)$$

The angle is always the angle of numerator $\varphi$ minus the angle of denominator $\theta$.

**Summation of Polar Numbers**

$$(A\angle\varphi) + (B\angle\theta) \rightarrow (A\cos\varphi + jA\sin\varphi) + (B\cos\theta + jB\sin\theta)$$
$$= (A\cos\varphi + B\cos\theta) + j(A\sin\varphi + B\sin\theta)$$

Both numbers must be converted to a rectangle and added as real and imaginary numbers.

**Note 3.19** The mathematical operation of a mix of polar and rectangle suggests a transformation from one form to another to ease the mathematical operations. It is more suitable for division and product in polar form and summation in rectangle form. However, as explained earlier, it is possible to do all mathematical operations in one form.

**Example 3.30 Do the following mathematical operations in a suitable form (either polar or rectangle):**

| | |
|---|---|
| $(1 + j20)(1 - j20)$. | Answer: 401 |
| $\frac{1+j20}{1-j1}$. | Answer: $-9.5 + j10.5$ |
| $\frac{2-j5}{3+j7}$. | Answer: $-0.5 - j0.5$ |
| $(10 \angle -90) + (10 \angle -30)$. | Answer: $8.66 - j15$ |
| $\frac{10\angle -60}{5\angle 30}$. | Answer: $-j2$ |

**Note 3.20** Conjugate of a complex number is shown by a star and is obtained as follows:

$$(a + jb)^* = a - jb$$

$$(A\angle\varphi)^* = A\angle - \varphi$$

**Example 3.31 Solve $\frac{1+j2}{4-j3}$.**

*Solution.* To simplify the function, the complex number of the denominator should be converted to a real number. That is possible by multiplying and dividing the entire function by the complex conjugate of the denominator as follows:

$$\frac{1+j2}{4-j3} = \frac{1+j2}{4-j3} \times \frac{(4-j3)^*}{(4-j3)^*} = \frac{1+j2}{4-j3} \times \frac{4+j3}{4+j3} = \frac{(4+j3)(1+j2)}{(4^2+3^2)}$$

$$= \frac{4 \times 1 + 4 \times j2 + j3 \times 1 + j3 \times j2}{16+9}$$

$$\frac{4+j8+j3-6}{25} = \frac{(4-6)+j(8+6)}{25} = \frac{-2+j11}{25}$$

**Note 3.21** $(a + jb)(a - jb) = a^2 + b^2$. For instance, with $a = 4$ and $b = 3$, the argument becomes

$$(4 - j3)(4 + j3) = 16 + 9 = 25$$

### Summation of Sinusoidal Functions

Functions with similar frequencies can be added using phasor, as follows:

$$f(t) = M \sin(\omega t + \alpha) + N \sin(\omega t + \beta)$$

**Note 3.22**  Check the frequencies to be the same.

**Note 3.23**  Check the form of functions to be the same. Both must be converted to sin or cos if they are not the same.

Converting the functions to phasor obtains

$$f(t) = M \angle \alpha + N \angle \beta$$

The rectangle form of this function is obtained by converting each term individually to a complex number as follows:

$$f(t) = (M \cos \alpha + jM \sin \alpha) + (N \cos \beta + jN \sin \beta)$$

Adding the real parts and imaginary parts separately results in

$$f(t) = (M \cos \alpha + N \cos \beta) + j(M \sin \alpha + N \sin \beta)$$

Considering

$$p \equiv M \cos \alpha + N \cos \beta$$
$$q \equiv M \sin \alpha + N \sin \beta$$

The function becomes

$$f(t) = p + jq$$

The rectangle to polar conversion can be accomplished as follows:

In polar form, there is a need for amplitude and a phase. The amplitude can be obtained by

$$A = \sqrt{p^2 + q^2}$$

The phase can be obtained by

$$\varphi = \tan^{-1} \frac{\text{Im}(f(t))}{\text{Re}(f(t))}$$

$$\varphi = \tan^{-1} \frac{q}{p}.$$

**Example 3.32 Find the result of** $e(t) = 10 \sin (50\, t + 30) + 20 \cos (50\, t + 20)$**.**

*Solution.* Since the frequencies are the same, the values can be added. Otherwise, they have to be added by calculating point-by-point values of each sine function.

However, the functions are sine and cosine. There is a need to convert one to another, i.e., both have to be sine or both have to be cosine. Let us convert both functions to sine. Using the transform, both in sine function can be written as

$$e(t) = 10 \sin (50t + 30) + 20 \sin (50t + 20 + 90)$$

Phasor presentation results in

$$e(t) = 10\angle 30 + 20\angle 110$$

Adding two polar numbers is better when both are converted to Cartesian or rectangle. This results in

$$e(t) = 10 \cos 30 + j10 \sin 30 + 20 \cos 110 + j20 \sin 110$$

$$e(t) = 10 \times 0.866 + j10 \times 0.5 + 20 \times -0.342 + j20 \times 0.939$$

$$e(t) = 8.66 + j5 - 6.84 + j18.78$$

$$e(t) = 1.82 - j13.78$$

Converting to polar results in

(continued)

**Example 3.32** (continued)

$$e(t) = \sqrt{1.82^2 + (-13.78)^2} \angle \tan^{-1} \frac{-13.78}{1.82}$$

$$e(t) = 13.899 \angle -82.47$$

In the time series function, considering the original frequency, the summation becomes

$$e(t) = 13.899 \sin(50t - 82.47)$$

**Example 3.33 Simplify these functions.**

$$f(t) = 20 \sin(100t + 10) + 150 \cos(100t - 26) - 120 \cos(100t + 30)$$

$$f(t) = 20 \cos(377t) - 50 \sin(377t - 30) + 20 \cos(150t) - 50 \sin(150t - 30)$$

## *Damped Sinusoidal Function*

A negatively damped exponential function shows a decay trajectory, which results in a decaying sinusoidal waveform if multiplied by a sinusoidal function. Figure 3.37 shows the function $f_1(t) = e^{-\alpha t} u(t)$ and $f_2(t) = \sin(\omega t)$. The product of these functions is sinusoidally wrapped around a decaying factor as $f_3(t) = e^{-\alpha t} \sin(\omega t)$ $u(t)$.

**Note 3.24**   Derivative of a function consisting of the product of two functions $f_1 f_2$ can be found as

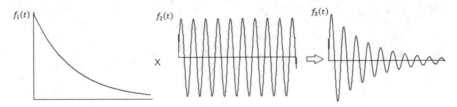

**Fig. 3.37**   The product of a sinusoidal and exponentially damping function results in a damped sinusoidal

$$\frac{d}{dt}(f_1 f_2) = \left(\frac{d}{dt}f_1\right)f_2 + f_1\left(\frac{d}{dt}f_2\right)$$

Therefore,

$$\frac{d}{dt}(e^{-at}\sin(\omega t)u(t)) = \left(\frac{d}{dt}e^{-at}\right)\sin(\omega t) + e^{-at}\left(\frac{d}{dt}\sin(\omega t)\right)$$

Considering the definitions done in this chapter yields

$$(-ae^{-at})\sin(\omega t) + e^{-at}(\omega\cos(\omega t))$$

**Example 3.34 Sketch.**

$$f(t) = 5e^{-t}\sin(10t)u(t).$$

*Solution.* The amplitude of this signal at $t = 0$ is $f(t = 0) = 5e^{-0}\sin(0)$ or $f(0) = 0$. The $u(t)$ function shows that the function starts at time $t \geq 0$. This is a decaying sinusoidal because the power of the exponential argument is negative, i.e., negative damping. As time increases, the function value decreases and reaches zero when $t \to \infty$. In this decaying function, a sinusoidal is bounded by the exponential. The amplitude drops as the exponential function continues to drop, limiting the sinusoidal amplitude but not its frequency. The sinusoidal function has a frequency of 10 rad/s. The function oscillates every $\frac{10}{2\pi}$ s. Figure 3.38 shows the time variation of the signal $f(t)$.

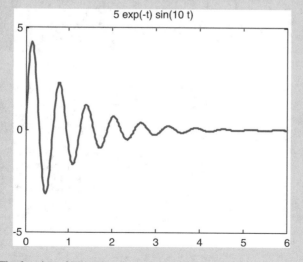

**Fig. 3.38** The function of Example 3.16. Note that the frequency is 10 rad/s

**Example 3.35 Sketch.**

$$f(t) = 5e^{-0.5t} \sin{(10t + 30)}u(t).$$

*Solution.* The function has initial value of $f(t = 0) = 5e^{0} \sin{(30)} = 2.5$. A phase $+30°$ shows a $30°$ shift of the signal to the left. The term $u(t)$ guarantees that the function appears starting at a time positive. Therefore, the negative time is eliminated. If the entire waveform existed for time positive and negative, the first zero crossing would have occurred when the argument was zero at $10t + \frac{\pi}{6} = 0$ or $t = -\frac{\pi}{60}$ s. Note that the phase shift angle must be in radians in this calculation. Figure 3.39 shows the waveform in a positive time.

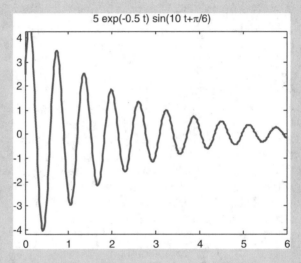

**Fig. 3.39**  The function of Example 3.35

**Example 3.36 Find the mathematical expression of the waveform shown in Fig. 3.40.**
*Solution.* Looking at the figure closely, some of the measured amplitudes and times can be used to determine the signal (Fig. 3.41).

The dotted line shows an exponential function with amplitude 3 and a damping factor of $\alpha$. The point of $(0.448, 1.916)$ can be used to determine the damping factor.

(continued)

**Example 3.36** (continued)

$$3e^{-\alpha t}|_{t=0.448} = 1.916$$

$$3e^{-0.448\alpha} = 1.916$$

$$e^{-0.448\alpha} = \frac{1.916}{3}$$

$$-0.448\alpha = \ln\left(\frac{1.916}{3}\right) = -0.448$$

Hence,

$$\alpha = 1$$

Now, we need to determine the sinusoidal function. One period of the sinusoidal function was approximately 0.2 s. Therefore,

$$T = 0.2 \rightarrow f = \frac{1}{0.2} = 5\ \text{Hz} \Rightarrow \omega = 2\pi f = 10\pi \frac{\text{rad}}{\text{s}}$$

The sinusoidal function becomes

$$\sin(10\pi t)$$

Combining the functions

$$f(t) = 3e^{-t}\sin(10\pi t)$$

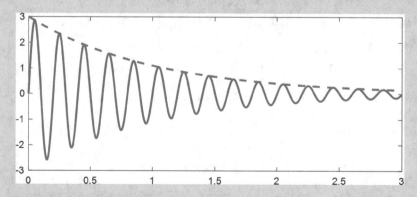

**Fig. 3.40** Waveform of Example 3.36

(continued)

**Example 3.36** (continued)

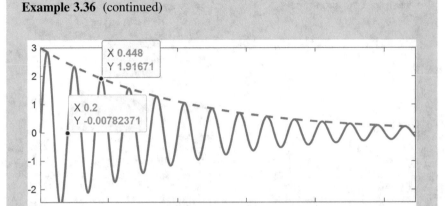

**Fig. 3.41** Measurement points to obtain the period and the damping factor

**Example 3.37 Sketch the waveform $f(t) = 10e^{-t} \sin(100\pi t + 60°)$.**
*Solution.* The frequency of the waveform is $100\pi \frac{rad}{s}$. That is $f = \frac{100\pi}{2\pi} = 50$ Hz.

The period of this waveform is $T = \frac{1}{f} = \frac{1}{50} = 20$ ms. This means that the waveform crosses zero every 10 ms and repeats itself every 20 ms.

The phase shift of 60° also translates into a time delay which can be obtained as follows:

Every period of this waveform is 20 ms which is 360°. Therefore, a 60° phase shift is equivalent to 3.33 ms obtained as follows:

$$\frac{360}{60} = \frac{20 \text{ m}}{?} \rightarrow ? = \frac{60}{360} \times 20 \text{ m} = 3.33 \text{ ms}$$

Therefore, the first zero crossing of the waveform occurs in negative time at $t = -3.33$ ms. The second zero crossing occurs at $t = 10$ m $-$ 3.33 m $=$ 6.67 ms. The waveform is completed in one cycle at a time $t = 6.67$ m $+$ 10 m $= 16.67$ ms.

The amplitude at the first peak can be obtained from the waveform evaluated at times of 90° shift from the first zero crossing and in intervals of 10 ms afterward. Ninety degree phase shift equals $\frac{90}{360} \times 20$ m $= 5$ ms. Therefore the first peak occurs at 5 $-$ 3.33 $=$ 1.67 ms. The second peak at 10 $+$ 1.67 $= 11.67$ ms, and so forth (Fig. 3.42).

(continued)

**Example 3.37** (continued)

The maximum of the sine waveform is 1, and the peak of the exponential is obtained by
Positive Peak

$$10e^{-t}|_{t=1.67 \text{ m}} = 10e^{-1.67 \text{ m}} = 9.99$$

Negative Peak

$$-10e^{-t}|_{t=11.67 \text{ m}} = -10e^{-11.67 \text{ m}} = -9.8$$

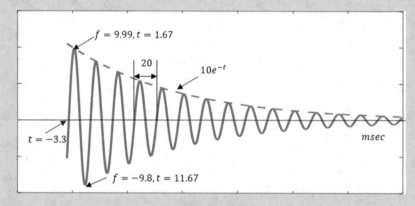

**Fig. 3.42** All the measured points are shown on the graph

## Periodic Waveform Mathematical Expression

Periodic waveforms repeat in a fixed amount of time, called the period. The key to expressing periodic waveforms may depend on how these waveforms are. Generally, writing the mathematical expression for a couple of periods of these waveforms should explain their mathematical structure and provide a lumped formula for the entire waveform. This is a working solution for simple waveforms.

**Example 3.38 Find the mathematical expression for the waveform shown in Fig. 3.43.**

*Solution.* As the waveform shows, it is a repeated pulse with 2 s at the amplitude of 10 followed by 2 s of amplitude 0. Therefore, the waveform repeats itself every 4 s. The first couple of periods is a pulse which can be written as

First period,

$$+10u(t) - 10u(t-2)$$

Second period,

$$+10u(t-4) - 10u(t-6)$$

Third period,

$$+10u(t-8) - 10u(t-10)$$

and so forth.

As mathematical expressions show, the amplitudes are alternating with a + and − , and the first period is a +. This seems like a pattern of $-1^0, -1^1, -1^2, -1^3, -1^4, -1^5 \ldots$ which is $+1, -1, +1, -1, +1, -1, \ldots$.

The time shift also has a pattern of 0, 2, 4, 6, 8, 10, … which can be expressed by $0 \times 2, 1 \times 2, 2 \times 2, 3 \times 2, 4 \times 2, 5 \times 2, \ldots$.

Therefore, the function $f(t) = -1^0 u(t - 0 \times 2) - 1^1 u(t - 1 \times 2) - 1^2 u(t - 2 \times 2) - 1^3 u(t - 3 \times 2) - 1^4 u(t - 4 \times 2) - 1^5 u(t - 5 \times 2)\ldots$.

The lumped formula can be expressed as

$$f(t) = \sum_{n=0}^{\infty} -1^n u(t - 2 \times n)$$

**Fig. 3.43** Waveform of
Example 3.38

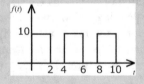

**Example 3.39 The periodic waveform of Example 3.38 represents the current of a 1 mH inductor. Find the voltage drop across the inductor.**

*Solution.* The current in lamped form is expressed as

$$i(t) = \sum_{n=0}^{\infty} -1^n u(t - 2 \times n)$$

The voltage drop across an inductor is

$$v = L \frac{di}{dt}$$

Therefore,

$$v = 1e - 3 \frac{d}{dt} \left( \sum_{n=0}^{\infty} -1^n u(t - 2 \times n) \right)$$

$$v = 1e - 3 \sum_{n=0}^{\infty} -1^n \delta(t - 2 \times n)$$

**Example 3.40 The current of a 0.1 F capacitor is $i(t) = \sum_{n=0}^{\infty} -1^n u(t - 2 \times n)$. Find the voltage drop across the capacitor.**

*Solution.* The voltage and current of a capacitor are

$$i = C \frac{dv}{dt}$$

Therefore, the voltage is

$$v = \frac{1}{C} \int i \, dt$$

$$v = \frac{1}{0.1} \int \left( \sum_{n=0}^{\infty} -1^n u(t - 2 \times n) \right)$$

(continued)

**Example 3.40** (continued)

The integral of a unit step function is a ramp function. If the unit step is shifted by $a$ seconds, the ramp will also be shifted by the same amount. Therefore,

$$\int ku(t-a) = k(t-a)u(t-a)$$

Hence,

$$v = 10 \sum_{n=0}^{\infty} -1^n(t - 2 \times n)u(t - 2 \times n)$$

The waveform of the voltage is shown in Fig. 3.44.

**Fig. 3.44** Waveform of Example 3.40

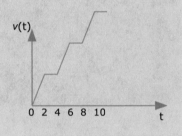

## Average of a Signal

A periodic waveform at period $T$ is presented by $f(t) = f(t + T)$. The average or DC value of this periodic signal can be obtained by $1/T$ times its integral over a period, as follows:

$$f_{dc} = f_{ave} = \frac{1}{T} \int_0^T f(t)dt$$

**Example 3.41 Find the DC value of the periodic waveform expressed in Fig. 3.45.**
*Solution.* The waveform shows a period of 4 s because it repeats itself every 4 s. This time can also be measured from the first zero crossing of the waveform to the next zero crossing or the time from one peak to the next.

(continued)

**Example 3.41** (continued)

According to Fig. 3.45, the function can be split into the time sections and written $f(t) = \begin{cases} 10\,u(t), & 0 < t < 2 \\ 0, & 2 < t < 4 \end{cases}$. Accordingly, the average of the signal can be found as

$$f_{dc} = f_{ave} = \frac{1}{T} \int_0^T f(t)dt$$

Replacing with the mathematical expression in time intervals results in

$$f_{dc} = f_{ave} = \frac{1}{4} \int_0^4 f(t)dt = \frac{1}{4}\left( \int_0^2 10u(t)dt + \int_2^4 0dt \right)$$

$$= \frac{1}{4}\left( 10t \Big|_0^2 + 0 \right) = \frac{1}{4}(10(2-0)) = 5$$

**Fig. 3.45** The function of Example 3.21

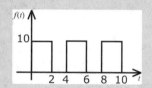

**Example 3.42 In the waveform of Fig. 3.45, the pulse width is decreased to 1 s. The waveform now has a 1 s on-time and an off-time of 3 s. Find the DC of the waveform.**

*Solution.* Since the period of the waveform is 4 s, and only the on-time has decreased, the DC content of the waveform can be calculated as

$$f_{dc} = f_{ave} = \frac{1}{4}\int_0^4 f(t)dt = \frac{1}{4}\left( \int_0^1 10u(t) + \int_1^3 0 \right) = \frac{1}{4} \times 10 = 2.5$$

It can be observed that the width in this pulse can directly control its DC.

**Example 3.43 Consider a pulse with an amplitude of $V$ and the on-time of $t_{on}$ at the period of $T$. Find the DC content of this signal.**

*Solution.* $t_{on}$ is the modulation of this signal. Integral over one-period results in

$$f_{dc} = f_{ave} = \frac{1}{T} \int_0^{t_{on}} Vu(t) = \frac{t_{on} \times V}{T} = \frac{t_{on}}{T} \times V$$

The ratio of $\frac{t_{on}}{T}$ is also known as duty cycle, $d$. Therefore, the DC of a pulse with duty cycle $d$ and amplitude $V$ is

$$f_{dc} = Vd$$

**Example 3.44 A pulse of 300 V DC is applied to a resistor. Calculate the DC voltage observed across the resistor at 40%, 50%, 80%, and 100% duty cycles.**

*Solution*

$$@d = 40\%, V_{dc} = dV = 0.4 \times 300 = 120 \text{ V}$$

$$@d = 50\%, V_{dc} = dV = 0.5 \times 300 = 150 \text{ V}$$

$$@d = 80\%, V_{dc} = dV = 0.8 \times 300 = 240 \text{ V}$$

$$@d = 100\%, V_{dc} = dV = 1 \times 300 = 300 \text{ V}$$

## Root Mean Square (RMS)

The root mean square of a periodic function $f(t) = f(t + T)$ at period $T$ can be obtained as follows:

$$f_{rms} = \sqrt{\frac{1}{T} \int_0^T f^2(t) dt}$$

**Note 3.25** The function must be squared first, and then the root of its average must be calculated.

**Example 3.45 Find the RMS of the function shown in Fig. 3.46.**
To calculate the RMS, the period of the waveform and mathematical expression or the value of the function in each time interval must be expressed. In this example, the period is $T = 4$ s, as the signal repeats itself every 4 s. The function has a value of 10 in time $0 < t < 2$ and is zero in time $2 < t < 4$, and this sequence is repeated. The rms can be obtained as follows:

$$f_{rms} = \sqrt{\frac{1}{T} \int_0^T f^2(t)dt} = \sqrt{\frac{1}{4} \left( \int_0^2 (10)^2 dt + \int_2^4 0 dt \right)}$$

$$f_{rms} = \sqrt{\frac{1}{4} \left( 100t \Big|_0^2 + 0 \right)} = \sqrt{\frac{1}{4}(100(2-0)+0)} = \sqrt{50} = 7.07$$

**Fig. 3.46** The function of
Example 3.45

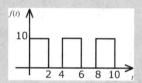

**Example 3.46 Find the RMS of the function $f(t) = A \sin \omega t$.**
*Solution.* The RMS of the function can be obtained as follows:

$$f_{rms} = \sqrt{\frac{1}{T} \int_o^T (A \sin \omega t)^2 d\omega t}$$

Since the period of this waveform is $T = 2\pi$, and the integral is taken over $d\omega t$, the RMS becomes

$$f_{rms} = \sqrt{\frac{1}{2\pi} \int_o^{2\pi} (A \sin \omega t)^2 d\omega t}$$

(continued)

**Example 3.46** (continued)

$$f_{\text{rms}} = \sqrt{\frac{A^2}{2\pi} \int_{o}^{2\pi} \frac{1 - \cos 2\omega t}{2} d\omega t}$$

$$f_{\text{rms}} = \sqrt{\frac{A^2}{2 \times 2\pi} 2\pi} = \frac{A}{\sqrt{2}}$$

**Example 3.47 Find the RMS of the voltage measured at the utility outlet of peak voltage** $110\sqrt{2}$ **V,60 Hz function** $f(t) = 110\sqrt{2} \sin 377t$.
*Solution.* The RMS of the function can be obtained as follows:

$$f_{\text{rms}} = \sqrt{\frac{1}{T} \int_{o}^{T} \left(110\sqrt{2} \sin 377t\right)^2 d\omega t}$$

Since the period of this waveform is $T = 2\pi$ and the integral is taken over $d\omega t$, the rms becomes

$$f_{\text{rms}} = \sqrt{\frac{1}{2\pi} \int_{o}^{2\pi} \left(110\sqrt{2} \sin 377t\right)^2 d\omega t}$$

$$f_{\text{rms}} = \sqrt{\frac{\left(110\sqrt{2}\right)^2}{2\pi} \int_{o}^{2\pi} \frac{1 - \cos 2 \times 377t}{2} d\omega t}$$

$$f_{\text{rms}} = \sqrt{\frac{\left(110\sqrt{2}\right)^2}{2 \times 2\pi} 2\pi} = \frac{110\sqrt{2}}{\sqrt{2}} = 110$$

## Circuit Simplification Techniques

### *Voltage Division*

Consider a circuit with a series of resistors connected to a voltage source. The voltage drop across each resistor is proportional to its resistance. Since the current passing through all elements is similar, the total current can be obtained as

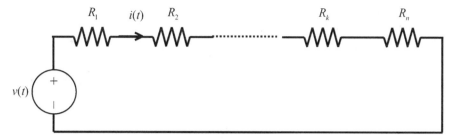

**Fig. 3.47** Series connection of resistors and voltage division

$$i = \frac{v}{\sum R}.$$

Therefore, the voltage drop of a resistor $R_k$ is $v_k = \frac{R_k}{\sum R} v$ (Fig. 3.47).

**Example 3.48 Find the voltage drop across all circuit elements shown in Fig. 3.48.**
*Solution.* The circuit has two sets of resistors in series. A $10\,\Omega$ and $2\|8 = 1.6\,\Omega$ connected in series to a 100 V voltage source. The voltage across 10 and $1.6\,\Omega$ resistors is

$$V_{10\ \Omega} = \frac{10}{10 + 1.6} 100 = 86.21 \text{ V}$$

$$V_{1.6\ \Omega} = \frac{1.6}{1.6 + 10} 100 = 13.79 \text{ V}$$

**Fig. 3.48** Circuit of
Example 3.48

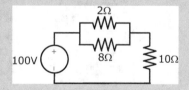

## Current Division

The current shared among several elements in parallel depends on the conductance of each parallel branch (Fig. 3.49).

Consider a circuit with $n$ resistors in parallel fed from a current source $i(t)$. The current drawn from the source is the summation of all branch currents, such that

**Fig. 3.49** Parallel
connection of resistors and
current division

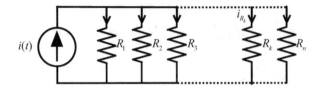

$$i = \sum_{m=1}^{n} i_{R_m}$$

$$i = \frac{v}{R_1} + \frac{v}{R_2} + \ldots + \frac{v}{R_n} = v\left(\frac{1}{R_1} + \frac{1}{R_2} + \ldots + \frac{1}{R_n}\right)$$

The voltage across the circuit is obtained from

$$v = \frac{i}{\left(\frac{1}{R_1} + \frac{1}{R_2} + \ldots + \frac{1}{R_n}\right)}$$

The current passing through any resistor is $i_{R_k} = \frac{v}{R_k} = \dfrac{\frac{1}{R_k}}{\left(\frac{1}{R_1} + \frac{1}{R_2} + \ldots + \frac{1}{R_n}\right)} \cdot i.$

**Example 3.49 Find the current of each branch in the circuit of Fig. 3.50.**
*Solution*

- The current of 4 Ω is $I_{R_4\ \Omega} = \dfrac{\frac{1}{4}}{\left(\frac{1}{4} + \frac{1}{5} + \frac{1}{10}\right)} 20 = 9.1$ A.

- The current of 5 Ω is $I_{R_5\ \Omega} = \dfrac{\frac{1}{5}}{\left(\frac{1}{4} + \frac{1}{5} + \frac{1}{10}\right)} 20 = 7.27$ A.

- The current of 10 Ω is $I_{R_{10}\ \Omega} = \dfrac{\frac{1}{10}}{\left(\frac{1}{4} + \frac{1}{5} + \frac{1}{10}\right)} 20 = 3.63$ A.

**Fig. 3.50** Circuit of
Example 3.49

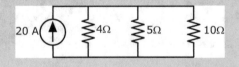

## Source Conversion

A voltage source in series to a resistor can be converted to a current source and
parallel resistance. The size of the current source is obtained by the voltage source
value divided by the resistance (Fig. 3.51).

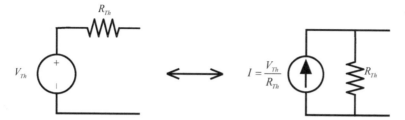

**Fig. 3.51** Thevenin and Norton equivalent conversions

**Note 3.26** The source conversion and Thevenin to Norton equivalent are also valid for dependent sources.

**Example 3.50 Convert the Thevenin to Norton and vice versa in circuits of Figs. 3.52, 3.53, and 3.54.**

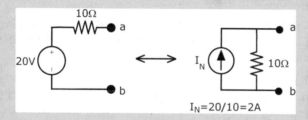

**Fig. 3.52** Circuit of Example 3.50

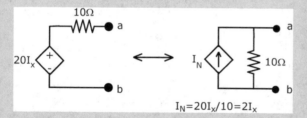

**Fig. 3.53** Another circuit of Example 3.50

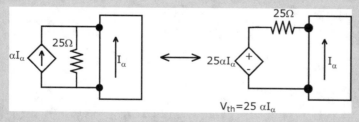

**Fig. 3.54** Circuit of Example 3.50. $I_\alpha$ is the current of an element in the circuit. The amount of current generated by the current source depends on $I_\alpha$

**Fig. 3.55** Thevenin
equivalent from points
*a* and *b*

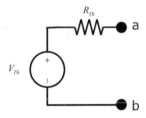

## Thevenin Equivalent Circuit

Any circuit from a desired set of terminals can be modeled as a voltage source and
a series resistance. This Thevenin equivalent circuit has two components $V_{th}$
(an independent voltage source with fixed value) and $R_{th}$ (or an impedance in the
case of the RLC circuit), as shown in Fig. 3.55.

Thevenin equivalent circuits are defined for all resistive circuits with inductors,
capacitors, and dependent sources.

**Note 3.27** To obtain the $V_{th}$ across terminal ports $a$ and $b$, the circuit must be
disconnected from the load at these terminals. The voltage difference built at the
terminals is measured as $V_{th}$.

**Note 3.28** Only independent sources must be turned off to obtain the $R_{th}$ at terminal
ports $a$ and $b$. This means a zero voltage source (short circuit) and a zero current
source (open circuit).

---

**Example 3.51 Find the Thevenin equivalent of the circuit in Fig. 3.56.**
*Solution.* Finding $V_{th}$ requires that $R_L$ be disconnected from the circuit. If so,
the 5 Ω resistor branch is disconnected and does not pass any current.
Therefore, the voltage drop across this resistor becomes zero, and the terminal
voltage becomes equivalent to the voltage drop across the 2 Ω resistor.

The KVL for loop suggests that $-10 + 8I_1 + 2I_1 = 0$. Therefore, $I_1 =$
$\frac{10}{10} = 1$ A. The voltage of 2 Ω resistor becomes $V_{th} = V_{2\Omega} = 2 \times 1 = 2$ V.

Finding $R_{th}$ needs a 0 V voltage source (short circuit) and zero A current
source (open source).

Removing the voltage source leaves the 8 and 2 Ω in parallel and results in a
series with 5 Ω. $R_{th} = \left(8 \| 2\right) + 5 = \frac{8 \times 2}{8+2} + 5 = 6.6$ Ω.

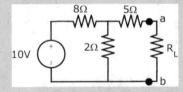

**Fig. 3.56** Figure Thevenin of Example 3.51

(continued)

**Example 3.51** (continued)

The equivalent Thevenin circuit is shown as follows (Fig. 3.57):

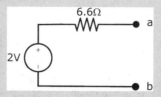

**Fig. 3.57** Thevenin equivalent of the circuit is shown in Fig. 3.56

**Example 3.52 Find the Thevenin equivalent of the circuit shown in Fig. 3.58.**
*Solution.* Since the load must be disconnected from the terminals, no current passes the 5 kΩ resistor. Hence, the voltage drop across this element is zero. Therefore, $V_{th} = V_x = V_1$, which is node 1. Writing a KCL for node 1 determines the voltage.

KCL ①. Passive element 1 kΩ resistor drains the current out of the node, and the dependent current source forces the current into the node (direction of the current sources). Therefore,

$$\frac{V_1 - 2}{3000} - \frac{V_x}{1000} = 0$$

Considering $V_x = V_1$ results in

$$\frac{V_x - 2}{3000} - \frac{V_x}{1000} = 0$$

$$\frac{-2V_x}{3000} - \frac{2}{1000} = 0$$

$$\frac{-2V_x}{3000} = \frac{2}{1000}$$

$$V_x = V_{th} = -3 \text{ V}.$$

To obtain $R_{th}$, the independent voltage source must be turned off, and the dependent current source must remain in the circuit. The circuit is shown in Fig. 3.59.

(continued)

**Example 3.52**  (continued)

As the circuit shows, the resistance measured at terminals depends on the value of the dependent current source. An external voltage source $V$ is connected to the terminals to excite this dependent source. The current drawn from the source $I$ is measured, and the resistance is shown at the terminal can be obtained $R_{th} = \frac{V}{I}$.

$$\text{KCL①} \frac{V_1}{3000} - \frac{V_x}{1000} + \frac{V_1 - V_x}{5000} = 0$$

From the circuit $V_x = V$, replacing in KCL results in $\frac{V_1}{3000} - \frac{V}{1000} + \frac{V_1 - V}{5000} = 0$.

$$V_1 \left( \frac{1}{3000} + \frac{1}{5000} \right) = V \left( \frac{1}{1000} + \frac{1}{5000} \right)$$

$$V_1 \left( \frac{1}{3000} + \frac{1}{5000} \right) = V \left( \frac{1}{1000} + \frac{1}{5000} \right)$$

$$V_1 \left( \frac{8}{15,000} \right) = V \left( \frac{6}{5000} \right), \text{or } V_1 = \frac{9}{4} V$$

$$I = \frac{V - V_1}{5000} = \frac{V - \frac{9}{4} V}{5000} = \frac{-5}{4} V \frac{1}{5000}$$

$$V = -4000 I \rightarrow R_{th} = -4000 \ \Omega.$$

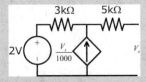

**Fig. 3.58**  Equivalent circuit of Example 3.52

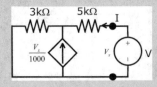

**Fig. 3.59**  All independent sources have been zeroed out, and since there is a dependable source in the circuit, an external source with voltage $V$ needs to be connected to terminal points to excite the circuit for impedance measurement

**Example 3.53 Find the Thevenin of the circuit shown in Fig. 3.60.**
*Solution.* To obtain the Thevenin impedance, an external source must be used. Connecting voltage source $V$ at the terminals creates a two-loop system.

- KVL ① $-7i + 3i + 5(i_1 - i) = 0$
- KVL ② $V + 5(i - i_1) = 0$

From the first KVL, $i_1 = \frac{12}{8} i$. Replacing in the second equation, $V = \frac{20}{8} i$. Therefore, the Thevenin equivalent circuit is a simple resistor with a resistance $\frac{20}{8}$ $\Omega$.

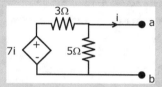

**Fig. 3.60** Circuit of Example 3.53

# Norton Equivalent Circuit

Any circuit from a set of the desired terminal can be represented by a current source parallel to a resistor. The current source shows the short circuit current that might have passed the terminals if it was shorted, hence called the short circuit current source, and the parallel resistance shows the equivalent resistance when all the independent sources are turned off (Fig. 3.61).

**Note 3.29** The load resistance across the terminals must become a short circuit in the Norton equivalent circuit. The current passing this short circuit is the equivalent of the Norton current source.

**Note 3.30** The Norton resistance is obtained similarly to the Thevenin resistance.

**Fig. 3.61** Norton
equivalent

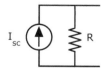

**Example 3.54 Find the Norton equivalent of the following circuit (Fig. 3.62).**

*Solution.* A short circuit of the load leaves the $R_L$ out of the circuit. Hence, the $I_{sc}$ is similar to the current passing through the 5 $\Omega$ resistor.

The total current of the circuit $I$ can be calculated by $I = \frac{10}{R_{eq}}$. The equivalent resistance is parallel to $5\|2$ in series with the 8 $\Omega$. $R_{eq} = 8 + (5\|2) = 9.428\ \Omega$, which makes the total current $I = 1.06$ A.

The current of the 5 $\Omega$ resistor is obtained from the current division as

$$I_{sc} = I_{5\,\Omega} = \frac{2}{2+5} 1.06 = 0.303 \text{ A}$$

Finding $R_{th}$ needs a zero $V$ voltage source (short circuit) and zero $A$ current source (open source). The circuit of Fig. 3.63 is obtained. Removing the voltage source leaves the 8 and 2 $\Omega$ in parallel and results in a series with 5 $\Omega$. $R_{th} = \left(8\|2\right) + 5 = \frac{8\times2}{8+2} + 5 = 6.6\ \Omega$.

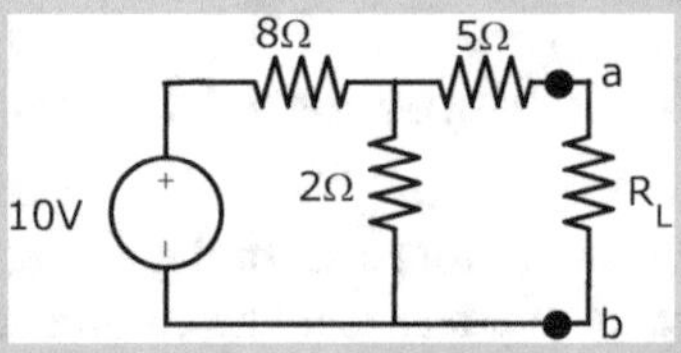

**Fig. 3.62**  Circuit of
Example 3.54

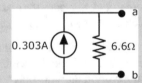

**Fig. 3.63**  Norton
equivalent of the circuit in
Fig. 3.62

# Norton and Thevenin Equivalent

Thevenin and Norton circuits can be converted to each other (Fig. 3.64).

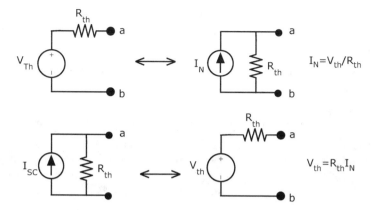

**Fig. 3.64** Thevenin and Norton's transformation

## Power Calculations

### *Consumption of Power*

Consider a resistor $R$ as part of a circuit. DC current $I$ passing through this resistor will generate a DC voltage drop $V$ across the resistor. In this condition, the resistor starts to consume power and dissipate heat. The amount of power loss is directly proportional to the amount of voltage and current and is obtained by

$$P_{\text{Loss}} = VI$$

The dissipated power is measured in watts. Considering Ohm's law in resistors, $V = RI$, the power dissipation can be obtained as

$$P_{\text{Loss}} = VI = (RI)I = RI^2 = \frac{V^2}{R}$$

The resistor is a passive element, which means the amount of energy stored over a cycle is zero. Resistors cannot store electric energy. However, they can store thermal energy in a particular type of ceramics.

**Note 3.31** Power consumption has a positive sign, e.g., a +100 W load consumes power equivalent to 100 W.

### *Generation of Power*

A DC power source feeds the circuit with a DC voltage and current. The product of the total current drawn from the source by the voltage of the source determines the amount of power the source has generated and fed to the circuit. Since the current is outgoing from the terminals of the source, the power is generated and is considered a negative value.

**Note 3.32** Power generation has a negative sign, e.g., a $-100$ W source generates and feeds the circuit by 100 W.

---

**Example 3.55  Consider a 100 W power load connected to a power source. The load consumes +100 W, and the source generates $-100$ W. If the voltage is 20 V, the load has 5 A current entering the terminal, and the source has 5 A current exiting the terminal.**

---

**Example 3.56 A battery unit is utilized to drive an electric vehicle, as shown in Fig. 3.65. The battery unit can both generate and absorb power. The battery is discharged to propel the vehicle forward on an uphill road. It is charged through regenerative braking when the vehicle's energy is harvested to be stored on a downhill road.**

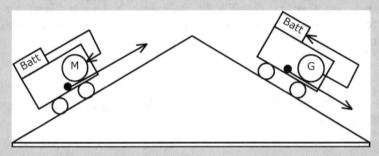

**Fig. 3.65** An electric vehicle discharges the battery uphill and charges it on the downhill road

## Maximum Power Transfer to Load in Pure Resistive Circuits

Consider a Thevenin equivalent of a resistive circuit connected to a load resistance, as shown in Fig. 3.66. A voltage source and a Thevenin resistance force the current $I$ through the circuit.

The load current is obtained from $I_L = \frac{V_{th}}{R_{th}+R_L}$ . Therefore, the power delivery to the load is

**Fig. 3.66** A Thevenin
equivalent circuit delivers
power to a resistive load.
The power delivery is
maximum when the
Thevenin and load
resistance are equal

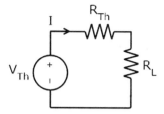

$$P_L = R_L I_L^2 = R_L \left( \frac{V_{th}}{R_{th} + R_L} \right)^2$$

To maximize the power delivery to the load, $\frac{dP_L}{dR_L} = 0$.

$$\frac{dP_L}{dR_L} = \frac{V_{th}^2 (R_{th} + R_L)^2 - 2(R_{th} + R_L)R_L V_{th}^2}{(R_{th} + R_L)^4} = \frac{V_{th}^2 (R_{th} + R_L) - 2R_L V_{th}^2}{(R_{th} + R_L)^3}$$

$$\frac{dP_L}{dR_L} = \frac{V_{th}^2 R_{th} - V_{th}^2 R_L}{(R_{th} + R_L)^3} = 0$$

Hence, the condition to transfer maximum power from the source to the load is

$$R_{th} = R_L$$

**Example 3.57 Find the load resistance that can absorb maximum power
from the circuit of Fig. 3.67.**
*Solution.* Thevenin equivalent of the circuit concerning load terminals can be
found by disconnecting the load and turning the independent sources off. This
results in

$$R_L = R_{th} = 4 + \left( 3 \| 9 \right) = \frac{25}{4} \ \Omega$$

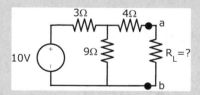

**Fig. 3.67** Circuit of Example 3.57

## Problems

3.1. Sketch the following functions:

(a) $f(t) = 10u(t)$
(b) $f(t) = 10u(t + 7)$
(c) $f(t) = 10u(t - 7)$
(d) $f(t) = 10u(3\ t - 15)$
(e) $f(t) = -10u(t)$
(f) $f(t) = 10u(-t)$

3.2. Sketch the following functions:

(a) $f(t) = 2tu(t)$
(b) $f(t) = 2(t - 3)u(t - 3)$
(c) $f(t) = 2(t - 3)u(t - 7)$
(d) $f(t) = -2tu(t)$
(e) $f(t) = 2tu(-t)$

3.3. Find the mathematical expression of the following signals:

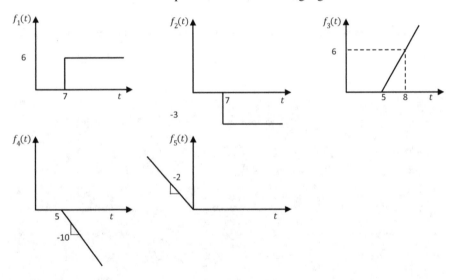

3.4. Sketch the following functions:

(a) $f(t) = u(t - 1) - u(t - 2)$
(b) $f(t) = u(t - 1) + u(t - 2) - u(t - 3)$
(c) $f(t) = 3u(t - 1) + 2u(t - 2) - 5u(t - 4)$
(d) $f(t) = tu(t) - (t - 2)u(t - 2)$
(e) $f(t) = tu(t) - (t - 3)u(t - 3) - (t - 4)u(t - 4) + (t - 5)u(t - 5)$
(f) $f(t) = 3(t - 2)u(t - 2) - 6(t - 4)u(t - 4) + 3(t - 6)u(t - 6)$
(g) $f(t) = 3tu(t - 5)$
(h) $f(t) = 5(t - 1)u(t - 3) + 3(t - 2)u(t - 4)$

3.5. Find the mathematical expression of the following signals:

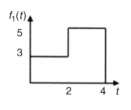

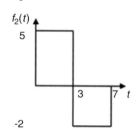

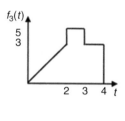

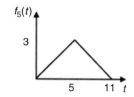

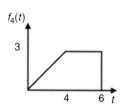

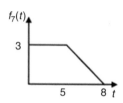

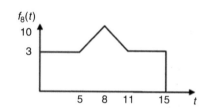

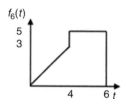

3.6. Sketch the following functions and determine their amplitude and damping factor.

(a) $f(t) = 3e^{-t}u(t)$
(b) $f(t) = 2e^{+5t}u(t)$
(c) $f(t) = 2e^{-3(t-1)}u(t-1)$
(d) $f(t) = e^{-t}u(t) - e^{-(t-5)}u(t-5)$
(e) $f(t) = 3e^{-(t-2)}u(t-1)$

3.7. In the following signals, determine the amplitude, phase shift, and frequency in (rad/s) and Hz.

(a) $f(t) = A \sin(\omega t + \phi)$
(b) $f(t) = 200 \sin(377t + 10)$
(c) $f(t) = 110\sqrt{2} \sin(120\pi t + 60)$
(d) $f(t) = \sin(t)$
(e) $f(t) = 20 \sin\left(100t + \frac{\pi}{6}\right)$

3.8. Do the following math operations:

(a) $\int u(t)dt$

(b) $\int 10u(t-1)dt$

(c) $\int 10u(t) - 10u(t-2)dt$

(d) $\int tu(t)dt$

(e) $\int 3(t-1)u(t-1)dt$

(f) $\int u(-t)dt$

(g) $\int -tu(-t)dt$

(h) $\int 10 \sin (377t + 20)dt$

(i) $\int 110\sqrt{2}\cos \left(100\pi t + \frac{\pi}{6}\right)dt$

(j) $\int e^{-5t}dt$

(k) $\int \frac{1}{5}e^{-3t} + \frac{3}{4}e^{2t}dt$

(l) $\int \delta(t)dt$

(m) $\int (t^2 + 5)\,\delta(t)dt$

(n) $\int (t^2 + 5)\,\delta(t-2)dt$

3.9. Do the following math operations:

(a) $\frac{d}{dt}(u(t))$

(b) $\frac{d}{dt}(u(t-1))$

(c) $\frac{d}{dt}(2u(t) + 3u(t-1))$

(d) $\frac{d}{dt}(3(t-1)u(t-1) + 5(t-3)u(t-3))$

(e) $\frac{d}{dt}(e^{-3t}u(t))$

(f) $\frac{d}{dt}(e^{-3t}u(t-2))$

(g) $\frac{d}{dt}\left(e^{-5t} + e^{5t}\ u(t)\right)$

(h) $\frac{d}{dt}(100 \sin 10\pi t)$

(i) $\frac{d}{dt}(200 \cos 1000t + 60)$

3.10. Do the following math operations:

(a) $2 \sin (10t + 60) + 3 \cos (10\,t + 30)$

(b) $\sin (100t + 10) - \sin (100t - 80)$

(c) $\cos (20\pi t) + \cos (20\pi t + 10) - \sin (20\pi t + 30)$

3.11. Find the amplitude, damping factor, frequency of oscillation, and initial phase in the following signals:

(a) $f(t) = 10 + \sin t$

(b) $f(t) = 5 + 100\sqrt{2} \sin 377t$

(c) $f(t) = \sin^2 5t$

(d) $f(t) = \sin^2 5t - \cos^2 5t$

(e) $f(t) = 1 + \sin^2 10t$

3.12. Find the average and rms of the following waveforms:

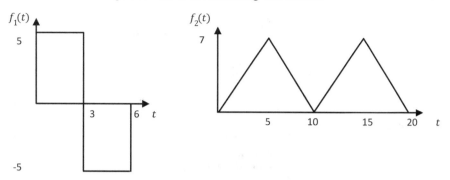

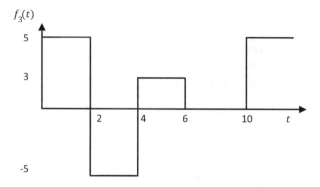

(a) $f_4(t) = 110 \sin 377t$
(b) $f_5(t) = 10 + \sin t$
(c) $f_6(t) = 5 + 100\sqrt{2} \sin 377t$
(d) $f_7(t) = \sin^2 5t$
(e) $f_8(t) = \sin^2 5t - \cos^2 5t$
(f) $f_9(t) = 1 + \sin^2 10t$

3.13. Find $I_1$, $I_2$.

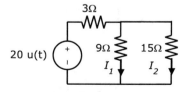

3.14. Find $I_1$, $V_o$.

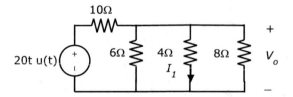

3.15. Find $V_1$, $V_2$, and $V_3$.

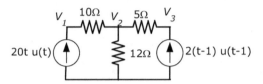

3.16. Find $V_1$ and $I$.

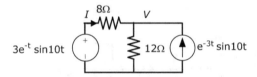

3.17. Find $I_1$, $I_2$.

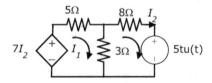

3.18. Find $V_1$.

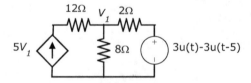

3.19. Find $I$ and $V_o$.

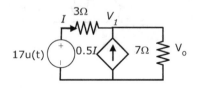

3.20. Find *I* and *V*.

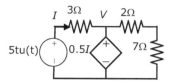

3.21. Find the voltage drop and current of all resistors. Find the power taken from the source and the power dissipated in each resistor.

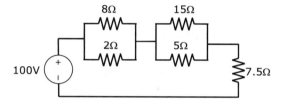

3.22. Find the voltage drop and current of all resistors. Find the power taken from the source and the power dissipated in each resistor.

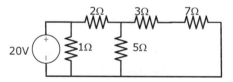

3.23. Find the voltage drop and current of all resistors. Find the power taken from the source and the power dissipated in each resistor.

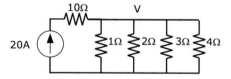

3.24. Find the voltage drop and current of all resistors. Find the power taken from the source and the power dissipated in each resistor.

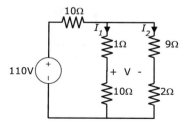

3.25. Find the Thevenin and Norton equivalent of the following circuit across the load terminals.

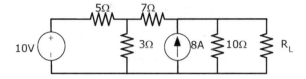

3.26. Find the Thevenin and Norton equivalent of the following circuit across the load terminals.

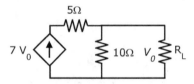

3.27. Find the Thevenin and Norton equivalent of the following circuit across the load terminals.

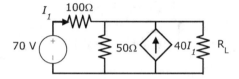

3.28. Find the Thevenin and Norton equivalent of the following circuit across the load terminals.

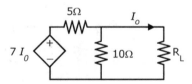

3.29. Find the load impedance at which the maximum power is transferred from the source to the load.

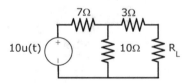

3.30. Find the load impedance at which the maximum power is transferred from the source to the load.

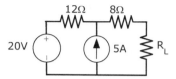

3.31. Find the load impedance at which the maximum power is transferred from the source to the load.

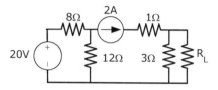

3.32. Find the load impedance at which the maximum power is transferred from the source to the load.

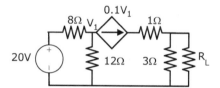

# Chapter 4
# Circuit Response Analysis

## Introduction

The flow of current in the circuit branches and the voltage drop across circuit elements depend on their behavior and ability to store energy. For instance, the voltage drop across a resistor is in phase with its current passing through. However, that is not the same in a capacitor or an inductor. This makes the circuit KVL and KCL equations integrodifferential equations. The order of these equations depends on the number of energy-storing elements. The circuit elements are introduced in this chapter, and their equations are discussed. The order of a circuit is discussed, and responses of first- and second-order circuits to their initial condition and external sources are analyzed.

## Resistors

Consider a resistor shown in Fig. 4.1 with current $i(t)$ passing through, resulting in a voltage drop $v(t)$. The relation of time-varying voltage and time-varying current to the resistance of the resistor follows Ohm's law as follows:

$$v(t) = Ri(t)$$

Ohm's law of a resistor indicates that the voltage drop across the resistor linearly depends on the current passing through. This also demonstrates that the voltage and current waveforms across a resistor are in-phase. Figure 4.2 shows the voltage and current of a resistor. As the figure shows, the zero-crossing of the two signals (voltage and current) is the same, and the peaks coincide. The current is a scaled waveform of the voltage.

A. Izadian, *Fundamentals of Modern Electric Circuit Analysis and Filter Synthesis*, https://doi.org/10.1007/978-3-031-21908-5_4

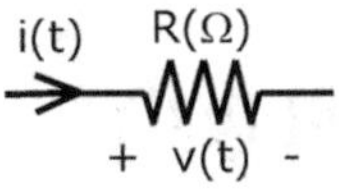

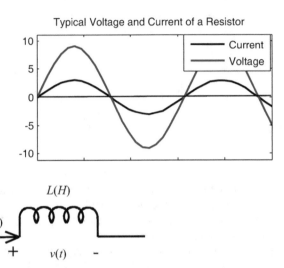

**Fig. 4.1** Schematic of a resistor. The voltage drop $v(t)$ dependency across a resistor $R$ when passing current $i(t)$

**Fig. 4.2** The voltage and current of a resistor are in-phase. They may have different values (peaks)

**Fig. 4.3** Circuit schematic of a charged inductor with initial current $I_0$. The voltage and current of the **inductor** are related through a differential equation as $v(t) = L\frac{di(t)}{dt}$

# Inductors

Consider an inductor with an inductance of $L(H)$ Henrys. The inductor also has an initial current of $I_0$, as shown in Fig. 4.3.

The relation between the voltage drop and the current of the inductor is obtained by

$$v(t) = L\frac{di(t)}{dt}$$

where $i(t)$ is the current passing through the inductor and $v(t)$ is the voltage drop across the inductor. As the equations demonstrate, the voltage and current in sinusoidal waveforms are 90° out of phase, where the current waveform lags the voltage waveform. Figure 4.4 shows the voltage and current of an inductor.

Considering an initial current of $I_0$ passing through the inductor, the current can be expressed as

**Fig. 4.4** The voltage and current of an inductor are 90° out of phase. The current lags the voltage. It is recommended to take voltage as a reference and then measure the phase shift of the current

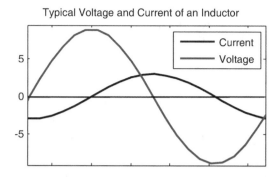

Typical Voltage and Current of an Inductor

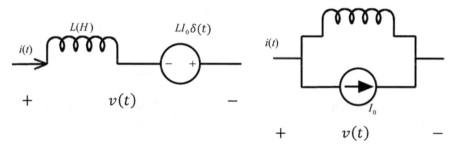

**Fig. 4.5** A charged inductor with initial current $I_0$. The initial condition can be modeled as a voltage source in series to the inductor. The value of the voltage source is $LI_0\delta(t)$, and the polarity of the voltage source is selected such that the current out of this source follows the same direction as the initial current. The series inductor and the voltage source model are best for KVL analysis. The model of a charged inductor can also be shown as a current source in parallel to the inductor. This model is best for KCL analysis

$$i(t) = I_0 u(t)$$

Replacing this current in the inductor's equation results in the initial voltage of

$$V_0 = L\frac{d(I_0 u(t))}{dt} = LI_0\delta(t)$$

The model of a charged inductor can be presented as an inductor without charge in parallel or series to a source that represents the initial condition (Fig. 4.5). The initial condition can also be modeled as a current source with the value of $I_0$, which is connected in parallel to the inductor. Figure 4.6 shows the inductor without charge and a current source representing the initial charge.

The inductor is an energy-storing element—the net energy stored in an ideal inductor when a sinusoidal voltage is applied to zero. The inductor is fully charged in a positive cycle and discharged in a negative cycle. The inductor is therefore called a passive element.

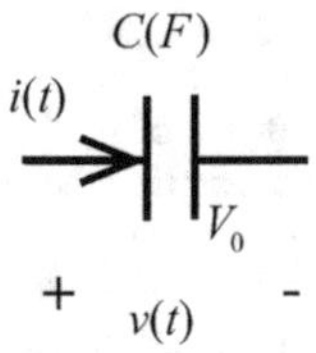

**Fig. 4.6** Circuit schematic
of a charged capacitor. The
voltage and current are
related through a differential
equation as $i(t) = C\frac{dv(t)}{dt}$

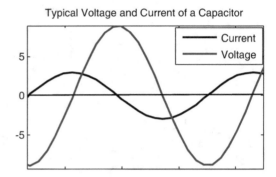

**Fig. 4.7** The voltage and
current of a capacitor are 90°
out of phase. The current
leads to the voltage. It is
recommended to take
voltage as a reference and
then measure the phase shift
of the current

The amount of stored energy (in Joules) in an inductor is proportional to its
inductance $L(H)$ and the square of current passing through $I^2$ as follows:

$$W = \frac{1}{2}LI^2 \ (J)$$

The energy stored in an inductor depends on the current passing through the
inductor. Therefore, the energy change in the inductor directly influences its current.
This means that

$$\frac{dW}{dt} \propto \frac{dI}{dt}$$

## Capacitors

Consider a capacitor with capacitance $C$ (F) Farads initially charged at voltage $V_0$ as
shown in Fig. 4.6. The current $i(t)$ generates voltage $v(t)$, which are related as

$$i(t) = C\frac{dv(t)}{dt}$$

This shows that the current and voltage across a capacitor, when sinusoidal
signals are applied, are 90° out of phase, with the current waveform leading the
voltage waveform. The phase shift is shown in Fig. 4.7.

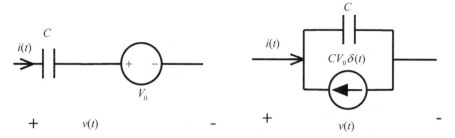

**Fig. 4.8** Model of a charged capacitor when the initial charge of the capacitor is shown as a series-connected voltage source. This model is best for KVL analysis. The initial charge can also be demonstrated as a current source with amplitude $CV_0\delta(t)$. The current direction is reversed when the capacitor starts to discharge; it sends the current out of its positive terminal. This model is suitable for KCL analysis

The initial charge of the capacitor can be modeled as a separate source in series or parallel to the no-charged capacitor. Considering a series voltage source, the model of the charged capacitor is shown in Fig. 4.8. Consider the voltage and current relation of a capacitor and its initial charge as

$$v(t) = V_0 u(t)$$

Then, the initial current of the capacitor becomes

$$I_0 = C\frac{d(V_0 u(t))}{dt} = CV_0\delta(t)$$

An ideal capacitor receives charge in a positive cycle and is fully charged. The negative cycle is fully discharged and charged with opposite polarity. Therefore, the net stored charge in this capacitor is zero in a period. For this reason, a capacitor is a passive element.

The amount of stored energy (in Joules) in a capacitor is proportional to its capacitance $C$ ($F$), and the square of applied voltage $V^2$ is as follows:

$$W = \frac{1}{2}CV^2 \text{ (J)}$$

The amount of energy stored in a capacitor depends on the voltage drop across the capacitor. Therefore, the energy variation in the capacitor directly influences its terminal voltage. This means that

$$\frac{dW}{dt} \propto \frac{dV}{dt}$$

## Order of a Circuit

Recalling circuit element definitions, it was determined that resistors just dissipate energy as heat, but inductors and capacitors store energy. The net energy storage of these elements over one cycle was zero. This makes these elements categorized as passive elements.

Differential and integral equations express the relation of the voltage and current of energy-storing elements. Therefore, each energy-storing element can potentially increase the order of a differential equation written for a circuit. These differential and integral equations are obtained through mesh and node analysis, i.e., KVL and KCL.

To determine the order of a given circuit:

- All possible simplifications of capacitors and inductors must be considered in the first step of determining a circuit's order. It means that the equivalent of a series connection of capacitors/inductors or parallel connection of capacitors/inductors must be obtained. The highest possible order is the count number of equivalent energy-storing elements.
- If a loop of the simplified circuit is formed by many capacitors and cannot be simplified further, the order is reduced by one for each capacitive loop.
- If a circuit node is only connected to many inductors and cannot be simplified further, the order is reduced by one for each of these inductive nodes.

This can be expressed as

$$order\ of\ a\ circuit = number\ of\ energy-storing\ elements,$$

$$capacitive\ loops,\ and\ inductive\ nodes$$

---

**Example 4.1 Find the order of the following circuits (Figs. 4.9, 4.10, 4.11, 4.12, 4.13, 4.14, 4.15, and 4.16)**

**Fig. 4.9** The circuit has one energy-storing element. Therefore, it is a first-order circuit. The differential equations that show the voltage or current relations are the first-order equation

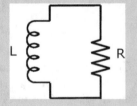

(continued)

**Example 4.1** (continued)

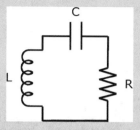

**Fig. 4.10** The circuit is simplified already and has two energy-storing elements. Therefore, it is a second-order circuit. The differential equations showing the circuit's voltage or current are second-order equations

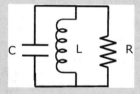

**Fig. 4.11** The circuit is simplified already and has two energy-storing elements. Therefore, it is a second-order circuit

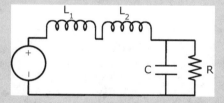

**Fig. 4.12** Two series inductors $L_1$ and $L_2$ can be reduced to one equivalent inductor. Then, the number of energy-storing elements becomes two. One is the equivalent inductor, and the other one is the capacitor. There is no capacitive loop or inductive node

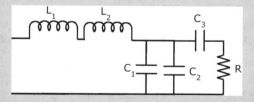

**Fig. 4.13** Two inductors $L_1$ and $L_2$ in series are equivalent to one inductor. Two capacitors $C_1$ and $C_2$ in parallel are equivalent to one capacitor. Then, the number of energy-storing elements becomes the equivalent inductor, equivalent capacitor, and $C_3$, a total of three. No inductive node and no capacitive loop exist. Therefore, the order is 3

(continued)

**Example 4.1** (continued)

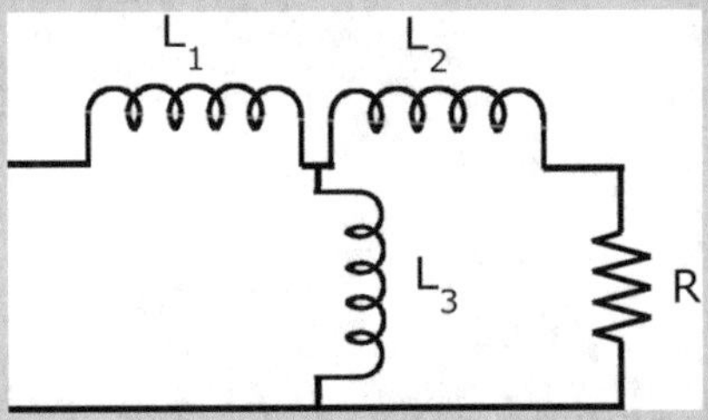

**Fig. 4.14** The inductors
form an inductive node
(only inductors are
connected). The order of the
circuit is 3 energy-storing
elements − 1 inductive node
− 0 capacitive loop =
2. The order is 2

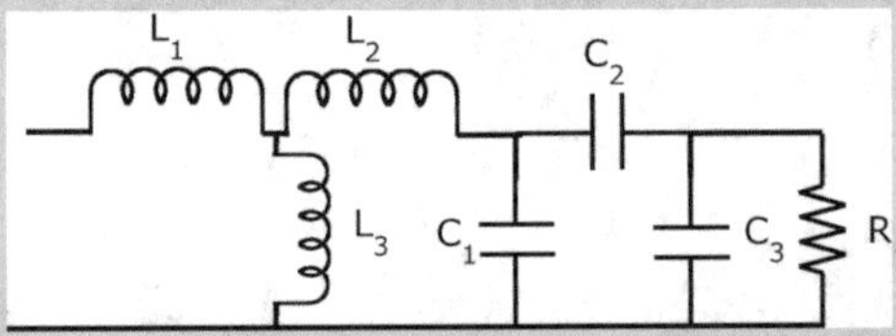

**Fig. 4.15** Inductors form an
inductive node; capacitors
form a capacitive loop. The
order of the circuit is
6 energy storing elements−
1 inductive node −1
capacitive loop = 4

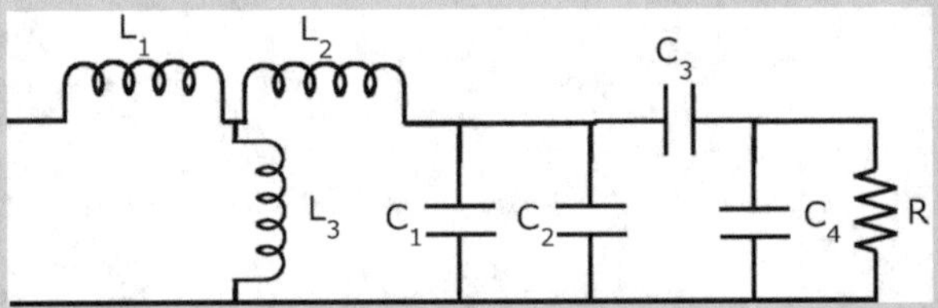

**Fig. 4.16** In parallel, two capacitors C1 and C2 are equivalent to one capacitor. The
inductors form an inductive node; the equivalent capacitor and the other two form a
capacitive loop. Therefore, the order of the circuit is 6 energy-storing elements ($C_1$ and $C_2$
are equivalent to one capacitor) − 1 inductive node − 1 capacitive loop = 4

## First-Order Circuits

A first-order circuit contains an equivalent of one energy-storing element as a
capacitor or an inductor. These circuits are analyzed in two conditions:

1. First is when its initial charge of energy-storing elements drives the circuit. The
   circuit response is identified as the voltage and current profiles. The response
   generated from initial charges is called a *natural response*.
2. The second is when the response is generated from an external source. The response
   is called a *forced response*, and depending on the type of the source applied, the

name is adapted too, e.g., a step function source generates the step response, an impulse source generates the impulse response, and a sinusoidal source generates the steady-state sinusoidal response (when the transients are damped).

## Natural Response: RL Circuits

Consider a series connection of a charged inductor and a resistor, as shown in Fig. 4.17. The initial charge of the inductor $I_0$ is the current of the inductor at the moment of switching. The inductor can be connected to a current source to create the initial conditions to establish the current $I_0$. The objective is to find the inductor current before and after the switching.

The inductor intends to keep the current constant even after the switching. Therefore, the current direction is the same before and after the switching. However, the inductor must change its polarity to feed the current in the same direction when discharged. Therefore, the polarity of the inductor before and after the switching event changes. Knowing this effect, the circuit must be analyzed in two events before and after switching.

*Before switching*, the switch has been closed for a long time, providing enough time for the energy-storing elements to be fully charged. The inductor, when charged with DC current, becomes a short circuit. The short circuit across the resistor takes all the sourced current (in this example). Therefore, the initial current is obtained.

*After switching*, the charged inductor is connected to the circuit, and without the source, it is discharged to the resistor (Fig. 4.18).

**Fig. 4.17** RL is the first-order circuit. The initial condition is obtained by charging the inductor with an external source and then disconnecting the source at the desired time

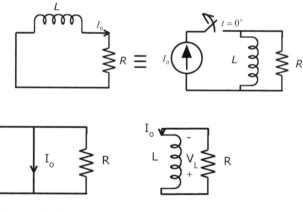

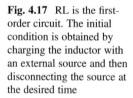

Before Switching                          After Switching

**Fig. 4.18** The circuit schematics before switching and after switching are shown. Before switching, the inductor is fully charged and becomes a short circuit. Once the source is disconnected, the inductor is discharged, and the current drops through the resistor. From the before switching circuit topology, the initial conditions are obtained. From the after-switching circuit topology, the time constant and the final value of the circuit parameters are obtained

The KVL (considering the polarity change in the inductor) can be written as

$$- V_L - V_R = 0$$

Ohm's law indicates $V_L = L\frac{di(t)}{dt}$ that $V_R = Ri(t)$. Therefore,

$$- L\frac{di(t)}{dt} - Ri(t) = 0$$

For simplicity, the time dependency of the current is removed from the equations.

$$- L\frac{di}{dt} - Ri = 0$$

Solving for $i$ results in

$$L\frac{di}{dt} = - Ri$$

$$L\frac{di}{i} = - Rdt \rightarrow \frac{di}{i} = - \frac{R}{L}dt$$

Taking integral of both sides $\int_{t_0}^{t} \frac{di}{i} = \int_{t_0}^{t} - \frac{R}{L}dt$ results in

$$\ln i(t)\Big|_{t_0}^{t} = - \frac{R}{L}t\Big|_{t_0}^{t}$$

$$\ln i(t) - \ln i(t_0) = - \frac{R}{L}(t - t_0)$$

Considering $t_0 = 0$ and the initial condition $i(t_0 = 0) = I_0$,

$$\ln i(t) - \ln I_0 = - \frac{R}{L}t = - \frac{t}{L/R} = - \frac{t}{\tau}$$

$$\ln a - \ln b = \ln \frac{a}{b}$$

Therefore,

$$\ln \frac{i(t)}{I_0} = - \frac{t}{\tau}$$

$\ln a = x \leftrightarrow a = e^x$. Therefore,

$$\frac{i(t)}{I_0} = e^{-\frac{t}{\tau}}$$

$$i(t) = I_0 e^{-\frac{t}{\tau}}, \ t \geq 0$$

This equation requires the initial condition $I_0$ obtained from the before switching circuit analysis and the time constant $\tau$, obtained from the circuit after the switching.

**Example 4.2 Consider an *RL* first-order circuit shown in Fig. 4.19. The switch has been closed for a long time, opening at $t = 0^+$ second. Find the current of the inductor $i(t)$.**

**Fig. 4.19** Circuit of
Example 4.2

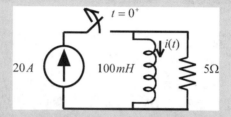

*Solution.* There is a need for the initial conditions and time constant.

The initial condition is obtained from the circuit before switching, shown in Fig. 4.20. The inductor, connected to a 20 A dc source, is charged and becomes a short circuit. Therefore, it bypasses the 5 $\Omega$ resistor from the circuit, letting all the current sources pass through the inductor. Therefore, the current of the inductor before switching reaches 20 A. Since the direction of current $i(t)$ and the initial current $I_0$ match, the initial current is a positive number.

**Fig. 4.20** The circuit
schematic before the
switching event. A fully
charged inductor becomes a
short circuit

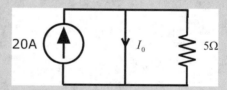

After switching, the circuit becomes a discharging RL circuit. The time constant from the ports of the inductor is measured to be $\tau = \frac{L}{R} = \frac{100 \ m}{5} = 0.02$ s.

Inserting the initial condition and the time constant into the inductor current template $i(t) = I_0 e^{-\frac{t}{\tau}}$ results in

$$i(t) = 20e^{-\frac{t}{0.02}} = 20e^{-50t}, \ t \geq 0 \ \text{(A)}$$

The voltage across the resistor is obtained as

(continued)

**Example 4.2** (continued)

$$V_R = -5i(t) = -100e^{-50t} \text{ (V)}$$

As seen from the voltage, a sudden discharge of inductors generates high voltages. Of course, this example analyzed the one-time charge discharge of the inductor. This process in real-world applications might be repeated periodically, hence, generating a train of high voltages.

**Example 4.3 Repeat Example 4.2 with $R = 10\,\Omega$, $L = 1$ H and the source current of 100 A. Find the voltage of the inductor $v(t)$.**
*Solution.* From the before switching circuit, the initial current is $I_0 = 100$ A. The equivalent resistance across the inductor is $R_{\text{th}} = 10\,\Omega$. Therefore, the time constant of the circuit that allows for the discharge of the inductor is $\tau = \frac{L}{R} = \frac{1}{10}$.

The inductor current is

$$i(t) = I_0 e^{-\frac{t}{\tau}} = 100e^{-\frac{t}{0.1}} = 100e^{-10t} \text{ A}$$

The voltage drop across the inductor is

$$v(t) = L\frac{di}{dt} = 1 \times \frac{d}{dt}\left(100e^{-10t}\right) = 100 \times -10 \times e^{-10t} = -1000e^{-10t} \text{ V}$$

If the inductor is charged every 10 ms, the voltage at the end of the discharge reaches

$$v(t = 10e-3) = -1000e^{-10 \times 10 \times 10^{-3}} = -900 \text{ V}$$

Therefore, the maximum and minimum voltage amplitude are 1000 V and 900 V. This maintains an average of $-950$ V.

**Example 4.4 Consider an RL circuit shown in Fig. 4.21. The switch has been closed for a long time and opens at $t = 0^+$. Find the inductor current $i(t)$ for positive time.**

**Fig. 4.21** Circuit schematic of Example 4.4

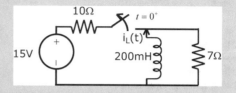

*Solution.* Inductor current before switching reaches a steady current, as shown in Fig. 4.22. The charged inductor becomes a short circuit bypassing the 7 $\Omega$ resistor. The initial current forced by the voltage source is measured against the desired direction of $i(t)$. Hence, it measures a negative value.

$$I_0 = -\frac{15}{10} = -1.5\,\text{A}$$

**Fig. 4.22** Circuit before the switching event. A fully charged inductor becomes a short circuit

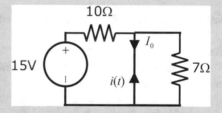

Figure 4.23 shows the circuit after the switching event. The switch opens and disconnects the source from the RL component. The circuit's time constant becomes $\tau = \frac{L}{R} = \frac{200}{7}$ ms.

**Fig. 4.23** Circuit after the switching event. The source is disconnected, and the inductor current is discharged through the resistor

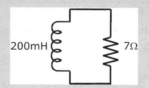

The current is $i(t) = -1.5e^{-\frac{t}{200e-3}} = -1.5e^{-35t}$, $t \geq 0$.
The voltage across the 7 $\Omega$ resistor becomes

$$V_R = 7i(t) = 10.5e^{-35t}\ (\text{V})$$

**Fig. 4.24** A first-order *RC* circuit with a switching event

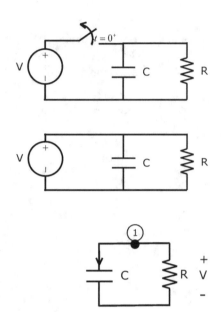

**Fig. 4.25** Before the switching event, the capacitor is fully charged and becomes an open circuit

**Fig. 4.26** After the switching, the source is disconnected, and the initial voltage stored in the capacitor is discharged through the resistor

## Natural Response: RC First-Order Circuit

Consider a circuit with a capacitor as an energy-storing element. A first-order RC circuit is shown in Fig. 4.24 in which, when the switch is closed, it forms a node involving a voltage source, a capacitor, and a resistor. The switch isolates the source from the rest of the circuit and allows for the charging of the capacitor.

Consider that the switch is closed for an extended time to charge the capacitor fully. In reality, charging for more than four times the time constant guarantees almost a full charge. The switch is open at time $t = 0^+$, disconnecting the source from the circuit. The capacitor is now discharged through the resistor. The objective is to obtain the voltage profile across the capacitor after the switching event. The circuit is analyzed both before switching and after switching topologies.

*Before switching*, the circuit is shown in Fig. 4.25. A fully charged capacitor becomes an open circuit forcing the current to pass through the resistor entirely (in this circuit). In this case, the voltage across the capacitor is the resistor's voltage (because they are connected in parallel). The initial voltage is, therefore, the source voltage $V_0$.

The charged capacitor keeps the voltage constant at the terminals before and after the switching event. The capacitor accomplishes this task by changing the current direction. Therefore, the charging current to the capacitor terminals can suddenly change direction to exit the terminal, keeping the voltage constant.

*After switching*, the source is disconnected, and the resistor drains the capacitor and converts the stored energy to heat. A KCL at node ① of the circuit shown in Fig. 4.26 results in

$$- i_C(t) - i_R(t) = 0$$

From Ohm's law, $i_C(t) = C \frac{dv(t)}{dt}$ and $i_R = \frac{v(t)}{R}$. For simplicity, the time dependency is omitted.

$$C \frac{dv}{dt} = -\frac{v}{R}$$

$$\frac{dv}{v} = -\frac{dt}{RC}$$

Taking the integral of the equation results in

$$\int_{t_0}^{t} \frac{dv}{v} = \int_{t_0}^{t} -\frac{dt}{RC}$$

Therefore,

$$\ln v(t) \Big|_{t_0}^{t} = -\frac{1}{RC} t \Big|_{t_0}^{t}$$

$$\ln v(t) - \ln v(t_0) = -\frac{1}{RC}(t - t_0)$$

Considering the initial time as zero $t_0 = 0$, $v(t_0 = 0) = V_0$, and the time constant, $\tau = RC$,

$$\ln v(t) - \ln V_0 = -\frac{1}{RC} t = \frac{-t}{\tau}$$

$$\ln a - \ln b = \ln \frac{a}{b}$$

Therefore,

$$\ln \frac{V(t)}{V_0} = -\frac{t}{\tau}$$

$$\ln a = x \leftrightarrow a = e^x$$

$$\frac{v(t)}{V_0} = e^{-\frac{t}{\tau}}$$

$$v(t) = V_0 e^{-\frac{t}{\tau}}, \ t \geq 0$$

Comparing the results obtained from RL and RC first-order circuits, it can be concluded that the natural response $x(t)$ of the first-order circuits is obtained by

$$x(t) = X_0 e^{-\frac{t}{\tau}}, \ t \geq 0.$$

where $X_0$ is the initial condition of the parameter $x(t)$ obtained from the before switching circuit and $\tau$ is the time constant obtained from the after switching circuit.

**Example 4.5** In the circuit shown in Fig. 4.27, the switch has been in a closed position for an extended time. At time $t = 0^+$ the switch is opened. Find the voltage $v(t)$ across the capacitor for $t \geq 0$. Find the current of the capacitor $i(t)$.

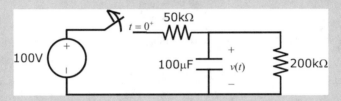

**Fig. 4.27**   Circuit of Example 4.5

*Solution.* Since the circuit has been connected to the voltage source for a long time, the capacitor is charged through the 50 kΩ resistor. However, as the capacitor is fully charged, it becomes an open circuit letting only the 200 kΩ resistor in the circuit. Since the capacitor and 200 kΩ resistor are parallel, they share the same voltage. Therefore, the initial voltage of the capacitor will be the same as the voltage drop across this resistor. Since these resistors are connected in series, a voltage divider between the 50 kΩ and 200 kΩ shows the desired voltage as

$$V_0 = V_{200 \text{ k}\Omega} = \frac{200 \text{ k}}{50 \text{ k} + 200 \text{ k}} \, 100 = 80 \text{ V}$$

After the switching event, the voltage source and the 50 kΩ resistor are disconnected from the circuit. The circuit topology is shown in Fig. 4.28.

**Fig. 4.28**   After the switching event, a charged capacitor is discharged through the resistor

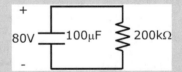

The time constant is obtained as

$$\tau = RC = 200 \text{ k} \times 100 \text{ μ} = 200e3 \times 100e - 6 = 20{,}000e - 3 = 20 \text{ s}$$

The voltage response across the capacitor is obtained by

$$v(t) = V_0 e^{-\frac{t}{\tau}}, \ t \geq 0$$

$$v(t) = 80e^{-\frac{t}{20}} = 80e^{-0.05t}, \ t \geq 0.\text{V}$$

$$i(t) = C\frac{dv}{dt} = 100 \text{ μ} \times \frac{d}{dt}\left(80e^{-0.05t}\right) = 100e - 6 \times 80 \times -0.05 \, e^{-0.05t} = -0.4e^{-0.05t} \text{ mA}$$

## Forced Response of First-Order Circuits

Forced response of circuits is obtained by utilizing external sources, which may or may not include initial charges. The forced response due to a step function is known as the step response, and a forced response due to an impulse is known as the impulse response. The circuit may switch between two charged elements, introducing a step in the circuit. However, when an input is applied in the circuit, the primary objective is to find the capacitors' voltage or inductors' current. Voltages, currents, power, and energy of other elements may be calculated from these values.

## Step Response of RL Circuit

Consider a first-order RL circuit connected to a voltage source through a switch. The initial condition of the inductor is known as $I_0$, which is obtained from the circuit topology before the switching event. The switch changes its position such that the RL circuit experiences a new driving force other than the initial charge of the inductor itself. Figure 4.29 shows a switch that is closed at $t = 0^+$.

A KVL in the loop can be written as follows:

$$-V_S + v_L(t) + v_R(t) = 0$$

The DC source forces a current through the circuit, and the charging inductor becomes a short circuit. However, this study aims to determine the inductor current profile variation that starts from the initial $I_0$ and reaches the final value $I_f = \frac{V_s}{R}$.

KVL is rewritten as follows:

$$-V_S + L\frac{di(t)}{dt} + Ri(t) = 0$$

Then,

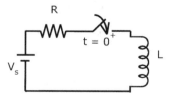

**Fig. 4.29** An *RL* circuit is connected to the source at $t = 0^+$. The source and the switch resemble a step function applied to the circuit. The response is the current of the inductor

$$\frac{di(t)}{dt} = \frac{1}{L}(V_S - Ri(t))$$

$$\frac{di(t)}{dt} = \frac{-R}{L}\left(i(t) - \frac{V_S}{R}\right)$$

$$\frac{di(t)}{\left(i(t) - \frac{V_S}{R}\right)} = \frac{-R}{L}dt$$

Integrating both sides from $t_0$ to $t$ yields

$$\int_{t_0}^{t} \frac{di(t)}{\left(i(t) - \frac{V_S}{R}\right)} = \int_{t_0}^{t} \frac{-R}{L}dt$$

$$\ln\left(i(t) - \frac{V_S}{R}\right)\bigg|_{t_0}^{t} = \frac{-R}{L}t\bigg|_{t_0}^{t}$$

$$\ln\left(i(t) - \frac{V_S}{R}\right) - \ln\left(I_0 - \frac{V_S}{R}\right) = \frac{-R}{L}(t - t_0)$$

$$\ln a - \ln b = \ln\frac{a}{b}$$

Considering $t_0 = 0$, and the $i(t = 0) = I_0$ results in

$$\ln\frac{\left(i(t) - \frac{V_S}{R}\right)}{\left(I_0 - \frac{V_S}{R}\right)} = -\frac{t}{\tau}$$

$$\ln a = x \leftrightarrow a = e^x$$

$$\frac{\left(i(t) - \frac{V_S}{R}\right)}{\left(I_0 - \frac{V_S}{R}\right)} = e^{-\frac{t}{\tau}}$$

Solving for $i(t)$.

$$i(t) = \frac{V_S}{R} + \left(I_0 - \frac{V_S}{R}\right)e^{-\frac{t}{\tau}}, \quad t \geq 0$$

Recalling circuit values $I_f = \frac{V_s}{R}$ and the initial condition, the step response can be obtained from

$$i(t) = I_f + (I_0 - I_f)e^{-\frac{t}{\tau}}, \quad t \geq 0.$$

**Example 4.6** In the circuit of Fig. 4.30, switch 1 has been closed for a long time. At time $t = 0^+$ it is opened, and switch 2 is closed. Find the current of the inductor $i(t)$.

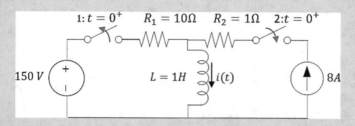

**Fig. 4.30** Circuit of Example 4.6.

*Solution.* Switch 1 has been closed for a long time, so the inductor had enough time to charge fully. A fully charged inductor becomes a short circuit allowing the entire current to pass through without adding any charge to its capacity. Therefore, the initial current is produced by the 150 V source through the current limiting resistor of $R_1 = 10 \ \Omega$. The current is

$$I_0 = \frac{150}{R_1} = \frac{150}{10} = 15 \text{ A}$$

After the switching, the 8 A current source is connected to the circuit and charges the inductor to its level. However, the amount of current in this case (switch 1 open and 2 closed) is determined by the 8 A current source. Therefore, the find current is

$$I_f = 8 \text{ A}$$

After switching, the time constant depends on the amount of $R_2$. The time constant is determined as

$$\tau = \frac{L}{R_2} = \frac{1}{1} = 1$$

The inductor current is

$$i(t) = I_f + \left(I_0 - I_f\right)e^{-\frac{t}{\tau}}$$

(continued)

**Example 4.6** (continued)

$$i(t) = 8 + (15 - 8)e^{-t} = 8 + 7e^{-t} \text{ A}$$

The initial current value at time $t = 0^+$ is

$$i(t = 0) = 8 + 7e^{-0} = 15 \text{ A}$$

The final current value at time $t = \infty$ is

$$i(t = \infty) = 8 + 7e^{-\infty} = 8 \text{ A}$$

**Example 4.7 The switch in the circuit shown in Fig. 4.31 has been in position $a$ for a long time. At time $t = 0^+$, it changes to position $b$. Find the current of the inductor $i(t)$.**

**Fig. 4.31** Circuit of
Example 4.7

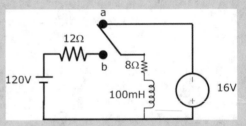

*Solution.* The circuit in position $a$ is connected to a $-16$ V source. The terminals of the source force the current through the inductor in the opposite direction of the desired $i(t)$. The initial current is obtained as

$$I_0 = -\frac{16}{8} = -2 \text{ A}$$

This calculation is that the inductor is charged and becomes a short circuit. Therefore, in position $a$, the circuit has an 8 $\Omega$ resistor and a $-16$ V source.

In position $b$, the circuit is connected to a voltage source of 120 V with an internal resistance of 12 $\Omega$. The circuit topology in switch position $b$ is shown in Fig. 4.32. The final/ultimate current of the circuit is obtained when the inductor is fully charged, i.e., short circuit.

(continued)

**Example 4.7** (continued)

**Fig. 4.32** The circuit schematic after the switching event

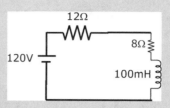

The final value of the current can be obtained as follows:

$$I_f = \frac{120}{12+8} = 6 \text{ A}$$

The inductor in position $b$ is also a short circuit. The inductor is not the short circuit when the current transient starts from $-2$ A to 6 A.

In position $b$, the circuit time constant is calculated to be $\tau = \frac{L}{R} = 5$ ms.

The transient inductor current is therefore obtained as

$$i(t) = I_f + (I_0 - I_f)e^{-\frac{t}{\tau}}, \ t \geq 0$$

$$i(t) = 6 + (-2 - 6)e^{-\frac{t}{5e-3}} = 6 - 8e^{-200t}, \ t \geq 0$$

**Forced Response of First-Order RC Circuit**

Consider an RC circuit shown in Fig. 4.33 with a switch to provide an option of connecting to a source. A sudden switching event introduces a step function into the system. The response is seen as a current changing voltage across the capacitor. The initial charge of the capacitor $V_0$ experiences a transient and shifts to a final value of $V_f$.

Before the switching event, the capacitor has an initial voltage. After the switching event, the circuit topology configures a node that has voltage $v(t)$. A KCL analysis at this node ① results in the voltage variation as follows:

$$\text{KCL①.} -I_s + i_R(t) + i_C(t) = 0$$

Considering Ohm's law results in

$$-I_s + \frac{v(t)}{R} + C\frac{dv(t)}{dt} = 0$$

**Fig. 4.33** Forced response
of an RC circuit. The current
source is connected to the
RC circuit through the
switch and forms a step in
the current

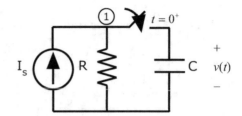

$$\frac{-1}{RC}(v(t) - RI_s) = \frac{dv(t)}{dt}$$

$$\frac{dv(t)}{(v(t) - RI_s)} = \frac{-1}{RC}dt$$

Integrating both sides from $t_0$ to $t$ yields

$$\int_{t_0}^{t} \frac{dv(t)}{(v(t) - RI_s)} = \int_{t_0}^{t} \frac{-1}{RC}dt$$

$$\ln(v(t) - RI_s)\Big|_{t_0}^{t} = \frac{-1}{RC}t\Big|_{t_0}^{t}$$

$$\ln(v(t) - RI_s) - \ln(V_0 - RI_s) = \frac{-1}{RC}(t - t_0)$$

$$\ln a - \ln b = \ln\frac{a}{b}$$

Considering $t_0 = 0$, and the initial voltage $v(t_0 = 0) = V_0$, results in

$$\ln\frac{(v(t) - RI_s)}{(V_0 - RI_s)} = -\frac{t}{\tau}$$

$$\ln a = x \leftrightarrow a = e^x$$

$$\frac{(v(t) - RI_s)}{(V_0 - RI_s)} = e^{-\frac{t}{\tau}}$$

Solve for $v(t)$:

$$v(t) = RI_s + (V_0 - RI_s)e^{-\frac{t}{\tau}}, \ t \geq 0$$

Recalling from circuit values, $V_f = RI_S$ and the initial condition, the step response
can be obtained from

$$v(t) = V_f + (V_0 - V_f)e^{-\frac{t}{\tau}}, \ t \geq 0$$

**Example 4.8 Consider a first-order RC circuit as shown in Fig. 4.34. The switch has been in position ① for a long time, and at time $t = 0^+$, it changes to position ②. Find the voltage $v(t)$ across the capacitor for all times $t \geq 0$.**

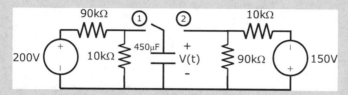

**Fig. 4.34** Circuit of Example 4.8

*Solution.* When the switch is in position ①, the capacitor is connected to a voltage source and receives a steady-state voltage equal to the value set by the voltage divider of 10 kΩ and 90 kΩ. When the capacitor is charged, it becomes an open circuit, and since it is connected in parallel to the 10 kΩ resistor, they share the same voltage.

Therefore, the initial voltage is

$$V_0 = \frac{10\ k}{10\ k + 90\ k} 200 = 20\ V$$

When the switch is in position ② for a long time, the capacitor voltage will reach another ultimate value set by the circuit connected to the 150 V source. A voltage divider in this circuit results in the final voltage $V_f$. As the source polarity is opposite the measured voltage across the capacitor, the final voltage becomes a negative value.

$$V_f = \frac{90\ k}{90\ k + 10\ k}(-150) = -135\ V$$

The circuit's time constant is obtained from the circuit topology after switching, and equivalent resistance across the resistor can obtain the time constant across the capacitor. From the circuit, two resistors of 10 kΩ and 90 kΩ are connected in parallel when the $-150$ V source is removed. Therefore, the time constant is

(continued)

**Example 4.8** (continued)

$$\tau = R_{eq}C = \left(10\ \text{k}\,\middle\|\,90\ \text{k}\right)450\ \mu = 4.05\ \text{ms}$$

The circuit response is therefore obtained as follows

$$v(t) = V_f + \left(V_0 - V_f\right)e^{-\frac{t}{\tau}},\ t \geq 0$$

Replacing the results of the calculated value in

$$v(t) = -135 + (20 - (-135))e^{-\frac{t}{4.05e-3}},\ t \geq 0$$
$$v(t) = -135 + 155e^{-246.9t},\ t \geq 0$$

The circuit time constant is always obtained from the equivalent resistance across the terminals of the capacitor or the terminals of the inductor.

## Second-Order Circuits

Second-order circuits have the equivalent of two energy-storing elements. As introduced earlier, the circuit elements have the relations of their voltage and current relations defined as either a linear function $v = Ri$ in resistors or differential or integral equations as $i = L\frac{di}{dt}$ and $i = \frac{1}{L}\int v\,dt$ in inductors $i = C\frac{dv}{dt}$ and $v = \frac{1}{C}\int i\,dt$ capacitors. One objective was to analyze circuits and find the voltage of capacitors and the current of inductors. From this analysis, other parameters can be identified. To analyze higher-order circuits, KVL for each loop and KCL for each node may be written to form differential equations based on the desired parameter $v(t)$ or $i(t)$.

Accordingly, circuits can be analyzed, and parameters can be identified. However, two typical *RLC* circuits where all elements are connected in series or parallel are of more interest because of the properties they show. Another analysis can be obtained by considering the elements' initial charges or analyzing an external source's effect.

### Natural Response of RLC Parallel Circuits

Consider a circuit consisting of a resistor, an equivalent inductor, and an equivalent capacitor connected in parallel, as shown in Fig. 4.35.

The inductor has an initial current $I_0$, and the capacitor has an initial voltage of $V_0$. There is an interest in finding the voltage across the terminals, $v(t)$. Since there is no external source and the voltage response is generated purely from initial charges, the voltage response is called a natural response.

**Fig. 4.35** Parallel
connection of *RLC*
elements. The initial
conditions induce a voltage
across the elements. The
response of interest is the
*v(t)*

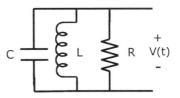

The parallel connection of these elements forms a voltage v(t) node. Therefore, at this node, the balance of currents must hold. KCL results in

$$i_R(t) + i_L(t) + i_C(t) = 0$$

These currents leave the node through passive elements and naturally drain the current out of the node. To find the voltage at the terminals, each of these currents must be written in v(t). Therefore, replacing the currents with the voltages results in

$$\frac{v(t)}{R} + \frac{1}{L}\int v(t)dt + C\frac{dv(t)}{dt} = 0$$

An integrodifferential equation has been obtained based on variable v(t). To solve this equation for v(t), a one-time differential must be taken to convert the entire equation to a differential equation. This is a one-time differential because the equation has a one-time integral. Taking differential from the equation results in a second-order differential equation as follows:

$$\frac{d}{dt}\left(\frac{v(t)}{R} + \frac{1}{L}\int v(t)dt + C\frac{dv(t)}{dt} = 0\right)$$

$$\frac{1}{R}\frac{dv(t)}{dt} + \frac{1}{L}v(t) + C\frac{d^2v(t)}{dt^2} = 0$$

Sorting the equation based on the order of derivative results in

$$C\frac{d^2v(t)}{dt^2} + \frac{1}{R}\frac{dv(t)}{dt} + \frac{1}{L}v(t) = 0$$

Dividing by the highest-order differential term $C\frac{d^2v(t)}{dt^2}$ (in this Eq. C) results in a *monic* polynomial. Making the polynomial monic results in

$$\frac{d^2v(t)}{dt^2} + \frac{1}{RC}\frac{dv(t)}{dt} + \frac{1}{LC}v(t) = 0$$

The circuit resulted in a second-order differential equation, which was expected because the circuit was a second-order circuit. A general solution of this linear second-order differential equation contains the terms $v(t) = e^{\lambda t}$ with two values for the $\lambda$. (Likewise, in a third-order circuit, it is expected to have three values for $\lambda$.)

Replacing this general solution in the equation helps find the two values for $\lambda$ as follows:

$$\frac{d^2\left(e^{\lambda t}\right)}{dt^2} + \frac{1}{RC}\frac{d\left(e^{\lambda t}\right)}{dt} + \frac{1}{LC}e^{\lambda t} = 0$$

Replacing the derivatives $\frac{d\left(e^{\lambda t}\right)}{dt} = \lambda e^{\lambda t}$ and $\frac{d^2\left(e^{\lambda t}\right)}{dt^2} = \lambda^2 e^{\lambda t}$ in the equation results in

$$\lambda^2 e^{\lambda t} + \frac{1}{RC}\lambda e^{\lambda t} + \frac{1}{LC}e^{\lambda t} = 0$$

Factoring the exponential term $e^{\lambda t}$ out results in

$$e^{\lambda t}\left(\lambda^2 + \frac{1}{RC}\lambda + \frac{1}{LC}\right) = 0$$

Since $\lambda$ has physical limitations and cannot reach $-\infty$, then $e^{\lambda t} \neq 0$. Therefore, in a parallel *RLC* circuit

$$\lambda^2 + \frac{1}{RC}\lambda + \frac{1}{LC} = 0$$

This is also called the *characteristics equation*.

The roots of the characteristics equations $\lambda_1$ and $\lambda_2$ determine the $v(t)$ response. Consider

$$\alpha = \frac{1}{2RC}$$

and

$$\omega_0^2 = \frac{1}{LC}$$

where $\alpha$ is the damping factor and $\omega_0$ $\left(\frac{\text{rad}}{\text{s}}\right)$ is the resonant frequency; therefore, the characteristics equation can be written as

$$\lambda^2 + 2\alpha\lambda + \omega_0^2 = 0$$

This quadratic equation has two roots, $\lambda_1$ and $\lambda_2$. These roots are obtained as follows:

$$\lambda_{1,2} = \frac{-2\alpha \pm \sqrt{(2\alpha)^2 - 4\omega_0^2}}{2} = -\alpha \pm \sqrt{\alpha^2 - \omega_0^2}$$

The value $(\alpha^2 - \omega_0^2)$ might be positive, zero, or negative. In each case, the value of the roots changes, ultimately changing the $v(t)$ response.

(a) If $\alpha^2 - \omega_0^2 > 0$, there are two distinct real roots as $\lambda_1 = -\alpha + \sqrt{\alpha^2 - \omega_0^2}$ and $\lambda_2 = -\alpha - \sqrt{\alpha^2 - \omega_0^2}$. The response is *overdamped* and becomes

$$v(t) = A_1 e^{\lambda_1 t} + A_2 e^{\lambda_2 t}$$

Initial conditions must be used to find $A_1$ and $A_2$. Since two parameters must be determined, two equations must be formed.

Considering the initial voltage of the capacitor, one of the equations can be found as follows:

$$v(t = 0) = V_0$$

$$v(t = 0) = A_1 + A_2 = V_0$$

Also evaluating the KCL $i_R(t) + i_L(t) + i_C(t) = 0$ at time $t = 0$ results in

$$i_R(t = 0) + i_L(t = 0) + i_C(t = 0) = 0$$

Since the initial voltage of the capacitor and initial current of the inductor is known, this results in

$$\frac{V_0}{R} + I_0 + C \frac{dv(0)}{dt} = 0$$

This equation can be used to determine the second condition

$$\frac{dv(0)}{dt} = \frac{-1}{C} \left( \frac{V_0}{R} + I_0 \right)$$

Therefore, the $v(t)$ derivative at time $t = 0$ must hold. Therefore,

$$\frac{dv(t = 0)}{dt} = \frac{d\left(A_1 e^{\lambda_1 t} + A_2 e^{\lambda_2 t}\right)}{dt}\bigg|_{t=0} = \left(A_1 \lambda_1 e^{\lambda_1 t} + A_2 \lambda_2 e^{\lambda_2 t}\right)\big|_{t=0}$$
$$= A_1 \lambda_1 + A_2 \lambda_2$$

Hence, the second equation is

$$A_1 \lambda_1 + A_2 \lambda_2 = \frac{-1}{C} \left( \frac{V_0}{R} + I_0 \right)$$

$A_1$ and $A_2$ can be found.

(b) If $\alpha^2 - \omega_0^2 = 0$, there are two equal real roots $\lambda_1 = \lambda_2 = -\alpha$. The response is *critically damped* and becomes

$$v(t) = B_1 t e^{-\alpha t} + B_2 e^{-\alpha t}$$

Initial conditions must be used to find $B_1$ and $B_2$. Since two parameters must be determined, two equations must be formed.

The first equation is formed from the initial voltage of the capacitor as follows:

$$v(t=0) = V_0$$
$$v(t=0) = B_1 = V_0$$

Writing a KCL at time $t = 0$ results in

$$i_R(t=0) + i_L(t=0) + i_C(t=0) = 0$$

Since the initial voltage of the capacitor and initial current of the inductor is known, this results in

$$\frac{V_0}{R} + I_0 + C\frac{dv(0)}{dt} = 0$$

This equation provides the second condition as

$$\frac{dv(0)}{dt} = \frac{-1}{C}\left(\frac{V_0}{R} + I_0\right)$$

The $v(t)$ derivative at time $t = 0$ must hold. Therefore

$$\frac{dv(t=0)}{dt} = \frac{d(B_1 t e^{-\alpha t} + B_2 e^{-\alpha t})}{dt}\Big|_{t=0}$$
$$= (-\alpha B_1 t e^{-\alpha t} + B_1 e^{-\alpha t} - \alpha B_2 e^{-\alpha t})\big|_{t=0} = B_1 - \alpha B_2$$

Math reminder: $\frac{d(f_1(t)f_2(t))}{dt} = \frac{df_1(t)}{dt}f_2(t) + f_1(t)\frac{df_2(t)}{dt}$.
Therefore,

$$B_1 - \alpha B_2 = V_0 - \alpha B_2 = \frac{-1}{C}\left(\frac{V_0}{R} + I_0\right)$$
$$B_2 = \frac{1}{\alpha}\left(V_0\left(1 + \frac{1}{RC}\right) + \frac{I_0}{C}\right)$$

(c) If $\alpha^2 - \omega_0^2 < 0$, there are two complex conjugate roots. Considering the damping frequency $\omega_d$ as $\omega_d^2 = \omega_0^2 - \alpha^2$, then $-\left(\omega_0^2 - \alpha^2\right) = -\omega_d^2 > 0$. Therefore, $\lambda_1 = -\alpha + j\omega_d$, $\lambda_2 = -\alpha - j\omega_d$. The response is *underdamped* and becomes

$$v(t) = C_1 e^{(-\alpha + j\omega_d)t} + C_2 e^{(-\alpha - j\omega_d)t}$$

Expanding the exponential functions, the response becomes

$$v(t) = e^{-\alpha t}(C_1 \cos(\omega_d t) + C_2 \sin(\omega_d t))$$

Initial conditions must be used to find $C_1$ and $C_2$. Since two parameters must be determined, two equations must be formed.

The first equation is formed from the initial voltage of the capacitor as follows:

$$v(t = 0) = V_0$$
$$v(t = 0) = C_1 = V_0.$$

Writing a KCL at time $t = 0$ results in

$$i_R(t = 0) + i_L(t = 0) + i_C(t = 0) = 0$$

Since the initial voltage of the capacitor and current of the inductor is known, this results in

$$\frac{V_0}{R} + I_0 + C\frac{dv(0)}{dt} = 0$$

This equation provides the second condition as

$$\frac{dv(0)}{dt} = \frac{-1}{C}\left(\frac{V_0}{R} + I_0\right)$$

The $v(t)$ derivative at time $t = 0$ must hold. Therefore

$$\frac{dv(t = 0)}{dt} = \frac{de^{-\alpha t}(C_1 \cos(\omega_d t) + C_2 \sin(\omega_d t))}{dt}\Big|_{t=0}$$

$$\frac{dv(t)}{dt} = \left(-\alpha e^{-\alpha t}(C_1 \cos(\omega_d t) + C_2 \sin(\omega_d t)) + e^{-\alpha t}(-C_1\omega_d \sin(\omega_d t) + C_2\omega_d \cos(\omega_d t))\right)\Big|_{t=0}$$

$$= -C_1\alpha + C_2\omega_d$$

Therefore,

$$- C_1\alpha + C_2\omega_{\mathrm{d}} = \frac{-1}{C}\left(\frac{V_0}{R} + I_0\right)$$

$C_1$ and $C_2$ can be found.

## Summary of RLC Parallel Circuit

**Note 4.1**  R, L, and C values will determine $\alpha$ and $\omega_0$. The sign $\alpha^2 - \omega_0^2$ determines the type of response.

**Note 4.2**  The response in overdamped circuits indicates that the rise of output voltage has slowed growth and has been damped so much that the oscillations are eliminated.

**Note 4.3**  The response in critically damped circuits is the limit of damping at which the system response starts to oscillate by epsilon decrement of damping factor concerning the resonant frequency, i.e., first signs of oscillations are about to start.

**Note 4.4**  The response in underdamped circuits starts to show the damped oscillations. The sign of oscillations in an underdamped circuit is the existence of the first peak.

| $\alpha^2 - \omega_0^2 > 0$ | Overdamped | $v(t) = A_1 e^{\lambda_1 t} + A_2 e^{\lambda_2 t}$ $A_1 + A_2 = V_0$ $A_1\lambda_1 + A_2\lambda_2 = \frac{-1}{C}\left(\frac{V_0}{R} + I_0\right)$ | |
|---|---|---|---|
| $\alpha^2 - \omega_0^2 = 0$ | Critically damped | $v(t) = B_1 t e^{-\alpha t} + B_2 e^{-\alpha t}$ $B_1 = V_0$ $B_2 = \frac{1}{\alpha}\left(V_0\left(1 + \frac{1}{RC}\right) + \frac{I_0}{C}\right)$ | |
| $\alpha^2 - \omega_0^2 < 0$ | Underdamped | $v(t) = e^{-\alpha}$ $^t(C_1 \cos(\omega_{\mathrm{d}} t) + C_2 \sin(\omega_{\mathrm{d}} t))$ $C1 = V0$ $- C_1\alpha + C_2\omega_{\mathrm{d}} = \frac{-1}{C}\left(\frac{V_0}{R} + I_0\right)s$ | |

**Example 4.9 Consider an RLC circuit in parallel, as shown in Fig. 4.35. At $R = 25\ \Omega$, $L = 10$ mH, and $C = 1\ \mu$F, find the voltage response $v(t)$ if the initial charge of capacitor is $V_0 = 150$ V and the initial current of the inductor $I_0 = 4$ A.**

*Solution.* In a parallel RLC circuit, the voltage response can be obtained using the characteristics equation $\lambda^2 + \frac{1}{RC}\lambda + \frac{1}{LC} = 0$. Considering the circuit element values, the characteristics equation becomes

$$\lambda^2 + \frac{1}{25 \times 1e-6}\lambda + \frac{1}{10\,m \times 1e-6} = 0$$

$$\alpha = \frac{1}{2RC} = \frac{1}{2 \times 25 \times 1e-6} = 20,000$$

$$\omega_0 = \frac{1}{\sqrt{LC}} = \frac{1}{\sqrt{10e-3 \times 1e-6}} = 10,000 \text{ rad/s}$$

These values indicate that $\alpha^2 - \omega_0^2 > 0$, therefore, the system is *overdamped*.

The response, according to the table, becomes

$$v(t) = A_1 e^{\lambda_1 t} + A_2 e^{\lambda_2 t}$$

where $\lambda_{1,2} = -\alpha \pm \sqrt{\alpha^2 - \omega_0^2} = -20,000 \pm \sqrt{20,000^2 - 10,000^2} = -20,000 \pm 17,320.5 = -2679.5 \text{ rad/s} - 37,320.5 \text{ rad/s}$.

To find the constants $A_1$ and $A_2$, the following equations can be used

$$A_1 + A_2 = V_0, \rightarrow A_1 + A_2 = 150$$

$$A_1\lambda_1 + A_2\lambda_2 = \frac{-1}{C}\left(\frac{V_0}{R} + I_0\right) \rightarrow -2679.5A_1 - 37,320.5A_2$$

$$= \frac{-1}{1e-6}\left(\frac{150}{25} + 4\right) = -10e6$$

This results in $A_1 = -127.07$ V and $A_2 = 277.072$ V. Hence,

$$v(t) = -127.07e^{-2679.5t} + 277.072e^{-37,320.5t} \text{ (V)}$$

**Example 4.10 Consider the previous example when the resistor is adjusted to $R = 100\ \Omega$. Find the damping coefficient, resonant frequency, and voltage response $v(t)$.**

*Solution.* In a parallel RLC circuit, the voltage response can be obtained using the characteristics equation $\lambda^2 + \frac{1}{RC}\lambda + \frac{1}{LC} = 0$. Considering the circuit element values, the characteristics equation becomes

$$\lambda^2 + \frac{1}{100 \times 1e-6}\lambda + \frac{1}{10e-3 \times 1e-6} = 0$$

$$\alpha = \frac{1}{2RC} = \frac{1}{2 \times 100 \times 1e-6} = 5000$$

$$\omega_0 = \frac{1}{\sqrt{LC}} = \frac{1}{\sqrt{10e-3 \times 1e-6}} = 10{,}000 \text{ rad/s}$$

These values indicate that $\alpha^2 - \omega_0^2 < 0$; therefore, the system is underdamped. This means the response would oscillate, and the amplitudes of oscillations are damped.

The damping frequency $\omega_d^2 = \omega_0^2 - \alpha^2$ can be obtained as $\omega_d^2 = 10{,}000^2 - 5000^2$ and $\omega_d = 8660.2$ rad/s.

The response, according to the table, becomes

$$v(t) = e^{-\alpha t}(C_1 \cos(\omega_d t) + C_2 \sin(\omega_d t))$$
$$v(t) = e^{-5000t}(C_1 \cos(8660.2t) + C_2 \sin(8660.2t))$$
$$C_1 = V_0 = 150 \text{ V}$$

$$-C_1\alpha + C_2\omega_d = \frac{-1}{C}\left(\frac{V_0}{R} + I_0\right)$$

$$-150 \times 5000 + C_2 8660.2 = \frac{-1}{1e-6}\left(\frac{150}{100} + 4\right)$$

$$C_2 = -721.69 \text{ V}$$

Therefore, the answer is

$$v(t) = e^{-5000t}(150\cos(8660.2t) - 721.69\sin(8660.2t)) \text{ V}$$

**Example 4.11** In the previous example, consider an unknown resistance value. Adjust the resistance to obtain a critically damped circuit, and using the initial conditions, find the system response.

*Solution.* To have critical damping $\alpha^2 - \omega_0^2 = 0$. Therefore, $\left(\frac{1}{2RC}\right)^2 = \left(\frac{1}{\sqrt{LC}}\right)^2$

$$\left(\frac{1}{2R \times 1e-6}\right)^2 = \left(\frac{1}{\sqrt{10e-3 \times 1e-6}}\right)^2$$

$$\frac{1}{2R \times 1e-6} = 10{,}000 \rightarrow R = 50\ \Omega$$

In critical damping $\alpha = \omega_0 = 10{,}000$ rad/s

$$v(t) = B_1 t e^{-\alpha t} + B_2 e^{-\alpha t}$$
$$v(t) = B_1 t e^{-10{,}000t} + B_2 e^{-10{,}000t}$$
$$B_1 = V_0 = 150\ \text{V}$$

$$B_2 = \frac{1}{\alpha}\left(V_0\left(1 + \frac{1}{RC}\right) + \frac{I_0}{C}\right) = \frac{1}{10{,}000}\left(150\left(1 + \frac{1}{50 \times 1e-6}\right) + \frac{4}{1e-6}\right)$$

$$= 70.015\ \text{V}$$

$$v(t) = 150t e^{-10{,}000t} + 70.015 e^{-10{,}000t}\ \text{V}$$

**Example 4.12** Consider the RLC parallel circuit in the past three examples. Knowing the voltage response $v(t)$, find the current of each element in case where $R = 25\ \Omega$, $R = 100\ \Omega$, $R = 500\ \Omega$.

*Solution.* Ohm's law should be imposed.

The current of resistors are as follows:

When $R = 25\ \Omega$, overdamp $i(t) = \frac{v(t)}{R} = \frac{-127.07 e^{-2679.5t} + 277.072 e^{-37320.5t}}{25}\ \text{A}$

When $R = 100\ \Omega$, underdamp

$$i(t) = \frac{v(t)}{R} = \frac{e^{-5000t}(150\cos(8660.2t) - 721.69\sin(8660.2t))}{100}\ \text{A}$$

When $R = 500\ \Omega$, critically damp $i(t) = \frac{v(t)}{R} = \frac{150t e^{-10{,}000t} + 430.015 e^{-10{,}000t}}{500}\ \text{A}$

(continued)

**Example 4.12** (continued)

The current inductor is

$$i_L(t) = \frac{1}{L} \int v(t) dt$$

When $R = 25\ \Omega$, overdamp

$$i_L(t) = \frac{1}{10e - 3} \int \left(-127.07e^{-2679.5t} + 277.072e^{-37,320.5t}\right) dt$$

$$= 100 \left(\frac{-127.07}{-2697.5} e^{-2679.5t} + \frac{277.072}{-37320.5} e^{-37,320.5t}\right)$$

$$i_L(t) = \left(4.71e^{-2679.5t} - 0.74e^{-37,320.5t}\right) \text{A}$$

When $R = 100\ \Omega$, underdamp

$$i_L(t) = \frac{1}{10e - 3} \int \left(e^{-5000t}(150\cos(8660.2t) - 721.69\sin(8660.2t))\right) dt$$

$$= \left(e^{-5000t}(6.202\cos(8660.2t) - 29.84\sin(8660.2t))\right) \text{A}$$

When $R = 500\ \Omega$, critically damp

$$i_L(t) = \frac{1}{10e - 3} \int \left(150te^{-10,000t} + 430.015e^{-10,000t}\right) dt$$

$$i_L(t) = 6.202te^{-10,000t} + 17.78e^{-10,000t}\ \text{A}$$

The current of the capacitor is

$$i_C(t) = C\frac{dv(t)}{dt}$$

When $R = 25\ \Omega$, overdamp

$$i_C(t) = (1e - 6)\frac{d}{dt}\left(-127.07e^{-2679.5t} + 277.072e^{-37,320.5t}\right)$$

$$i_C(t) = (1e - 6)\left(-127.07 \times -2679.5e^{-2679.5t} + 277.072 \times -37,320.5e^{-37,320.5t}\right)$$

$$i_C(t) = \left(0.34e^{-2679.5t} - 10.34e^{-37,320.5t}\right)\ \text{A}$$

(continued)

**Example 4.12** (continued)

When $R = 100\ \Omega$, underdamp

$$i_C(t) = (1e-6)\frac{d}{dt}\left(e^{-5000t}(150\cos(8660.2t) - 721.69\sin(8660.2t))\right)$$

$$i_C(t) = (1e-6)\left(-5000e^{-5000t}(150\cos(8660.2t) - 721.69\sin(8660.2t))\right.$$
$$\left. + e^{-5000t}(150\times -8660.2\sin(8660.2t) - 721.69\times 8660.2\cos(8660.2t))\right)$$

$$i_C(t) = \left(e^{-5000t}(-7\cos(8660.2t) + 2.318\sin(8660.2t))\right)\text{A}$$

When $R = 500\ \Omega$, critically damp

$$i_C(t) = (1e-6)\frac{d}{dt}\left(150te^{-10,000t} + 430.015e^{-10,000t}\right)$$

$$i_C(t) = (1e-6)$$
$$\times \left(150e^{-10,000t} - 150\times 10,000te^{-10,000t} + 430.015\times -10,000e^{-10,000t}\right)$$

$$i_C(t) = (1e-6)$$
$$\times \left(150e^{-10,000t} - 150\times 10,000te^{-10000t} + 430.015\times -10,000e^{-10,000t}\right)$$

$$i_C(t) = \left(4.3e^{-10,000t} - 1.5te^{-10,000t}\right)\text{A}$$

---

**Example 4.13 Characteristic roots of an RLC parallel circuit are given as $\lambda_{1,2} = -1000 - 4000\ \frac{\text{rad}}{\text{s}}$. Determine the damping and the natural frequency of the circuit. Assuming $C = 1\ \mu\text{F}$, find the $R$ and $L$ for designing this circuit.**

*Solution.* The circuit has two distinct real roots, yielding an overdamp circuit. The characteristic roots are $\lambda_1 = -\alpha + \sqrt{\alpha^2 - \omega_0^2}$ and $\lambda_2 = -\alpha - \sqrt{\alpha^2 - \omega_0^2}$. Therefore,

$$\lambda_1 + \lambda_2 = -2\alpha$$

$$-1000 - 4000 = -2\alpha \rightarrow \alpha = 2500\ \frac{\text{rad}}{\text{s}}$$

(continued)

**Example 4.13** (continued)

Replacing one of the characteristic roots, the value of $\omega_0$ can be found:

$$\lambda_2 = -4000 = -2500 - \sqrt{2500^2 - \omega_0^2} \rightarrow \omega_0 = 2000 \; \frac{\text{rad}}{\text{s}}$$

$$\alpha = \frac{1}{2RC} = 2500, C = 1 \; \mu\text{F (given)} \rightarrow R = \frac{1}{2 \times 1 \; \mu \times 2500} = 200 \; \Omega$$

$$\omega_0 = \frac{1}{\sqrt{LC}} \rightarrow 2000^2 = \frac{1}{LC} \rightarrow L = \frac{1}{1 \; \mu \times 2000^2} = 0.25 \; \text{H}$$

**Example 4.14 Characteristic roots of an RLC parallel circuit are given as $\lambda_{1,2} = -1000 \pm j\, 5000 \; \frac{\text{rad}}{\text{s}}$. Determine the damping and the natural frequency of the circuit. Assuming $C = 1 \; \mu\text{F}$, find the $R$ and $L$ for designing this circuit.**

*Solution.* Since the characteristic roots of the circuit are complex conjugate, the circuit is under-damped. The characteristic roots are

$$\lambda_{1,2} = -\alpha \pm j\omega_d = -\alpha \pm j\sqrt{\omega_0^2 - \alpha^2}$$

Compared with the roots, the damping factor and the damping frequency can be found as

$$\lambda_{1,2} = -\alpha \pm j\omega_d = -1000 \pm j5000$$

$$\alpha = 1000, \omega_d = 5000$$

$$\omega_d = \sqrt{\omega_0^2 - \alpha^2} \rightarrow 5000 = \sqrt{\omega_0^2 - 1000^2} \rightarrow \omega_0 = 5099 \; \frac{\text{rad}}{\text{s}}$$

$$\alpha = \frac{1}{2RC} = 1000, C = 1 \; \mu\text{F (given)} \rightarrow R = \frac{1}{2 \times 1 \; \mu \times 1000} = 500 \; \Omega$$

$$\omega_0 = \frac{1}{\sqrt{LC}} \rightarrow 5099^2 = \frac{1}{LC} \rightarrow L = \frac{1}{1 \; \mu \times 5099^2} = 38.46 \; \text{mH}$$

**Fig. 4.36** *RLC* series circuit

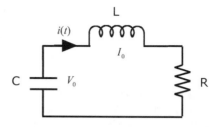

## Natural Response of RLC Series Circuits

Consider a circuit consisting of a loop of R, L, and C where the initial charge of the capacitor is $V_0$, and the initial charge of the inductor is $I_0$. The circuit is shown in Fig. 4.36. The loop has current $i(t)$ that drops a voltage across each element.

KVL in this loop indicates

$$v_R(t) + v_L(t) + v_C(t) = 0$$

Ohm's law indicates the voltage drop across each element as follows:

$$R\,i(t) + L\frac{di(t)}{dt} + \frac{1}{C}\int i(t)dt = 0$$

The KVL equation results in an integrodifferential equation over $i(t)$. To solve for the current, there is a need to take a one-time differential because a single integral exists in the equation.

Taking a one-time differential from the equation results in

$$\frac{d}{dt}\left(R\,i(t) + L\frac{di(t)}{dt} + \frac{1}{C}\int i(t)dt = 0\right)$$

$$R\frac{di(t)}{dt} + L\frac{d^2 i(t)}{dt^2} + Ci(t) = 0$$

Sorting the equation based on the order of derivative results in

$$L\frac{d^2 i(t)}{dt^2} + R\frac{di(t)}{dt} + Ci(t) = 0$$

Dividing by the highest-order differential term (eq. L) results in a *monic* polynomial. Dividing the characteristics equation by $L$ results in

$$\frac{d^2 i(t)}{dt^2} + \frac{R}{L} \frac{di(t)}{dt} + \frac{1}{LC} i(t) = 0$$

The circuit has resulted in a second-order differential equation which was expected because the circuit was second-order (two energy-storing elements). A general solution of this linear second-order differential equation is $i(t) = e^{\lambda t}$ with two roots for the variable $\lambda$. To find the roots, the general solution needs to be replaced into the equation, which results in

$$\frac{d^2 (e^{\lambda t})}{dt^2} + \frac{R}{L} \frac{d(e^{\lambda t})}{dt} + \frac{1}{LC} e^{\lambda t} = 0$$

Considering the derivatives $\frac{d(e^{\lambda t})}{dt} = \lambda e^{\lambda t}$ and $\frac{d^2 (e^{\lambda t})}{dt^2} = \lambda^2 e^{\lambda t}$ replacing these in the differential equation results in

$$\lambda^2 e^{\lambda t} + \frac{R}{L} \lambda e^{\lambda t} + \frac{1}{LC} e^{\lambda t} = 0$$

Factoring the exponential term out results in

$$e^{\lambda t} \left( \lambda^2 + \frac{R}{L} \lambda + \frac{1}{LC} \right) = 0$$

Since $\lambda$ has physical limitations and cannot reach $-\infty$, $e^{\lambda t} \neq 0$. Therefore, the characteristics equation needs to be zero

$$\lambda^2 + \frac{R}{L} \lambda + \frac{1}{LC} = 0$$

The roots of this equation determine templates of $i(t)$ response. Considering

$$\alpha = \frac{R}{2L}$$

and

$$\omega_0^2 = \frac{1}{LC}$$

where $\alpha$ is the damping factor and $\omega_0 \left( \frac{rad}{s} \right)$ is the resonant frequency; therefore, the characteristics equation can be written as

$$\lambda^2 + 2\alpha\lambda + \omega_0^2 = 0$$

This quadratic equation has two roots, $\lambda_1$ and $\lambda_2$. These roots are obtained as follows:

$$\lambda_{1,2} = \frac{-2\alpha \pm \sqrt{(2\alpha)^2 - 4\omega_0^2}}{2} = -\alpha \pm \sqrt{\alpha^2 - \omega_0^2}$$

The value $(\alpha^2 - \omega_0^2)$ might be positive, zero, or negative. In each case, the value of the roots changes, ultimately changing the $i(t)$ response.

(a) If $\alpha^2 - \omega_0^2 > 0$,s there are two distinct real roots for $\lambda_1 = -\alpha + \sqrt{\alpha^2 - \omega_0^2}$ and $\lambda_2 = -\alpha - \sqrt{\alpha^2 - \omega_0^2}$. The response is *overdamped* and becomes

$$i(t) = A_1 e^{\lambda_1 t} + A_2 e^{\lambda_2 t}$$

Initial conditions must be used to find $A_1$ and $A_2$. Since two parameters must be determined, two equations must be formed.

The first equation is formed from the initial current of the inductor as follows:

$$i(t=0) = I_0$$

$$i(t=0) = A_1 + A_2 = I_0$$

Writing a KVL at time $t = 0$ results in

$$v_R(t=0) + v_L(t=0) + v_C(t=0) = 0$$

Since the initial voltage and current are known, this results in

$$RI_0 + L\frac{di(0)}{dt} + CV_0 = 0$$

This equation provides the second condition as

$$\frac{di(0)}{dt} = \frac{-1}{L}(RI_0 + V_0)$$

The $i(t)$ derivative at time $t = 0$ must hold. Therefore,

$$\frac{di(t=0)}{dt} = \frac{d\left(A_1 e^{\lambda_1 t} + A_2 e^{\lambda_2 t}\right)}{dt}\bigg|_{t=0} = \left(A_1\lambda_1 e^{\lambda_1 t} + A_2\lambda_2 e^{\lambda_2 t}\right)\big|_{t=0} = A_1\lambda_1 + A_2\lambda_2$$

Therefore,

$$A_1\lambda_1 + A_2\lambda_2 = \frac{-1}{L}(RI_0 + V_0)$$

$A_1$ and $A_2$ can be found.

(b) If $\alpha^2 - \omega_0^2 = 0$, there are two real repeated roots $\lambda_1 = \lambda_2 = -\alpha$. The response is *critically damped* and becomes

$$i(t) = B_1 t e^{-\alpha t} + B_2 e^{-\alpha t}$$

Initial conditions must be used to find $B_1$ and $B_2$. Since two parameters must be determined, two equations must be formed.

The first equation is formed from the initial current of the inductor as follows:

$$i(t=0) = I_0$$
$$i(t=0) = B_1 = I_0$$

Writing a KVL at time $t = 0$ results in

$$v_R(t=0) + v_L(t=0) + v_C(t=0) = 0$$

Since the initial voltage of the capacitor and current of the inductor is known, this results in

$$RI_0 + L\frac{di(0)}{dt} + CV_0 = 0$$

This equation provides the second condition as

$$\frac{di(0)}{dt} = \frac{-1}{L}(RI_0 + V_0)$$

Derivative of $i(t)$ at time $t = 0$ must hold. Therefore,

$$\frac{di(t=0)}{dt} = \frac{d(B_1 t e^{-\alpha t} + B_2 e^{-\alpha t})}{dt}\Big|_{t=0}$$
$$= (-\alpha B_1 t e^{-\alpha t} + B_1 e^{-\alpha t} - \alpha B_2 e^{-\alpha t})\Big|_{t=0} = B_1 - \alpha B_2$$

Therefore,

$$B_1 - \alpha B_2 = I_0 - \alpha B_2 = \frac{-1}{L}(RI_0 + V_0)$$

$$B_2 = \frac{1}{\alpha}\left(I_0\left(1 + \frac{R}{L}\right) + \frac{V_0}{L}\right)$$

(c) If $\alpha^2 - \omega_0^2 < 0$, there are two complex conjugate roots. Consider $\omega_d^2 = \omega_0^2 - \alpha^2$, then $-(\omega_0^2 - \alpha^2) = -\omega_d^2 > 0$. Therefore, $\lambda_1 = -\alpha + j\omega_d$ and $\lambda_2 = -\alpha - j\omega_d$. The response is *underdamped* and becomes

$$i(t) = C_1 e^{(-\alpha + j\omega_d)t} + C_2 e^{(-\alpha - j\omega_d)t}$$

Expanding the exponential functions, the response becomes

$$i(t) = e^{-\alpha t}(C_1 \cos(\omega_d t) + C_2 \sin(\omega_d t))$$

Initial conditions must be used to find $C_1$ and $C_2$. Since two parameters must be determined, two equations must be formed.

The first equation is formed from the initial current of the inductor as follows:

$$i(t = 0) = I_0$$
$$i(t = 0) = C_1 = I_0$$

Writing a KVL at time $t = 0$ results in

$$v_R(t = 0) + v_L(t = 0) + v_C(t = 0) = 0$$

Since the initial voltage of the capacitor and current of the inductor is known, this results in

$$RI_0 + L\frac{di(0)}{dt} + CV_0 = 0$$

This equation provides the second condition as

$$\frac{di(0)}{dt} = \frac{-1}{L}(RI_0 + V_0)$$

The $i(t)$ derivative at time $t = 0$ must hold. Therefore,

$$\frac{di(t=0)}{dt} = \frac{de^{-at}(C_1 \cos(\omega_d t) + C_2 \sin(\omega_d t))}{dt}\Big|_{t=0}$$
$$= (-\alpha e^{-at}(C_1 \cos(\omega_d t) + C_2 \sin(\omega_d t))$$
$$+ e^{-at}(-C_1 \omega_d \sin(\omega_d t) + C_2 \omega_d \cos(\omega_d t)))|_{t=0}$$
$$= -C_1 \alpha + C_2 \omega_d$$

Therefore,

$$-C_1 \alpha + C_2 \omega_d = \frac{-1}{L}(RI_0 + V_0)$$

$C_1$ and $C_2$ can be found.

## Summary of RLC Series Circuit

| $\alpha^2 - \omega_0^2 > 0$ | Overdamped | $i(t) = A_1 e^{\lambda_1 t} + A_2 e^{\lambda_2 t}$ <br> $A_1 + A_2 = I_0$ <br> $A_1 \lambda_1 + A_2 \lambda_2 = \frac{-1}{L}(RI_0 + V_0)$ | |
|---|---|---|---|
| $\alpha^2 - \omega_0^2 = 0$ | Critically damped | $i(t) = B_1 t e^{-at} + B_2 e^{-at}$ <br> $B_1 = I_0$ <br> $B_2 = \frac{1}{\alpha}\left(I_0\left(1 + \frac{R}{L}\right) + \frac{V_0}{L}\right)$ | |
| $\alpha^2 - \omega_0^2 < 0$ | Underdamped | $i(t) = e^{-\alpha t}(C_1 \cos(\omega_d t) + C_2 \sin(\omega_d t))$ <br> $C_1 = I_0$ <br> $-C_1 \alpha + C_2 \omega_d = \frac{-1}{L}(RI_0 + V_0)$ | |

**Example 4.15 Characteristics roots of an RLC series circuit are given as $\lambda_{1,2} = -1000 - 4000 \frac{\text{rad}}{\text{s}}$. Determine the damping and the natural frequency of the circuit. Assuming $L = 10$ mH, find the $R$ and $C$ for designing this circuit.**

*Solution.* The circuit has two distinct real roots, yielding an overdamp circuit. The characteristics roots are $\lambda_1 = -\alpha + \sqrt{\alpha^2 - \omega_0^2}$ and $\lambda_2 = -\alpha - \sqrt{\alpha^2 - \omega_0^2}$. Therefore,

$$\lambda_1 + \lambda_2 = -2\alpha$$

$$-1000 - 4000 = -2\alpha \rightarrow \alpha = 2500 \; \frac{\text{rad}}{\text{s}}$$

Replacing one of the characteristics roots, the value of $\omega_0$ can be found:

$$\lambda_2 = -4000 = -2500 - \sqrt{2500^2 - \omega_0^2} \rightarrow \omega_0 = 2000 \; \frac{\text{rad}}{\text{s}}$$

$$\alpha = \frac{R}{2L} = 2500, L = 10 \text{ mH (given)} \rightarrow R = 2L \times 2500 = 50 \; \Omega$$

$$\omega_0 = \frac{1}{\sqrt{LC}} \rightarrow 2000^2 = \frac{1}{LC} \rightarrow C = \frac{1}{10 \text{ m} \times 2000^2} = 25 \; \mu\text{F}$$

**Example 4.16 Characteristics roots of an RLC series circuit are given as $\lambda_{1,2} = -1000 \pm j\, 5000 \frac{\text{rad}}{\text{s}}$. Determine the damping and the natural frequency of the circuit. Assuming $L = 10$ mH, find the $R$ and $C$ for designing this circuit.**

*Solution.* Since the characteristic roots of the circuit are complex conjugate, the circuit is underdamped. The characteristics roots are

$$\lambda_{1,2} = -\alpha \pm j\omega_d = -\alpha \pm j\sqrt{\omega_0^2 - \alpha^2}$$

Compared with the roots, the damping factor and the damping frequency can be found as

$$\lambda_{1,2} = -\alpha \pm j\omega_d = -1000 \pm j5000$$

$$\alpha = 1000, \omega_d = 5000$$

(continued)

**Example 4.16** (continued)

$$\omega_d = \sqrt{\omega_0^2 - \alpha^2} \rightarrow 5000 = \sqrt{\omega_0^2 - 1000^2} \rightarrow \omega_0 = 5099 \; \frac{rad}{s}$$

$$\alpha = \frac{R}{2L} = 1000, L = 10 \text{ mH (given)} \rightarrow R = 2L \times 1000 = 20 \; \Omega$$

$$\omega_0 = \frac{1}{\sqrt{LC}} \rightarrow 5099^2 = \frac{1}{LC} \rightarrow C = \frac{1}{10 \text{ m} \times 5099^2} = 3.84 \; \mu F$$

## Problems

4.1. Find the voltage across a 10 $\Omega$ resistor if the current flowing through is as follows:

(a) $i(t) = 10u(t)$
(b) $i(t) = 10tu(t)$
(c) $i(t) = 200 \sin (60\pi t + 10)$
(d) $i(t) = t^2 u(t)$
(e) $i(t) = e^{-3t} \sin 10\pi t$

4.2. Find the current of a 100 $\Omega$ resistor if the applied voltage is as follows:

(a) $v(t) = 120\sqrt{2} \sin (100\pi t + 10)$
(b) $v(t) = u(t) + u(t - 2) - 2u(t - 3)$
(c) $v(t) = 10tu(t) - 20(t - 1)u(t - 1) + 10(t - 3)u(t - 3)$
(d) $v(t) = e^{-3t} \sin 100\pi t$
(e) $v(t) = 100te^{-10t}u(t)$

4.3. Find the voltage induced across an $L = 100$ mH inductor when the current is as follows:

(a) $i(t) = 10u(t)$
(b) $i(t) = 10tu(t)$
(c) $i(t) = 200 \sin (60\pi t + 10)$
(d) $i(t) = t^2 u(t)$
(e) $i(t) = e^{-3t} \sin 10\pi t$

4.4. Find the current through an $L = 100$ mH inductor if the voltage applied across it is as follows:

(a) $v(t) = 120\sqrt{2} \sin (100\pi t + 10)$
(b) $v(t) = u(t) + u(t - 2) - 2u(t - 3)$
(c) $v(t) = 10tu(t) - 20(t - 1)u(t - 1) + 10(t - 3)u(t - 3)$
(d) $v(t) = e^{-3t} \sin 100\pi t$
(e) $v(t) = 100te^{-10t}u(t)$

4.5. Find the voltage across a $C = 100\ \mu F$ capacitor when the current is as follows:

(a) $i(t) = 10u(t)$
(b) $i(t) = 10tu(t)$
(c) $i(t) = 200 \sin (60\pi t + 10)$
(d) $i(t) = t^2 u(t)$
(e) $i(t) = e^{-3t} \sin 10\pi t$

4.6. Find the current through a $C = 100\ \mu F$ capacitor if the voltage applied across it is as follows:

(a) $v(t) = 120\sqrt{2} \sin (100\pi t + 10)$
(b) $v(t) = u(t) + u(t - 2) - 2u(t - 3)$
(c) $v(t) = 10tu(t) - 20(t - 1)u(t - 1) + 10(t - 3)u(t - 3)$
(d) $v(t) = e^{-3t} \sin 100\pi t$
(e) $v(t) = 100te^{-10t}u(t)$

4.7. The switch has been closed for a long time. At time $t = 0^+$, it is opened. Find $v(t)$ and $i(t)$.

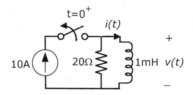

4.8. The switch has been closed for a long time. At time $t = 0^+$, it is opened. Find $v(t)$ and $i(t)$.

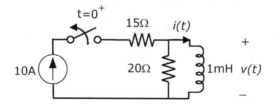

4.9. The switch has been closed for a long time. At time $t = 0^+$, it is opened. Find $v(t)$ and $i(t)$.

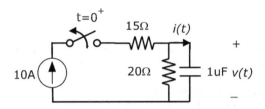

4.10. The switch has been closed for a long time. At time $t = 0^+$ it is opened. Find $v(t)$ and $i(t)$.

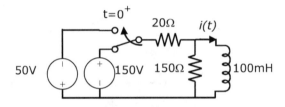

4.11. The switch has been closed for a long time. At time $t = 0^+$, it is opened. Find $v(t)$, $i_{c1}(t)$, $i_{c2}(t)$.

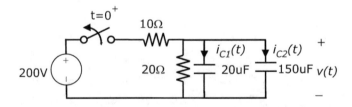

4.12. The switch has been closed for a long time. At time $t = 0^+$ it is opened. Find $i(t)$, $v_1(t)$, and $v_2(t)$.

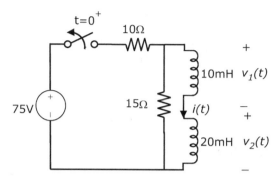

4.13. Find the characteristics equations, characteristics roots, damping conditions, and the response in the following circuit.

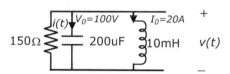

4.14. Find the characteristics equations, characteristics roots, damping conditions, and the response in the following circuit.

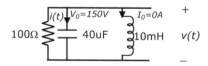

4.15. Find the characteristics equations, characteristics roots, damping conditions, and the response in the following circuit.

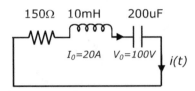

4.16. Characteristic roots of a parallel RLC circuit are $\lambda_{1,2} = -1000 \pm j5000 \ \frac{rad}{s}$. Find the system's natural voltage response if the initial conditions are $I_0 = -25$ A and $V_0 = 150$ V.

4.17. Characteristic roots of a parallel RLC circuit are $\lambda_1 = -1000 \ \frac{rad}{s}$, $\lambda_2 = -5000 \ \frac{rad}{s}$. Find the system voltage response if the circuit has initial conditions as $V_0 = 100$ V, $\frac{d}{dt}v(0^+) = -12{,}000 \ \frac{V}{s}$.

4.18. Characteristic roots of a series RLC circuit: Find the current system response if the initial conditions are $I_0 = -250$ A and $V_0 = -200$ V.

4.19. The response of a parallel RLC circuit recorded from the oscilloscope follows. Find the characteristic roots and characteristic equations.

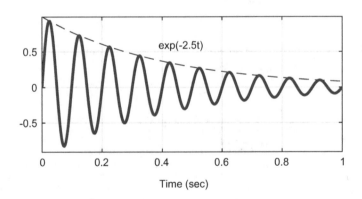

4.20. Natural responses of some RLC circuits are as follows. Find the characteristic roots, characteristic equations, and initial conditions of the circuit.

(a) $i(t) = 150e^{-200t} \cos 1000t$
(b) $v(t) = 100e^{-2000t} \sin (30{,}000t + 30)$
(c) $i(t) = 200e^{-300t} + 150e^{-120t}$
(d) $v(t) = 200te^{-2000t} + 20e^{-2000t}$

4.21. Design an RLC series circuit such that the current natural response becomes

$$i(t) = 100e^{-100t} \cos 750t \ \text{A}$$

Select the R, L, and C values and the initial conditions that result in the desired response.

Note that there might be multiple solutions for this design. Therefore, select the range smaller than mH,µF in the inductor and capacitor.

# Chapter 5
# Steady-State Sinusoidal Circuit Analysis

## Introduction

Sinusoidal waveforms, as explained in Chap. 3, have an amplitude $r$, a frequency $\omega$, and a phase shift or phase angle $\theta$ and are expressed as

$$f(t) = r \sin(\omega t + \theta)$$

The same function can be presented as a phasor. The phasor conveys essential information regarding a signal, *amplitude*, and *phase angle*, considering that the frequency throughout the operation is fixed. The amplitude and phase information resemble polar coordinates. A polar coordinate can be transformed into rectangle coordinates as well. This reciprocal transformation can be achieved as follows:

- $R \to P$. Rectangular to polar conversion. Consider a complex value $a + jb$ in rectangle coordinates of real and imaginary. This value in polar coordinate has an amplitude of $r = \sqrt{a^2 + b^2}$ and an angle of $\theta = \tan^{-1}\frac{b}{a}$.
- $P \to R$. Polar to rectangular conversion. Consider a number $r \angle \theta$ in polar coordinates. This number in rectangle coordinates has a real axis value of $a = r \cos \theta$ and an imaginary value of $b = r \sin \theta$.

$$r\angle\theta \leftrightarrow r\cos\theta + jr\sin\theta$$

$$r\cos\theta + jr\sin\theta \leftrightarrow re^{j\theta}$$

$$r\angle\theta \leftrightarrow re^{j\theta}$$

The conversion of the rectangle and polar coordinates is shown in Fig. 5.1.

A. Izadian, *Fundamentals of Modern Electric Circuit Analysis and Filter Synthesis*,
https://doi.org/10.1007/978-3-031-21908-5_5

**Fig. 5.1** The polar coordinate of $r \angle \theta$ can be projected on the earl and imaginary axes to identify the real and imaginary values in a rectangular coordinate

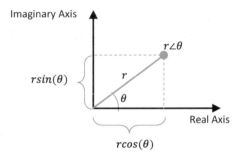

Example 5.1 Consider a sinusoidal voltage source $v(t) = 169.7 \sin (377t + 30°)$ being connected to a circuit. Find the frequency in $\left(\frac{rad}{s}\right)$,Hz, amplitude, and the phase shift of the signal.

*Solution.* This source voltage amplitude has a peak value of 169.7 V. Over time, the amplitude swings between **+169.7** and **−169.7** V following a sinusoidal waveform. The angular frequency $\omega = 377\left(\frac{rad}{s}\right)$ translates into the frequency in Hertz as $f(\text{Hz}) = \frac{\omega(\text{rad/s})}{2\pi} = \frac{377}{2\pi} = 60\,\text{Hz}$. This means that the waveform period is $T = \frac{1}{f} = \frac{1}{60} = 16.6$ ms. Therefore, it reaches its half-cycle in 8.4 ms. In the polar coordinate, the voltage will have an amplitude and a phase of 169.7 $\angle$ 30°.

Example 5.2 Express the function $v(t) = 169.7 \sin (377t + 30°)$ in phasor.
*Solution.* The polar coordinate is 169.7 $\angle$ 30. Therefore, the phasor becomes $169.7e^{j30}$.

Example 5.3 Find the phasor expression of the following numbers:
(a) $10 + j10$
(b) $2 + j\sqrt{3}$

   *Solution.*

(a) $10 + j10 \leftrightarrow \sqrt{10^2 + 10^2} \angle \tan^{-1}\frac{10}{10} = \sqrt{200}\angle 45 \leftrightarrow \sqrt{200}e^{j45}$

(b) $2 + j\sqrt{3} \leftrightarrow \sqrt{2^2 + \sqrt{3}^2} \angle \tan^{-1}\frac{\sqrt{3}}{2} = \sqrt{7}\angle 50.76 \leftrightarrow \sqrt{7}e^{j50.76}$

**Example 5.4 Find the polar expression of the following quantities:**
(a) $10 \angle 30°$
(b) $110\sqrt{2}\angle -60°$

*Solution.*

(a) The real and imaginary part of the number can be obtained by projecting the vector on the real and imaginary axes. The real part is $10\cos(30)$ and the imaginary part is $10\sin(30)$. Therefore,

$$10\angle30° = 10\cos(30) + j10\sin(30) = 8.66 + j5$$

(b) The real part can be obtained by $110\sqrt{2}\cos(-60)$, and the imaginary part can be obtained by $110\sqrt{2}\sin(-60)$. The rectangle coordinate presentation becomes

$$110\sqrt{2}\angle -60° = 110\sqrt{2}\cos(-60) + 110\sqrt{2}\sin(-60)$$
$$77.78 - j134.72$$

# How to Use Phasor in Circuit Analysis

An electric circuit might be excited by a variety of waveforms. The representation of quantities in phasor is helpful because it preserves the frequency information and can be independent of the time and trigonometric functions. The circuit quantities are either voltage, current, or impedance values expressed in the phasor. The impedance of components once excited by a sinusoidal waveform might be frequency-dependent. Following are some steps to analyze the circuit response (i.e., the voltage drops and current flows) under sinusoidal excitations.

The first step is to determine the stage of the response. This means determining whether the response has reached a stable operation or its amplitude is changing due to the nature of the circuit. The response stages are discussed in detail in the next section.

The second step is to identify the equivalent of circuit elements and impedance when a sinusoidal source at the frequency $\omega = 2\pi f$ is utilized. This is explained in detail in this chapter.

The third step uses circuit analysis laws, such as KVL and KCL, to calculate the circuit values. Several examples are provided in this chapter to show the circuit analysis under sinusoidal excitation.

**Fig. 5.2** Resistive-
inductive-capacitive or RLC
circuit with all components
connected in series

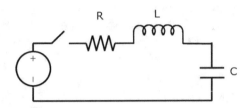

## Circuit Response Stages

Consider an RLC circuit consisting of a sinusoidal source and a switch to connect the circuit to the source at the desired time (Fig. 5.2).

The circuit is considered to have no initial condition, meaning that the capacitors and inductors are fully discharged. The switch is closed at $t = 0^+$, connecting the source to the circuit. The voltage source characteristics suggest that potentially an unlimited amount of current is available to fill the capacitors and inductors while maintaining the voltage. Therefore, the current $i(t)$ is influenced by the circuit topology and the switching angle representing the voltage amplitude. In general, and depending on the switching time, colliding with the instantaneous amplitude of the sinusoidal waveform, the current response can be divided into three periods:

- The first 1–3 cycles show the sub-transient response.
- The next 10–15 cycles show the transient response.
- The steady-state response starts once the amplitude settles to a fixed peak value.

The waveform frequency is fixed during the sub-transient, transient, and steady-state responses. However, only their amplitude is different due to the nature of the circuit components. Figure 5.3 shows sub-transient, transient, and steady-state parts of the current response.

This chapter analyzes the steady-state system response, where all the switching transient responses are already damped and a fixed amplitude of voltage and current is reached. Under this condition, the equivalents of resistors, inductors, and capacitors are calculated.

## Resistors in Steady State

Consider a voltage source $v(t) = V_m \sin(\omega t + \theta)$ at the peak value of $V_m$, and the angular frequency of $\omega$ (rad/s) is applied across a resistor. The current is obtained according to Ohm's law as follows (Fig. 5.4):

$$v(t) = Ri(t)$$

and

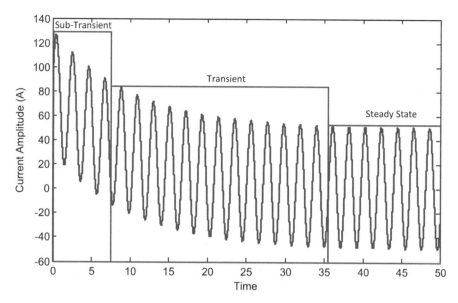

**Fig. 5.3** Transient response to sinusoidal excitation. Three main stages may exist sub-transient, transient, and steady-state

**Fig. 5.4** The equivalent of a pure resistor, when excited by a sinusoidal source at frequency $\omega$, remains a resistor with the same resistance value

$$i(t) = \frac{v(t)}{R}$$

With the given sinusoidal voltage source, the current becomes

$$i(t) = \frac{V_m \sin(\omega t + \theta)}{R} = \frac{V_m}{R} \sin(\omega t + \theta) = I_m \sin(\omega t + \theta).$$

where $I_m$ is the current. The current in phasor form $I$ can also be obtained as follows:

$$I = \frac{V}{R} = \frac{V_m \angle \theta}{R} = \frac{V_m}{R} \angle \theta = I_m \angle \theta.$$

As can be seen from the voltage and current phasors, resistors cannot change the current phase. Both voltage and current had a phase angle of $\theta$ but different amplitudes. A resistor scales the current according to the amount of its resistance. Therefore, a resistor under steady-state conditions shows a purely resistive impedance with zero-degree phase shift from voltage to current.

**Fig. 5.5** The voltage and current waveforms in a pure resistive circuit. The zero crossings are the same for both the voltage and the current, which means they are in phase. Only their amplitudes are scaled by the resistance value as $V = RI$

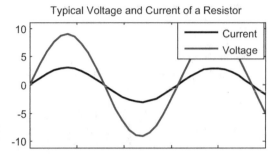

## Power Factor of Resistive Circuits

Power factor is defined as the cosine of the phase shift between voltage and current passing through the circuit, i.e., $PF = \cos\left(\widehat{v, i}\right)$. Therefore, the power factor can be $0 \leq PF \leq 1$, where 1 shows a pure resistive circuit and 0 shows a pure inductive or capacitive circuit. However, since in a pure resistive circuit, the phase shift in current with respect to voltage (reference) is zero (Fig. 5.5); the power factor (PF) of a resistive circuit is 1. $PF = \cos(0) = 1$. It is also called that the voltage and current in a pure resistive circuit are in-phase, meaning that the zero crossing and peaks of the voltage and current waveforms occur simultaneously.

> **Example 5.5 Find the current phasor and power factor in the circuit shown in Fig. 5.6.**
> *Solution.* The voltage source has an amplitude of 20 V and an initial phase of 20°. The voltage is applied across a 5Ω resistor. Therefore, the current is
>
> $$I = \frac{V}{R}$$
>
> $$I = \frac{10\angle 20}{5} = \frac{10}{5}\angle 20 - 0 = 2\angle 20 \text{ A}$$
>
> It can be seen that the phase of the current and the phase of the voltage are the same. In other words, the voltage and current are in-phase in a resistor.
>
>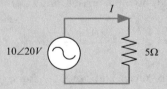
>
> **Fig. 5.6** Circuit of Example 5.5

(continued)

**Example 5.5** (continued)

The power factor in this circuit is the $\cos\left(\widehat{V,I}\right)$. Since the voltage and current have the same phase, their angle is zero. Therefore, the power factor of this circuit is

$$PF = \cos(0) = 1$$

**Example 5.6 In the circuit of Fig. 5.7, find the current phasor in each branch.**

*Solution.* Each resistor is connected to the source, and each resistor's current can be found by dividing the source voltage by the resistance of the branch.

$$I_1 = \frac{30\angle 20}{30} = 1\angle 20 \text{ A}$$

$$I_2 = \frac{10\angle 20}{30} = \frac{1}{3}\angle 20 \text{ A}$$

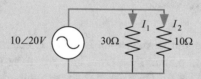

**Fig. 5.7** Circuit of Example 5.6

**Example 5.7 In the circuit of Fig. 5.8, find the current phasor in each branch.**

*Solution.* First, the current $I$ needs to be calculated. The current is then divided between $20\Omega$ and $15\Omega$ resistors.

The equivalent resistance across the source is a parallel of $20\Omega$ and $15\Omega$, in series with $10\Omega$.

$$R_{eq} = (20\|15) + 10 = \frac{20 \times 15}{20 + 15} + 10 = 18.57 \ \Omega$$

(continued)

**Example 5.7** (continued)

Therefore, the current $I$ can be found as

$$I = \frac{10\angle 60}{18.57} = 0.538\angle 60 \text{ A}$$

$$I_1 = \frac{20}{15+20}I = \frac{20}{35}\times 0.538\angle 60 = 0.307\angle 60 \text{ A}$$

$$I_2 = \frac{15}{15+20}I = \frac{15}{35}\times 0.538\angle 60 = 0.230\angle 60 \text{ A}$$

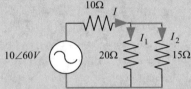

**Fig. 5.8** Circuit of Example 5.7

## Inductors in Steady State

The relation of voltage and current in an inductor is determined by $v(t) = L\frac{di(t)}{dt}$. Considering a sinusoidal current through an inductor under the steady-state condition as $i(t) = I_m \sin(\omega t + \theta)$, the voltage drop across the inductor is obtained as follows:

$$v(t) = L\frac{di(t)}{dt} = L\frac{d(I_m \sin(\omega t + \theta))}{dt} = LI_m\omega \cos(\omega t + \theta)$$
$$= L\omega I_m \sin(\omega t + \theta + 90°) = L\omega I_m \angle(\theta + 90°)$$

This shows that the current is lagging the voltage by 90°. This phase shift is shown by a $j$ factor in Fig. 5.9 and is presented in the following equations.

$$V = LI_m\omega e^{j(\theta+90°)} = LI_m\omega e^{j\theta}e^{j90°} = j\omega L(I_m e^{j\theta}) = j\omega L(I_m e^{j\theta}) = j\omega L(I_m\angle\theta)$$

According to these equations, inductors impede the flow of current in the presence of sinusoidal (time-varying) waveform excitations. The impedance of an inductor at inductance $L$ (H) operating at an angular frequency of $\omega$ (rad/s) is $j\omega L$ measured in ($\Omega$). Phasor representation of Ohm's law for an inductor under steady-state sinusoidal condition is as follows:

**Fig. 5.9** The voltage and current are 90° apart in a purely inductive circuit, which means that the peak occurs at the zero of the other. Since the current is lagging the voltage and considering voltage as a reference, the current occurs with a time delay equal to $t_{delay} = \frac{\pi}{2}T$

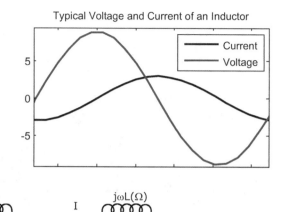

Typical Voltage and Current of an Inductor

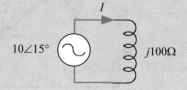

**Fig. 5.10** The equivalent of an inductor with inductance $L$ (H), when excited by a sinusoidal source at frequency $\omega$, becomes an inductor with impedance $j\omega L$ measured in ($\Omega$)

$$V = j\omega LI = X_L I$$

It can be concluded that an inductor $L$ (H) in steady-state sinusoidal shows a reactance of $X_L = j\omega L$ ($\Omega$) (Fig. 5.10).

**Note 5.1** The inductance $L$ is measured in Henrys (H), but $j\omega L$ is measured in Ohm ($\Omega$).

## Power Factor of Inductive Circuits

Since the current phase shift with respect to voltage is 90°, the power factor becomes zero as PF $= \cos 90° = 0$. It can be concluded that a pure inductive circuit has a power factor of zero.

**Example 5.8 Find the current phasor in the circuit of Fig. 5.11.**
*Solution.* At the angular frequency of $\omega = 100 \frac{rad}{s}$, the reactance of the inductor becomes $j\omega L = j100 \ \Omega$.
 The circuit with the ohmic values of the reactance is shown as

$10\angle15°$     $I$     $j100\Omega$

(continued)

**Example 5.8** (continued)

The current phasor is

$$I = \frac{10\angle15}{j100} = \frac{10\angle15}{100\angle90} = 0.1\angle15 - 90 = 0.1\angle - 75 \text{ A}$$

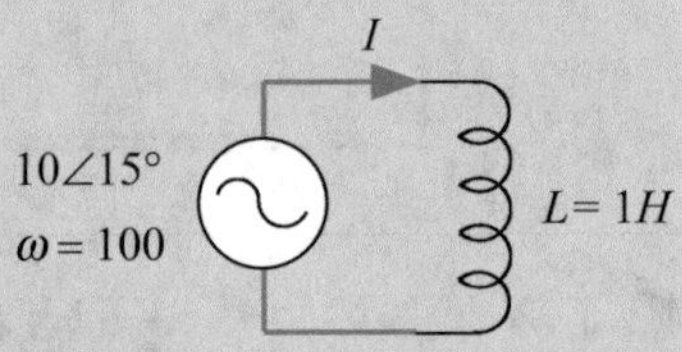

$$10\angle15°$$
$$\omega = 100$$
$$L = 1H$$

**Fig. 5.11** Circuit of Example 5.8

## Capacitors in Steady-State Sinusoidal

The relation of voltage and current in a capacitor is determined by $v(t) = \frac{1}{C} \int i(t) dt$. Considering a sinusoidal current through the capacitor under the steady-state condition as $i(t) = I_m \sin(\omega t + \theta)$, the voltage drop across the capacitor is obtained as follows:

$$v(t) = \frac{1}{C} \int i(t) dt = \frac{1}{C} \int (I_m \sin(\omega t + \theta)) dt = \frac{-I_m}{\omega C} \cos(\omega t + \theta)$$
$$= \frac{I_m}{\omega C} \sin(\omega t + \theta - 90°) = \frac{I_m}{\omega C} \angle(\theta - 90°)$$

This shows that the current is leading the voltage by 90°. This phase shift is shown by the $-j$ factor in Fig. 5.12 and is presented in the following equations.

**Fig. 5.12** The voltage and current are 90° apart in a purely capacitive circuit, which means that the peak of one waveform occurs at the zero of the other waveform. Since the current is leading the voltage and considering voltage as a reference, the current occurs with a time ahead of the voltage equal to $t_{lead} = \frac{\pi}{2} T$ s

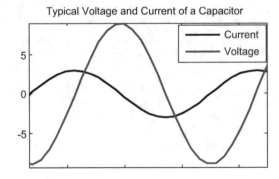

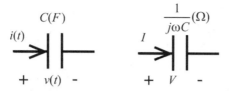

**Fig. 5.13** The equivalent of a capacitor with capacitance $C$ (F), when excited by a sinusoidal source at frequency $\omega$, becomes a capacitor with impedance $\frac{1}{j\omega C}$ measured in ($\Omega$)

$$V = \frac{I_m}{\omega C} e^{j(\theta+90°)} = \frac{I_m}{\omega C} e^{j\theta} e^{-j90°} = -j\frac{1}{\omega C}\left(I_m e^{j\theta}\right) = \frac{1}{j\omega C}\left(I_m e^{j\theta}\right) = \frac{1}{j\omega C}(I_m \angle\theta)$$

According to these equations, the capacitor impedes the flow of current in the presence of sinusoidal (periodic) waveform excitations. The impedance of a capacitor at capacitance $C$ (F) operating at an angular frequency of $\omega$ (rad/s) is $\frac{1}{j\omega C}$ measured in ($\Omega$). Phasor representation of Ohm's law for a capacitor under steady-state sinusoidal conditions is as follows:

$$V = \frac{1}{j\omega C} I$$

It can be concluded that a capacitor $C$ (F) in steady-state sinusoidal has a reactance of $X_C = \frac{1}{j\omega C}$ ($\Omega$) (Fig. 5.13).

## Power Factor of Capacitive Circuits

Since the phase shift in current with respect to the voltage is $-90°$, the power factor $PF = \cos(-90°) = 0$. It can be concluded that a pure capacitive circuit has $PF = 0$.

**Example 5.9 Find the current in the circuit of Fig. 5.14.**
*Solution.* The reactance of the capacitor at frequency $\omega = 100$ can be calculated as

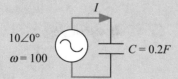

**Fig. 5.14** Circuit of Example 5.9

(continued)

**Example 5.9** (continued)

$$X_c = \frac{1}{j\omega C} = \frac{1}{j100 \times 0.2} = -j0.05\Omega$$

The circuit, therefore, can be shown as

$$10\angle 0° \qquad \frac{1}{j\omega C} = -j0.05\Omega$$

The current through this reactance is

$$I = \frac{10\angle 0}{-j0.05} = \frac{10\angle 0}{0.05\angle - 90} = 200\angle 90 \text{ A}$$

Note that $j = 1 \angle 90$, $-j = 1 \angle -90$.

## Resistive-Inductive Circuits

Consider a series connection of an inductor $L$ (H) and a resistor $R(\Omega)$ to a voltage source $v(t) = V_m \sin(\omega t)$ (Fig. 5.15). This circuit operating at frequency $\omega$ shows the total impedance of $R(\Omega)$ in the resistor and $j\omega L$ ($\Omega$) in the inductor. The total ohmic value of impedance $z(j\omega)$ in this series circuit is the summation of both elements (because of series connection) as follows:

$$z(j\omega) = R + j\omega L = R + jX_L(\Omega)$$

**Fig. 5.15** A resistive-inductive RL circuit excited by external sinusoidal source $v(t)$

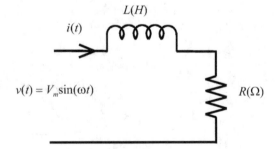

The impedance of this circuit has a real part known as *resistance R* and an imaginary part known as *reactance* $X_L$. In polar coordinates, the impedance of this circuit can be written as

$$Z = \sqrt{R^2 + X_L^2} \angle \left( \tan^{-1} \frac{X_L}{R} \right) = z \angle \varphi$$

where $\varphi$ is the phase angle of the impedance, which determines the amount of phase shift in current with respect to voltage. This phase is a positive value because the values of resistance and reactance observed from an inductor are positive. As the phase angle is positive for an RL circuit, the circuit generates a lag type of behavior. This means the current lags the voltage when the voltage is set as a reference. Ohm's law indicates that the voltage drop across the impedance $Z = z \angle \varphi$ when current $I = I_m \angle \theta$ passes through the impedance can be calculated as $V = ZI$.

This can be expanded using the phasor as follows:

$$V = ZI = z \angle \varphi I_m \angle \theta = z I_m \angle (\varphi + \theta)$$

Therefore, the current knowing the voltage of the source can be obtained by dividing the voltage over the impedance as follows:

$$I = \frac{V}{Z}$$

Considering the phasors values,

$$I = \frac{V_m \angle 0}{z \angle \varphi} = \frac{V_m}{z} \angle - \varphi$$

This current can also be obtained by utilizing the circuit element values as follows:

$$I = \frac{V_m}{\sqrt{R^2 + X_L^2}} \angle - \left( \tan^{-1} \frac{X_L}{R} \right)$$

## Power Factor of Resistive-Inductive Circuits

Consider the circuit of Fig. 5.15. The phase shift in current with respect to voltage is

$$\varphi = - \left( \tan^{-1} \frac{X_L}{R} \right)$$

Therefore, the power factor of the RL circuit is

$$PF = \cos \left( - \tan^{-1} \frac{X_L}{R} \right)$$

Since the circuit generates a lagging current, the power factor can be read as $PF = \cos \left( - \tan^{-1} \frac{X_L}{R} \right)$ Lag.

The resistance ratio over the impedance amplitude can also be obtained power factor. Therefore,

$$PF = \frac{R}{|Z|} = \frac{R}{\sqrt{R^2 + X_L^2}}$$

**Power factor can also be obtained from the division of the voltage measured across the resistor over the entire circuit voltage drop as follows:**

$$PF = \frac{V_R}{V_m}$$

The outcome of this discussion is as follows:

- The phase of the current in an RL circuit is a negative value.
- The phase of impedance in an RL circuit is a positive value.
- An RL circuit shows a phase lag in current with respect to voltage.

**Example 5.10 A resistive-inductive circuit is excited at the frequency of 60 Hz. The equivalent resistance is $R = 10\ \Omega$, and the inductance of $L = 100$ mH. What is the power factor of the circuit?**
*Solution*

$$PF = \frac{R}{|Z|} = \frac{R}{\sqrt{R^2 + X_L^2}} = \frac{10}{\sqrt{10^2 + (2\pi 60 100e - 3)^2}} = \frac{10}{39} = 0.256$$

**Example 5.11 In the circuit of Fig. 5.10 (check the figure number), the resistance and reactance are $R = 10\ \Omega$, $X = j20\ \Omega$. Find the impedance of the circuit in polar form. Find the power factor of the circuit.**

*Solution.* The resistor and the inductor are connected in series. Therefore, the circuit impedance is a series of resistance and reactance. It is

$$Z = R + jX = 10 + j20\ \Omega$$

Converting the impedance from rectangular coordinates to polar coordinates results in

$$|Z| = \sqrt{R^2 + X^2} = \sqrt{100 + 400} = 22.36\ \Omega$$

$$\angle Z = \tan^{-1}\frac{X}{R} = \tan^{-1}\frac{20}{10} = 63.43\,^{\circ}$$

The diagram of impedance can be shown on the real and imaginary axes. There is a resistance of $20\Omega$ on the real axis and a reactance of $20\Omega$ perpendicular to form a rectangle.

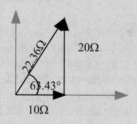

**Example 5.12 The impedance of a circuit is measured to be $Z = 100\ \angle\ 60\ \Omega$. Find the resistance and reactance of the circuit. How much is the inductance at the frequency of $f = 60$ Hz?**

*Solution.* The impedance can be converted from polar form to rectangular coordinates as follows:

$$Z = 100\angle 60 = 100\cos 60 + j100\sin 60$$

$$Z = 50 + j86.6\ \Omega$$

Therefore, $R = 50\ \Omega$, $X = 86.6\ \Omega$. The reactance at the frequency $f$ (Hz) can be calculated as

(continued)

**Example 5.12** (continued)

$$X = 2\pi f\, L$$

Therefore, the inductance of the inductor at $f = 60$ Hz can be found as

$$L = \frac{X}{2\pi f} = \frac{86.6}{2\pi \times 60} = 229 \text{ mH}$$

## Vector Analysis of RL Circuits

The rectangle form impedance $z(j\omega) = R + jX_L$ has a real part $R$ and an imaginary part $X_L$. The factor $j$ indicates a right-angle (90°) rotation from the real axis. These impedance values show the amplitude of unit vectors on the real and the imaginary axis. In polar coordinates, this impedance shows an amplitude $z$ and a phase $\varphi$. The impedance in both rectangle and polar coordinates is shown (Fig. 5.15).

This analysis can be expanded to KVL in the loop as follows:

$$-V_s + V_R + V_L = 0$$

Note that these voltages are presented as vectors or phasors.

Consider the source voltage $V_m \angle 0$ as a reference. The current of an RL circuit has a time delay in reaching peak values with respect to the voltage by angle $\varphi$. Vectors of voltage and current are shown in Fig. 5.16.

The phase shift between two vectors generates shifted rectangle coordinates on the current vector. Therefore, to obtain the KVL in the loop, the voltage drop across the resistor generates an in-phase value of $V_R = RI$ with the current and a vector with 90° rotation (perpendicular to the current vector) that demonstrates the voltage drop across the inductor as $V_L = j\omega L I$ (Fig. 5.17).

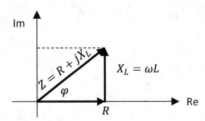

**Fig. 5.16** A series RL circuit impedance is split into a real part resistance $R$ and an imaginary part reactance of $\omega L$ vertical to the resistance on the imaginary axis. Factor $j$ also confirms the angle of reactance with respect to the resistance

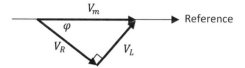

**Fig. 5.17** Summation of voltage drops in the RL circuit. The voltage drop across the resistor has a phase shift with respect to the reference because the current phase passes through the resistor. The voltage drop across the inductor has a 90° phase shift from the vector of the resistor voltage drop. The factor $j$ in this voltage drop makes the phase shift

**Example 5.13 The rms voltage drop measured across elements of a series RL circuit is $V_R = 25$ V, $V_L = 10$ V. Find the power factor of the circuit.**
*Solution.* The source voltage is the vector summation of the resistive and inductive voltages. Considering that the inductive voltage has a $j$ factor, the source voltage is

$$V_m = V_R + jV_L$$

$$V_m = 25 + j10$$

Therefore,

$$PF = \frac{V_R}{\sqrt{V_R^2 + V_L^2}} = \frac{25}{\sqrt{25^2 + 10^2}} = \frac{25}{26.92} = 0.928$$

Since the circuit has an inductive effect, the power factor becomes PF = 0.928 lag.

**Example 5.14 In the circuit of Fig. 5.18, find the circuit's impedance, resistance, reactance, and power factor.**
*Solution.* The resistor and the inductor are connected in parallel. The impedance can be calculated as follows:

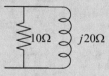

**Fig. 5.18** Circuit of Example 5.15

(continued)

**Example 5.14** (continued)

$$Z = R \| jX = \frac{R \times jX}{R + jX} = \frac{10 \times j20}{10 + j20} = \frac{j200}{10 + j20} = \frac{j200}{10 + j20} \times \frac{10 - j20}{10 - j20}$$

$$= \frac{4000 + j2000}{100 + 400} = \frac{4000}{500} + \frac{j2000}{500} = 8 + j4 \ \Omega$$

Therefore, $R = 8$, $X = 4$. Converting the impedance to polar form results in

$$Z = \sqrt{8^2 + 4^2} \angle \tan^{-1} \frac{4}{8} = 8.9 \angle 26.58 \ \Omega$$

Power factor can be obtained from the known angle of the impedance or the resistance and reactance of the impedance.

$$PF = \cos(\angle Z) = \cos 26.58 = 0.89$$

or

$$PF = \frac{R}{|Z|} = \frac{8}{8.9} = 0.89$$

**Example 5.15** The impedance of the two circuits was measured as $Z_1 = 1 + j5 \ \Omega$ and $Z_2 = 5 + j1 \ \Omega$. What is the expected phase shift of the current with respect to the voltage in each of these circuits? Find the power factor of these circuits and discuss your findings.
*Solution.*
  Circuit-1

$$PF_1 = \frac{R_1}{|Z_1|} = \frac{1}{\sqrt{1^2 + 5^2}} = \frac{1}{\sqrt{26}} = 0.19$$

The impedance $Z_1$ has small resistance and a considerable reactance value. Therefore, the power factor is expected to be close to zero for highly inductive inductors.
  In this circuit, the phase shift of the current with respect to the voltage is

$$\phi_1 = \cos^{-1} 0.19 = 79°$$

(continued)

**Example 5.15** (continued)

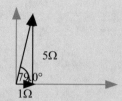

Circuit-2

$$PF_2 = \frac{R_1}{|Z_1|} = \frac{5}{\sqrt{5^2 + 1^2}} = \frac{5}{\sqrt{26}} = 0.98$$

The impedance $Z_2$ has considerable resistance and a small inductance value. Therefore, the power factor is expected to be close to 1 toward highly resistive circuits.

In this circuit, the phase shift of the current with respect to the voltage is

$$\phi_1 = \cos^{-1} 0.89 = 11.4°$$

## Resistive-Capacitive Circuits

Consider a parallel connection of a capacitor $C$ (F) at reactance of $X_C = \frac{1}{\omega C}$ ($\Omega$) and a resistance $R(\Omega)$ to a voltage source $v(t) = V_m \sin(\omega t)$, as shown in Fig. 5.19. This circuit operating at frequency $\omega$ shows total admittance $\frac{1}{R}$ ($\Omega^{-1}$) from the resistor

**Fig. 5.19** A parallel resistive-capacitive circuit excited by a sinusoidal voltage source $v(t)$. The current $i(t)$ is flowing in the circuit

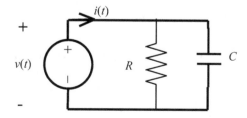

and $j\omega C$ $(\Omega^{-1})$ in a capacitor. The total ohmic value of admittance $Y(j\omega)$ in this parallel circuit is the summation of both elements as follows (Fig. 5.19):

$$Y(j\omega) = \frac{1}{R} + \frac{1}{jX_C} = G - jB \ (\Omega^{-1})$$

## Using Admittance

Admittance of this circuit has a real part known as *conductance G* and an imaginary part known as *susceptance* $B = \omega C$. A transformation to polar coordinates, the admittance of this circuit can be written as

$$Y = \sqrt{G^2 + B^2} \angle \left( \tan^{-1} \frac{B}{G} \right) = y \angle \psi$$

where $\psi$ is the phase angle of the circuit, which shows a shift in current with respect to voltage. This phase is positive because the value of susceptance observed from a capacitor is positive. As the phase angle is positive for an RC circuit, the circuit generates a lead type of behavior. This means the current leads the voltage when the voltage is set as a reference. Ohm's law indicates that the voltage drop across the admittance $Y = y \angle \psi$ when current $I = I_m \angle \theta$ passes through the admittance can be calculated $V = \frac{I}{Y}$. Therefore, the current knowing the voltage of the source can be obtained by the product of the voltage and the admittance.

$$I = YV$$

Note that these values are expressed in phasors.

$$I = y \angle \psi \ V_m \angle 0 = y V_m \angle \psi$$

This current can also be obtained utilizing the rectangle coordinate values as follows:

$$I = \frac{V_m}{\sqrt{G^2 + B^2}} \angle \left( \tan^{-1} \frac{B}{G} \right)$$

## Using Impedance

The current phasor using the impedance values can be obtained as follows:

$$Z(j\omega) = R \| \frac{1}{jC\omega} = \frac{R\frac{1}{jC\omega}}{R + \frac{1}{jC\omega}} = \frac{R}{1 + j\omega RC} = \frac{R\angle 0}{\sqrt{1 + (\omega RC)^2} \angle \tan^{-1}(\omega RC)}$$

$$= \frac{R}{\sqrt{1 + (\omega RC)^2}} \angle - \tan^{-1}(\omega RC)$$

$$Z(j\omega) = z\angle - \varphi$$

where $\varphi$ is the phase angle of the circuit, which shows a shift in current with respect to voltage. The phase of a resistive-capacitive circuit is a negative value because the value of reactance observed from a capacitor is negative, resembling a lead circuit. Ohm's law indicates that the voltage drop across the impedance $Z = z \angle - \varphi$ when current $I = I_m \angle \theta$ passes through the admittance can be calculated as $V = ZI$. Therefore, the current knowing the voltage of the source can be obtained by dividing the voltage by the impedance:

$$I = \frac{V}{Z}$$

Note that these values are expressed in phasors. Therefore,

$$I = \frac{V_m \angle 0}{z \angle - \varphi} = \frac{V_m}{z} \angle \varphi$$

## Power Factor of Resistive-Capacitive Circuits

The phase shift in current with respect to voltage is $(-\tan^{-1}(\omega RC))$. Therefore, the power factor of an RC circuit is $PF = \cos(-\tan^{-1}(\omega RC))$. Since the circuit generates a leading current with respect to the voltage, the power factor can be read as $PF = \cos(-\tan^{-1}(\omega RC))$ lead.

The outcome of this discussion is as follows:

- The phase angle of current in an RC circuit is positive.
- The phase angle of admittance in an RC circuit is positive.
- The phase angle of impedance in an RC circuit is negative.
- An RC circuit shows a phase lead in current with respect to a reference voltage.

## Vector Analysis of RC Circuits

The impedance of an RC circuit shown in Fig. 5.20 from the source terminals is $z(j\omega) = R - jX_C$ in rectangle form. This has a real part, $R$, and an imaginary part $-X_C$. The factor $-j$ indicates a right angle $(-90°)$ rotation from the real axis. In polar coordinates, this impedance shows an amplitude $z$ and a phase $\varphi$. The impedance in both rectangle and polar coordinates is shown (Fig. 5.20).

This analysis can be expanded to KVL written as

$$- V_s + V_R + V_C = 0$$

Considering the vectors of these voltages, their balance becomes

$$V_s = \sqrt{V_R^2 + (-V_C)^2}$$

or

$$V_s = \sqrt{V_R^2 + V_C^2}$$

Considering the source voltage $V_m \angle 0$ as a reference, the current of an RC circuit has a time lead in reaching peak values with respect to the voltage by angle $-\varphi$. Vectors of voltage and current are shown in Fig. 5.21.

The phase shift between two vectors generates shifted rectangle coordinates on the current vector. In the KVL, the voltage drop across the resistor $V_R = RI$ is

**Fig. 5.20** RC circuit excited by an external source. The current also has a sinusoidal form but at a different phase angle

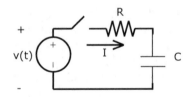

**Fig. 5.21** The impedance of an RC circuit has two components of $R$ on the reference axis and a capacitive component as $-\frac{1}{j\omega C}$, or $-jX_C$. The equivalent impedance is a vector summation of $R$ and $-jX_C$ as $Z = R - jX_C$

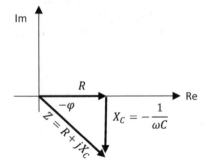

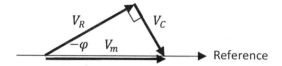

**Fig. 5.22** The voltage drop across the resistor and the capacitor is $-90°$ apart. Consider that the voltage of the resistor and the source is $-\varphi$ apart

in-phase with the current. The voltage drop across the capacitor has a phase shift of $-90°$ (perpendicular to the current vector), as $V_C = -j\frac{1}{\omega C}I$ shown in Fig. 5.22.

---

**Example 5.16** In the circuit shown in Fig. 5.20, the resistance and reactance are $R = 10\ \Omega$, $X = -j15\ \Omega$. Find the impedance and power factor. Determine whether the circuit is lead or lag.

*Solution.* In a series connection of two elements, the impedance becomes

$$Z = R + jX = 10 + (-j15)\ \Omega$$

Converting it to the polar coordinate, the amplitude and phase become

$$Z = \sqrt{10^2 + 15^2} \angle \tan^{-1} - \frac{15}{10} = 18.02 \angle -56.3°\ \Omega$$

There are two signs in this circuit that can help determine the lead or lag characteristics.

1. The reactance part of the impedance is negative.
2. The angle of the impedance is negative.

**Note:** It should be noted that the real part of the impedance is a positive value and the imaginary part of the impedance is negative. Therefore, the impedance vector is in the fourth quarter, similar to Fig. 5.21. Therefore, the negative angle $-56.3$ makes sense.

*However, if the real part of a complex number is negative, the vector will fall either in the second or the third quarter, depending on the sign of the imaginary part. When the angle of the vector is calculated through $\tan^{-1}\frac{Im}{Re}$, the angle falls in the first or the fourth quarter. The angle must then be added by $180°$ to align with its true direction correctly.*

**Example 5.17 The impedance of a circuit is assumed to be** $Z = -10 + j15 \ \Omega$**. Convert the impedance to the polar coordinates.**
*Solution.*

$$|Z| = \sqrt{10^2 + 15^2} = 18.02 \ \Omega$$

The impedance vector is located in the third quarter, having a negative real and a positive imaginary part. However, the angle is calculated to be

$$\angle Z = \tan^{-1} - \frac{15}{10} = -56.3\,°$$

The $-56.3°$ falls in the fourth quarter does not match the impedance. Therefore, $180 + (-56.3) = 123.7°$. So the correct polar coordinates are

$$Z = 18.02 \angle + 123.7\,° \ \Omega$$

**Example 5.18 In the circuit of Fig. 5.13 (RC Parallel),** $R = 10 \ \Omega$ **and** $\frac{1}{j\omega C} = -j15 \ \Omega$ **. Find the circuit's impedance, resistance, and reactance observed at the terminals. Find the power factor and determine if it is lead or lag.**
*Solution.*

$$Z = 10 \| -j15 = \frac{10 \times -j15}{10 + (-j15)} = \frac{-j150}{10 - j15} = \frac{150 \angle -90}{\sqrt{10^2 + 15^2} \angle \tan^{-1} - \frac{15}{10}}$$

$$= \frac{150 \angle -90}{18.02 \angle -56.8} = 8.32 \angle -33.2\,° \ \Omega$$

$$Z = 8.32 \cos(-33.2) + j8.32 \sin(-33.2) = 6.94 - j4.55 \ \Omega$$

Therefore, the resistance is $R = 6.94 \ \Omega$ and the reactance is $X = -4.55 \ \Omega$. The circuit is lead as the imaginary part of the impedance is negative, or the impedance angle is negative.

$$PF = \frac{R}{|Z|} = \frac{R}{\sqrt{R^2 + X^2}} = \frac{6.96}{\sqrt{6.94^2 + 4.55^2}} = 0.836$$

or

$$PF = \cos(-33.2) = 0.836$$

**Example 5.19** In the circuit of Fig. 5.10, the voltage source is $V = 100 \angle 0$ V, $R = 10\ \Omega$, and $X = 20\ \Omega$. Find the current in the circuit and the voltage drop across each element. Draw the voltage drop diagram.

*Solution.* The current can be calculated as

$$I = \frac{V}{Z} = \frac{100\angle 0}{10 + j20} = \frac{100\angle 0}{22.36\angle 63.43} = 4.47\angle - 63.43 \text{ A}$$

The voltage drop across each element is its impedance times the current passing through. Therefore,

$$V_R = RI = 10 \times 4.47\angle - 63.43 = 44.7\angle - 63.43 \text{ V}$$

$$V_L = jXI = j20 \times 4.47\angle - 63.43 = 89.4\angle 26.57 \text{ V}$$

It can be seen that in this circuit (series connection or $R$, $L$), the source voltage and the voltage drop across the resistor and the inductor form a triangle as

$$V_S^2 = V_R^2 + V_L^2$$

$$100^2 = 44.7^2 + 89.4^2$$

**Example 5.20** The circuit shown in Fig. 5.20 shows that the voltage drop across the resistor is 25% of the source voltage. What is the power factor of the circuit?

*Solution.* Power factor can be found as follows:

$$\text{PF} = \frac{V_R}{V_s} = \frac{0.25V_s}{V_s} = 0.25$$

**Example 5.21 Find the impedance of the circuit shown in Fig. 5.23 at frequency $\omega$.**

*Solution.*

$$z(j\omega) = R + \left( \frac{1}{j\omega C_1} \,\middle\|\, \left( j\omega L + \frac{1}{j\omega C_2} \right) \right)$$

$$z(j\omega) = R + \left( \frac{\frac{1}{j\omega C_1} \left( j\omega L + \frac{1}{j\omega C_2} \right)}{\frac{1}{j\omega C_1} + \left( j\omega L + \frac{1}{j\omega C_2} \right)} \right)$$

$$z(j\omega) = R + \left( \frac{-\frac{1}{C_1 C_2 \omega^2} \left( 1 - L C_2 \omega^2 \right)}{\frac{-1}{C_1 C_2 \omega^2} \left( j C_2 \omega - L C_1 C_2 \omega^2 + j\omega C_1 \right)} \right)$$

$$z(j\omega) = \left( R - \frac{L C_1 \omega^2 (1 - L C_2 \omega^2)}{\left( - L C_1 C_2 \omega^2 \right)^2 + (\omega (C_1 + C_2))^2} \right)$$

$$- j \left( \frac{\omega (C_1 + C_2)(1 - L C_2 \omega^2)}{\left( L C_1 C_2 \omega^2 \right)^2 + (\omega (C_1 + C_2))^2} \right)$$

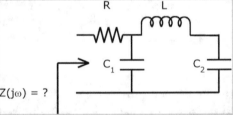

**Fig. 5.23** Circuit of Example 5.22

**Example 5.22 Find impedance of the circuit shown in Fig. 5.24 at frequency $\omega$.**

$$Z(j\omega) = R_1 + \frac{1}{\frac{1}{R_2 + j\omega L} + \frac{1}{R_3 + \frac{1}{j\omega C}}}$$

(continued)

**Example 5.22** (continued)

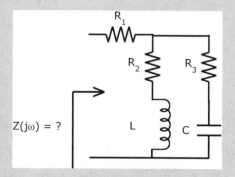

**Fig. 5.24** Circuit of Example 5.23

$$Z(j\omega) = R_1 + \cfrac{1}{\cfrac{R_2 - j\omega L}{R_2^2 + \omega^2 L^2} + \cfrac{R_3 + \frac{j}{\omega C}}{R_3^2 + \frac{1}{\omega^2 C^2}}}$$

$$Z(j\omega) = R_1 + \cfrac{1}{\cfrac{R_2 - j\omega L}{R_2^2 + \omega^2 L^2} + \cfrac{R_3 + \frac{j}{\omega C}}{R_3^2 + \frac{1}{\omega^2 C^2}}}$$

Consider

$$A = R_2^2 + \omega^2 L^2$$

$$B = R_3^2 + \frac{1}{\omega^2 C^2}$$

Therefore,

$$Z(j\omega) = R_1 + \cfrac{1}{\cfrac{R_2 - j\omega L}{A} + \cfrac{R_3 + j\frac{1}{\omega C}}{B}}$$

$$Z(j\omega) = R_1 + \cfrac{1}{\cfrac{BR_2 - jB\omega L + AR_3 + j\frac{A}{\omega C}}{AB}}$$

$$Z(j\omega) = R_1 + \frac{AB}{BR_2 - jB\omega L + AR_3 + j\frac{A}{\omega C}}$$

(continued)

**Example 5.22** (continued)

$$Z(j\omega) = R_1 + \frac{AB}{(BR_2 + AR_3) + j\left(\frac{A}{\omega C} - B\omega L\right)}$$

Consider

$$M = BR_2 + AR_3$$

$$N = \frac{A}{\omega C} - B\omega L$$

Therefore,

$$Z(j\omega) = R_1 + \frac{AB}{M + jN}$$

$$Z(j\omega) = R_1 + \frac{AB(M - jN)}{M^2 + N^2}$$

$$Z(j\omega) = \left(R_1 + \frac{ABM}{M^2 + N^2}\right) - j\frac{ABN}{M^2 + N^2}$$

**Example 5.23** Find impedance of the circuit shown in Fig. 5.25 at frequency $\omega$.

*Solution.*

$$Z(j\omega) = \frac{1}{Y(j\omega)}$$

$$Y(j\omega) = \frac{1}{R_1} + \frac{1}{R_2 + j\omega L} + \frac{1}{R_3 + \frac{1}{j\omega C}}$$

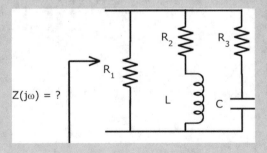

**Fig. 5.25** Circuit of Example 5.24

(continued)

**Example 5.23** (continued)

Each part of $Y(j\omega)$ is recommended to be simplified first, and then the real and imaginary parts are separated and determined.

Therefore,

$\frac{1}{R_1}$ is already simplified.

$\frac{1}{R_2+j\omega L}$ becomes

$$\frac{1}{R_2+j\omega L} = \frac{1}{R_2+j\omega L}\frac{R_2-j\omega L}{R_2-j\omega L} = \frac{R_2-j\omega L}{R_2^2+\omega^2 L^2}$$

$\frac{1}{R_3+\frac{1}{j\omega C}}$ becomes

$$\frac{1}{R_3+\frac{1}{j\omega C}} = \frac{1}{R_3-j\frac{1}{\omega C}} = \frac{1}{R_3-j\frac{1}{\omega C}}\frac{R_3+j\frac{1}{\omega C}}{R_3+j\frac{1}{\omega C}} = \frac{R_3+j\frac{1}{\omega C}}{R_3^2+\frac{1}{\omega^2 C^2}}$$

As a result, the admittance of the circuit becomes

$$Y(j\omega) = \frac{1}{R_1} + \frac{R_2-j\omega L}{R_2^2+\omega^2 L^2} + \frac{R_3+j\frac{1}{\omega C}}{R_3^2+\frac{1}{\omega^2 C^2}}$$

$$Y(j\omega) = \left(\frac{1}{R_1} + \frac{R_2}{R_2^2+\omega^2 L^2} + \frac{R_3}{R_3^2+\frac{1}{\omega^2 C^2}}\right) + j\left(\frac{-\omega L}{R_2^2+\omega^2 L^2} + \frac{\frac{1}{\omega C}}{R_3^2+\frac{1}{\omega^2 C^2}}\right)$$

$$Z(j\omega) = \frac{1}{\left(\dfrac{1}{R_1} + \dfrac{R_2}{R_2^2+\omega^2 L^2} + \dfrac{R_3}{R_3^2+\frac{1}{\omega^2 C^2}}\right) + j\left(\dfrac{-\omega L}{R_2^2+\omega^2 L^2} + \dfrac{\frac{1}{\omega C}}{R_3^2+\frac{1}{\omega^2 C^2}}\right)}$$

$$Z(j\omega) = \frac{\left(\dfrac{1}{R_1} + \dfrac{R_2}{R_2^2+\omega^2 L^2} + \dfrac{R_3}{R_3^2+\frac{1}{\omega^2 C^2}}\right) - j\left(\dfrac{-\omega L}{R_2^2+\omega^2 L^2} + \dfrac{\frac{1}{\omega C}}{R_3^2+\frac{1}{\omega^2 C^2}}\right)}{\left(\dfrac{1}{R_1} + \dfrac{R_2}{R_2^2+\omega^2 L^2} + \dfrac{R_3}{R_3^2+\frac{1}{\omega^2 C^2}}\right)^2 + \left(\dfrac{-\omega L}{R_2^2+\omega^2 L^2} + \dfrac{\frac{1}{\omega C}}{R_3^2+\frac{1}{\omega^2 C^2}}\right)^2}$$

## Steady-State Analysis of Circuits

Circuits operating in steady-state sinusoidal conditions show specific impedance determining the branches' current and nodes' voltage. The circuit equivalent in a given angular frequency must be evaluated to obtain these parameters. Then KVL and KCL can be used to analyze the circuit. It is important to note that all voltages, currents, and impedances follow vector and phasor analysis such that pure ohmic values are in phase with the reference, and the inductive impedance is projected on a positive imaginary. The capacitive impedance is projected on the negative imaginary axis.

## RLC Series

Consider a series connection of $R$, $L$, and $C$ to a voltage source $v(t) = V_m \sin(\omega t)$, as shown in Fig. 5.26. The impedance of the circuit observed from the source terminals is measured to be (Fig. 5.26)

$$z(j\omega) = R + jX_L - jX_C = R + j\omega L + \frac{1}{j\omega C} = R + j\omega L - \frac{j}{\omega C} = R + j\left(\omega L - \frac{1}{\omega C}\right)$$

Vector representation of the impedance is obtained by having $R$ on the positive real axis, $\omega L$ on the positive imaginary axis because of $+j$, and $\frac{1}{\omega C}$ on the negative imaginary axis because of $-j$.

The impedance amplitude is obtained as $|z| = \sqrt{R^2 + (X_L - X_C)^2} = \sqrt{R^2 + \left(L\omega - \frac{1}{C\omega}\right)^2}$. The impedance phase is obtained as $\angle z = \varphi = \tan^{-1}\frac{X_L - X_C}{R}$.

The impedance of the inductor and capacitor cancel each other to some extent, depending on their values. If $|L\omega| > |\frac{1}{C\omega}|$, the circuit becomes more inductive, and if $|L\omega| < |\frac{1}{C\omega}|$, the circuit becomes more capacitive.

When the overall impedance becomes more inductive, the phase of impedance becomes a positive value, and when the overall impedance becomes more capacitive, the phase of impedance becomes negative.

**Fig. 5.26** An RLC series is excited by a sinusoidal voltage source

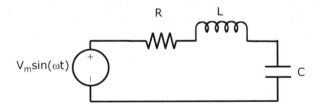

The amplitude of impedance is also influenced by the inductor and capacitor reactance. The impedance value is minimum (pure resistive) when the inductor impedance equals the impedance of the capacitor or $X_L = X_C$.

$$|z|_{\min}\Big|_{X_L=X_C} = R$$

At this point ($X_L = X_C$), the impedance becomes purely resistive as the impedance value decreases, and the circuit current increases to its maximum value. This operating point at which the energy stored in capacitors and inductors of the circuit cancels each other is called *resonance*.

KVL indicates that the summation of voltage drops in a loop is zero. It should be noted that the voltages in the steady-state analysis represent vectors either on the real axis or on the imaginary axis with positive and negative values.

For instance, the KVL in loop ① is written as follows:

$$-V_m\angle 0 + V_R + jV_L - jV_C = 0$$

$$\sqrt{V_R^2 + (V_L - V_C)^2} = V_m$$

Considering the loop current phasor $I$, the KVL can be written as

$$RI + jX_LI - jX_CI = V_m\angle 0$$

The summation of amplitudes suggests that

$$\sqrt{(RI)^2 + (X_LI - X_CI)^2} = |V_m|$$

Factoring $I$ out results in

$$|I|\sqrt{R^2 + (X_L - X_C)^2} = |V_m|$$

Therefore, the current amplitude of the loop is obtained from dividing the voltage by impedance amplitude as

$$|I| = \frac{|V_m|}{\sqrt{R^2 + (X_L - X_C)^2}}$$

The phase of current, as explained earlier, can be obtained from the circuit as

$$\theta = -\tan^{-1}\frac{X_L - X_C}{R}$$

**Example 5.24 In the circuit of Fig. 5.26, the elements are $R = 10\,\Omega$, $L = 1.5$ H, $C = \frac{1}{4}$ F. The operating frequency is $\omega = 1\ \frac{\text{rad}}{\text{s}}$. Find the impedance and the power factor of the circuit.**

*Solution.* In a series connection of the elements, the impedance is

$$Z = R + j\omega L + \frac{1}{j\omega C}$$

$$Z = 10 + j \times 1 \times 1.5 + \frac{1}{j \times 1 \times \frac{1}{4}} = 10 + j1.5 - j4 = 10 - j2.5\ \Omega$$

The power factor can be calculated as follows:

$$\text{PF} = \frac{R}{|Z|} = \frac{10}{\sqrt{10^2 + 2.5^2}} = 0.969\ \text{lead}$$

The circuit is lead as the imaginary part of the impedance is negative, which means that there is more capacitive than inductive reactance.

**Example 5.25 In the circuit of Fig. 5.26, find the amount of capacitive reactance to make the power factor PF = 1. The other circuit values are $R = 10\,\Omega$, $X_L = 1.5\,\Omega$.**

*Solution.*

$$\text{PF} = \frac{R}{\sqrt{R^2 + (X_L - X_C)^2}} = \frac{10}{\sqrt{10^2 + (1.5 - X_C)^2}} = 1$$

$$10^2 + (1.5 - X_C)^2 = 10^2$$

$$(1.5 - X_C)^2 = 0$$

$$X_C = 1.5$$

This means that at PF = 1, the entire reactance of the capacitor must cancel the reactance of the inductor, i.e., @PF = 1; $|X_L| = |X_C|$.

**Example 5.26 In the circuit of Fig. 5.20, $R = 10\ \Omega$, $X_L = 1.5\ \Omega$. Find the capacitive reactance such that the power factor of the circuit becomes PF $= 0.9$.**

*Solution.* The power factor of the circuit with an unknown $X_C$ becomes

$$\text{PF} = \frac{R}{\sqrt{R^2 + (X_L - X_C)^2}} = \frac{10}{\sqrt{10^2 + (1.5 - X_C)^2}} = 0.9$$

$$\frac{10^2}{10^2 + (1.5 - X_C)^2} = 0.81$$

$$100 = 81 + 0.81(1.5 - X_C)^2$$

$$(1.5 - X_C)^2 = 23.45$$

$$1.5 - X_C = \pm 4.84$$

$$X_C = 1.5 \pm 4.84 = 6.43, \text{or}, -3.34$$

Since the negative sign of the capacitive reactance is already considered in the equations, only the positive value is acceptable; therefore, $-jX_C = -j6.43\ \Omega$.

**Example 5.27 Find the inductor's voltage and the power factor in the following circuit.**

*Solution.* As Fig. 5.27 shows, the voltage drop on each element is known except for the inductor. The balance of voltages can be written as

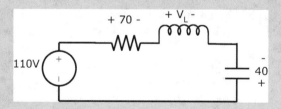

**Fig. 5.27** Circuit of Example 5.28

(continued)

**Example 5.27** (continued)

$$\sqrt{V_R{}^2 + (V_L - V_C)^2} = V_m$$

$$\sqrt{70^2 + (V_L - 40)^2} = 110$$

$$(V_L - 40)^2 = 110^2 - 70^2 = 7200$$

$$V_L - 40 = 84.85$$

$$V_L = 124.85 \ \text{V}$$

Power factor can be obtained from

$$\mathbf{PF} = \frac{\mathbf{V_R}}{\mathbf{V}} = \frac{\mathbf{70}}{\mathbf{110}} = \mathbf{0.63} \ \textbf{lag}.$$

*The circuit lags because the voltage drop across the inductor is larger than the voltage drop across the capacitor. This has resulted from larger inductive reactance concerning the capacitive reactance for the same circuit current.* Therefore, $X_L > X_C$. The voltage across the inductor can be higher than the amplitude of the voltage source, which may occur due to energy exchange between the inductor and capacitor. The phase delay naturally occurs in the capacitor and inductor, causing the capacitor and inductor voltages to reach peak value with a time delay. Therefore, a simple summation of voltages without considering their phase angle is not an accurate KVL.

**Example 5.28 Considering a series connection of RLC circuit to a voltage source of $v(t) = 167 \sin (377t + 10)$ V draws a current as $i(t) = 7.07 \sin (377t + 70)$ A. Find the impedance, resistance, and reactance of the circuit. Is this circuit more capacitive or more inductive? Find the power factor of the circuit.**
*Solution.* The voltage and current phasors are obtained as $V = 167 \angle 10$ and $I = 7.07 \angle 70$. Knowing that the impedance is $Z = \frac{V}{I}$:

$$Z = \frac{167 \angle 10}{7.07 \angle 70} = 23.62 \angle - 60 \ \Omega$$

Since the phase of impedance is negative, the circuit is more capacitive. Therefore, the impedance in rectangle coordinates shows the resistance and reactance as follows:

(continued)

**Example 5.28** (continued)

$$Z = \frac{167\angle 10}{7.07\angle 70} = 23.62\angle -60 = 23.62\cos(-60) + j23.62\sin(-60)$$
$$= \underbrace{11.81}_{R} - \underbrace{j20.45}_{-jX_C}\Omega$$

The real part of impedance shows the resistance, and the imaginary part shows the reactance. Since the reactance is negative, the circuit is more capacitive.

The power factor of the circuit is obtained from the resistance ratio of 11.81 $\Omega$ over the impedance amplitude of 23.62 $\Omega$. PF $= \frac{11.81}{23.62} = 0.5$ lead. The circuit is lead because it shows more capacitive behavior.

Power factor can also be obtained PF $= \cos{(\widehat{V, I})} = \cos{(-60)} = 0.5$.

## RLC Parallel

Consider a parallel connection of $R$, $L$, and $C$ to a current source $i(t) = I_m \sin(\omega t)$. The circuit forms a node considered $v(t)$ volts (Fig. 5.28).

A KCL at node ① can be written as follows:

$$-i(t) + i_R + i_L + i_C = 0$$

Phasor current values can be replaced in the KCL. This yields

$$-I_m\angle 0 + \frac{V}{R} + \frac{V}{j\omega L} + \frac{V}{1/j\omega C} = 0$$

Solving for $V$ (phasor representation of voltage) can be found as

$$V\left(\frac{1}{R} + \frac{1}{j\omega L} + j\omega C\right) = I_m$$

**Fig. 5.28** Parallel RLC circuit excited by a sinusoidal current source

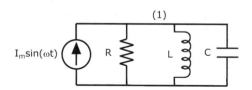

$$V = \frac{I_m}{\left(\frac{1}{R} + \frac{1}{j\omega L} + j\omega C\right)} = \frac{I_m}{\left(\frac{1}{R} + \frac{-j}{\omega L} + j\omega C\right)}$$

$$V = \frac{I_m}{\sqrt{\left(\frac{1}{R}\right)^2 + \left(\omega C - \frac{1}{\omega L}\right)^2}} \angle \left(- \tan^{-1} R\left(\omega C - \frac{1}{\omega L}\right)\right)$$

*The power factor* of this circuit is obtained by

$$PF = \cos\left(\tan^{-1} R\left(\frac{1}{\omega L} - \omega C\right)\right)$$

The circuit has maximum voltage when the admittance reaches a minimum value. At this point, the $\omega C - \frac{1}{\omega L} = 0$ results in $\omega^2 CL = 1$ or $\omega = \frac{1}{\sqrt{LC}}$. Power factor at resonance reaches unity, $PF = 1$. The maximum voltage can be obtained as $V = RI_m$.

**Example 5.29** In the circuit of Fig. 5.26, the source and the circuit parameters are as follows:

$$V_m = 20, \omega = 10 \ \frac{rad}{s}, L = 1 \ H, C = \frac{1}{20} \ F.$$

**Find the current in the circuit $i(t)$, PF lead or lag? Find the voltage drop across all the circuit elements.**
*Solution.* The reactance of the inductor and capacitor at this frequency is

$$jX_L = j\omega L = j10 \times 1 = j10 \ \Omega$$

$$jX_C = \frac{1}{j\omega C} = \frac{1}{j \times 10 \times \frac{1}{20}} = -j2 \ \Omega$$

The impedance of the circuit from the source terminals is

$$Z = 10 + j10 - j2 = 10 + j8 \ \Omega$$

The current is

$$I = \frac{V}{Z} = \frac{20\angle 0}{10 + j8} = \frac{20\angle}{12.8\angle 38.65} = 1.56\angle - 38.65 \ A$$

The angle between the voltage and the current is

(continued)

**Example 5.29** (continued)

$$(\widehat{V,\,I}) = 0 - (-38.65) = +38.65\,°$$

$$\mathrm{PF} = \cos(\widehat{V,\,I}) = \cos 38.65 = 0.78\ \mathrm{Lag}$$

The circuit impedance shows a more inductive reactance as the imaginary part is a positive quantity. Therefore, the current is lagging the voltage, so the circuit is lagging.

Since the current in the circuit is known, the voltage drop across all elements can be found as

$$V_R = RI = 10 \times 1.56\angle - 38.65 = 15.6\angle - 38.65\,°\ \mathrm{V}$$

$$V_L = jX_L I = j10 \times 1.56\angle - 38.65 = 15.6\angle 90 - 38.65 = 15.6\angle 51.35\,°\ \mathrm{V}$$

$$V_C = jX_c I = -j2 \times 1.56\angle - 38.65 = 3.12\angle - 90 - 38.65 = 3.12\angle - 128.65\,°\ \mathrm{V}$$

As the voltages show, their vector summation must add to the source value. This means that the

$$V_m = V_R + V_L + V_C$$

That is

$$20\angle 0 = 15.6\angle - 38.65 + 15.6\angle 51.35 + 3.13\angle - 128.65$$

Converting all values to rectangular coordinates yields

$$20 + j0 = 15.6\cos 38.65 - j15.6\sin 38.65 + 15.6\cos 51.35 + j15.6\sin 51.35$$
$$+ 3.31\cos 128.65 - j3.13\sin 128.65 = 12.18 - j9.74 + 9.74 + j12.18$$
$$- 2.06 - j2.58$$

$$20 + j0 \cong 19.86 - j0.14$$

The results are very close, and the difference comes from the round-off errors in the current and the algebraic process.

**Example 5.30 In the circuit of Fig. 5.28, the values are given as follows:**

$$I_m = 100\angle 0° \, A, R = 10 \, \Omega, X_L = j10 \, \Omega, X_C = -j2\Omega$$

**Find the voltage of the node and the current in each branch.**
*Solution.* A KCL at the node yields

$$\sum I = 0 \rightarrow -I_m + I_R + I_L + I_C = 0$$

The source current is negative because of the direction of the current that is entering the node. Other currents are positive as they are considered to be leaving the node. The value of these currents can be calculated based on the node voltage $V$.

Therefore,

$$-100 + \frac{V}{R} + \frac{V}{X_L} + \frac{V}{X_C} = 0$$

$$-100 + \frac{V}{10} + \frac{V}{j10} + \frac{V}{-j2} = 0$$

Simplifying results in

$$-100 + 0.1V + (-j0.1)V + (j0.5)V = 0$$

$$V(0.1 + j0.4) = 100$$

$$V = \frac{100}{0.1 + j0.4} = \frac{100}{0.41\angle 75.96}$$

$$V = 243.9 \angle -75.96 \, V$$

The current of each branch is the voltage divided by the impedance of that branch, as follows:

$$I_R = \frac{V}{10} = \frac{243.9\angle -75.96}{10} = 24.39 \angle -75.96 \, A$$

$$I_L = \frac{V}{j10} = \frac{243.9\angle -75.96}{10\angle 90} = 24.39\angle(-75.96 - 90) = 24.39\angle -165.95 \, A$$

(continued)

**Example 5.30** (continued)

$$I_C = \frac{V}{-j2} = \frac{243.9\angle -75.96}{2\angle -90} = 121.95\angle(-75.96 + 90)$$

$$= 121.95\angle 14.04 \text{ A}$$

However, the phase shift of the voltage of 243.9 $\angle -75.96$ and the source current of 100 $\angle 0$ can be found as follows:

Try to make the voltage phase zero by shifting both waveforms by +75.97. Therefore, after the shift, the voltage of the waveforms becomes

$$V_{\text{shifted}} = 243.9\angle 0 \& I_{m_{\text{shifted}}} = 100\angle(0 + 75.96) = 100\angle 75.96$$

As the waveforms show, the current is ahead of the voltage, meaning that the system is lead.

**Example 5.31 An inductor and a capacitor are connected in parallel (Fig. 5.29). Determine a condition that the equivalent of these elements becomes a lead or a lag circuit.**
*Solution.* The parallel equivalent of these two elements becomes

$$Z_{\text{eq}} = jX_L \| -jX_C = \frac{jX_L \times -jX_C}{jX_L - jX_C} = -j\frac{X_L X_C}{X_L - X_C}$$

The impedance is lead or more capacitive if the $Im\{Z_{\text{eq}}\} < 0$. This yields

$$\text{Lead} : X_L - X_C > 0 \rightarrow X_C < X_L$$

The impedance is lag or more inductive if the $Im\{Z_{\text{eq}}\} > 0$. This yields

$$\text{Lag} : X_L - X_C < 0 \rightarrow X_L < X_C$$

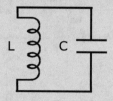

**Fig. 5.29** Circuit of Example 5.32

**Example 5.32 An inductor and a capacitor are connected in series (Fig. 5.30). Determine a condition that the equivalent of these elements becomes a lead or a lag circuit.**

*Solution.* Series equivalent of these two elements becomes

$$Z_{eq} = jX_L - jX_C = j(X_L - X_C)$$

The impedance is lead or more capacitive if the $Im\{Z_{eq}\} < 0$. This yields:

$$\text{Lead} : X_L - X_C < 0 \rightarrow X_C > X_L$$

The impedance is lag or more inductive if the $Im\{Z_{eq}\} > 0$. This yields

$$\text{Lag} : X_L - X_C > 0 \rightarrow X_L > X_C$$

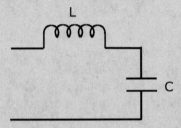

**Fig. 5.30**   Circuit of Example 5.33

**Example 5.33 Find the current drawn from the source in the given circuit in Fig. 5.31.**

*Solution.* The source is operating at an angular frequency of $\omega = 3000$ rad/s. Therefore, the ohmic value of the inductor becomes $jX_L = j\omega L = j3000 \times \frac{1}{3} = j1000$ $\Omega$, and the ohmic value of capacitor impedance becomes $-jX_c = -j\frac{1}{\omega C} = -j\frac{1}{3000 \times \frac{1}{6}e - 6} = -j2000$ $\Omega$.

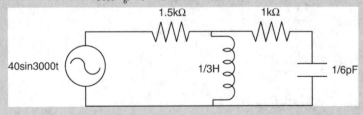

**Fig. 5.31**   Circuit of Example 5.34

(continued)

**Example 5.33** (continued)

The circuit is shown in Fig. 5.32.

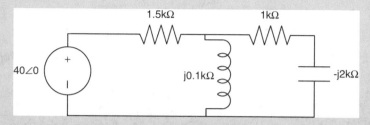

**Fig. 5.32** Circuit of Example 5.32 when impedance values are calculated at the frequency of the source, 3000 rad/s

The source's current is the voltage division over the impedance measured at the terminals of the source. This is

$$
I = \frac{40\angle 0}{1500 + j1000 \| (1000 - j2000)} = \frac{40\angle 0}{1500 + \dfrac{j1000 \times (1000 - j2000)}{j1000 + (1000 - j2000)}}
$$

$$
= \frac{40\angle 0}{1500 + \dfrac{j1000 \times (1000 - j2000)}{(1000 - j1000)}} = \frac{40\angle 0}{1500 + \dfrac{j1 \times (1000 - j2000)}{(1 - j1)}}
$$

$$
= \frac{40\angle 0}{1500 + \dfrac{(1 + j1)j1 \times (1000 - j2000)}{(1 + j1)(1 - j1)}} = \frac{40\angle 0}{2000 + j1500} = \frac{40\angle 0}{2500\angle 36.87}
$$

$$
= 16\angle(-36.87) \text{ mA}
$$

**Example 5.34 Consider the circuit shown in Fig. 5.33. Using node KCL, find the voltage of each node.**
*Solution.* The circuit has two nodes, ① and ②, at the voltage of $v_1(t)$ and $v_2(t)$, respectively, in the time domain and $V_1$ and $V_2$ as phasors. Node ① involves five elements, including the 1A source forcing the current to the node and four passive elements that drain the current out of the node. Node ② has five elements, including a $0.5\angle-90°$ leaving the node. Considering that all currents leaving the node through passive elements are positive and all currents entering the node are negative, KCLs for these nodes can be written as follows

(continued)

**Example 5.34** (continued)

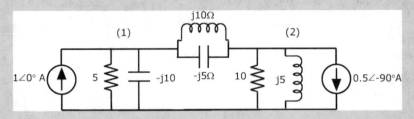

**Fig. 5.33** Circuit of Example 5.35

$$\text{KCL}①. \quad -1\angle0 + \frac{V_1}{5} + \frac{V_1}{-j10} + \frac{V_1 - V_2}{-j5} + \frac{V_1 - V_2}{-j10} = 0$$

$$\text{KCL}②. \quad 1.2\angle - 90 + \frac{V_2}{10} + \frac{V_2}{j5} + \frac{V_1 - V_2}{-j5} + \frac{V_1 - V_2}{-j10} = 0$$

Simplifying these equations results in

$$(0.2 + j0.2)V_1 - j0.1V_2 = 1\angle0$$

$$-j0.1V_1 + (0.1 - j0.1)V_2 = -0.5\angle -90 = -(-j)0.5 = j0.5$$

Solving for $V_1$ and $V_2$ results in

$$V_1 = (1 - j2) = 2.24\angle -63.4°\,\text{V}$$

$$V_2 = (-2 + j4) = 4.47\angle116.6°\,\text{V}$$

Considering the angular frequency of $\omega = 377$ rad/s, the time representation of these voltages becomes

$$v_1(t) = 2.24\sin(377t - 63.4°)\,\text{V}$$

$$v_1(t) = 4.47\sin(377t + 116.6°)\,\text{V}$$

**Example 5.35 In the circuit shown in Fig. 5.34, using nodal analysis, find**
$v_1(t)$ **and** $v_2(t)$, **knowing the angular frequency** $\omega = 1000$ **rad/s.**
*Solution.* The circuit has two nodes shown in Fig. 5.34 as ① and ②. Node ①
involves two current sources of 20 and 50 mA, where one leaves the node (+50
$\angle - 90$ mA) and enters the node (−20 mA). The (50 mA) source enters the
node ② and becomes a (−50 $\angle - 90$ mA). The current through each passive
element follows Ohm's law as $I = YV$.

$$\text{KCL①} : -20m + 50m\angle - 90 + V_1 j50m + (V_1 - V_2) \times -j25m = 0$$

$$\text{KCL②} : -50m\angle - 90 + (V_1 - V_2) \times -j25m + V_2 40m = 0$$

Solving for $V_1$ and $V_2$ results in

$$V_1 = 1.062\angle 23.3\,\text{V} \rightarrow v_1(t) = 1.062 \sin (1000t + 23.3)\,\text{V}$$
$$V_2 = 1.593\angle - 50.0\,\text{V} \rightarrow v_2(t) = 1.593 \sin (1000t - 50)\,\text{V}$$

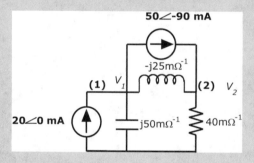

**Fig. 5.34** Circuit of Example 5.36

**Example 5.36 In the given circuit, find the current of each loop**
**(Fig. 5.35).**
*Solution.* The voltage source feeds the circuit at 10 V and an angular frequency
of 1000 rad/s. The 4 mH inductor shows the impedance of $j4\,\Omega$, $(j\omega L)$, and the
500 μF capacitance shows $-j2\,\Omega$, $(1/j\omega C)$. The current-dependent voltage
source has a dependency on the current in loop ①. Therefore, writing KVL is
treated as an independent voltage source with value as a function of $I_1$ as $2I_1$.
The circuit has two loops with current phasors $I_1$ and $I_2$ circulating clockwise
(the direction is optional). In each loop, the KVL suggests some voltage drops
starting from the * sign, as follows:

(continued)

**Example 5.36** (continued)

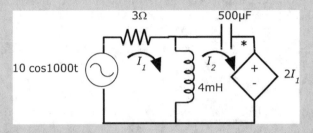

**Fig. 5.35** Circuit of Example 5.37

$$\text{KCL}\textcircled{1} : 3I_1 + j4(I_1 - I_2) - 10\angle 0 = 0$$
$$\text{KCL}\textcircled{2} : j4(I_1 - I_2) - j2I_2 + 2I_1 = 0$$

Equations can be simplified into

$$(3 + j4)I_1 - j4I_2 = 10$$
$$(2 - j4)I_1 + j2I_2 = 0$$

Solving for $I_1$ and $I_2$ results in

$$I_1 = 1.24\angle 29.7\,\text{A} \rightarrow i_1(t) = 1.24\cos(3000t + 29.7)\,\text{A}$$
$$I_2 = 2.77\angle 56.3\,\text{A} \rightarrow i_2(t) = 2.77\cos(3000t + 56.3)\,\text{A}$$

**Example 5.37 Find the voltage across the current source terminals (Fig. 5.36).**
*Solution.* The current source is connected to two parallel impedances of $-j5\,\Omega$ and $10 + j5\,\Omega$. The voltage at the terminals is $V = ZI$. Therefore,

$$V = 3\angle 30 \left(-j5 \| (10 + j5)\right)$$

(continued)

**Example 5.37** (continued)

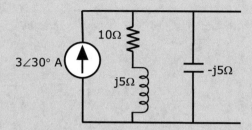

**Fig. 5.36** Circuit of Example 5.38

$$V = 3\angle 30 \left( \frac{-j5(10 + j5)}{-j5 + (10 + j5)} \right) = 3\angle 30 \left( \frac{-j(10 + j5)}{2} \right)$$

$$= 3\angle 30 \left( \frac{1\angle - 90 \times 11.183\angle 26.56}{2} \right)$$

$$V = 3 \times 11.183 \times \frac{1}{2} \angle (-30 - 90 + 26.56) = 16.773\angle - 33.4 \, \text{V}$$

## Resonance

Consider a circuit (RLC) connected to a variable/adjustable frequency source. In each half-cycle, the capacitors and inductors are charged and will be discharged in their opposite cycles. A phase delay in this charge and discharge is observed in the inductor and capacitor. As the frequency changes, the time delay in this charge and discharge process changes. This also demonstrates different amounts of impedance observed from inductors and capacitors. All inductive effects cancel all capacitive effects at a specific frequency, resulting in a purely resistive circuit.

At resonant, the voltage or current, depending on the circuit topology, reaches an extreme amount (maximum or minimum). For instance, it was observed that the current in a series RLC circuit reached the maximum at the resonance frequency $\omega = \frac{1}{\sqrt{LC}}$. It was also observed that the voltage of a parallel RLC circuit reached its maximum value at the resonant frequency $\omega = \frac{1}{\sqrt{LC}}$.

To obtain the resonant frequency, a frequency that makes the imaginary part of the impedance or imaginary part of the reactance zero must be identified.

**Example 5.38 Find the input impedance of the following circuit, and calculate the resonant frequency (Fig. 5.37).**

*Solution.* The circuit shows a $C$ and $R_2$ in series with $R_1$ and $L$. At the operating frequency of $\omega$, the input impedance from the terminal of this circuit is measured as

$$Z(j\omega) = R_1 + j\omega L + \left( \frac{1}{j\omega C} \middle\| R_2 \right)$$

In this case, each term is simplified individually. A common denominator does not help, so it can be avoided. Hence,

$$Z(j\omega) = R_1 + j\omega L + \left( \frac{\frac{1}{j\omega C} R_2}{\frac{1}{j\omega C} + R_2} \right)$$

$$Z(j\omega) = R_1 + j\omega L + \left( \frac{R_2}{1 + j\omega C R_2} \right)$$

$$Z(j\omega) = R_1 + j\omega L + \frac{1 - j\omega C R_2}{1 - j\omega C R_2} \left( \frac{R_2}{1 + j\omega C R_2} \right)$$

$$Z(j\omega) = R_1 + j\omega L + \left( \frac{R_2(1 - j\omega C R_2)}{1 + (\omega C R_2)^2} \right) = R_1 + j\omega L$$

$$+ \left( \frac{R_2}{1 + (\omega C R_2)^2} - \frac{j\omega C R_2{}^2}{1 + (\omega C R_2)^2} \right)$$

Now, the fractions can be separated into real and imaginary, and each part can be collected as follows:

$$Z(j\omega) = \left( R_1 + \frac{R_2}{1 + (\omega C R_2)^2} \right) + j\omega \left( L - \frac{C R_2^2}{1 + (\omega C R_2)^2} \right)$$

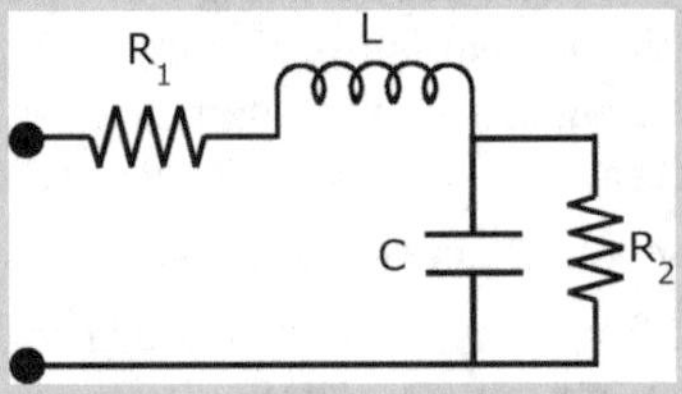

**Fig. 5.37** Circuit of Example 5.39

(continued)

**Example 5.38** (continued)

To operate at resonance, the imaginary part of the impedance must be zero. Therefore,

$$\text{Im}(Z(j\omega))\big|_{\omega=\omega_0} = 0$$

$$j\omega_0 \left( L - \frac{CR_2{}^2}{1 + (\omega_0 CR_2)^2} \right) = 0$$

Solving for $\omega_0$ and considering $\omega_0 \neq 0$ yields

$$L - \frac{CR_2{}^2}{1 + (\omega_0 CR_2)^2} = 0$$

$$L\left(1 + (\omega_0 CR_2)^2\right) = CR_2{}^2$$

$$\omega_0{}^2 = \frac{1}{(CR_2)^2} \left( \frac{CR_2{}^2}{L} - 1 \right) = \frac{1}{LC} - \frac{1}{C^2 R_2{}^2}$$

$$\omega_0 = \sqrt{\frac{1}{LC} - \frac{1}{C^2 R_2^2}} \quad \left(\frac{\text{rad}}{\text{s}}\right).$$

**Example 5.39 Find the input admittance of the circuit shown in Fig. 5.38 and the resonant frequency.**
*Solution.* Admittance of the circuit from the terminal at the frequency $\omega$ has $R_1$, $R_2 + j\omega L$, and $\frac{1}{j\omega C}$ parallel.

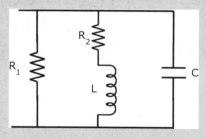

**Fig. 5.38** Circuit of Example 5.40

(continued)

**Example 5.39** (continued)

$$Y(j\omega) = \frac{1}{R_1} + \frac{1}{R_2 + j\omega L} + \frac{1}{\frac{1}{j\omega C}}$$

Simplifying each term yields

$$Y(j\omega) = \frac{1}{R_1} + \frac{R_2 - j\omega L}{R_2 - j\omega L} \frac{1}{R_2 + j\omega L} + j\omega C$$

$$Y(j\omega) = \frac{1}{R_1} + \left(\frac{R_2 - j\omega L}{R_2{}^2 + \omega^2 L^2}\right) + j\omega C$$

Splitting the fraction into real-imaginary sections yields

$$Y(j\omega) = \frac{1}{R_1} + \left(\frac{R_2}{R_2{}^2 + \omega^2 L^2} - \frac{j\omega L}{R_2{}^2 + \omega^2 L^2}\right) + j\omega C$$

$$Y(j\omega) = \frac{1}{R_1} + \frac{R_2}{R_2{}^2 + \omega^2 L^2} + j\omega \left(C - \frac{L}{R_2{}^2 + \omega^2 L^2}\right)$$

At the resonant frequency of $\omega = \omega_0$,

$$I_m(Y(j\omega))\big|_{\omega = \omega_0} = 0$$

Considering $\omega_0 \neq 0$, the resonant frequency becomes

$$C - \frac{L}{R_2{}^2 + \omega_0{}^2 L^2} = 0$$

Solving for $\omega_0$ yields,

$$\omega_0 = \sqrt{\frac{1}{LC} - \frac{R_2^2}{L^2}} \left(\frac{\text{rad}}{\text{s}}\right)$$

## Power in Sinusoidal Steady-State Operation

The term power is defined as the product of voltage and current. Voltage and current can be instantaneous, RMS, or DC. Therefore, instantaneous power (measured in volt-amps (VA)) or apparent power and DC power (measured in watts (W)) can be measured as

$$S = \text{instantaneous power} = \text{apparent power} = v(t).i(t)$$

## Apparent Power

Consider an RLC circuit where the current has a phase shift $\varphi$ with respect to the voltage, as shown in Fig. 5.39. The current can be projected to in-phase and vertical components with respect to the voltage. The in-phase current component with respect to voltage has the value of $I \cos \varphi$, and the vertical components have the value of $I \sin \varphi$. Therefore, it can be represented as $\left(V, \widehat{I \cos \phi}\right) = 0$ and $\left(V, \widehat{I \sin \phi}\right) = 90$. This angle is +90 for a lag system and −90 for a lead system (Fig. 5.39).

$$S = v(t)i(t) \ (\text{VA})$$

$$S = V_p \sin (\omega t) I_p \sin (\omega t - \varphi)$$

$V_p$ and $I_p$ are the peak values of voltage and current, respectively.

$$S = \frac{1}{2} V_p I_p (\cos \varphi - \cos (2\omega t - \varphi))$$

**Fig. 5.39** Projection of current with a phase shift on its reference

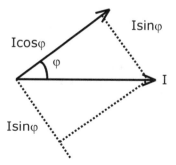

$$S = \frac{1}{2} V_p I_p \cos \varphi - \frac{1}{2} V_p I_p \cos (2\omega t - \varphi)$$

Utilizing RMS values as $V = \frac{V_p}{\sqrt{2}}$ and $I = \frac{I_p}{\sqrt{2}}$, the apparent power becomes

$$S = VI \cos \varphi - VI \cos (2\omega t - \varphi)$$

The apparent power has two components:

- $VI \cos \varphi$ is a constant value power and contributes to what is known as *active power*.
- $VI \cos (2\omega t - \varphi)$ is a pulsating power at a frequency twice the voltage and current frequency. This power has an average value of zero over one cycle; therefore, it does not contribute to the actual work done by the circuit.

*As a reminder*, the voltage waveform is considered a reference, and the current might be leading, in-phase, or lagging the voltage.

- *The phase of a lead current is positive.*
- *The phase of an in-phase current is zero.*
- *The phase of a lag current is negative.*

In phasors' presentation, the apparent power is calculated by

$$S = VI^* = \frac{1}{2} V_p I_p^* \ (\text{VA})$$

Considering real and imaginary parts of the current, apparent power can be converted as follows:

$$S = VI( \cos \varphi + j \sin \varphi)$$

$$S = VI \cos \varphi + jVI \sin \varphi = P + jQ$$

Hence, active power $P$(W) and reactive power $Q$(VAR) can be found as follows:

$$P = VI \cos \varphi, \text{or,} P = \frac{1}{2} V_p I_p \cos \varphi \ (\text{W})$$

$$Q = VI \sin \varphi, \text{or,} Q = \frac{1}{2} V_p I_p \sin \varphi \ (\text{VAR})$$

Where $V_p$ and $I_p$ are the peak values.

Apparent power is a vector summation of active and reactive power, as shown in Fig. 5.40.

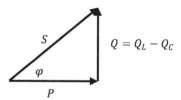

**Fig. 5.40** Power triangle. The balance of the active, reactive, and apparent power forms a triangle known as a power triangle. Considering that $Q_L$ is the consumption of reactive power in an inductor, $Q_C$ is the generation of reactive power in a capacitor, and $Q = Q_L - Q_C$ is the balance of reactive power; then the $Q$ value might be more consumed, i.e., $Q > 0$; more generation, i.e., $Q < 0$; or balanced, i.e., $Q = 0$

The amplitude of apparent power is $|S| = \sqrt{P^2 + Q^2}$, and the phase is obtained as

$$\frac{Q}{P} = \frac{VI \ \sin \varphi}{VI \ \cos \varphi} = \tan \varphi$$

$$\varphi = \tan^{-1} \frac{Q}{P}$$

**Example 5.40** In a circuit, the voltage and current phasors are $V = 10 \angle 20 \degree$ V, $I = 15 \angle 65 \degree$ A. Find the apparent power, active power, and reactive power.

*Solution.* The apparent power can be calculated as

$$S = VI^* = 10 \angle 20 \ (15 \angle 65)^*$$

$$S = 10 \times 15 \angle 20 - 65$$

$$S = 150 \angle - 45 \text{ VA}$$

Converting to rectangular coordinates,

$$S = 150 \cos (-45) + j150 \sin (-45)$$

$$S = 106.05 - j106.05 \text{ VA}$$

However, the powers are also presented as

$$S = P + jQ$$

(continued)

**Example 5.40** (continued)

Therefore,

$$P = 106.05 \text{ W}, Q = -106.05 \text{ VAR}$$

The negative sign in power means a generation status, and a positive sign means the power is consumed.

**Example 5.41 A circuit consumes apparent power as $S = 45 + j10$ VA. Find the active, reactive phase shift in the circuit, PF, lead, or lag.**
*Solution.*

$$S = P + jQ = 45 + j10$$

This means

$$P = 45 \text{ W}, Q = 10 \text{ VAR}$$

The phase angle can be obtained from

$$\varphi = \tan^{-1} \frac{Q}{P}$$

Therefore,

$$\varphi = \tan^{-1} \frac{10}{45} = 12.52^\circ$$

The phase is positive, meaning that the reactive power is being consumed. As inductors consume the reactive power, the circuit is Lag.
The power factor can be obtained from this phase shift as follows:

$$PF = \cos \varphi = \cos 12.52 = 0.976$$

## Active Power

The mean power consumed in a circuit is obtained by the product of voltage and the in-phase component of the current. Measured in watts (W), the mean power is also called active power and is shown as

$$P = V(I \cos \varphi) = (VI) \cos \varphi = S \cos \varphi \ \ (\text{W})$$

Considering an impedance in resistive and reactive parts, the power consumed in the resistive part is active. This power is converted into torque in electric machines or generates heat in space heaters. The active power does the work in electromechanical systems. For this reason, $\cos\varphi$ is called the power factor, meaning the portion of the apparent power that does the active work.

Consider an impedance of an RL circuit shown in Fig. 5.41.

The angle between the impedance and the resistance determines the current phase shift as follows:

$$\cos \varphi = \frac{R}{Z}$$

Therefore, the active power can be calculated as

$$P = VI \cos \varphi = VI \frac{R}{Z} = \frac{V}{Z} IR = IIR = RI^2 \ \ (\text{W})$$

## Reactive Power

Perpendicular to the in-phase component of the current, the power developed into the imaginary part of the impedance is known as reactive power. Measured in volt-ampere reactive (VAR), the value for this power can be obtained as

$$Q = V(I \sin \varphi) = (VI) \sin \varphi = S \sin \varphi \ \ (\text{VAR})$$

Considering the impedance shown in Fig. 5.41, the value of $\sin\varphi$ can be obtained as follows:

$$\sin \varphi = \frac{X}{Z}$$

Therefore, the reactive power can be calculated as

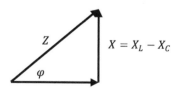

**Fig. 5.41** The balance of capacitive and inductive reactance and resistance values determine the impedance. The balance of inductive and capacitive reactance is $X = X_L - X_C$

$$Q = VI \sin\varphi = VI\frac{X}{Z} = \frac{V}{Z}IX = IIX = XI^2 \quad (\mathbf{VAR})$$

As the equation illustrates, only the reactance generates or consumes the reactive power in an impedance consisting of resistance and reactance.

**Note 5.2** Power consumption has a positive value, and power generation has a negative value.

**Note 5.3** Apparent power can also be obtained from the product of $Z(j\omega)$ and the current $I$ as follows:

$$S = Z(j\omega)I^2 = (R + jX)I^2 = RI^2 + jXI^2 = P + jQ$$

**Note 5.4** As the sign of each element, $RI^2$ and $XI^2$ are positive; the circuit is a resistive-inductive circuit that consumes active power (in resistor) and reactive power (an inductor).

## Reactive Power of an RC Circuit

Consider an RC series circuit that has resistance $R$ and reactance $-X$; therefore, the impedance is obtained as

$$Z(j\omega) = R - jX$$

The apparent power of this circuit can be obtained as follows:

$$S = Z(j\omega)I^2 = (R - jX)I^2 = RI^2 - jXI^2 = P - jQ$$

It can be observed that the resistive-capacitive circuits consume active power $+RI^2$ and generate reactive power $-XI^2$ in the capacitor.

**Note 5.5** Reactive power is generated in capacitors and is consumed in inductors.

**Note 5.6** Ideal reactive elements $L$ and $C$ do not consume or generate active power.

**Note 5.7** Resistors do not consume or generate reactive power.

**Example 5.42 A 100 kW load has a power factor of _PF_ = 0.7. Find the active power, apparent power, and the phase shift between the voltage and current of this load.**

*Solution.* The apparent power can be obtained as follows:

$$P = S \cos \varphi \rightarrow S = \frac{P}{\cos \varphi} = \frac{100k}{0.7} = 142.85 \, \text{kVA}$$

Since PF = 0.7, the phase shift between the voltage and the current can be obtained as

$$\cos \varphi = 0.7, \varphi = 45.57^{\circ}$$

The reactive power can be obtained as follows:

$$Q = P \tan \varphi = 100k \tan 45.57 = 102.02 \, \text{kVAR}$$

**Example 5.43 An electric motor converts 80% of its 100 kVA power to a useful torque and heat. What is the power factor of the electric motor? How much power does the motor need to build up the magnetic fields, charge the windings, and build a magnetic field in the air gap?**

*Solution.* An electric motor generates torque directly utilizing the active power and builds magnetic and airgap fields by consuming the reactive power. Therefore, the power factor is the portion of the power that generates torque (80%) of the total apparent power. Meaning

$$PF = 0.8$$

Therefore, the reactive power is the rest of the apparent power in the power triangle. That means

$$S^2 = P^2 + Q^2$$

So

$$100k^2 = 80k^2 + Q^2$$

$$Q = 60 \, \text{kVAR}$$

**Example 5.44 In Fig. 5.42, find $I$, $P$, $Q$, $S$, PF. Find a suitable amount of capacitor reactance to improve the PF to 0.98.**
*Solution.*

$$I = \frac{V}{Z} = \frac{110\angle 0}{100 + j50} = \frac{110\angle 0}{111.8\angle 26.57} = 0.98\angle - 26.57°\,\text{A}$$

$$S = VI^* = 110\angle 0(0.98\angle - 26.57)^* = 110 \times 0.98\angle + 26.57$$
$$= 107.8\angle 26.57\,\text{VA}$$

In the rectangular coordinates, the power is

$$S = 107.8\cos 26.57 + j\,107.8\sin 26.57 = 96.42 + j48.21$$

$$S = P + jQ$$

Therefore,

$$P = 96.42\,\text{W},\ Q = 48.22\,\text{VAR}$$

The power factor can be obtained as follows:

$$\text{PF} = \frac{P}{S} = \frac{96.42}{107.8} = 0.9$$

To correct the power factor to 0.98, the capacitor's reactance needs to be calculated. Considering the impedance of the circuit with a capacitor $-jX_C$. The corrected power factor can be found as

$$\text{PF} = \frac{R}{\sqrt{R^2 + (X_L - X_C)^2}}$$

$$0.98 = \frac{100}{\sqrt{100^2 + (50 - X_C)^2}} \rightarrow X_C = 29.69\,\Omega$$

**Fig. 5.42** Circuit of Example 5.45

(continued)

**Example 5.44** (continued)

At frequency $f = 60$ Hz, the amount of capacitance can be calculated in $F$ as

$$X_C = \frac{1}{C\omega} = \frac{1}{2\pi f C} \rightarrow C = \frac{1}{X_C 2\pi f} = \frac{1}{29.69 \times 2\pi \times 60} = 89 \ \mu F$$

## Nonideal Inductors

Ideal inductors have no internal resistance, and therefore, they only demonstrate a reactance. However, the resistance of wires used to make the inductors sometimes cannot be ignored. Therefore, the nonideal inductors have impedance similar to an RL circuit. To demonstrate the internal resistance of a nonideal inductor, a quality factor $Q_f$ is defined (Fig. 5.43).

## Quality Factor ($Q_f$)

The reciprocal of the power factor is known as the quality factor $Q_f = \frac{1}{PF}$. This value shows the merit of a coil. The power factor of an ideal inductor was calculated to be PF $= 0$, as the resistive part was zero. Therefore, the quality factor of an ideal inductor is $Q_f = \frac{1}{0} = \infty$. Actual (nonideal) inductors have internal resistance, making their power factors non-zero PF $= \frac{R}{Z} = \frac{R}{\sqrt{R^2 + X_L^2}}$ (Fig. 5.44).

Considering the circuit of Fig. 5.44, the quality factor is obtained as

$$Q_f = \frac{1}{PF} = \frac{Z}{R} = \frac{\sqrt{R^2 + X_L^2}}{R}$$

**Fig. 5.43** A nonideal inductor is shown by its internal resistance and inductance

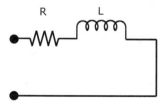

**Fig. 5.44** The small amount of internal resistance $R$ creates a phase angle $\varphi$. Voltage drops of $RI$ and $XI$ are perpendicular

For inductors that have large reactance, i.e., $L\omega \gg R$, the quality factor can be obtained as

$$Q_f = \frac{\sqrt{R^2 + X_L{}^2}}{R} \simeq \frac{\sqrt{X_L{}^2}}{R} = \frac{L\omega}{R}$$

**Example 5.45** In an **RL** series circuit, the current and voltage are measured to be $i(t) = 7.5 \sin(377t + 120)$ A and $v(t) = 150 \sin(377t + 150)$ V. Find:

- *The impedance of the circuit*
- *The value of the resistance*
- *The value of reactance and the inductance in (H)*
- *The amount of active power*
- *The amount of reactive power*
- *The power factor*

*Solution.* According to Ohm's law, the voltage and current phasor are related as follows:

$$V = ZI$$

$$Z = \frac{150\angle 150}{7.5\angle 120} = 20\angle 30$$

As the impedance has a positive phase, the circuit is resistive-inductive. The real part of the impedance shows the resistance, and the imaginary part of the impedance shows the reactance as follows:

$$Z = 20\angle 30 = 20\cos 30 + j20\sin 30 = 17.32 + j10 \ \Omega$$

$$R = 17.32 \ \Omega$$

(continued)

**Example 5.45** (continued)

$$X_L = 10 \ \Omega$$

$$X_L = L\omega = 10$$

$$L = \frac{10}{377} = 26.5 \ \text{mH}$$

Apparent power utilizing the peak values of voltage and current can be calculated as

$$S = \frac{1}{2}VI^* = \frac{1}{2}(150\angle 150)(7.5\angle 120)^* = \frac{1}{2}(150\angle 150)7.5\angle - 120$$

$$= 562.5\angle 30\,\text{VA}$$

Converting to rectangle coordinates yields

$$S = 562.5\cos 30 + j562.5\sin 30 = 487.1 + j281.25$$

$$P = 487.1\,\text{W}, Q = 281.25\,\text{VAR}$$

The power factor is

$$\text{PF} = \cos 30 = 0.86\,\text{lag}.$$

The quality factor of the inductor becomes

$$Q_f = \frac{1}{\text{PF}} = \frac{1}{0.86} = 1.15$$

**Example 5.46** The voltage drop of a coil when a DC of 9 A passes through is measured to be 4.5 V. The same coil when an AC sinusoidal of 9 A at 25 Hz is passing through drops 24 V. Find the impedance, power, PF, and $Q_f$ at a voltage of 150 V, 60 Hz.
*Solution.* The DC and voltage drop result in resistance value as

$$R = \frac{V}{I} = \frac{4.5}{9} = 0.5\,\Omega$$

At 25 Hz, the impedance value is obtained as

(continued)

**Example 5.46** (continued)

$$Z = \frac{V}{I} = \frac{24}{9} = 2.66\,\Omega$$

The reactance value can be calculated as

$$Z = \sqrt{R^2 + X_L^2}$$

$$X_L = \sqrt{Z^2 - R^2} = \sqrt{2.66^2 - 0.5^2} = 2.61\,\Omega$$

$$L = \frac{X_L}{\omega} = \frac{2.61}{2\pi 25} = 17.4\,\text{mH}$$

At 60 Hz frequency,

$$X_L = 17.4m \times 2\pi \times 60 = 6.56\,\Omega$$

$$Z = \sqrt{R^2 + X_L^2} = \sqrt{0.5^2 + 6.56^2} = 6.57\,\Omega$$

$$I = \frac{V}{Z} = \frac{150}{6.57} = 22.8\,\text{A}$$

The impedance phase is $\varphi = \tan\frac{6.56}{0.5} = 85.64°$. Therefore, the current phase angle is $-85.64°$.

Apparent power utilizing RMS parameters becomes

$$S = VI^* = (150\angle 0)(22.8\angle - 85.64) * = (150\angle 0)\,22.8\angle + 85.64$$
$$= 3420\angle 85.64\,\text{VA}$$

$$S = 3420\cos 85.64 + j3420\sin 85.64 = 256 + j3410$$

$$P = 256\,\text{W}, Q = 3410\,\text{VAR}$$

$$\text{PF} = \cos 85.64 = 0.076$$

$$Q_f = \frac{1}{0.076} = 13.15$$

**Example 5.47 A black box series circuit has the following voltage and current measurements** $v(t) = 200\sqrt{2} \sin{(377t + 10)}$ **V** $i(t) = 10\sqrt{2} \cos{(377t - 35)}$ **A. Find the circuit elements and their values.**

*Solution.* The circuit current is given in cos form, which needs to be converted to a sin function as follows:

$$i(t) = 10\sqrt{2}\cos{(377t - 35)} = 10\sqrt{2}\sin{(377t - 35 + 90)}$$
$$= 10\sqrt{2}\sin{(377t + 55)}$$

The impedance of the circuit is obtained as follows:

$$Z = \frac{V_P}{I_P} = \frac{200\sqrt{2}\angle 10}{10\sqrt{2}\angle 55} = 20\angle - 45 \ \Omega$$

The impedance angle is a negative value, so the circuit is an RC circuit. The circuit is more capacitive than inductive. The resistance becomes $R = 20\cos{(-45)} = 14.14 \ \Omega$. The amount of reactance is $-X_C = 20\sin{(-45)} = -14.14 \ \Omega$.

The operating frequency $\omega = 377$ rad/s results in

$$X_C = \frac{1}{C\omega} \rightarrow C = \frac{1}{\omega X_C} = \frac{1}{377 \times 14.14} = 1.87e - 4\,\mathrm{F} = 187\,\mu\mathrm{F}.$$

**Example 5.48 A circuit when the current of** $4 - j5$ **A is passing through drops a voltage of** $180 + j90$ **V. Find the impedance of the circuit, power consumption, or generation of each element, and determine whether the system is lead or lag and the power factor.**

*Solution.* The circuit impedance can be obtained by

$$Z = \frac{V}{I} = \frac{180 + j90}{4 - j5} = \frac{201.24\angle 26.56}{6.40\angle - 51.34} = 31.444\angle 77.9\,\Omega$$

Power is calculated by

$$S = VI^* = 201.24\angle 26.56 \times 6.40\angle + 51.34 = 1287.9\angle 77.9\,\mathrm{VA}$$

(continued)

**Example 5.48** (continued)

$$S = 1287.9 \cos 77.9 + j1287.9 \sin 77.9 \, \text{VA}$$

$$S = 268.08 + j1259.28 \, \text{VA}$$

**Example 5.49** An electric motor takes a $2 \angle -60°$ A current at the voltage of $200 \angle 0°$ at $f = 50$ Hz. Find $R$, $X$, PF, $L$ ($H$), $S$, $P$, $Q$ of the load. How much does the apparent, active, and reactive power change if the voltage and frequency are cut in half? Note that the ratio of the $\frac{V}{f}$ remains the same to generate the same amount of torque at the load.

*Solution.*

$$Z = \frac{V}{I} = \frac{200 \angle 0}{2 \angle -60} = \frac{200}{2} \angle 60 = 100 \angle 60 \; \Omega$$

$$Z = 100 \cos 60 + j100 \sin 60 = 50 + j86.6 \equiv R + jX$$

Therefore,

$$R = 50 \; \Omega, X = 86.6 \; \Omega$$

$$\text{PF} = \cos 60 = 0.5$$

$$@f = 50 \; \text{Hz}, X = L\omega = 86.6 \rightarrow L = \frac{86.6}{2\pi f} = \frac{86.6}{2\pi \times 50} = 0.275 \; \text{H}$$

$$S = VI^* = 200 \angle 0 \, (2 \angle -60)^* = 400 \angle +60 \, \text{VA}$$

$$S = 400 \cos 60 + j400 \sin 60 = 200 + j346.4 \equiv P + jQ$$

Therefore,

$$@f = 50 \; \text{Hz}, P = 200 \; \text{W}, Q = 346.4 \; \text{VAR}$$

At frequency $f = 25$ Hz and voltage $V = 100$ V, the amount of power will depend on the amount of current passing the circuit and the resistance and reactance of the load.

(continued)

**Example 5.49** (continued)

$$@f = 25 \text{ Hz}, V = 100$$

$$R = 50 \ \Omega, X = \frac{86.6}{2} = 43.3 \ \Omega$$

The current is

$$I = \frac{100}{50 + j43.3} = \frac{100}{66.14 \angle 40.89} = 1.5 \angle - 40.89 \text{ A}$$

$$S = VI^* = 100(1.5 \angle - 40.89)^* = 150 \angle 40.89 \text{ VA}$$

The active and reactive powers are

$$P = 150 \cos 40.89 = 112.5 \text{ W}$$

$$Q = 150 \sin 40.89 = 99 \text{ VAR}$$

It can be observed that the motor's load still has the same amount of torque, but as the frequency and voltage dropped, the amount of active power and reactive power drawn from the source dropped as well. This is utilized when a load needs to rotate at a slower speed, but the motor has to maintain the torque. This saves energy in the form of variable frequency drives.

## Nonideal Capacitors

The dielectric material used in capacitors is ideally loss-free. It means that the charge applied to terminals of capacitors (i.e., plates) stays on the plates for an infinite time. There is no internal current leak or internal discharge. However, that might not be true for existing dielectric materials, with minimal current passing through the material and hence discharge.

The model for this internal discharge is a resistor that can be added in series or parallel to an ideal capacitor forming an RC circuit. The current in this circuit deviates from the ideal 90° and generates an in-phase and a perpendicular component (Fig. 5.45).

## Model as RC Series

Consider the internal resistance $R_{se}$ in series to an ideal capacitor. The current angle with respect to the voltage as a reference is $\varphi$ as opposed to 90°. Consider the angle deviation from 90° as $\beta = 90 - \varphi$.

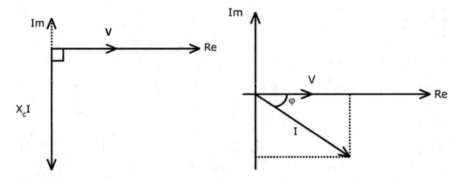

**Fig. 5.45** Nonideal capacitors. The internal resistance creates a phase angle $\varphi$ between the voltage across the terminals of the capacitor and the current flown in the capacitor

**Fig. 5.46** The internal resistance of a nonideal capacitor when presented in series

$$\tan\beta = \frac{I\ R_{se}}{I\ X_C} = \frac{R_{se}}{X_C} = R_{se}C\omega$$

$$R_{se} = X_C\tan\beta$$

Power loss in this model can be calculated as

$$P = R_{se}I^2 = X_CI^2\tan\beta\ \ (\mathrm{W})$$

Power factor can be obtained as follows:

$$\mathrm{PF} = \sin\beta = \cos\varphi$$

**Proof** From Fig. 5.46, it can be written as

$$\cos\varphi = \frac{R_{se}}{\sqrt{R_{se}^2 + X_C^2}} = \frac{X_C\tan\beta}{\sqrt{(X_C\tan\beta)^2 + X_C^2}} = \frac{\tan\beta}{\sqrt{\tan^2\beta + 1}}$$

$$= \frac{\frac{\sin\beta}{\cos\beta}}{\sqrt{\left(\frac{\sin\beta}{\cos\beta}\right)^2 + 1}} = \frac{\frac{\sin\beta}{\cos\beta}}{\sqrt{\frac{\sin^2\beta + \cos^2\beta}{\cos^2\beta}}} = \sin\beta$$

$$\tan \beta = \frac{\sin \beta}{\cos \beta} = \frac{\cos \varphi}{\sqrt{1 - \cos^2 \varphi}} = \frac{PF}{\sqrt{1 - PF^2}}$$

$$R_{se} = \frac{PF}{C\omega}$$

**Example 5.50** A 10 µF capacitor has a power factor of PF $= 0.1$ at $f = 60$ Hz. Find the equivalent model of this capacitor in series form (Fig. 5.47).

*Solution.*

$$R_{se} = \frac{PF}{C\omega} = \frac{0.1}{10\mu \times 2\pi f} = \frac{0.1}{10e - 6 \times 2\pi \times 60} = 265.2 \, \Omega$$

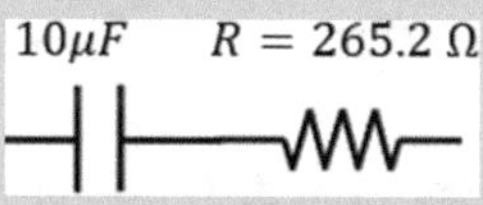

$$10\mu F \qquad R = 265.2 \, \Omega$$

**Fig. 5.47** Equivalent circuit of capacitor in Example 5.51

## Model as RC Parallel

Consider the internal resistance $R_{sh}$ in parallel to an ideal capacitor. Figure 5.48 shows the current balance in each branch and the real and imaginary values of the current phasor.

$$\tan \beta = \frac{I_1}{I_2} = \frac{V/R_{sh}}{V/X_C} = \frac{X_C}{R_{sh}} = \frac{1}{R_{sh}C\omega}$$

$$\tan \beta = \frac{1}{R_{sh}C\omega}$$

$$R_{sh} = \frac{1}{C\omega \tan \beta} = \frac{X_c}{\tan \beta}$$

Power loss in this model can be calculated as

**Fig. 5.48** The internal resistance of a nonideal capacitor when presented in parallel

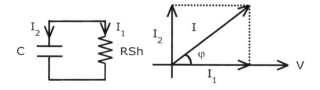

$$P = \frac{V^2}{R_{sh}} = \frac{V^2}{X_C} \tan\beta \ \ (\text{W})$$

$$R_{sh} = \frac{1}{C\omega PF}$$

**Example 5.51 A 10 μF capacitor has a power factor of PF = 0.1 at $f = 60$ Hz. Find the equivalent model of this capacitor in parallel form (Fig. 5.49).**

$$R_{sh} = \frac{1}{C\omega PF} = \frac{1}{10e - 6 \times 2\pi \times 60 \times 0.1} = 2652.5 \ \Omega$$

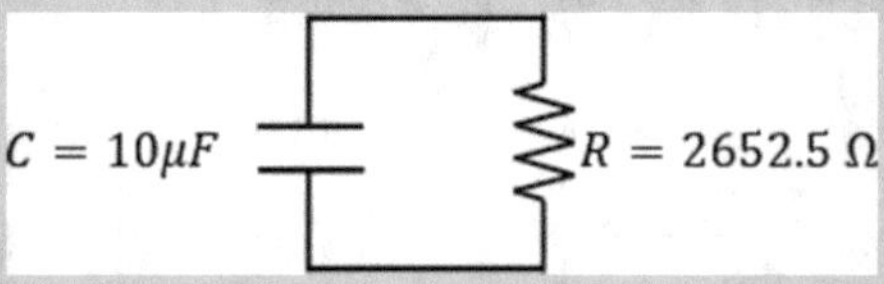

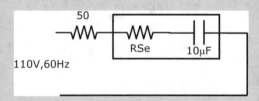

$C = 10\mu F$  $R = 2652.5 \ \Omega$

**Fig. 5.49** Equivalent circuit of capacitor in Example 5.52

**Example 5.52 A nonideal capacitor with a capacitance of 10 μF and phase deviation of $\beta = 10°$ is connected in series with a 50 Ω resistance shown in Fig. 5.50. The RC circuit is connected to a 120 V, 60 Hz source.**
Find:

1. The increase in resistance due to the insertion of the capacitor
2. The power loss of the capacitor
3. Power factor PF

50
RSe    10μF
110V,60Hz

**Fig. 5.50** Circuit of Example 5.53

(continued)

**Example 5.52** (continued)

*Solution.* The impedance of the capacitor at 60 Hz is

$$X_C = \frac{1}{C\omega} = \frac{1}{10e - 6 \times 2\pi 60} = 265.25\,\Omega$$
$$\varphi = 90 - \beta = 90 - 10 = 80°$$

Considering a series RC circuit,

$$R_{se} = X_C \tan\beta = 265.25 \tan 10 = 46.77\,\Omega$$

Therefore, the circuit has 50 Ω in series with an internal resistance of 46.77 Ω, which becomes a total of 96.77 Ω. Total impedance becomes

$$Z(j\omega) = 96.77 - j265.26\,\Omega$$

The impedance in polar form shows the amplitude (to calculate current) and the angle (to calculate the PF).

$$Z(j\omega) = 282.36\angle - 69.95\,\Omega$$

To find the current (needed for loss calculations),

$$I = \frac{V}{|Z|} = \frac{110}{282.36} = 0.389\,\text{A}$$

The power loss of the capacitor is obtained by

$$P = R_{se}I^2 = 46.77\,(0.389)^2 = 7.098\,(\text{W})$$

The power factor is obtained by

$$PF = \cos\varphi = \cos 80 = 0.173\,\text{lead}$$

## Dielectric Heating

The formation of nonideal capacitors and loss in dielectric can be used in industry for heating. Voltage and frequency of operation can be adjusted to generate the amount of heat needed. The power loss, required voltage, and frequency using the parallel model is obtained as follows:

$$P = \frac{V^2}{R_{sh}}$$

$$V = \sqrt{PR_{sh}}$$

Considering $R_{sh} = \frac{1}{C\omega \ \tan \beta}$, the power loss is a function of $V^2$ and frequency $\omega$.

$$P = V^2 C\omega \tan \beta$$

Adjusting the frequency and voltage from $V_1$, $\omega_1$ to $V_2$, $\omega_2$ for similar material to generate a desired amount of heat can be identified as follows:

$$P \propto V_1{}^2\omega_1 = V_2{}^2\omega_2$$

**Example 5.53 Design a dielectric heating oven that can generate 200 W of power in a block material with $d = $ 1-inch thickness, area of $A = 1$ square foot, with relative permittivity of $\epsilon = 5$, at PF $= 0.05$ and $f = 30$ MHz. If the voltage is limited to 200 V, find the required frequency to generate the same power.**
*Solution*

The block of material forms a capacitor with a capacitance of $C = \epsilon_0 \epsilon \frac{A}{d}$:

$$d = 1 \, \text{in} = 2.54 \, \text{cm} = 2.54e - 2 \, \text{m}$$

$$A = 30.48e - 2 \times 30.48e - 2 \, \text{m}^2$$

The capacitance of the system is

$$C = \epsilon_0 \epsilon \frac{A}{d} = 5 \times 8.85e - 12 \times \frac{0.0929}{2.54e - 2} = 161.84 \, \text{pF} = 161.84e - 12 \, \text{F}$$

$$PF = \cos \varphi = \sin \beta = 0.05$$

(continued)

**Example 5.53** (continued)

$$\tan \beta \cong 0.05$$

$$R_{sh} = \frac{1}{C\omega \tan \beta} = \frac{1}{161.84e - 12 \times 2\pi \times 30e6 \times 0.05} = 655.60\,\Omega$$

$$V = \sqrt{PR_{sh}} = \sqrt{200 \times 655.60} = 362.1\,V$$

When the voltage is limited to 200 V, the frequency and voltage balance are as follows:

$$P \propto V_1{}^2\omega_1 = V_2{}^2\omega_2$$

$$362.1^2 \times 2\pi 30\,MHz = 200^2 \times 2\pi f_2$$

The adjusted frequency becomes

$$f_2 = 98.337\,MHz$$

## Thevenin Equivalent Circuits in Sinusoidal Steady State

Similar to the Thevenin equivalent defined in RLC circuits, the impedance value of elements is considered when the circuit operates at the steady-state sinusoidal. Therefore, the circuit can be presented as a voltage source in series to an impedance (Fig. 5.51).

The value of the voltage source and the impedance is obtained as follows. When the load is disconnected from the circuit, the voltage measured at the terminals becomes the Thevenin voltage. To obtain the Thevenin impedance, the value of independent sources must become zero. It means an independent voltage source becomes a short circuit, and an independent current source becomes an open circuit. Then, Thevenin impedance can be measured at the terminals.

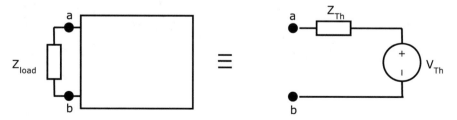

**Fig. 5.51** Thevenin equivalent circuit from ports $a$ and $b$ across the rest of the circuit is known as a load

**Example 5.54 Find the Thevenin equivalent of the circuit shown in Fig. 5.52.**

*Solution.* The load must be disconnected to find the Thevenin impedance, and all the independent sources must be turned off. This leads to a short circuit equivalent for the voltage source. The Thevenin impedance is

$$Z_{th} = Z_3 + (Z_1 \| Z_2)$$

or

$$Z_{th} = Z_3 + \frac{Z_1 Z_2}{Z_1 + Z_2}$$

The Thevenin voltage is obtained when the load is disconnected across terminals a and b. Since the load is disconnected, the current through the impedance $Z_3$ is zero. Therefore, the voltage measured at the terminal is the voltage across the impedance $Z_2$.

$$V_{th} = V_{Z_2}$$

A voltage division between $Z_1$, $Z_2$ results in the Thevenin voltage.

$$V_{th} = V_{Z_2} = \frac{Z_2}{Z_1 + Z_2} V_s$$

Therefore, the Thevenin equivalent can be shown as the following circuit (Fig. 5.53).

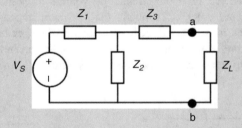

**Fig. 5.52** Circuit of Example 5.55

(continued)

**Example 5.54** (continued)

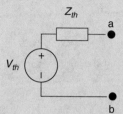

**Fig. 5.53** Thevenin equivalent of circuit in Example 5.55

**Example 5.55 Find the Thevenin equivalent of the circuit shown in Fig. 5.54.**

*Solution.* The Thevenin impedance is (Fig. 5.55)

$$Z_{th} = -j10 + (10\|j15)$$

or

$$Z_{th} = -j10 + \frac{10 \times j15}{10 + j15} = -j10 + \frac{j150\,(10 - j15)}{10^2 + 15^2}$$

$$Z_{th} = -j10 + \frac{j150(10 - j15)}{325} = -j10 + j4.61 + 6.92 = 6.92 - j5.39\ \Omega$$

$$V_{th} = V_{j15\Omega} = \frac{j15}{10 + j15} \times 110\angle30° = \frac{15\angle90 \times 110\angle30}{101.12\angle56.31} = 16.31\angle63.69°\,V$$

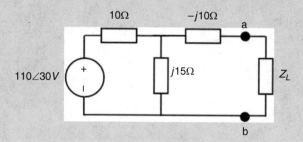

**Fig. 5.54** Circuit of Example 5.56

(continued)

**Example 5.55** (continued)

**Fig. 5.55** Thevenin
equivalent circuit of
Example 5.56

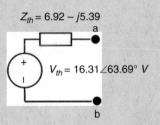

$Z_{th} = 6.92 - j5.39$

$V_{th} = 16.31\angle 63.69°\ V$

**Example 5.56 Find the Thevenin equivalent of the following circuit measured at terminals $a$ and $b$ (Fig. 5.56).**

*Solution.* The ZL must be disconnected from the circuit to obtain the Thevenin voltage. The voltage measured at terminals $a$ and $b$ is the same as the voltage drop across the impedance $j6\ \Omega$ because no current passes through the impedance $-j7\ \Omega$; hence, it drops zero volts.

The voltage of $j6\ \Omega$ is obtained through a voltage divider between $9 + j3\ \Omega$ and $j6\ \Omega$, as follows:

$$V_{th} = \frac{j6}{j6 + (9 + j3)}\ 12\angle 30 = \frac{6\angle 90\ \ 12\angle 30}{12.72\angle 45} = \frac{6 \times 12}{12.72}\angle(90 + 30 - 45)$$

$$= 5.66\angle 75\ V$$

To obtain the Thevenin impedance, the $12\angle 30\ V$ source is turned off and becomes zero volts or a short circuit (Fig. 5.57).

The impedance observed at the terminal is a parallel connection of $9 + j3\ \Omega$ and $j6\ \Omega$ in series with $-j7\ \Omega$ as follows:

$$Z_{th} = \left((9 + j3)\|j6\right) - j7 = \frac{(9 + j3)j6}{(9 + j3) + j6} - j7 = \frac{(3 + j1)j2}{1 + j1} - j7$$

$$Z_{th} = \frac{1 - j1}{1 - j1}\frac{j6 - 2}{1 + j1} - j7 = \frac{4 + j8}{2} - j7 = 2 - j3\ \Omega$$

Therefore, the circuit from terminals has an equivalent, as shown in Fig. 5.58.

(continued)

**Example 5.56** (continued)

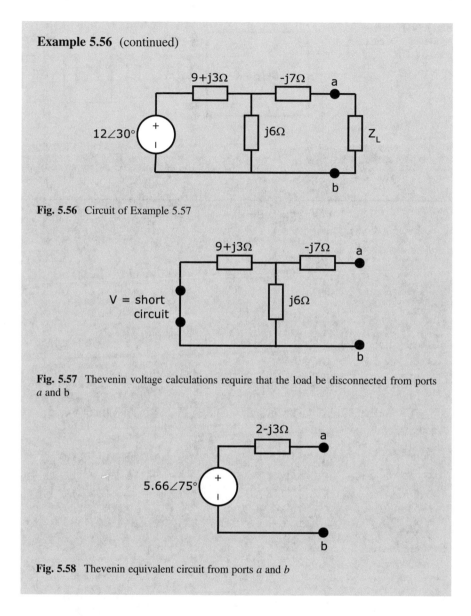

**Fig. 5.56** Circuit of Example 5.57

**Fig. 5.57** Thevenin voltage calculations require that the load be disconnected from ports *a* and b

**Fig. 5.58** Thevenin equivalent circuit from ports *a* and *b*

## Norton Equivalent and Source Conversion

The Norton model presents any circuit's equivalent at the desired terminal through a current source parallel to an impedance. The amount of current source (also known as short circuit current source) is obtained by measuring the current that passes through when shorting the circuit terminals. The impedance is obtained similarly to the Thevenin equivalent. Figure 5.59 shows a Norton equivalent of a network.

**Fig. 5.59** Norton
equivalent of a circuit from
ports *a* and *b*

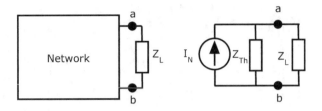

**Example 5.57 Find the Norton equivalent of the following circuit (Fig. 5.57).**

*Solution.* Circuit load must be replaced with a short circuit at terminals *a* and *b*. The short circuit current is the current through $-j7\ \Omega$ impedance. The source current is calculated to be

$$I = \frac{12\angle 30}{(9+j3) + \left(j6 \middle\| -j7\right)} = \frac{12\angle 30}{(9+j3) + \left(\frac{j6 \times -j7}{j6 - j7}\right)}$$

$$= \frac{12\angle 30}{(9+j3) + \left(\frac{j6 \times -j7}{j6 - j7}\right)}$$

$$= \frac{12\angle 30}{(9+j3) + j42} = \frac{12\angle 30}{45.89\angle 78.69} = 0.26\angle -48.69\ \text{A}$$

The impedance observed at the terminal is a parallel connection of $9 + j3\ \Omega$ and $j6\ \Omega$ in series to $-j7\ \Omega$ as follows

$$Z_{\text{th}} = \left((9+j3)\middle\| j6\right) - j7 = \frac{(9+j3)j6}{(9+j3) + j6} - j7 = \frac{(3+j1)j2}{1+j1} - j7$$

$$Z_{\text{th}} = \frac{1-j1}{1-j1}\frac{j6-2}{1+j1} - j7 = \frac{4+j8}{2} - j7 = 2 - j3\ \Omega$$

Therefore, the circuit from terminals has an equivalent, as shown in Fig. 5.60.

**Fig. 5.60** Norton equivalent circuit

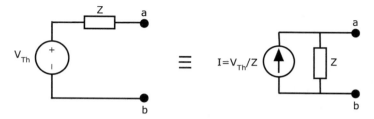

**Fig. 5.61** Conversion of Thevenin to Norton equivalent circuit

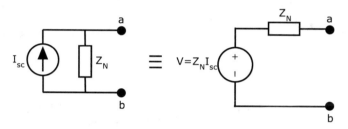

**Fig. 5.62** Conversion of Norton to Thevenin equivalent circuit

**Thevenin to Norton Conversion** Equivalent circuits of Thevenin and Norton can be converted to each other. Figure 5.61 shows the Thevenin to Norton conversion. Figure 5.62 shows the Norton to Thevenin conversion.

**Example 5.58 Convert all Thevenin to Norton and all Norton to Thevenin (Fig. 5.63).**

*Solution.* The current can be calculated by $I_{sc} = \frac{120 - j150}{1 + j2} = -36 - j78\,\text{A}$ (Fig. 5.64).

*Solution*

The voltage can be calculated by $V = (1 - j5)(70 + j15) = 145 - j335\,\text{V}$.

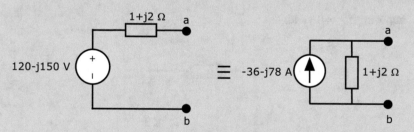

**Fig. 5.63** Conversion of Thevenin to Norton in Example 5.59

(continued)

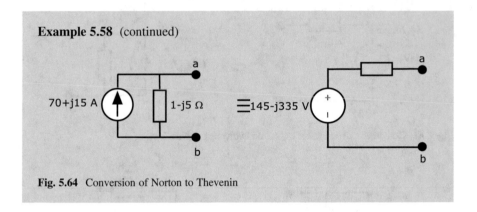

**Example 5.58** (continued)

**Fig. 5.64** Conversion of Norton to Thevenin

## Maximum Power Transfer

Consider a circuit that is shown by its equivalent Thevenin model ($V_{th}$, $Z_{th}(j\omega)$) and is connected to a load impedance of $Z_L(j\omega)$. The power delivered to the load from the source can be calculated as follows:

Considering each impedance's real and imaginary parts as $Z_{th} = R_{th} + jX_{th}$ and $Z_L = R_L + jX_L$, the amount of power delivered to the load is calculated as

$$P = R_L I^2 = R_L \left( \frac{|V|}{|Z_{th} + Z_L|} \right)^2 = R_L \left( \frac{|V|}{|R_{th} + R_L + j(X_{th} + X_L)|} \right)^2$$

$$= \frac{R_L |V|^2}{|(R_{th} + R_L)^2 + (X_{th} + X_L)^2|}$$

To maximize the power delivery to the load (Fig. 5.65),

$$\frac{dP}{dR_L} = 0 \text{ and } \frac{dP}{dX_L} = 0$$

Imposing $\frac{dP}{dR_L} = 0$ yields

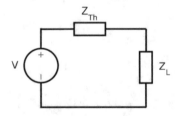

**Fig. 5.65** Thevenin equivalent circuit is feeding a load. The maximum power is transferred from the source to the load when the Thevenin impedance and the load impedance become pairs of the complex conjugate. This means $Z_L = Z_{th}^*$

$$\frac{|V|^2|(R_{th}+R_L)^2+(X_{th}+X_L)^2|-R_L\;|V|^2|2(R_{th}+R_L)|}{|(R_{th}+R_L)^2+(X_{th}+X_L)^2|^2}=0$$

$$R_L=R_{th}$$

Imposing $\frac{dP}{dX_L}=0$ yields

$$\frac{-2R_L\;|V|^2|(X_{th}+X_L)|}{|(R_{th}+R_L)^2+(X_{th}+X_L)^2|^2}=0$$

$$X_L=-X_{th}$$

It can be concluded that to transfer maximum power in AC sinusoidal steady-state conditions, the load impedance must be the complex conjugate of the Thevenin impedance as

$$@P_{max}\;:Z_L=Z_{th}{}^*$$

**Note 5.8** To obtain the load impedance that transfers maximum power from the source to the load through an existing circuit, the load impedance must be a complex conjugate of the Thevenin impedance.

**Example 5.59 Considering the following circuit, find the load impedance to cause maximum power transfer from the source to the load (Fig. 5.66).**
*Solution.* To transfer maximum power to the load, $Z_L = Z_{th}{}^*$ must hold. Therefore, finding the Thevenin equivalent of the circuit is required from the load terminals.

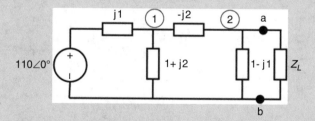

**Fig. 5.66** Circuit of Example 5.60

The Thevenin voltage can be obtained from the voltage across the $1-j\,\Omega$ impedance at node ② (when the load is disconnected). $V_2 = V_{th} = V_{1-j}\,\Omega.$

(continued)

**Example 5.59** (continued)

KCL ①. All elements connected to node (1) are passive, draining the node's current.

$$\frac{V_1 - 110\angle 0}{j} + \frac{V_1}{1 + j2} + \frac{V_1 - V_{th}}{j2} = 0$$

KCL ②. Since the load is disconnected (to get the Thevenin impedance), there are two elements in the current balance as follows:

$$\frac{V_{th} - V_1}{j2} + \frac{V_{th}}{1 - j} = 0$$

From ①

$$V_1\left(\frac{1}{j} + \frac{1}{1 + j2} + \frac{1}{j2}\right) - \frac{V_{th}}{j2} = \frac{110\angle 0}{j} \rightarrow V_1\left(-j + \frac{1 + j2}{5} + \frac{-j}{2}\right) + j\frac{V_{th}}{2} = -j110$$

$$V_1(0.2 - j1.1) + j0.5V_{th} = -j110$$

From ②

$$V_{th}\left(\frac{1}{j2} + \frac{1}{1 - j}\right) - \frac{V_1}{j2} = 0 \rightarrow V_{th}\left(\frac{-j}{2} + \frac{1 + j}{2}\right) + j\frac{V_1}{2} = 0$$

$$0.5V_{th} + j0.5V_1 = 0 \rightarrow V_{th} = -jV_1$$

Replacing in ①

$$V_1(0.2 - j1.1) + j0.5(-jV_1) = -j110$$

$$V_1(0.7 - j1.1) = -j110 \rightarrow V_1 = \frac{-j110}{0.7 - j1.1} = 71.17 - j45.29$$

$$V_1 = 84.36\angle - 32.47 \text{ V}$$

$$V_{th} = 84.36\angle - 122.47 \text{ V}$$

The Norton impedance can be obtained from the circuit when the independent sources are turned off. From the terminals, the circuit shows

$$Z_{th} = \left(j \| (1 + j2) - j2\right) \| (1 - j) = 0.29 - j0.66 \ \Omega$$

Therefore, the load impedance should be $Z_L = Z_{th}^* = 0.29 + j0.66 \ \Omega$.

## Problem

5.1. Find the impedance and admittance of the circuit.

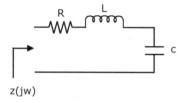

z(jw)

5.2. Find the impedance and admittance of the circuit.

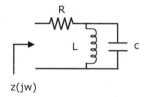

z(jw)

5.3. Find the impedance and admittance of the circuit.

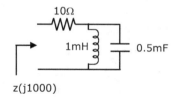

z(j1000)

5.4. Find the impedance and admittance of the circuit.

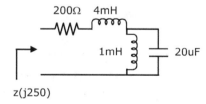

z(j250)

5.5. Find the impedance and admittance of the circuit.

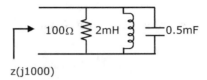

z(j1000)

5.6. Find the admittance of the circuit.

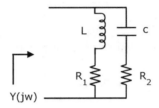

Y(jw)

5.7. Find the admittance of the circuit.

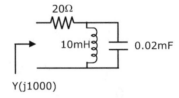

Y(j1000)

5.8. Find the admittance of the circuit.

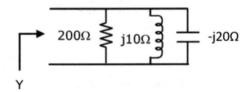

Y

5.9. Find the admittance of the circuit.

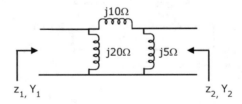

5.10. Find the admittance of the circuit.

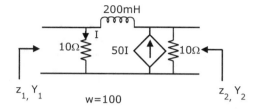

5.11. Find $I$, $V_1$, and $V_2$.

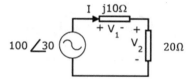

5.12. Find $I$, $V_1$, and $V_2$.

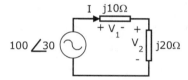

5.13. Find $I_1$, $I_2$, $I_3$, and $V$.

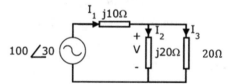

5.14. Find $I_1$, $I_2$, $I_3$, and $V$.

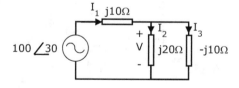

5.15. Find $I_1$, $I_2$, $I_3$, and $V$.

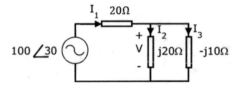

5.16. Find $I_1$, $I_2$, $I_3$, and $V$.

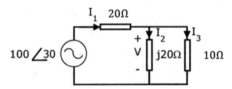

5.17. Find $I_1$, $I_2$, and $I_3$.

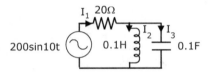

5.18. Find $I_1$, $I_2$, and $I_3$.

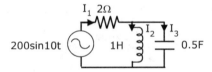

5.19. Find $I_1$, $I_2$, and $I_3$.

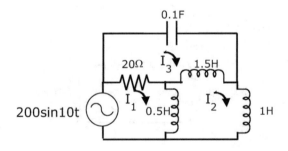

5.20. Find *I* and *V*.

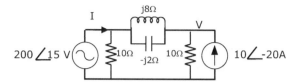

5.21. Find the current of each branch.

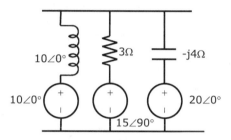

5.22. The current $i(t) = 300 \sin (377t + 50)$ when passing through an impedance shows a voltage drop of $v(t) = 480 \sin (377t + 10)$.

(a) Find the impedance of the element.
(b) Find the resistance and reactance of the element.
(c) Determine whether the impedance is more capacitive or more inductive.

5.23. Find the current through a $Z = 100 \angle 30\ \Omega$ impedance when it is connected to a voltage of $V = 120\sqrt{2}\angle10$ V.

5.24. A circuit creates a 60° phase shift in the current regarding the voltage. What is the power factor?

5.25. The voltage and current of an element are recorded as

$$v(t) = 110\sqrt{2} \sin (100\pi t + 30°)\, V$$
$$i(t) = 20 \sin (100\pi t - 30°)\, A$$

(a) Find the impedance
(b) Find the resistance
(c) Find the reactance
(d) Is the circuit more inductive or more capacitive?

5.26. The power factor of a circuit is $PF = 0.6$ lag. If the voltage phase is +25°, what is the phase of its current?

5.27. The power factor of a circuit is $PF = 0.6$ lead. If the voltage phase is +25°, what is the phase of its current?

5.28. In a series RLC circuit, the voltage drop across each element is measured as

$$V_R = 20\,\text{V}, V_L = 75\,\text{V}, V_C = 50\,\text{V}.$$

    (a) How much is the source voltage?
    (b) What is the circuit's power factor?

5.29. In the following circuit, find the inductor's voltage for different values of the capacitor $C = 1$ mF, $C = 10$ mF, and $C = 100$ mF.

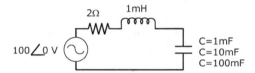

5.30. Find $V_C$.

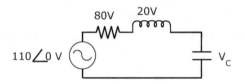

5.31. Find $V_C$.

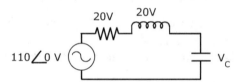

5.32. Find PF.

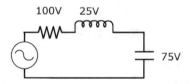

5.33. Find PF.

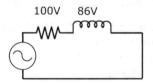

5.34. Find PF.

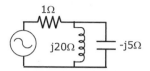

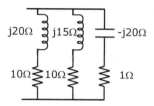

5.35. Find PF.

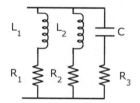

5.36. Find PF.

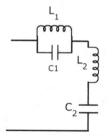

5.37. Find PF.

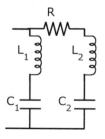

5.38. Find PF.

5.39. Find the apparent, active, reactive, and power factors in the following cases. Determine the generation or consumption of the power.

(a) $Z = 1 + j5 \ \Omega$, $I = 15 \ \angle -30$ A
(b) $Z = 1 + j5 \ \Omega$, $I = 15 \ \angle +30$ A
(c) $V_{rms} = 100 \ \angle \ 30 \ V_{rms}$, $I_{rms} = 20 \ \angle -40$ A
(d) $V_{rms} = 100 \ \angle \ 30 \ V_{rms}$, $I_{rms} = 20 \ \angle +40$ A
(e) $Z = 75 \ \angle \ 20 \ \Omega$, $V_{rms} = 220 \ \angle +10$ V
(f) $Z = 50 \ \angle -20 \ \Omega$, $V_{rms} = 110 \ \angle +35$ V
(g) $V = 100 \ \angle \ 30$ V, $I = 20 \ \angle -40$ A
(h) $V = 100 \ \angle \ 30$ V, $I = 20 \ \angle +40$ A
(i) $Z = 75 \ \angle \ 20 \ \Omega$, $V = 220 \ \angle +10$ V
(j) $Z = 50 \ \angle -20 \ \Omega$, $V = 110 \ \angle +35$ V

5.40. Find the requested variable in each case.

(a) $P = 1000$ W, $Q = 1500$ VAR, $S = ?$, PF $= ?$, lead or lag?
(b) $P = 1000$ W, $Q = -1500$ VAR, $S = ?$, PF $= ?$, lead or lag?
(c) $S = 100$ kVA, $P = 80$ kW, $Q = ?$, PF $= ?$, lead or lag?
(d) $S = 200$ kVA, $Q = 50$ kVA, $Q = ?$, PF $= ?$, lead or lag?
(e) $S = 500$ kVA, PF $= 0.6$ lead, $Q = ?$, $P = ?$
(f) $S = 500$ kVA, PF $= 0.6$ lag, $Q = ?$, $P = ?$
(g) $V_p = 200 \ \angle \ 30$ V, $I_p = 10 \ \angle - 40$, $S = ?$, $P = ?$, $Q = ?$, PF $= ?$, lead or lag?
(h) $V_p = 200 \ \angle \ 30$ V, $I_p = 10 \ \angle \ 40$, $S = ?$, $P = ?$, $Q = ?$, PF $= ?$, lead or lag?

5.41. Multiple loads take current from the grid in parallel in a household operating at 110 $V_{rms}$, 60 Hz. The current taken by these loads are

$$I_1 = 50\angle0 \text{A}, I_2 = 20\angle - 45 \text{ A}, I_3 = 12\angle - 30 \text{ A}, I_4 = 35\angle0 \text{ A}$$

(a) Find the apparent, active, and reactive power of each load.
(b) Find the total active power and total reactive power of the loads.
(c) Find the total apparent power of the household.
(d) Find the power factor of the house.
(e) Is this load lead or lag? Why?

5.42. One phase of an industrial load operating at 220 $V_{rms}$ has a load of $S = 2000 + j15,000$ VA.

(a) Find the power factor of this load. Lead or lag? Why?
(b) A capacitor compensates for the power factor in parallel to this load. Find the capacitor power required to change the power factor to 0.8 and 0.97.

5.43. An induction motor shows an inductive load of $S = 300$ kVA at power factor
PF $= 0.67$.

(a) Find the right amount of capacitor to be connected in parallel at the
terminal of this machine to bring the power factor to PF $= 0.97$.
(b) Find the active and reactive power taken from the grid before and after the
power factor correction.

5.44. A mix of wood chips and glue is pressed between two parallel plates at the
voltage of 600 V. The structure forms a plywood structure at a capacity of
1 μF. Find the required frequency at which a power loss of 500 W is applied to
the glue and woodchips to melt the glue. Consider the phase shift in the current
with respect to the voltage of the capacitor to be $\varphi = 10°$. Find the frequency at
which the power loss is 1200 W. At 1200 W, find the equivalent voltage if the
frequency is fixed at 1 kHz.

5.45. Find Thevenin and Norton.

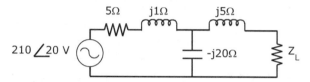

5.46. Find Thevenin and Norton.

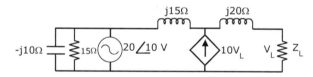

5.47. Find Thevenin and Norton.

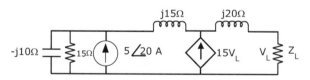

5.48. Find Thevenin and Norton.

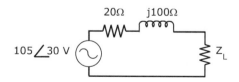

5.49. Find Thevenin and Norton.

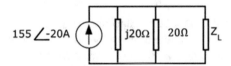

5.50. Find Thevenin and Norton.

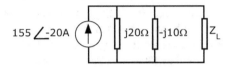

5.51. Find Thevenin and Norton.

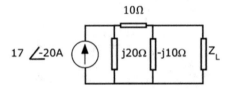

5.52. In Problems 5.45–5.51, find the load impedance at which the circuit transfers maximum power from the source to the load.

# Chapter 6
# Mutual Inductance

## Introduction

Electric circuits, specifically when excited by AC sources, can transfer energy by direct electric connection or magnetic coupling. Consider an inductor with $N$ turns of winding. The current $i$ passing through this inductor generates a magnetic flux $\phi$ around the windings. This flux creates a magnetic field that starts from the north pole and ends at the south pole. When the direction of current changes, the location of the north and south poles changes which causes a change in the direction of flux but still from the North pole to the south Pole (Fig. 6.1). The variation flux over generates a voltage in the coil that is measured by

$$v = N \frac{d\phi}{dt}$$

As explained earlier, the flux is a current dependent variable. Therefore,

$$v = N \frac{d\phi}{di} \frac{di}{dt}$$

The total flux variation accounted for the number of turns in a coil that determines its inductance. This means that a higher number of turns or higher area to generate more flux increases the inductance value. This can be calculated by

$$L = N \frac{d\phi}{di}$$

Replacing the inductance in the voltage equation reveals Ohm's law discussed earlier. Hence,

© The Author(s), under exclusive license to Springer Nature Switzerland AG 2023
A. Izadian, *Fundamentals of Modern Electric Circuit Analysis and Filter Synthesis*,
https://doi.org/10.1007/978-3-031-21908-5_6

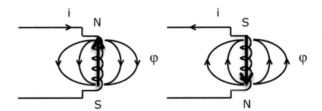

**Fig. 6.1** Magnetic field generated because of the current passing through a single inductor. The direction of the field is always from the north pole to the south pole. When the current direction changes, the magnetic field's direction also changes

$$v = L\frac{di}{dt}$$

## Self-Inductance and Mutual Inductance

The inductance value utilized in this equation is known as *self-inductance*. This means that the voltage induced in a coil is solely generated by the time-varying current of the same coil. However, if the flux generated by another coil passes through the windings of the inductor, a voltage is induced in the coil due to *mutual inductance* between coils.

Consider two adjacent coils (Fig. 6.2) with self-inductances of $L_1$ and $L_2$ Henrys and $N_1$ and $N_2$ turns. Coil 1 is connected to the $v_1$ source that establishes current $i_1$. The flux $\phi_1$ generated by coil 1 has two parts: part 1 links coil 1, $\phi_{11}$, due to self-inductance, and part 2 links coil 2, $\phi_{12}$, due to mutual inductance. Hence,

$$\phi_1 = \phi_{11} + \phi_{12}$$

## Induced Voltage

The induced voltages at the terminal of each coil depend on the total amount of flux linking the coil and the number of turns the coil has as follows:

Total flux $\phi_1 = \phi_{11} + \phi_{12}$ links coil 1 and is generated by current $i_1$:

$$v_1 = N_1\frac{d\phi_1}{dt} \rightarrow v_1 = N_1\left(\frac{d\phi_1}{di_1}\frac{di_1}{dt}\right) = \left(N_1\frac{d\phi_1}{di_1}\right)\frac{di_1}{dt} = L_1\frac{di_1}{dt}$$

The voltage in the second coil is generated by the current in the first coil through the linking flux $\phi_{12}$. Considering $M_{21}$ as the mutual inductance on coil 2 influenced by coil 1,

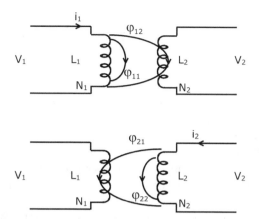

**Fig. 6.2** When the magnetic fields generated by one coil are cut by the windings of another adjacent coil, the voltage is induced in the second coil. The direction of the current entering the first coil and the direction of windings in the second coil determines the polarity of the induced voltage. The split of magnetic flux passing through the air and circulating the main coil generates the self-inductance, and the part passing the adjacent coil builds the mutual inductance

$$v_2 = N_2 \frac{d\phi_{12}}{dt} \rightarrow v_2 = N_2 \left( \frac{d\phi_{12}}{di_1} \frac{di_1}{dt} \right) = \left( N_2 \frac{d\phi_{12}}{di_1} \right) \frac{di_1}{dt} = M_{21} \frac{di_1}{dt}$$

Hence, $v_2 = M_{21} \frac{di_1}{dt}$ determines that the induced voltage in coil 2 is influenced by the current variation in coil 1 through mutual inductance, $M_{21}$. It also indicates that the voltage is induced in coil 2 due to current passing coil 1.

Now consider that the current is passing coil 2 and coil 1 voltage is measured (Fig. 6.2). The current $i_2$ generates a total flux of $\phi_2$ which has two parts of $\phi_{22}$ that link coil 2 and $\phi_{21}$ that links coil 1. The total flux is

$$\phi_2 = \phi_{22} + \phi_{21}$$

The induced voltages at the terminal of each coil depend on the total amount of flux linking the coil and the number of turns the coil has as follows:

Total flux $\phi_2 = \phi_{22} + \phi_{21}$ links coil 2 and is generated by current $i_2$:

$$v_2 = N_2 \frac{d\phi_2}{dt} \rightarrow v_2 = N_2 \left( \frac{d\phi_2}{di_2} \frac{di_2}{dt} \right) = \left( N_2 \frac{d\phi_2}{di_2} \right) \frac{di_2}{dt} = L_2 \frac{di_2}{dt}$$

The voltage induced in the first coil is generated by the current in the second coil through the linking flux $\phi_{21}$. Therefore,

$$v_1 = N_1 \frac{d\phi_{21}}{dt} \rightarrow v_1 = N_1 \left( \frac{d\phi_{21}}{di_2} \frac{di_2}{dt} \right) = \left( N_1 \frac{d\phi_{21}}{di_2} \right) \frac{di_2}{dt} = M_{12} \frac{di_2}{dt}$$

$$v_1 = M_{12} \frac{di_2}{dt}$$

Mutual inductance is a reciprocal quantity, meaning that the same voltage will be induced in coil 1 if the current is passed through coil 2. This means $M_{21} = M_{12} = M$, all measured in Henrys (H).

Mutual inductance exists when two or more coils are physically located such that the flux generated by one coil finds an appropriate path to link the adjacent coils. If this path does not exist or the flux is not time-varying, the mutual inductance disappears.

Mutual inductance is often shown by ● signs on one of the terminals at each port. The dot's location and the direction of current in and out of the dot determine the polarity of the induced voltage due to mutual inductance.

- The current entering the dotted terminal of a coil induces positive voltage at the dotted terminal of the second coil (Fig. 6.3).
- The current leaving the dotted terminal of a coil induces a negative voltage in the dotted terminal of the second coil (Figs. 6.4, 6.5, and 6.6).

Therefore, to determine the polarity of induced voltages in each coil, the location of dotted terminals, and the direction of current in the other coil must be considered.

**Fig. 6.3** Direction of both currents entering the dot ● induces the polarity that supports the direction of current $i_2$

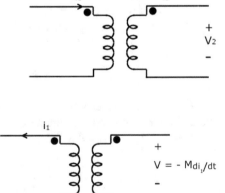

**Fig. 6.4** Reverse direction of the current $i_1$ induces the reverse polarity of voltage $v_2$

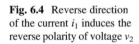

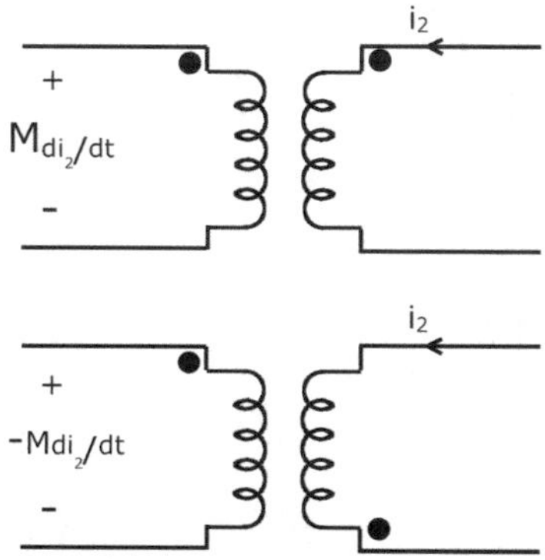

**Fig. 6.5** The amount of
voltage induced is directly
proportional to the value of
the mutual inductance $m$ and
the rate of the change of the
voltage-inducing current, $i_1$,
or $v = -M\frac{di_1}{dt}$

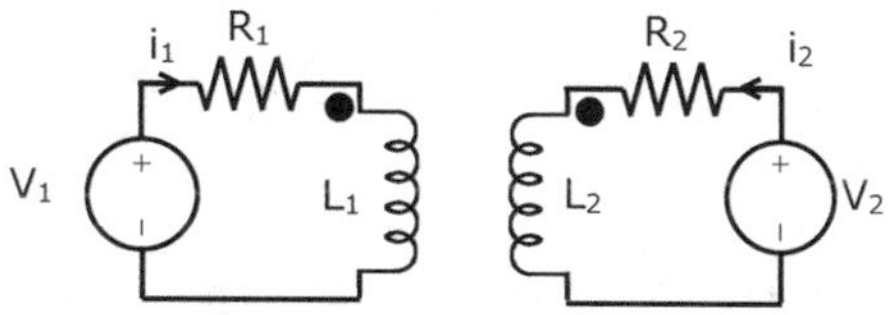

**Fig. 6.6** Direction of
currents is in the doted
mutual inductance, and the
voltages they induce must
support the direction of
these currents

**Example 6.1 Determine the polarity of the induced voltage in the circuit
of Fig. 6.6.**

*Solution.* Induced voltage due to mutual inductance in the first loop can be
modeled as a voltage source with positive polarity in the loop circulating
clockwise (Fig. 6.7).

**Fig. 6.7** The circuit of
Example 6.1

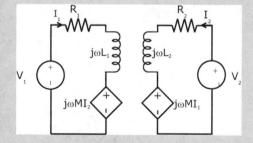

The amount of induced voltage source in the first loop is $+j\omega MI_2$ because
the current enters the dotted terminal in the second loop. The second loop has
an induced voltage of $+j\omega MI_1$ with positive polarity because the current $I_1$
enters the dotted terminal of the first loop.

Mutual inductance generates a current-controlled voltage source. In two
loops, when the dotted terminals are removed, the circuit can be shown as
Fig. 6.7 and calculated as follows:

(continued)

**Example 6.1** (continued)

KVL ①

$$-V_1 + R_1 I_1 + j\omega L_1 I_1 + j\omega M I_2 = 0$$

KVL ②

$$-V_2 + R_2 I_2 + j\omega L_2 I_2 + j\omega M I_1 = 0$$

Solving for $I_1$ and $I_2$ using Cramer's method is as follows:
First, the equations have to be simplified as follows:

$$\begin{bmatrix} R_1 + j\omega L_1 & j\omega M \\ j\omega M & R_2 + j\omega L_2 \end{bmatrix} \begin{bmatrix} I_1 \\ I_2 \end{bmatrix} = \begin{bmatrix} V_1 \\ V_2 \end{bmatrix}$$

$\begin{bmatrix} R_1 + j\omega L_1 & j\omega M \\ j\omega M & R_2 + j\omega L_2 \end{bmatrix}$ is the matrix of coefficients (let us consider it as $A$),

$$A = \begin{bmatrix} R_1 + j\omega L_1 & j\omega M \\ j\omega M & R_2 + j\omega L_2 \end{bmatrix}$$

$\begin{bmatrix} I_1 \\ I_2 \end{bmatrix}$ is the matrix of unknown variables (let us consider it as $X$),

$$X = \begin{bmatrix} I_1 \\ I_2 \end{bmatrix}$$

Moreover, $\begin{bmatrix} V_1 \\ V_2 \end{bmatrix}$ is the matrix of known inputs (let us consider it as $B$).

$$B = \begin{bmatrix} V_1 \\ V_2 \end{bmatrix}$$

Therefore, the equation reads as

$$AX = B$$

Solving for $X$ results in

(continued)

**Example 6.1** (continued)

$$X = A^{-1}B$$

Matrix analysis of circuits will be covered in Chap. 12. Unknown variables are obtained as

$$\begin{bmatrix} I_1 \\ I_2 \end{bmatrix} = \begin{bmatrix} R_1 + j\omega L_1 & j\omega M \\ j\omega M & R_2 + j\omega L_2 \end{bmatrix}^{-1} \begin{bmatrix} V_1 \\ V_2 \end{bmatrix}$$

$$\begin{bmatrix} I_1 \\ I_2 \end{bmatrix} = \frac{1}{(R_1 + j\omega L_1)(R_2 + j\omega L_2) - (j\omega M)(j\omega M)} \begin{bmatrix} R_2 + j\omega L_2 & -j\omega M \\ -j\omega M & R_1 + j\omega L_1 \end{bmatrix}$$

$$\times \begin{bmatrix} V_1 \\ V_2 \end{bmatrix}$$

$$I_1 = \frac{R_2 V_1 + j\omega(L_2 V_1 - MV_2)}{M^2\omega^2 + R_1 R_2 - L_1 L_2\omega^2 + j\omega(L_1 R_2 + L_2 R_1)}$$

$$I_2 = \frac{R_1 V_2 + j\omega(L_1 V_2 - MV_1)}{M^2\omega^2 + R_1 R_2 - L_1 L_2\omega^2 + j\omega(L_1 R_2 + L_2 R_1)}$$

**Example 6.2 Consider two mutually coupled inductors connected in series as shown in Fig. 6.8. Find the equivalent inductance of the circuit.**
*Solution.* As shown, the current enters the dotted terminal of inductor 1; therefore, the voltage induced in inductor 2 shows a positive voltage with respect to the dotted terminal of the second inductor. As the current of the first inductor enters the dotted terminal, the inductor shows a positive polarity voltage (Fig. 6.9).
A KVL of the circuit shows

$$V = j\omega L_1 I + j\omega MI + j\omega L_2 I + j\omega MI$$

$$V = j\omega(L_1 + M + L_2 + M)I = j\omega L_{eq}I$$

$$L_{eq} = L_1 + L_2 + 2M$$

**Fig. 6.8** Circuit of Example 6.2. The same current passes through the series inductors and enters each dotted node

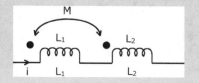

(continued)

**Example 6.2** (continued)

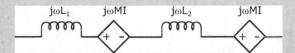

**Fig. 6.9** Equivalent circuit of Example 6.2. Induced voltages are shown as voltage sources. One should pay attention to the direction of the current and the polarity of the dependent voltage sources (showing the induced voltage)

**Example 6.3 Consider two mutually coupled inductors connected in series as shown in Fig. 6.3. Find the equivalent inductance of the circuit (Fig. 6.10).**

*Solution.* As shown, the current enters the dotted terminal of inductor 1; therefore, the voltage induced in inductor 2 shows a positive voltage with respect to the dotted terminal of the second inductor. The positive source terminal is connected to the dotted terminal. The same principle applies to the first inductor. As the current leaves the dotted terminal of the second inductor, it induces a negative voltage in the first inductor (Fig. 6.11).

A KVL of the circuit shows

$$V = j\omega L_1 I + (-j\omega MI) + j\omega L_2 I - j\omega MI$$

$$V = j\omega(L_1 - M + L_2 - M)I = j\omega L_{eq}I$$

$$L_{eq} = L_1 + L_2 - 2M$$

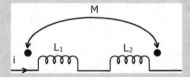

**Fig. 6.10** Circuit of Example 6.3. The same current passes through the series inductors, but the current has the opposite direction to each dotted node, enters one, and exits the other

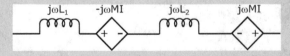

**Fig. 6.11** Equivalent circuit of Example 6.2. Induced voltages are shown as voltage sources. One should pay attention to the direction of the current and the polarity of the dependent voltage sources (showing the induced voltage)

**Example 6.4 Consider the circuit shown in Fig. 6.12 with mutual inductance between the inductors. Find the current in each circuit.**
*Solution.* Considering the mutual inductance, the dotted terminals, and the direction of currents, the voltage induced in the first circuit shows a negative voltage because the current $i_2$ leaves the dotted terminal of the second inductor.

In the second loop, the induced voltage is a positive polarity with respect to the dotted line because the current in the first loop enters the dotted terminal of the first inductor. The equivalent circuit is shown as follows (Fig. 6.13).
KVL ①

$$-100 - j10I_1 + j5I_1 - j3I_2 = 0$$

KVL ②

$$10I_2 - j3I_1 + jI_2 = 0$$

Using matrix approach,

$$\begin{bmatrix} -j5 & -j3 \\ -j3 & 10+j \end{bmatrix} \begin{bmatrix} I_1 \\ I_2 \end{bmatrix} = \begin{bmatrix} 100 \\ 0 \end{bmatrix}$$

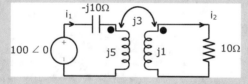

**Fig. 6.12** Circuit of Example 6.4

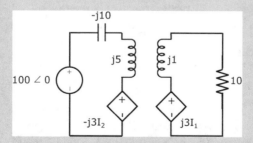

**Fig. 6.13** The induced voltage of the circuit in Fig. 6.12 is expanded to the dependent voltage sources. The mutual inductance previously shown as dots is now presented at the circuit level by voltage sources

(continued)

**Example 6.4** (continued)

$$\begin{bmatrix} I_1 \\ I_2 \end{bmatrix} = \begin{bmatrix} -j5 & -j3 \\ -j3 & 10+j \end{bmatrix}^{-1} \begin{bmatrix} 100 \\ 0 \end{bmatrix}$$

$$I_1 = 3.33 + j19.06\,\text{A} = 19.34\angle 80\,\text{A}$$

$$I_2 = -5.56 + j1.55\,\text{A} = 5.77\angle 164.4\,\text{A}$$

**Example 6.5 Consider the circuit of the previous example with a reverse connection of mutual inductance (dotted terminals are connected in reverse). Find the current of each circuit (Fig. 6.14).**

*Solution.* Since the current enters the dotted terminal in both inductors, they induce positive voltage to the mutually coupled inductors. This polarity is measured positively with respect to the dotted terminal. The equivalent circuit is shown in Fig. 6.15.

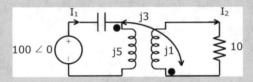

**Fig. 6.14** Circuit of Example 6.5

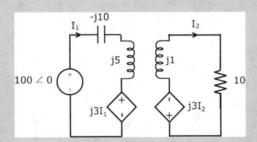

**Fig. 6.15** The induced voltage of the circuit in Fig. 6.14 is expanded to the dependent voltage sources. The mutual inductance previously shown as dots is now presented at the circuit level by voltage sources. Pay close attention to the direction of the currents entering or leaving the mutual inductance dots and the polarity of the voltage sources

(continued)

**Example 6.5**  (continued)

Therefore, the KVLs can be written as follows:
KVL ①

$$-100 - j10I_1 + j5I_1 + j3I_2 = 0$$

KVL ②

$$10I_2 + j3I_1 + jI_2 = 0$$

Using matrix approach,

$$\begin{bmatrix} -j5 & j3 \\ j3 & 10+j \end{bmatrix} \begin{bmatrix} I_1 \\ I_2 \end{bmatrix} = \begin{bmatrix} 100 \\ 0 \end{bmatrix}$$

$$\begin{bmatrix} I_1 \\ I_2 \end{bmatrix} = \begin{bmatrix} -j5 & j3 \\ j3 & 10+j \end{bmatrix}^{-1} \begin{bmatrix} 100 \\ 0 \end{bmatrix}$$

$$I_1 = 3.33 + j19.06\,\text{A} = 19.34\angle 80\,\text{A}$$

$$I_2 = 5.56 - j1.55\,\text{A} = 5.77\angle -15.57\,\text{A}$$

## Energy Stored in Coupled Circuits

Inductors store energy depending on the current amplitude and their self-inductance calculated $W = \frac{1}{2}LI^2$. Considering the power of a coil as a product of voltage and current, the energy is obtained as follows:

$$P(t) = v(t)i(t) = L\frac{di(t)}{dt}i(t)$$

$$W = \int P(t)dt = \int L\frac{di(t)}{dt}i(t)dt = L\int i(t)d(t) = \frac{1}{2}LI^2$$

Consider a mutually coupled coil, as shown in Fig. 6.16. Mutual inductance $M$ induces an additional voltage, which changes the power equation as follows:
In loop ①, the voltage is measured as

$$v_1 = L_1\frac{di_1(t)}{dt} + M\frac{di_2(t)}{dt}$$

The power measured from loop ① is obtained as

**Fig. 6.16** Mutual
inductance $M$ between two
coils at inductance $L_1$ and $L_2$

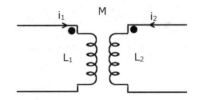

$$P_1(t) = v_1(t)i_1(t) = \left(L_1\frac{di_1(t)}{dt} + M\frac{di_2(t)}{dt}\right)i_1(t)$$

$$W_1 = \int \left(L_1\frac{di_1(t)}{dt} + M\frac{di_2(t)}{dt}\right)i_1(t)dt = \frac{1}{2}L_1I_1{}^2 + MI_1I_2$$

Considering the overall circuit, there are three energy-storing elements: self-inductance $L_1$, self-inductance $L_2$, and mutual inductance $M$.

$W_1$ considers the energy stored in self-inductance $L_1$ and the mutual inductance $M$. The energy stored in the self-inductance $L_2$ is measured as

$$W_2 = \int L_2\frac{di_2(t)}{dt}i_2(t)dt = \frac{1}{2}L_2I_2{}^2$$

The energy stored in the entire circuit is obtained as

$$W = W_1 + W_2$$

$$W = \frac{1}{2}L_1I_1{}^2 + \frac{1}{2}L_2I_2{}^2 + MI_1I_2$$

**Example 6.6 Find the stored energy in the circuit of Fig. 6.17.**
*Solution.* In loop ①, the voltage is measured as

$$v_1 = L_1\frac{di_1(t)}{dt} - M\frac{di_2(t)}{dt}.$$

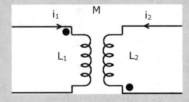

**Fig. 6.17** Mutual inductance $M$ with opposite dotted terminals between two coils of $L_1$ and $L_2$

(continued)

**Example 6.6** (continued)

The power measured from loop ① is obtained as

$$P_1(t) = v_1(t)i_1(t) = \left( L_1 \frac{di_1(t)}{dt} - M \frac{di_2(t)}{dt} \right) i_1(t)$$

$$W_1 = \int \left( L_1 \frac{di_1(t)}{dt} - M \frac{di_2(t)}{dt} \right) i_1(t)dt = \frac{1}{2}L_1I_1{}^2 - MI_1I_2$$

Considering the overall circuit, there are three energy-storing elements: self-inductance $L_1$, self-inductance $L_2$, and mutual inductance $M$.

$W_1$ considers the energy stored in self-inductance $L_1$ and the mutual inductance $M$. The energy stored in the self-inductance $L_2$ is measured as

$$W_2 = \int L_2 \frac{di_2(t)}{dt} i_2(t)dt = \frac{1}{2}L_2I_2{}^2$$

The energy stored in the entire circuit is obtained as

$$W = W_1 + W_2$$

$$W = \frac{1}{2}L_1I_1{}^2 + \frac{1}{2}L_2I_2{}^2 - MI_1I_2$$

## Limit of Mutual Inductance

Considering that the energy stored in passive inductors cannot be negative, $\frac{1}{2}L_1I_1{}^2 + \frac{1}{2}L_2I_2{}^2 - MI_1I_2 \geq 0$. As a complete square expression, the maximum limit of the mutual inductance is shown as

$$M \leq \sqrt{L_1L_2}$$

As mentioned earlier, the mutual inductance depends on the geometry of the coils with respect to each other, the magnetic core, and their orientation. A maximum mutual inductance is reached with the linking flux of one coil entirely passing through the second coil. This may occur when two coils are concentric, e.g., one coil is wrapped around the other. As the geometries move away from each other, the linking flux between the coils is reduced. The ratio of

$$k = \frac{\phi_{12}}{\phi_{11} + \phi_{12}} = \frac{\phi_{21}}{\phi_{21} + \phi_{22}}$$

From Coil 1    From Coil 2

$k$ is a factor between 0 and 1, $0 \le k \le 1$. The mutual inductance is obtained as

$$M = k\sqrt{L_1 L_2}$$

Considering the minimum and maximum values of $k$, the mutual inductance can reach

$$0 \le M \le \sqrt{L_1 L_2}$$

## Turn Ratio

Consider two coils wrapped around a core. The ratio of their inductance is

$$\frac{L_2}{L_1} \propto \frac{N_2{}^2}{N_1{}^2} = n^2$$

where $N_{1,2}$ is the number of turns a coil has and $n$ is the turn ratio. Therefore, the inductance ratio is proportional to the square of turn ratio.

## Equivalent Circuit of Mutual Inductance

Consider the following circuit in which two inductors $L_1$ and $L_2$ have a mutual inductance of $M$. This mutual inductance might be positive or negative, depending on the dotted terminals and the current direction, which also determines the polarity of the induced voltage. Therefore, the mutual inductance can generally be either a positive or negative number. This mutual inductance can be shown as an equivalent $T$ or $\Pi$ inductive circuit.

## *T Equivalent Circuit*

Considering the circuit shown in Fig. 6.18, KVLs in loops ① and ② show

$$V_1 = j\omega L_1 I_1 + j\omega M I_2$$
$$V_2 = j\omega M I_1 + j\omega L_2 I_2$$

In matrix form, these equations can be written as

$$\begin{bmatrix} V_1 \\ V_2 \end{bmatrix} = \begin{bmatrix} j\omega L_1 & j\omega M \\ j\omega M & j\omega L_2 \end{bmatrix} \begin{bmatrix} I_1 \\ I_2 \end{bmatrix}$$

Considering the same current direction and terminal voltages, a $T$ equivalent circuit is shown in Fig. 6.19. KVLs in loops ① and ② can be written as

$$V_1 = j\omega(L_a + L_b)I_1 + j\omega L_c I_2$$
$$V_2 = j\omega L_c I_1 + j\omega(L_b + L_c)I_2$$

In matrix form, the equivalent circuit becomes

$$\begin{bmatrix} V_1 \\ V_2 \end{bmatrix} = \begin{bmatrix} j\omega(L_a + L_b) & j\omega L_c \\ j\omega L_c & j\omega(L_b + L_c) \end{bmatrix} \begin{bmatrix} I_1 \\ I_2 \end{bmatrix}$$

Comparing the matrices, the equivalent $L_a$, $L_b$, and $L_c$ can be obtained as follows

**Fig. 6.18** Mutual inductance for $T$, $\Pi$ equivalent circuit

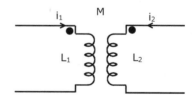

**Fig. 6.19** $T$ equivalent circuit of mutual inductance

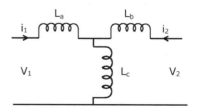

$$\begin{bmatrix} j\omega L_1 & j\omega M \\ j\omega M & j\omega L_2 \end{bmatrix} \equiv \begin{bmatrix} j\omega(L_a + L_b) & j\omega L_c \\ j\omega L_c & j\omega(L_b + L_c) \end{bmatrix}$$

Therefore,

$$L_a = L_1 - M$$
$$L_b = L_2 - M$$
$$L_c = M$$

## Π Equivalent Circuit

The admittance matrix can be obtained from the matrix form KVL written or the mutual inductance circuit Fig. 6.18, as follows (solving for current in the KVL):

$$\begin{bmatrix} I_1 \\ I_2 \end{bmatrix} = \begin{bmatrix} j\omega L_1 & j\omega M \\ j\omega M & j\omega L_2 \end{bmatrix}^{-1} \begin{bmatrix} V_1 \\ V_2 \end{bmatrix}$$

This yields

$$\begin{bmatrix} I_1 \\ I_2 \end{bmatrix} = \begin{bmatrix} \dfrac{L_2}{j\omega(L_1 L_2 - M^2)} & \dfrac{-M}{j\omega(L_1 L_2 - M^2)} \\ \dfrac{-M}{j\omega(L_1 L_2 - M^2)} & \dfrac{L_1}{j\omega(L_1 L_2 - M^2)} \end{bmatrix} \begin{bmatrix} V_1 \\ V_2 \end{bmatrix}$$

Consider a Π equivalent circuit as shown in Fig. 6.20. KCL in nodes ① and ② shows

$$I_1 = \frac{V_1}{j\omega L_A} + \frac{V_1 - V_2}{j\omega L_C}$$

$$I_2 = \frac{V_2}{j\omega L_B} + \frac{V_2 - V_1}{j\omega L_C}$$

In matrix form, these equations can be written as

**Fig. 6.20** Π equivalent circuit of mutual inductance

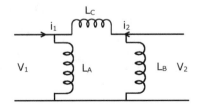

$$\begin{bmatrix} I_1 \\ I_2 \end{bmatrix} = \begin{bmatrix} \dfrac{1}{j\omega L_A} + \dfrac{1}{j\omega L_C} & \dfrac{-1}{j\omega L_C} \\ \dfrac{-1}{j\omega L_C} & \dfrac{1}{j\omega L_B} + \dfrac{1}{j\omega L_C} \end{bmatrix} \begin{bmatrix} V_1 \\ V_2 \end{bmatrix}$$

The equivalent admittance matrix and the original circuit admittance must be equal. This yields

$$\begin{bmatrix} \dfrac{L_2}{j\omega(L_1 L_2 - M^2)} & \dfrac{-M}{j\omega(L_1 L_2 - M^2)} \\ \dfrac{-M}{j\omega(L_1 L_2 - M^2)} & \dfrac{L_1}{j\omega(L_1 L_2 - M^2)} \end{bmatrix} \equiv \begin{bmatrix} \dfrac{1}{j\omega L_A} + \dfrac{1}{j\omega L_C} & \dfrac{-1}{j\omega L_C} \\ \dfrac{-1}{j\omega L_C} & \dfrac{1}{j\omega L_B} + \dfrac{1}{j\omega L_C} \end{bmatrix}$$

Therefore,

$$L_A = \frac{L_1 L_2 - M^2}{L_2 - M}$$

$$L_B = \frac{L_1 L_2 - M^2}{L_1 - M}$$

$$L_C = \frac{L_1 L_2 - M^2}{M}.$$

**Note 6.1** The loop current directions are preserved in the equivalent circuit.

**Note 6.2** The terminal voltage polarities are preserved in the equivalent circuit.

**Note 6.3** Positive or negative values of $\pm M$ must be considered depending on the original circuit.

**Note 6.4** Depending on the original circuit, $T$ or $\Pi$ circuits can be utilized.

**Example 6.7 Consider the circuit shown in Fig. 6.21. Using the $T$ equivalent circuit of the mutual inductance, find the current of each circuit (Fig. 6.21).**
*Solution.* The mutual inductance part of the circuit shown with terminals $a$, $b$, $c$, and $d$ can be replaced with its $T$ equivalent. The arrangement of dotted terminals and the direction of currents result in a positive mutual inductance value. The circuit and current directions are shown in Fig. 6.22.

(continued)

**Example 6.7** (continued)

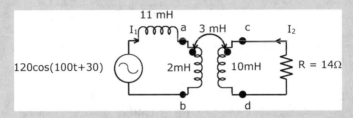

**Fig. 6.21** Circuit of Example 6.7. The mutual inductance can be replaced by its $T$ equivalent

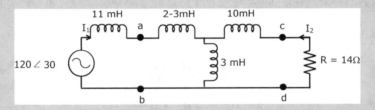

**Fig. 6.22** Mutual inductance is replaced by the $T$ equivalent. The values of the inductance are shown

The equivalent inductance observed at terminals $a$ and $b$ becomes (2 mH − 3 mH) = −1 mH—the impedance at $\omega = 1000$ rad/s shows equivalent impedance of a capacitor as $-j1\ \Omega$.

KVL in loop ①:

$$-120\angle 30 + j10I_1 + j3(I_1 + I_2) = 0$$

KVL in loop ②:

$$(14 + j7)I_2 + j3(I_1 + I_2) = 0$$

Using the matrix approach, the currents are obtained as

$$\begin{bmatrix} j13 & j3 \\ j3 & 14+j10 \end{bmatrix}\begin{bmatrix} I_1 \\ I_2 \end{bmatrix} = \begin{bmatrix} 120\angle 30 \\ 0 \end{bmatrix} = \begin{bmatrix} 103.92 + j60 \\ 0 \end{bmatrix}$$

$$\begin{bmatrix} I_1 \\ I_2 \end{bmatrix} = \begin{bmatrix} j13 & j3 \\ j3 & 14+j10 \end{bmatrix}^{-1}\begin{bmatrix} 103.92 + j60 \\ 0 \end{bmatrix}$$

$$\begin{bmatrix} I_1 \\ I_2 \end{bmatrix} = \begin{bmatrix} 4.99 - j8.01 \\ -1.64 + j0.103 \end{bmatrix} = \begin{bmatrix} 9.43\angle -58.07 \\ 1.64\angle -266.4 \end{bmatrix} \text{ A}$$

## Ideal Mutual Inductance

Consider the circuit shown in Fig. 6.23, in which, as mentioned earlier, the mutual inductance is influenced by the geometry of the windings and the core material. Coil ① (primary) has $N_1$ turns, and coil ② (secondary) has $N_2$ turns. If the magnetic material used in the circuit provides an ideal path such that the flux of one coil fully passes the second coil, the induced voltages can be written as (Figs. 6.23 and 6.24):

$$v_1 = N_1 \frac{d\phi}{dt}$$

and

$$v_2 = N_2 \frac{d\phi}{dt}$$

## Ideal Transformer

As the value of self-inductance in both coils and the value of mutual inductance reach very large numbers, the core has to provide a path for the flux to link both coils equally. The loading on the secondary increases the current at the secondary circuit. This generates flux in the core, increasing the current taken from the primary (Fig. 6.25).

Considering the induced voltage in both primary and secondary, the ratios become

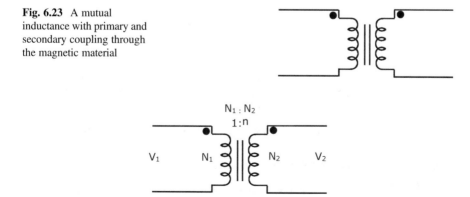

**Fig. 6.23** A mutual inductance with primary and secondary coupling through the magnetic material

**Fig. 6.24** The turn ratio is $N_1:N_2$ or normalized as $1:n$. Accordingly, the applied voltage ratios are proportional to the turn ratio, and the current flowing is reverse proportional to the turn ratio

**Fig. 6.25** Ideal transformer
with the load

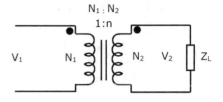

$$\frac{v_2}{v_1} = \frac{N_2}{N_1} = n$$

In an ideal transformer, the loss in each coil is negligible. The power input at primary $P_{\text{in}} = v_1 i_1$ is delivered at the secondary $P_{\text{out}} = v_2 i_2$. This yields

$$v_1 i_1 = v_2 i_2$$

Therefore, in phasor form, the ratio of voltages becomes

$$\frac{V_2}{V_1} = \frac{I_1}{I_2} = n$$

The impedance $Z_L$ at the secondary can be calculated as

$$V_2 = Z_L I_2$$

Replacing the primary voltage and current ratios results in

$$n V_1 = Z_L \frac{1}{n} I_1$$

This impedance observed (transferred) to primary is measured as

$$V_1 = \frac{Z_L}{n^2} I_1 \Rightarrow Z_1 = \frac{Z_L}{n^2}$$

**Note 6.5** Turn ratio $n = 1$ shows an isolation transformer. The voltage at the primary and secondary is the same, and the primary and secondary currents are similar. The impedance is observed similarly on both sides.

**Note 6.6** Turn ratio $n > 1$ shows a step-up transformer. In this case, the secondary voltage is increased by $n$ times, and at constant power, the current is scaled down by $\frac{1}{n}$ times. The impedance transferred to the secondary is $n^2$ times larger.

**Note 6.7** Turn ratio $n < 1$ shows a step-down transformer. In this case, the secondary voltage is dropped by $n$ times, and the current is increased at secondary $\frac{1}{n}$ times. The impedance at the secondary is still $n^2$ times that of the primary.

**Fig. 6.26** The equivalent
circuit of an ideal
transformer is seen from
the primary side

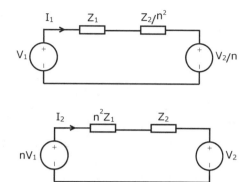

**Fig. 6.27** The equivalent
circuit of a transformer is
seen from the
secondary side

**Note 6.8** The impedance value is higher on the high voltage side.

**Note 6.9** Negative mutual inductance is observed as a $180°$ out of phase signal.

Equivalent circuit of an ideal transformer. Consider an ideal transformer with the impedance of primary and secondary windings. The equivalent circuit when the transformer is seen from the primary side, meaning that all voltages and impedances are transferred to the primary, is shown in Fig. 6.26. The impedance of the secondary $Z_2$ when transferred to the primary becomes $\frac{Z_2}{n^2}$, and the voltage of secondary $V_2$ when transferred to the primary becomes $\frac{V_2}{n}$.

The transformer equivalent circuit can also be seen from the secondary side. The values of the primary must be transferred to the secondary. Therefore, when transferred to the secondary, the impedance of the primary $Z_1$ becomes $n^2 Z_1$, and the voltage $V_1$ becomes $nV_1$. The equivalent circuit seen from the secondary side is shown in Fig. 6.27.

**Example 6.8 In the mutual inductance circuit shown in Fig. 6.28, find the currents in each loop. Find $i_1$ and $i_2$.**

$$KVL : -V_1 + R_1 I_1 + j\omega L_1 I_1 + j\omega M I_2 = 0$$

$$KVL : j\omega L_2 I_2 + j\omega M I_1 - V_2 + R_2 I_2 = 0$$

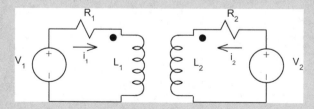

**Fig. 6.28** Circuit of Example 6.8

(continued)

**Example 6.8** (continued)

$$\begin{bmatrix} R_1 + j\omega L_1 & j\omega M \\ j\omega M & j\omega L_2 + R_2 \end{bmatrix}\begin{bmatrix} I_1 \\ I_2 \end{bmatrix} = \begin{bmatrix} V_1 \\ V_2 \end{bmatrix}$$

$$I_1 = \frac{\begin{vmatrix} V_1 & j\omega M \\ V_2 & j\omega L_2 + R_2 \end{vmatrix}}{\begin{vmatrix} R_1 + j\omega L & j\omega M \\ j\omega M & j\omega L_2 + R_2 \end{vmatrix}}$$

$$I_2 = \frac{\begin{vmatrix} R_1 + j\omega L\_1 & V_1 \\ j\omega M & V_2 \end{vmatrix}}{\begin{vmatrix} R_1 + j\omega L & j\omega M \\ j\omega M & j\omega L_2 + R_2 \end{vmatrix}}$$

**Example 6.9 Find the total energy stored in the mutual inductance circuit shown in Fig. 6.29.**

$$V_1(t) = L_1\frac{di_1}{dt} + M\frac{di_2}{dt}$$

$$P_1(t) = V_1(t) \cdot i_1(t) = \left(L_1\frac{di_1}{dt} + M\frac{di_2}{dt}\right) \cdot i_1$$

$$W_1(t) = (L_1, M) = \int\left(L_1\frac{di_1}{dt} + M\frac{di_2}{dt}\right) i_1\, dt$$

$$\text{Let}: \quad i_1(t) = I_1, \quad i_2(t) = I_2$$

$$W_1(t) = \int L_1 i_1 d_i + M i_1 di_2$$

$$= \frac{1}{2}l_1 I_1^2 (\text{energy in } L_1, \text{ self inductance}) + M I_1 I_2 (\text{energy in mutual inductance})$$

$$W_2(\text{energy in } L_2 \text{ only}) = \frac{1}{2}L_2 I_2^2$$

$$W_{\text{Total}} = W_1 + W_2$$

$$= \frac{1}{2}L_1 I_1^2 + \frac{1}{2}L_1 I_2^2 + M I_1 I_2$$

(continued)

**Example 6.9**  (continued)

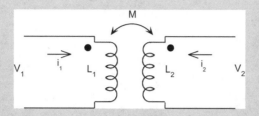

**Fig. 6.29**  Mutual inductance circuit of Example 6.9

**Example 6.10 In the circuit of Fig. 6.30, find the equivalent inductance from ports A and B.**

*Solution.* Using the $T$ equivalent circuit and considering the connection of two mutual inductances, the equivalent inductance becomes

$$L_{eq} = (L_1 - M) \| (L_2 - M) + M$$

$$= \frac{(L_1 - M)(L_2 - M)}{(L_1 - M) + (L_2 - M)} + M$$

$$L_{eq} = \frac{L_1 L_2 - M^2}{L_1 + L_2 - 2M}$$

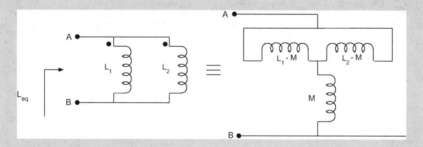

**Fig. 6.30**  Circuit of Example 6.10

**Example 6.11 In the circuit shown in Fig. 6.31, find $I_1$ and $I_2$ using equivalent circuits.**

*Solution.* The mutual inductance can be replaced by its $T$ equivalent circuit. The circuit with replaced $T$ model becomes (Fig. 6.32)

KVL in loop ①:

$$-120\angle30° + j10I_1 + j3(I_1 + I_2) = 0$$

KVL in loop ②:

$$j7I_2 + j3(I_1 + I_2) + 14I_2 = 0$$

$$\begin{bmatrix} j13 & j3 \\ j3 & j10 + 14 \end{bmatrix}\begin{bmatrix} I_1 \\ I_2 \end{bmatrix} = \begin{bmatrix} 120\angle30° \\ 0 \end{bmatrix}$$

$$\begin{bmatrix} I_1 \\ I_2 \end{bmatrix} = \begin{bmatrix} j13 & j3 \\ j3 & j10 + 14 \end{bmatrix}^{-1}\begin{bmatrix} 120\angle30° \\ 0 \end{bmatrix}$$

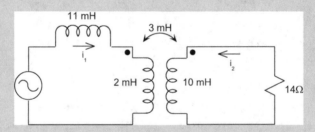

**Fig. 6.31** Circuit of Example 6.11

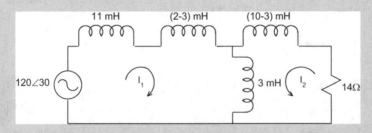

**Fig. 6.32** Circuit 6.31 is when the mutual inductance is replaced by the $T$ equivalent

(continued)

**Example 6.11** (continued)

$$I_1 = \frac{\begin{vmatrix} 120\angle 30° & j3 \\ 0 & j10+14 \end{vmatrix}}{\begin{vmatrix} j13 & j3 \\ j3 & j10+14 \end{vmatrix}}$$

$$= \frac{(60\sqrt{3}+j60)(14+j10)}{j13(14+j10)-(j3)^2}$$

$$= \frac{840\sqrt{3}+j600\sqrt{3}+j840-600}{j182-130+9}$$

$$= \frac{854.92+1879.23j}{j182-121} = \frac{2064.56\angle 65.54°}{218.55\angle 123.62°}$$

$$= 9.45\angle -58.07° \text{ A}$$

$$I_2 = \frac{\begin{vmatrix} j13 & 60\sqrt{3}+j60 \\ j3 & 0 \end{vmatrix}}{\begin{vmatrix} j13 & j3 \\ j3 & j10+14 \end{vmatrix}}$$

$$= \frac{-j3(60\sqrt{3}+j60)}{218.55\angle 123.62°} = \frac{-j180\sqrt{3}+180}{218.55\angle 123.62°}$$

$$= \frac{360\angle -120°}{218.55\angle 123.62°} = 1.54\angle -243.62° \text{ A}$$

## Problems

6.1. A 2 mH inductor has $N = 100$ turns. If the coil current is $i(t) = 3 \sin 120\pi t$, find the flux variations in the core and the induced voltage at the terminals.

6.2. Two inductors are located so that maximum flux linkage may occur. If the inductance of the coils is $L_1 = 25H$, $L_2 = 4H$, what is the maximum possible mutual inductance between the coils?

6.3. Determine the polarity of the induced voltage at the terminals of any of the following mutual inductance circuits.

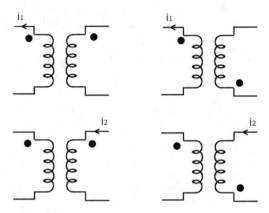

6.4. Calculate the total inductance of the following circuits.

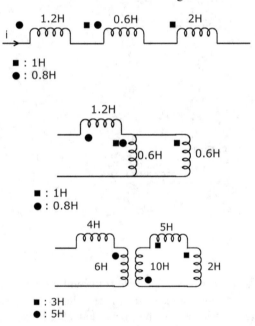

6.5. Find $V_o$ across the capacitor in the following circuit.

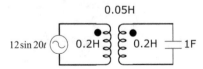

6.6. Find the input impedance in the following circuit.

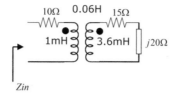

6.7. Find the *PF* of the circuit.

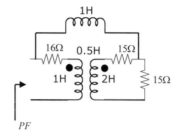

6.8. Find the reactance *X* such that the maximum power is transferred to the 10Ω load.

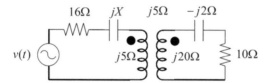

6.9. Find the current in the circuit.

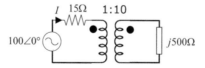

6.10. Find the currents $I_1$, $I_2$, and $I_3$.

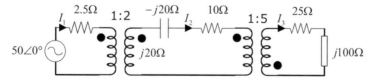

# Chapter 7
# Laplace Transform and Its Application in Circuits

## Introduction

Most of the circuits introduced have been analyzed in the time domain. This means that the input to the circuit, the circuit variables, and the responses have been presented as a function of time. All the input functions, such as unit step, ramp, impulse, exponential, and sinusoidal, have been introduced as a time-dependent variable, and their effects on circuits have been identified directly as a function of time. This required utilization of differential equations and solutions in the time domain. However, high-order circuits result in high-order differential equations, which are sometimes hard to solve considering the initial conditions. In addition, the time domain analysis is a limiting factor for circuits exposed to a spectrum of frequencies, such as filters.

Time domain analysis can be transformed into a frequency domain analysis to simplify the analysis of high-order circuits and incorporate the variable frequency nature of some circuits in effect. The Laplace transform is one of the transformations that can take the circuits from the time domain to the frequency domain. In the frequency domain, the input to the circuit, the circuit itself, and the results are obtained using algebraic equations. The system response can be transformed back into the time domain through the inverse of the Laplace transform.

The process of using the Laplace transform to analyze circuits is as follows:

1. Find the Laplace of the input functions.
2. Represent the circuit in the frequency domain.
3. Then find the desired response in the frequency domain.
4. Transform back to the time domain.

Figure 7.1 demonstrates a set of differential equations represented in the time domain. These equations are transformed to the frequency domain using the Laplace transform and are presented by algebraic equations, and the Laplace inverse takes the circuit back to the time domain.

A. Izadian, *Fundamentals of Modern Electric Circuit Analysis and Filter Synthesis*, https://doi.org/10.1007/978-3-031-21908-5_7

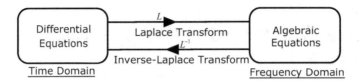

**Fig. 7.1** Laplace transform from time to the frequency domain and inverse Laplace transform from the frequency domain to the time domain

## Mathematical Background

Laplace transform is defined for a given function $f(t)$ to be represented in the frequency domain using the Laplace operator $s$. It reads Laplace of $f(t)$ is $F(s)$ and is written as

$$L\{f(t)\} = F(s)$$

Laplace transform is a two-sided transform that can be defined as the time from $-\infty$ to $+\infty$. The Laplace transform is defined as

$$L\{f(t)\} \triangleq \int_{-\infty}^{+\infty} f(t)e^{-st}dt$$

In this transform, the time domain representation of the function is transformed to a complex frequency domain $s = \sigma + j\omega$. However, this frequency remains a variable in all presentations of functions.

Because most electric circuits are defined and operated in positive time, the Laplace transform in this book is defined as one-sided, e.g., in positive time. Therefore, a one-sided transform is defined as follows:

$$L\{f(t)\} = F(s) \triangleq \int_0^{+\infty} f(t)e^{-st}dt.$$

The Laplace transform exists for a given function $f(t)$ if and only if the integral $\int_0^{+\infty} f(t)e^{-st}dt \neq \infty$ exists.

The inverse of the Laplace transform takes the functions from the frequency domain to the time domain. It is defined as

$$f(t) = L^{-1}\{F(s)\}$$

**Note 7.1** The functions in the time domain are presented in lowercase alphabets such as $f$, and when the frequency domain is presented, capital letters are used, such as $F$. This is helpful when the time variable $t$ and frequency operator $s$ are not shown in the transformations.

## Laplace of Unit Step Function

$$f(t) = u(t)$$

According to the definition,

$$U(s) = L\{u(t)\} \triangleq \int_0^\infty u(t) e^{-st} dt$$

However, the value of $u(t)$ for time-positive $t \geq 0$ is 1. Therefore,

$$U(s) = \int_0^\infty 1 e^{-st} dt = \frac{1}{-s} e^{-st} \Big|_0^\infty = \frac{1}{-s} \left( e^{-\infty} - e^0 \right) = \frac{1}{s}$$

Therefore,

$$L\{u(t)\} = \frac{1}{s}$$

**Example 7.1 Find the Laplace transform of $10u(t)$.**
*Solution*

$$U(s) = \int_0^\infty 10 e^{-st} dt = \frac{10}{-s} e^{-st} \Big|_0^\infty = -\frac{10}{s} \left( e^{-\infty} - e^0 \right) = \frac{10}{s}$$

## Laplace of Impulse Function

$$f(t) = \delta(t)$$

According to the definition,

$$\Delta(s) = L\{\delta(t)\} \triangleq \int_0^\infty \delta(t) e^{-st} dt$$

The impulse function has a value at the time that makes the argument zero. Therefore, this function has a value of $t = 0$. The integral becomes

$$\Delta(s) = \int_0^\infty \delta(t)e^{-st}dt\big|_{t=0} = 1$$

Therefore,

$$L\{\delta(t)\} = 1$$

## Laplace of Ramp Function

$$f(t) = r(t) = ktu(t)$$

According to the definition,

$$R(s) = L\{r(t)\} \triangleq \int_0^\infty r(t)e^{-st}dt$$

The value of function is $r(t) = ktu(t)$. Therefore,

$$R(s) = \int_0^\infty ktu(t)e^{-st}dt = k\left(t\frac{1}{-s}e^{-st} - \frac{1}{s^2}e^{-st}\right)\big|_0^\infty = k\frac{1}{s^2}$$

**Example 7.2 *Laplace*. $L\{5tu(t)\}$.**
*Solution.* The Laplace of a ramp with slope 5 is $F(s) = \frac{5}{s^2}$.

**Example 7.3 *Laplace inverse*. Find the original time-domain function $f(t)$ if the Laplace transform is $F(s) = \frac{10}{s^2}$.**
*Solution.* The Laplace function matches the ramp function $tu(t)$, and 10 is just an amplitude/coefficient/slope. Therefore, the original time domain function has been $f(t) = 10tu(t)$.
   Therefore,

$$L^{-1}\left\{\frac{10}{s^2}\right\} = 10tu(t).$$

## Laplace of Exponential Function

$$f(t) = e^{at}\ u(t)$$

The $u(t)$ part of the function represents that the function is defined in positive time, i.e., for $t \geq 0$.

According to the definition,

$$F(s) = L\{f(t)\} \triangleq \int_0^\infty f(t)e^{-st}dt$$

Therefore,

$$F(s) = \int_0^\infty e^{at}e^{-st}dt = \int_0^\infty e^{-t(s-a)}dt = \frac{1}{-(s-a)}e^{-(s-a)t}\Big|_0^\infty = \frac{1}{-(s-a)}$$

$$\times \left(e^{-\infty} - e^0\right) = \frac{1}{(s-a)}$$

Hence,

$$L\{e^{at}\ u(t)\} = \frac{1}{(s-a)}$$

---

**Example 7.4 *Laplace*. $L\{10e^{2t}u(t)\}$.**
*Solution.* The Laplace of the exponential function defined in positive time at the amplitude of 10 is $F(s) = \frac{10}{s-2}$.

---

**Example 7.5 *Laplace inverse*. Find the original time-domain function $f(t)$ if the Laplace transform is $F(s) = \frac{-3}{s+5}$.**
*Solution.* The form of the Laplace function $\frac{1}{s+a}$ matches that of the exponential function, and $-3$ is just an amplitude/coefficient. Therefore, the original time domain function has been $f(t) = -3e^{-5t}u(t)$. In this function, $u(t)$ means that the function is defined in positive time.

(continued)

**Example 7.5** (continued)

Since the time domain function is obtained from the frequency domain function, it can be written as

$$L^{-1}\left\{\frac{-3}{s+5}\right\} = -3e^{-5t}u(t).$$

**Example 7.6 *Laplace*. $L\{e^{-t}u(t)\}$.**
*Solution.* This is exponential with a damping factor of $-1$, and the Laplace transform is $F(s) = \frac{1}{s+1}$.

**Note 7.2** Note that if the exponential function has positive damping, the Laplace function will have a negative shift in frequency and vice versa.

**Example 7.7 *Laplace inverse*. Find the original time domain function $f(t)$ if the Laplace transform is $F(s) = \frac{3}{s+1}$.**
*Solution.* The form of Laplace $\frac{1}{s+1}$ matches with that of the exponential function $e^{\alpha t}u(t)$, where $\alpha = -1$ and 3 is just a coefficient. Therefore, the original time domain function has been $f(t) = 3e^{-t}u(t)$.
   Therefore,

$$L^{-1}\left\{\frac{3}{s+1}\right\} = 3e^{-t}u(t).$$

**Example 7.8 *Laplace*. $L\{10e^{20t}u(t)\}$.**
*Solution.* This is an exponential with damping of $+20$ and an amplitude of 10. Therefore, $F(s) = \frac{10}{s-20}$.

**Example 7.9 *Laplace inverse*. Find the original time domain function $f(t)$ if the Laplace transform is $F(s) = \frac{5}{s+20}$.**
*Solution.* The form of Laplace $\frac{1}{s+20}$ matches with that of the exponential function $e^{\alpha t}u(t)$, where $\alpha = -20$ and 5 is just a coefficient. Therefore, the original time domain function has been $f(t) = 5e^{-20t}u(t)$.

(continued)

**Example 7.9** (continued)

Therefore,

$$L^{-1}\left\{\frac{5}{s+20}\right\} = 5e^{-20t}u(t).$$

## Laplace of Sinusoidal Function

$$f(t) = \textbf{sin}\,(\omega t)$$

According to the definition,

$$F(s) = L\{f(t)\} \triangleq \int_0^\infty f(t)e^{-st}dt$$

Therefore,

$$F(s) = \int_0^\infty \sin\,(\omega t)e^{-st}dt$$

However, $\sin\,(\omega t) = \frac{e^{j\omega t} - e^{-j\omega t}}{j2}$. Therefore,

$$\infty \int_0^\infty \frac{e^{j\omega t} - e^{-j\omega t}}{j2}e^{-st}dt = \infty \int_0^\infty \frac{e^{-(s-j\omega)t} - e^{-(s+j\omega)t}}{j2}dt$$

$$= \frac{1}{j2}\left(\frac{1}{-(s-j\omega)}e^{-(s-j\omega)t} - \frac{1}{-(s+j\omega)}e^{-(s+j\omega)t}\right)\Big|_0^\infty$$

$$= \frac{-1}{j2}\left(\frac{1}{(s-j\omega)} - \frac{1}{(s+j\omega)}\right) = \frac{\omega}{s^2 + \omega^2}$$

Therefore,

$$L\{\,\textbf{sin}\,(\omega t)\} = \frac{\omega}{s^2 + \omega^2}$$

## Laplace of Co-sinusoidal Function

$$f(t) = \cos(\omega t)$$

According to the definition,

$$F(s) = L\{f(t)\} \triangleq \int_0^\infty f(t)e^{-st}dt$$

Therefore,

$$F(s) = \int_0^\infty \cos(\omega t)e^{-st}dt$$

However, $\cos(\omega t) = \frac{e^{j\omega t}+e^{-j\omega t}}{2}$. Therefore,

$$\infty \int_0^\infty \frac{e^{j\omega t}+e^{-j\omega t}}{2}e^{-st}dt = \infty \int_0^\infty \frac{e^{-(s-j\omega)t}+e^{-(s+j\omega)t}}{2}dt$$

$$= \frac{1}{2}\left(\frac{1}{-(s-j\omega)}e^{-(s-j\omega)t} - \frac{1}{(s+j\omega)}e^{-(s+j\omega)t}\right)\Big|_0^\infty$$

$$= \frac{1}{2}\left(\frac{1}{(s-j\omega)} + \frac{1}{(s+j\omega)}\right) = \frac{s}{s^2+\omega^2}$$

Therefore,

$$L\{\cos(\omega t)\} = \frac{s}{s^2+\omega^2}$$

**Example 7.10 Find {sin10t u(t)}.**
*Solution.* The sinusoidal function at an angular frequency of $\omega = 10$ is

$$L\{\sin 10t u(t)\} = \frac{10}{s^2+10^2} = \frac{10}{s^2+100}$$

**Example 7.11 Find the original time domain function that resulted in $F(s) = \frac{50}{s^2+9}$.**

*Solution.* Looking at the format of the Laplace transform $\frac{50}{3} \frac{3}{s^2+3^2}$ indicates a sinusoidal function at a frequency of $\omega = 3$ and amplitude of $\frac{50}{3}$. Therefore, $f(t) = \frac{50}{3} \sin 3tu(t)$. Here, $u(t)$ function indicates that the function $f(t)$ is defined at a time positive.

**Example 7.12 Find $L\{cos5t\ u(t)\}$.**

*Solution.* $L\{\cos 5tu(t)\} = \frac{s}{s^2+5^2} = \frac{s}{s^2+25}$

**Example 7.13 Find the original time domain function that resulted in $F(s) = \frac{10s}{s^2+16}$.**

*Solution.* The format of the Laplace transform $10\frac{s}{s^2+4^2}$ indicates a co-sinusoidal function at a frequency of $\omega = 4$ and an amplitude of 10. Therefore, $f(t) = 10 \cos4t\ u(t)$. The $u(t)$ part indicates that the function $f(t)$ is defined at a time positive.

**Example 7.14 Find the original time domain function that resulted in $F(s) = \frac{10s+3}{s^2+16}$.**

*Solution.* Looking at the numerator of the Laplace transform, if it is split into $10\frac{s}{s^2+4^2} + \frac{3}{4}\frac{4}{s^2+4^2}$, it shows a summation of two parts: (1) cos function at a frequency of $\omega = 4$ and an amplitude of 10 and (2) sin function at a frequency of $\omega = 4$ and amplitude of $\frac{3}{4}$. Therefore, $f(t) = \left(10\cos 4t + \frac{3}{4}\sin 4t\right)\ u(t)$.

**Example 7.15 Find $L\{10 \sin(5t + 30)\}$.**

*Solution*

$$L\{10\sin(5t+30)\} = L\{10(\sin 5t \cos 30 + \cos 5t \sin 30)\}$$

$$= 10\frac{5}{s^2+25}0.866 + 10\frac{s}{s^2+25}0.5$$

$$= \frac{43.3}{s^2+25} + \frac{5s}{s^2+25} = \frac{5s+43.3}{s^2+25}$$

## Laplace of Hyperbolic Sinusoidal Function

$$f(t) = \sin h(\omega t)$$

According to the definition,

$$F(s) = L\{f(t)\} \triangleq \int_0^\infty f(t)e^{-st}dt$$

Therefore,

$$F(s) = \int_0^\infty \sin h(\omega t)e^{-st}dt$$

However, $\sin h(\omega t) = \frac{e^{\omega t} - e^{-\omega t}}{2}$. Therefore,

$$\infty \int_0^{} \frac{e^{\omega t} - e^{-\omega t}}{2} e^{-st}dt = \infty \int_0^{} \frac{e^{-(s-\omega)t} - e^{-(s+\omega)t}}{2} dt$$

$$= \frac{1}{2}\left(\frac{1}{-(s-\omega)}e^{-(s-\omega)t} - \frac{1}{-(s+\omega)}e^{-(s+\omega)t}\right)\Big|_0^\infty$$

$$= \frac{-1}{2}\left(\frac{1}{(s-\omega)} - \frac{1}{(s+\omega)}\right) = \frac{\omega}{s^2 - \omega^2}$$

Therefore,

$$L\{\sin h(\omega t)\} = \frac{\omega}{s^2 - \omega^2}$$

## Laplace of Hyperbolic Co-sinusoidal Function

$$f(t) = \cos h(\omega t)$$

According to the definition,

$$F(s) = L\{f(t)\} \triangleq \int_0^\infty f(t)e^{-st}dt$$

Therefore,

$$F(s) = \int_0^\infty \cos h(\omega t) e^{-st} dt$$

However, $\cos h(\omega t) = \frac{e^{\omega t} + e^{-\omega t}}{2}$. Therefore,

$$\infty \int_0^{} \frac{e^{\omega t} + e^{-\omega t}}{2} e^{-st} dt = \infty \int_0^{} \frac{e^{-(s-\omega)t} + e^{-(s+\omega)t}}{2} dt$$

$$= \frac{1}{2}\left( \frac{1}{-(s-\omega)} e^{-(s-\omega)t} - \frac{1}{(s+\omega)} e^{-(s+\omega)t} \right)\Big|_0^\infty$$

$$= \frac{1}{2}\left( \frac{1}{(s-\omega)} + \frac{1}{(s+\omega)} \right) = \frac{s}{s^2 - \omega^2}$$

Therefore,

$$L\{\cos h(\omega t)\} = \frac{s}{s^2 - \omega^2}$$

**Example 7.16 Find $L\{\sinh 10 tu(t)\}$.**
*Solution.* $L\{\sin h10tu(t)\} = \frac{10}{s^2 - 10^2} = \frac{10}{s^2 - 100}$.

**Example 7.17 Find the original time domain function that resulted**
**in $F(s) = \frac{50}{s^2 - 9}$.**
*Solution.* Looking at the format of the Laplace transform $\frac{50}{3} \frac{3}{s^2 - 3^2}$ indicates a
hyperbolic sinusoidal function at a frequency of $\omega = 3$ and amplitude of $\frac{50}{3}$.
Therefore, $f(t) = \frac{50}{3} \sin h3tu(t)$.

**Example 7.18 Find $\{\cosh 5tu(t)\}$.**
*Solution*

$$L\{\cos h5tu(t)\} = \frac{s}{s^2 - 5^2} = \frac{s}{s^2 - 25}$$

**Example 7.19 Find the original time domain function that resulted in** $F(s) = \frac{10s}{s^2-16}$.

*Solution.* Looking at the format of the Laplace transform $10\frac{s}{s^2-4^2}$ indicates a hyperbolic sinusoidal function at a frequency of $\omega = 4$ and an amplitude of 10. Therefore, $f(t) = 10cosh4tu(t)$.

**Example 7.20 Find the original time domain function that resulted in** $F(s) = \frac{10s+3}{s^2-16}$.

*Solution.* Looking at the format of the Laplace transform, if the function is split into $10\frac{s}{s^2-4^2} + \frac{3}{4}\frac{4}{s^2-4^2}$, indicates the summation of two parts: (1) a cos$h$ function at a frequency of $\omega = 4$ and an amplitude of 10 and (2) a sin$h$ function at a frequency of $\omega = 4$ and an amplitude of $\frac{3}{4}$. Therefore, $f(t) = \left(10\cos h4t + \frac{3}{4}\sin h4t\right)u(t)$.

**Example 7.21 Find** $L\{10\ \text{sin}h(5t + 3)\}$.
*Solution*

$$L\{10 \sin h(5t+3)\} = L\{10(\sin h5t \cos h3 + \cos h5t \sin h3)\}$$

$$= 10\frac{5}{s^2+25}10.06 + 10\frac{s}{s^2+25}10.01$$

$$= \frac{503}{s^2+25} + \frac{101.1s}{s^2+25} = \frac{101.1s+503}{s^2+25}$$

## Laplace of Derivatives of Impulse

$$L\{\dot{\delta}\,(t)\}$$

Each derivative of the impulse function generates an $s$ factor. Therefore,

$$L\{\dot{\delta}(t)\} = s$$

$$L\{\ddot{\delta}(t)\} = s^2$$

In general,

$$L\left\{\delta^{(n)}(t)\right\} = s^n$$

## Laplace of Differential Functions

$$L\left\{\frac{df(t)}{dt}\right\} = sF(s) - f(0^+)$$

where $F(s)$ is the Laplace of function $f(t)$ in frequency domain and $f(0^+)$ is the time domain initial condition at $t = 0^+$.

**Note 7.3**

$$\frac{df(t)}{dt} \equiv \dot{f}(t)$$

$$L\left\{\frac{d^2f(t)}{dt^2}\right\} = s^2 F(s) - sf(0^+) - f'(0^+)$$

where $f'(0^+)$ is the differential of the initial condition in the time domain at $t = 0^+$.

**Note 7.4**

$$\frac{d^2f(t)}{dt^2} \equiv \ddot{f}(t)$$

$$L\left\{\frac{d^n f(t)}{dt^n}\right\} = s^n F(s) - s^{n-1} f(0^+) - s^{n-2} f'(0^+) - \cdots - f^{(n-1)}(0^+)$$

*Solve the following differential equations using Laplace transforms.*

**Example 7.22** Solve for $v(t).\dot{v} + 3v = (t)$, $v(0^+) = 1$.
*Solution.* Taking Laplace of both sides results in

$$L\{\dot{v} + 3v\} = L\{\delta(t)\}$$

$$L\{\dot{v}\} + 3L\{v\} = L\{\delta(t)\}$$

(continued)

**Example 7.22** (continued)

$$sV(s) - v(0^+) + 3V(s) = 1$$

Replacing from the initial condition,

$$sV(s) - 1 + 3V(s) = 1$$

Solving for $V(s)$,

$$V(s) = \frac{2}{s+3}$$

Original time domain function $v(t)$ has the format of exponential at damping factor $-3$ and coefficient 2 as follows:

$$v(t) = 2e^{-3t}u(t)$$

$u(t)$ determines the function, defined as a positive time.

**Example 7.23 Solve for** $v(t).\ddot{v} + 4\dot{v} + 3v = 0$, $v(0^+) = -1, \dot{v}(0^+) = 2$.
*Solution.* Taking Laplace from both sides of the equation results in

$$L\{\ddot{v} + 4\dot{v} + 3v = 0\}$$
$$(s^2 V(s) - sv(0^+) - \dot{v}(0^+)) + 4(sV(s) - v(0^+)) + 3V(s) = 0$$
$$(s^2 V(s) - s(-1) - (2)) + 4(sV(s) - (-1)) + 3V(s) = 0$$
$$V(s)(s^2 + 4s + 3) + s - 2 + 4 = 0$$
$$V(s) = \frac{-s+2}{s^2 + 4s + 3} = \frac{-s-2}{(s+1)(s+3)}$$

Splitting the fraction into two and assuming coefficients $A$ and $B$ result in

$$V(s) = \frac{A}{s+1} + \frac{B}{s+3} \equiv \frac{-s-2}{(s+1)(s+3)}$$

The common denominator and simplification yield

(continued)

**Example 7.23** (continued)

$$V(s) = \frac{(A+B)s + 3A + B}{(s+1)(s+3)} \equiv \frac{-s-2}{(s+1)(s+3)}$$

This results in $A + B = -1$ and $3A + B = -2$.
This gives $A = -\frac{1}{2}, B = -\frac{1}{2}$. Therefore,

$$V(s) = \frac{-\frac{1}{2}}{s+1} + \frac{-\frac{1}{2}}{s+3}$$

The original time domain function $v(t)$ has two exponential structures at damping factors $-1$ and $-3$ with coefficients 1 and $-2$, respectively. Therefore,

$$v(t) = \left( -\frac{1}{2}e^{-t} - \frac{1}{2}e^{-3t} \right)u(t)$$

## Laplace Operations

Several operations can simplify the Laplace transformation from frequency to time and vice versa. These operations have indicators discussed in this section, and note that these indicators trigger specific operations that need to be carefully considered. Otherwise, the transformation may become wrong.

### *Linear Combination of Functions*

$$L\{\alpha f(t) \pm \beta g(t)\}$$

Laplace of the linear combination of several functions is the summation of the Laplace of those functions.

Consider $L\{f(t)\} = F(s)$, and $L\{g(t)\} = G(s)$; then

$$L\{\alpha f(t) \pm \beta g(t)\} = \alpha L\{f(t)\} \pm \beta L\{g(t)\} = \alpha F(s) \pm \beta G(s)$$

**Note 7.5** This is true only for the summation of functions and *not* the product. Therefore,

$$L\{f(t)g(t)\} \neq L\{f(t)\}L\{g(t)\}$$

**Example 7.24 Find $L\{e^{-5t} + \sin 3t\}$.**
*Solution*

$$L\{e^{-5t} + \sin 3t\} = L\{e^{-5t}\} + L\{\sin 3t\} = \frac{1}{s+5} + \frac{3}{s^2 + 9}$$

**Example 7.25 Find $L\{\cos 5t + 2\sin 3t\}$.**
*Solution*

$$L\{\cos 5t + 2\sin 3t\} = L\{\cos 5t\} + 2L\{\sin 3t\} = \frac{s}{s^2 + 25} + 2\frac{3}{s^2 + 9}$$

$$= \frac{7s^2 + 159}{(s^2 + 25)(s^2 + 9)}$$

**Example 7.26 Find $L\{-\cos 5t + 2\sin 5t\}$.**
*Solution*

$$L\{-\cos 5t + 2\sin 5t\} = -L\{\cos 5t\} + 2L\{\sin 5t\} = \frac{-s}{s^2 + 25}$$

$$+ 2\frac{5}{s^2 + 25} = \frac{-s + 10}{(s^2 + 25)}$$

**Example 7.27 Find the original time-domain (Laplace inverse)**
**of $L^{-1}\left\{\frac{-s+10}{(s^2+25)}\right\}$.**
*Solution.* Splitting the function reveals two known structures for cos as $\frac{s}{s^2+\omega^2}$
and sin as $\frac{\omega}{s^2+\omega^2}$. Therefore,

$$L^{-1}\left\{\frac{-s}{(s^2 + 25)} + \frac{10}{(s^2 + 25)}\right\} = -\cos 5t + \frac{10}{5}\sin 5t, \, t \geq 0.$$

## *Shift in Time*

$$L\{f(t+a)\}$$

A shift in the time domain becomes exponential in the frequency domain.

$$L\{f(t+a)\} = e^{as}F(s)$$

---

**Example 7.28 Find the Laplace of the function shown in Fig. 7.2.**

**Fig. 7.2** Pulse function of
Example 7.28

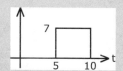

*Solution.* The function is a summation of two shifted unit steps by 5 and 10*s*, respectively, at an amplitude of 7 and −7. Therefore, $f(t) = 7u(t − 5) − 7u(t − 10)$. The Laplace becomes

$$L\{7u(t-5) - 7u(t-10)\} = 7L\{u(t-5)\} - 7L\{u(t-10)\}$$

$$= 7e^{-5s}\frac{1}{s} - 7e^{-10s}\frac{1}{s} = \frac{7}{s}\left(e^{-5s} - e^{-10s}\right)$$

---

**Example 7.29 Find the Laplace of the function** $f(t) = 2tu(t − 3)$.
*Solution.* As the function shows, only the step function is shifted by 3*s*. To use the time shift operations, all-time functions must have been shifted by the same 3*s*. To obtain such a function, the part that is not shifted will be shifted manually as follows:

$$f(t) = 2(t - 3 + 3)u(t - 3)$$

$$f(t) = 2(t - 3)u(t - 3) + 2(3)u(t - 3)$$

Now the function is shifted by the same amount of 3*s*, and the Laplace operation can be used as follows:

$$L\{2(t-3)u(t-3) + 2(3)u(t-3)\} = \frac{2}{s^2}e^{-3s} + \frac{6}{s}e^{-3s}$$

## Product by an Exponential

$$L\{e^{-at}f(t)\}$$

The appearance of an exponential and function $f(t)$ triggers a shift in frequency as follows:

$$L\{e^{-at}f(t)\} = F(s+a)$$

**Note 7.6** This means that after taking the Laplace of $f(t)$ as $F(s)$, convert all $s \to (s+a)$.

**Example 7.30** Find $L\{e^{-5t}u(t)\} = ?$
*Solution.* The exponential $e^{-5t}$ triggers a shift in frequency over the $L\{u(t)\} = \frac{1}{s}$ by $s \to s + 5$. This results in

$$L\{e^{-5t}u(t)\} = \frac{1}{s+5}$$

**Example 7.31** Find $L\{e^{-5t}\sin 3t\} = ?$
*Solution.* The exponential $e^{-5t}$ triggers a shift in frequency over the $L\{\sin 3t\} = \frac{3}{s^2+9}$ by $s \to s + 5$. This results in

$$L\{e^{-5t}\sin 3t\} = \frac{3}{(s+5)^2+9}$$

**Example 7.32** Find $L\{e^{-2t}\cosh tu(t-5)\} = ?$
*Solution.* Only one part of the function is shifted by $5s$. The rest should be manually shifted as follows:

$$f(t) = e^{-2(t-5+5)}\cos h(t-5+5)u(t-5)$$

(continued)

**Example 7.32** (continued)

$$f(t) = e^{-2(t-5)}e^{-2(5)}[\cos h(t-5)\cos h(5) - \sin h(t-5)\sin h(5)]u(t-5)$$

$$f(t) = \left[e^{-2(5)}\cos h(5)e^{-2(t-5)}\cos h(t-5) - e^{-2(5)}\sin h(5)e^{-2(t-5)}\sin h(t-5)\right]u(t-5)$$

Laplace transform of $f(t)$ becomes

$$F(s) = e^{-10}\cos h(5)\frac{s+2}{(s+2)^2+1}e^{-5s} + e^{-10}\sin h(5)\frac{1}{(s+2)^2+1}e^{-5s}$$

## *Product by Time Factors*

$$L\{tf(t)\}$$

Factors of $t$ trigger a differential in the frequency domain as follows:

$$L\{tf(t)\} = -\frac{d}{ds}F(s)$$

In general, when multiplied by $t^n$,

$$L\{t^n f(t)\} = (-1)^n\frac{d^n}{ds^n}F(s)$$

**Example 7.33 Find $L\{tu(t)\}$.**
*Solution.* $t$ indicates a derivative operation $-\frac{d}{ds}$ of the Laplace of $u(t)$. Considering that $\{u(t)\} = \frac{1}{s}$, it can be concluded that $L\{tu(t)\} = -\frac{d}{ds}\left(\frac{1}{s}\right) = \frac{1}{s^2}$.

**Example 7.34 Find $L\{t^2 u(t)\}$.**
*Solution.* $t^2$ indicates a second derivative of the Laplace of the function $u(t)$. Therefore,

$$L\{t^2 u(t)\} = -\frac{d}{ds}\left(-\frac{d}{ds}\left(\frac{1}{s}\right)\right) = -\frac{d}{ds}\left(\frac{1}{s^2}\right) = \frac{2}{s^3}$$

**Example 7.35 Find $L\{t^n u(t)\}$.**
*Solution.* Accordingly,

$$L\{t^n u(t)\} = \frac{n!}{s^{n+1}}$$

**Example 7.36 Find $L\{t^3 e^{-5t} u(t)\}$.**
*Solution.* The exponential $e^{-5t}$ triggers a shift in frequency over the Laplace of the function $t^3$. Therefore, $L\{t^3\} = \frac{3!}{s^{3+1}} = \frac{3!}{s^4}$ shifted by $s \to s + 5$ results in

$$L\{t^3 e^{-5t}\} = \frac{3!}{(s+5)^4}$$

## *Divide by Time Factors*

$$L\left\{\frac{1}{t} f(t)\right\}$$

The factor $\frac{1}{t}$ indicates an integral of the Laplace transform of the function $f(t)$ as follows:

$$L\left\{\frac{1}{t} f(t)\right\} = \int_0^s F(x) dx$$

## Complementary Laplace Inverse Techniques

To obtain the time domain transform of the Laplace functions, it is best to convert the function to individual functions or templates with known Laplace inverse transforms. Some of the techniques that can help identify known templates from given frequency domain functions are introduced in this section.

## Long Division

In fractions with a polynomial in the numerator with higher order than the polynomial in the denominator, a long division results in simpler functions to break into known templates.

**Example 7.37 Find the Laplace inverse of the function** $L^{-1}\left\{\frac{s^5+2s^4+s+1}{s^2+2}\right\}$.

*Solution.* Long division results in

$$\frac{s^5+2s^4+s+1}{s^2+2} = s^3 + 2s^2 - 2s - 4 + \frac{5s+9}{s^2+2}$$

Split the fraction into two terms:

$$\frac{s^5+2s^4+s+1}{s^2+2} = s^3 + 2s^2 - 2s - 4 + \left(\frac{5s+9}{s^2+2}\right)$$

$$= s^3 + 2s^2 - 2s - 4 + \left(5\frac{s}{s^2+\left(\sqrt{2}\right)^2} + \frac{9}{\sqrt{2}}\frac{\sqrt{2}}{s^2+\left(\sqrt{2}\right)^2}\right)$$

Now, all terms are presented in known templates:

$$L^{-1}\left\{s^3 + 2s^2 - 2s - 4 + 5\frac{s}{s^2+\left(\sqrt{2}\right)^2} + \frac{9}{\sqrt{2}}\frac{\sqrt{2}}{s^2+\left(\sqrt{2}\right)^2}\right\}$$

$$= \ldots \dddot{\delta} + 2\ddot{\delta} - 2\dot{\delta} - 4\delta + 5\cos\sqrt{2}t + \frac{9}{\sqrt{2}}\sin\sqrt{2}t, t \geq 0.$$

## Partial Fraction Expansion

This technique allows fractions with high-order polynomials to be broken into simpler functions and known templates to find the Laplace inverse transforms. Consider a fraction with numerator and denominator polynomials where the order of numerator is less than the order of denominator:

$$F(s) = \frac{\text{Num}(s)}{\text{Den}(s)}$$

Several cases may exist depending on the form and order of the denominator.

## Case 1. Real Distinct Roots

Simple distinct roots. In this case, the denominator has distinct roots. The numerator coefficients can be calculated by

$$F(s) = \frac{\text{Num}(s)}{\prod\limits_{i=1}^{n}(s+p_i)}$$

$$F(s) = \sum\limits_{i=1}^{n}\frac{A_i}{(s+p_i)}$$

Each $A_i$ coefficient is obtained by a product of the original function by the denominator of the fraction $\frac{A_i}{(s+p_i)}$ or $(s+p_i)$ and evaluating the entire product at the root of the denominator as follows:

$$A_i = ((s+p_i).F(s))|_{s=-p_i}$$

**Example 7.38 Do partial fraction expansion of a function $F(s) = \frac{s-1}{s^2+5s+6}$.**
*Solution.* The denominator $s^2 + 5s + 6$ has two roots: $s = -2$ and $s = -3$. Therefore,

$$F(s) = \frac{s-1}{s^2+5s+6} = \frac{A_1}{s+2} + \frac{A_2}{s+3}$$

To find each of the coefficients $A_1$ and $A_2$, the procedure is as follows:

$$A_1 = (s+2)\frac{s-1}{s^2+5s+6}\Big|_{s=-2} = \frac{-2-1}{-2+3} = -3$$

$$A_2 = (s+3)\frac{s-1}{s^2+5s+6}\Big|_{s=-3} = \frac{-3-1}{-3+2} = 4$$

$$F(s) = \frac{s-1}{s^2+5s+6} = \frac{-3}{s+2} + \frac{4}{s+3}$$

## Case 2. Repeated Roots

The denominator has repeated roots, root $p_j$ is repeated $m$ times, and some other roots are repeated.

$$F(s) = \frac{\text{Num}(s)}{(s+p_j)^m \prod\limits_{i=1}^{n-m}(s+p_i)}$$

The partial fraction expansion must include all power of $s + p_j$ from 1 to $m$ and all distinct roots $p_i$ as follows:

$$F(s) = \frac{B_1}{(s+p_j)^m} + \frac{B_2}{(s+p_j)^{m-1}} + \ldots + \frac{B_m}{s+p_j} + \sum_{i=1}^{n-m} \frac{A_i}{(s+p_i)}$$

To find $B$ coefficients:

$$B_1 = (s+p_j)^m F(s)\big|_{s=-p_j}$$

$$B_2 = \frac{d(s+p_j)^m F(s)}{ds}\big|_{s=-p_j}$$

In general form,

$$B_m = \frac{1}{(m-1)!} \frac{d^{m-1}(s+p_j)^m F(s)}{ds^{m-1}}\big|_{s=-p_j}$$

**Example 7.39 Do partial fraction expansion of a function $F(s) =$**
$\frac{s-1}{(s+1)^3(s^2+5s+6)}$.
*Solution.* The denominator has two distinct roots of $s = -2$ and $s = -3$ and three repeated roots of $s = -1$.

Therefore,

$$F(s) = \frac{s-1}{(s+1)^3(s^2+5s+6)} = \frac{B_1}{(s+1)^3} + \frac{B_2}{(s+1)^2} + \frac{B_3}{s+1} + \frac{A_1}{s+2}$$

$$+ \frac{A_2}{s+3}$$

$$B_1 = (s+1)^3 \frac{s-1}{(s+1)^3(s^2+5s+6)}\big|_{s=-1} = \frac{-1-1}{(-1^2-5+6)} = \frac{-2}{2}$$

$$B_1 = -1$$

$$B_2 = \frac{d}{ds}\left((s+1)^3 \frac{s-1}{(s+1)^3(s^2+5s+6)}\right)\big|_{s=-1} = \frac{d}{ds}\left(\frac{s-1}{(s^2+5s+6)}\right)\big|_{s=-1}$$

$$= \frac{(s^2+5s+6)-(s-1)(2s+5)}{(s^2+5s+6)^2}\big|_{s=-1}$$

$$= \frac{(-1^2-5+6)-(-1-1)(-2+5)}{(-1^2-5+6)^2} = \frac{2+6}{4}$$

$$B_2 = 2$$

(continued)

**Example 7.39** (continued)

$$B_3 = \frac{1}{2!}\frac{d^2}{ds^2}\left((s+1)^3\frac{s-1}{(s+1)^3(s^2+5s+6)}\right)\Big|_{s=-1}$$

$$= \frac{d}{ds}\left(\frac{(s^2+5s+6)-(s-1)(2s+5)}{(s^2+5s+6)^2}\right)\Big|_{s=-1}$$

$$= \frac{d}{ds}\left(\frac{(-s^2+2s+11)}{(s^2+5s+6)^2}\right)\Big|_{s=-1}$$

$$= \left(\frac{(-2s+2)(s^2+5s+6)^2-2(2s+5)(s^2+5s+6)(-s^2+2s+11)}{(s^2+5s+6)^4}\right)\Big|_{s=-1}$$

$$= \left(\frac{(-2s+2)(s^2+5s+6)-2(2s+5)(-s^2+2s+11)}{(s^2+5s+6)^3}\right)\Big|_{s=-1}$$

$$= \frac{4\times2-2\times3\times10}{2\times8} = \frac{-52}{2\times8}$$

$$B_3 = -3.75$$

$$A_1 = (s+2)\frac{s-1}{(s+1)^3(s+2)(s+3)}\Big|_{s=-2} = \frac{-2-1}{-1\times+1} = 3$$

$$A_2 = (s+3)\frac{s-1}{(s+1)^3(s+2)(s+3)}\Big|_{s=-3} = \frac{-3-1}{-8\times-1} = -0.5$$

$$F(s) = \frac{-1}{(s+1)^3} + \frac{2}{(s+1)^2} + \frac{-3.75}{s+1} + \frac{3}{s+2} + \frac{-0.5}{s+3}$$

### Case 3. Complex Conjugate Roots

In the case the denominator has a set of repeated complex conjugate roots as $s = -\alpha \pm j\beta$ repeated $m$ times as follows,

$$F(s) = \frac{\text{Num}(s)}{(s+\alpha-j\beta)^m(s+\alpha+j\beta)^m}$$

$$F(s) = \frac{k_1}{(s+\alpha-j\beta)^m} + \frac{k_1^*}{(s+\alpha+j\beta)^m} + \frac{k_2}{(s+\alpha-j\beta)^{m-1}} + \frac{k_2^*}{(s+\alpha+j\beta)^{m-1}}$$

$$+\cdots+\frac{k_m}{(s+\alpha-j\beta)} + \frac{k_m^*}{(s+\alpha+j\beta)}$$

The process is similar to the one in real repeated roots. However, the coefficients $k_1 = |k_1| \angle \theta_1, \ldots, k_m = |k_m| \angle \theta_m$ are complex conjugate values of $k_1^* = |k_1| \angle -\theta_1, \ldots, k_m^* = |k_m| \angle -\theta_m$.

*In a second-order system*, this process can be explained in the context of the roots through the following examples.

**Example 7.40 Find** $L^{-1}\left\{\frac{k}{s^2+2\zeta\omega_n s+\omega_n^2}\right\}$?

*Solution.* Convert the denominator to a complete square as follows:

$$s^2 + 2\zeta\omega_n s + \omega_n^2 = (s+\zeta\omega_n)^2 + \omega_n^2 - (\zeta\omega_n)^2 = (s+\zeta\omega_n)^2 + (1-\zeta^2)\omega_n^2$$

Considering

$$\omega_d^2 = (1-\zeta^2)\omega_n^2$$

or

$$\omega_d = \omega_n\sqrt{(1-\zeta^2)}$$

yields

$$L^{-1}\left\{\frac{k}{(s+\zeta\omega_n)^2 + (1-\zeta^2)\omega_n^2}\right\} = L^{-1}\left\{\frac{k}{(s+\zeta\omega_n)^2 + \omega_d^2}\right\}$$

This suggests a core sinusoidal function with an exponential presented in positive time as follows:

$$f(t) = \frac{k}{\omega_d} e^{-\zeta\omega_n t} \sin\omega_d t\, u(t)$$

Analyzing the roots of the denominator in the frequency domain yields

$$(s+\zeta\omega_n)^2 + \omega_d^2 = 0$$
$$s = -\zeta\omega_n \pm j\omega_d$$

It can be concluded that the roots of the denominator dictate the type of response such that the real part shows the amount of the exponential decay and the imaginary part of the root determines the damping frequency.

**Example 7.41 The roots of the denominator of a function in the frequency domain are** $s = -10 \pm j300$. **Find the core function in the time domain.**

*Solution.* The roots are

$$s = -10 \pm j300 = -\zeta\omega_n \pm j\omega_d$$

(continued)

**Example 7.41** (continued)

Therefore, the response in the time domain without consideration of the coefficients and the roots of the numerator can be expressed as

$$f(t) = \frac{1}{\omega_d} e^{-\zeta\omega_n t} \sin \omega_d t u(t) = \frac{1}{300} e^{-10t} \sin 300t \, u(t)$$

Example 7.42 Find $L^{-1}\left\{\frac{k(s+\zeta\omega_n)}{s^2+2\zeta\omega_n s+\omega_n^2}\right\}$.

Solution. Convert the denominator to a complete square as follows:

$$s^2 + 2\zeta\omega_n s + \omega_n^2 = (s + \zeta\omega_n)^2 + \omega_n^2 - (\zeta\omega_n)^2 = (s + \zeta\omega_n)^2 + \left(1 - \zeta^2\right)\omega_n^2$$

Considering

$$\omega_d^2 = \left(1 - \zeta^2\right)\omega_n^2$$

yields

$$L^{-1}\left\{\frac{k(s + \zeta\omega_n)}{(s + \zeta\omega_n)^2 + \left(1 - \zeta^2\right)\omega_n^2}\right\} = L^{-1}\left\{\frac{k(s + \zeta\omega_n)}{(s + \zeta\omega_n)^2 + \omega_d^2}\right\}$$

This suggests a core co-sinusoidal function with an exponential presented in positive time as follows:

$$f(t) = k e^{-\zeta\omega_n t} \cos \omega_d t \, u(t)$$

**Example 7.42** Find $L^{-1}\left\{\frac{ks}{s^2+2\zeta\omega_n s+\omega_n^2}\right\}$.

*Solution.* Convert the denominator to a complete square as follows:

$$s^2 + 2\zeta\omega_n s + \omega_n^2 = (s + \zeta\omega_n)^2 + \omega_n^2 - (\zeta\omega_n)^2 = (s + \zeta\omega_n)^2 + \left(1 - \zeta^2\right)\omega_n^2$$

Considering

$$\omega_d^2 = \left(1 - \zeta^2\right)\omega_n^2$$

or

(continued)

**Example 7.42** (continued)

$$\omega_d = \omega_n \sqrt{\left(1 - \zeta^2\right)}$$

yields

$$L^{-1}\left\{\frac{ks}{(s + \zeta\omega_n)^2 + (1 - \zeta^2)\omega_n^2}\right\} = L^{-1}\left\{\frac{k(s + \zeta\omega_n - \zeta\omega_n)}{(s + \zeta\omega_n)^2 + \omega_d^2}\right\}$$

Separating the fractions results in

$$L^{-1}\left\{\frac{k(s + \zeta\omega_n)}{(s + \zeta\omega_n)^2 + \omega_d^2} - \frac{k\zeta\omega_n}{(s + \zeta\omega_n)^2 + \omega_d^2}\right\}$$

This suggests a core co-sinusoidal and sinusoidal functions with an exponential presented in positive time as follows:

$$f(t) = \left(ke^{-\zeta\omega_n t}\cos\omega_d t + \frac{k\zeta\omega_n}{\omega_d}e^{-\zeta\omega_n t}\sin\omega_d t\right)u(t)$$

**Example 7.43** Find $L^{-1}\left\{F(s) = \frac{5}{s^2 + 2s + 50}\right\}$.

*Solution.* The denominator has two complex conjugate roots. One way to solve these problems is to convert them into a complete square, as follows:

$$s^2 + 2s + 50 = (s + 1)^2 + 49$$

The shift in frequency $s + 1$ produces an exponential in time with one check that all the $s$ in the function must be shifted by the same amount. Since there is only one $s$ and it is shifted, the rest of the Laplace inverse is to determine the main function in the form of $\frac{1}{s^2 + \omega^2}$ as $\frac{1}{s^2 + 49}$ with a frequency shift of $+1$ and a coefficient of $\frac{5}{\omega}$. Therefore, the time function is a sinusoidal function multiplied by an exponential function as follows:

$$F(s) = \frac{5}{7}\frac{7}{(s + 1)^2 + 7^2}$$

(continued)

**Example 7.43** (continued)

Therefore, the Laplace resembles a function that is sinusoidal at the core but shifted by +1 and presented in the time positive $t = 0^+$ shown by the $u(t)$. The answer is

$$f(t) = \frac{5}{7} e^{-t} \sin 7t \, u(t)$$

**Example 7.44 Find** $L^{-1}\left\{ F(s) = \frac{5s}{s^2 + 14s + 58} \right\}.$

*Solution.* The denominator can be converted to a complete square as follows:

$$s^2 + 14s + 58 = \left( s + \frac{14}{2} \right)^2 + 58 - \left( \frac{14}{2} \right)^2 = (s + 7)^2 + 9$$

As the denominator dictates, the frequency $s$ is shifted by +7. Therefore, all the frequencies must be shifted by the same amount. This suggests that the numerator expression must be shifted exactly +7. The numerator's shift can be accomplished as

$$s = s + 7 - 7$$

Hence,

$$F(s) = 5 \frac{s + 7 - 7}{(s + 7)^2 + 9}$$

Now, the expression of $s + 7$ needs to be kept in one unit. Hence, the function can be separated into two fractions as follows:

$$F(s) = 5 \frac{s + 7}{(s + 7)^2 + 9} + 5 \frac{-7}{(s + 7)^2 + 9}$$

Now the functions in each fraction show the same amount of frequency shift, meaning all $s$'s are shifted by +7, in the form of $(s + 7)$.

The form in each fraction suggests a core $\cos 3t$ and $\sin 3t$ functions, accompanied by an exponential function defined in time positive. Therefore,

$$f(t) = \left( 5e^{-7t} \cos 3t - \frac{35}{3} e^{-7t} \sin 3t \right) u(t)$$

R

**Fig. 7.3** A resistor is represented in the frequency domain

## Application of Laplace in Electric Circuits

As illustrated earlier, Laplace is utilized to express and solve circuit parameters in the frequency domain. Laplace transforms of all circuit components, including the input sources, the elements, and the responses, must be completely expressed in the frequency domain. The transformation of sources in the Laplace domain directly applies the methods introduced at the beginning of this chapter. In this section, the Laplace transform of the circuit elements is introduced.

### Resistors in Frequency Domain

Consider a resistor shown in Fig. 7.3 with current $i(t)$ passing through and a voltage drop $v(t)$ across the element. Ohm's law indicates that

$$v(t) = Ri(t)$$

The Laplace transform of this equation, considering no initial condition (because resistors do not store energy), can be obtained as follows:

$$L\{v(t) = Ri(t)\}$$
$$V(s) = RI(s)$$

Therefore, the resistance value in the frequency domain remains similar to its time-domain value, measured in ohm $\Omega$.

### Inductors in Frequency Domain

Consider an inductor with an inductance of $L$ Henrys. Current $i(t)$ drops voltage $v(t)$, which is related as follows:

$$v(t) = L\frac{di(t)}{dt}$$

Considering an initial current of $I_o$ through the inductor, the Laplace transform of the equation is obtained as follows:

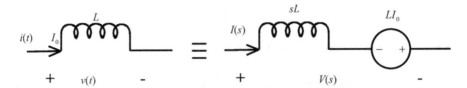

**Fig. 7.4** Laplace transformation of a charged inductor from the time to the frequency domain. The initial charge of the inductor is shown as a voltage source in a series connection to the inductor. This model is best for KVL analysis. Note that the polarity of the voltage source is reversed

$$L\left\{v(t)=L\frac{di(t)}{dt}\right\}$$

$$V(s)=L(sI(s)-I_o)$$

$$V(s)=sLI(s)-LI_o$$

Interpreting this equation in a mesh and through KVL, the voltage drop $V(s)$ equals the voltage drop across an impedance $sL$ times the current $I(s)$ in series with a voltage source generated by the existence of initial current $LI_o$. The inductor tends to keep the current constant by changing the voltage polarity. For this reason, the source indicating the initial current demonstrates a negative polarity at time $t=0^+$. Figure 7.4 shows the frequency domain equivalent of an inductor charged with initial current $I_o$.

**Note 7.7** The inductance $L(H)$ in the time domain is measured in Henrys, but the Laplace transform in the frequency domain is measured in ohm $sL(\Omega)$.

Solving for $I(s)$ results in admittance equivalent of an inductor in the frequency domain:

$$V(s)=sLI(s)-LI_o$$

$$I(s)=\frac{1}{sL}V(s)+\frac{1}{s}I_o$$

This equation demonstrates that the current passing a charged inductor is equivalent to a parallel of the inductor with admittance $\frac{1}{sL}$ and a current source generated because of the initial current $\frac{1}{s}I_o$. Figure 7.5 shows the parallel equivalent of a charged inductor in the frequency domain.

## Capacitors in Frequency Domain

Consider a capacitor at capacitance $C$ that is initially charged at voltage $V_0$ as shown in Fig. 7.6. The current $i(t)$ generates voltage $v(t)$, which is related as

**Fig. 7.5** Laplace
transformation of a charged
inductor when the initial
charge of the inductor is
shown as a current source
parallel to the inductor. This
model is best for KCL
analysis

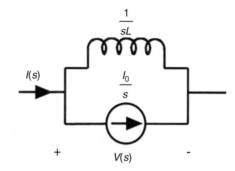

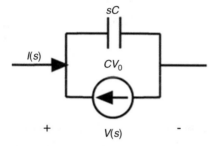

**Fig. 7.6** Laplace transformation of a charged capacitor when the initial charge of the capacitor is
shown as a current source parallel to the capacitor. This model is best for KCL analysis, and note
that the direction of the current source is reversed

$$i(t) = C \frac{dv(t)}{dt}$$

Taking Laplace of this equation considering the initial charge at time $v(t = 0^+) = V_0$
results in

$$L\left\{ i(t) = C \frac{dv(t)}{dt} \right\}$$

$$I(s) = C(sV(s) - v(0^+)) = CsV(s) - CV_0$$

Considering a parallel equivalent circuit, the charged capacitor in the frequency
domain is presented as

$$I(s) = CsV(s) - CV_0$$

The equivalent circuit is shown in Fig. 7.6. The voltage $V(s)$ can be obtained to
show a charged capacitor in series form. The voltage is obtained as follows:

$$V(s) = \frac{1}{Cs} I(s) + \frac{1}{s} CV_0$$

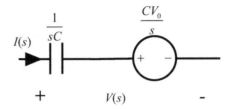

**Fig. 7.7** Laplace transformation of a charged capacitor from the time to the frequency domain. The initial charge of the capacitor is shown as a voltage source in a series connection to the inductor. This model is best for KVL analysis

**Table 7.1** Table of circuit element conversion

| Circuit elements | Time domain (unit) | Frequency domain impedance ($\Omega$) | Frequency domain admittance ($\Omega^{-1}$) |
|---|---|---|---|
| Resistor | $R(\Omega)$ | $R$ | $\frac{1}{R}$ |
| Inductor | $L(H)$ | $sL$ | $\frac{1}{sL}$ |
| Capacitor | $C(F)$ | $\frac{1}{sC}$ | $sC$ |

The circuit is obtained by writing the KVL across the uncharged capacitor and the source representing the initial charge as a voltage source. A parallel circuit represents the initial charge once this source is presented as a current source. Considering a KVL, as shown in the above equation, the equivalent circuit can be obtained in Fig. 7.7.

## Circuit Analysis Using Laplace Transform

Laplace can be used in circuit analysis in two ways, (1) solving differential equations obtained from circuits and (2) writing circuit KVL and KCL equations directly in Laplace. This section studies techniques to write circuit equations directly in Laplace.

The following steps are recommended to write KVL and KCL equations in the Laplace domain, although some of these steps can be skipped depending on the circuit.

- *Step 1*: The circuit elements must be converted to their Laplace equivalent. Table 7.1 lists the equivalent of circuit elements in the frequency domain.
- *Step 2*: Convert current and voltage sources from the time domain to the Laplace domain. These sources preserve the source type, but their functions or values are transformed into the frequency domain. For instance, a current source of $i(t) = 10u(t)$ remains a current source with the same direction, but its value becomes $I(s) = 10\frac{1}{s}$.
- *Step 3*: Dependent sources remain dependent, and their functions are converted to the Laplace domain. Their dependent parameter from the time domain is converted to the Laplace of the same parameter.

- *Step 4*: The voltage drop across a Laplace represented impedance is $V(s) = Z(s)$ $I(s)$ and can be used in KVL.
- *Step 5*: The current of a branch written in KCL is obtained from $I(s) = \frac{V(s)}{Z(s)}$.
- *Step 6*: Writing KVL and KCL in the Laplace domain should result in algebraic equations. Solve these equations for the desired parameters.
- *Step 7*: Take the Laplace inverse of the circuit responses, and find the time domain functions and values.

---

**Example 7.45 Find the current flowing through the circuit of Fig. 7.8 in the Laplace domain.**

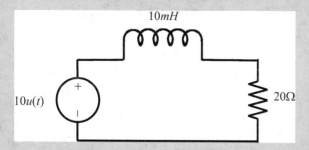

**Fig. 7.8**  Circuit of Example 7.45 in the time domain

*Solution.* Steps 1 and 5. Converting the circuit elements and the source function in the Laplace domain requires the inductor 10 mH to be presented as $10e - 3s$ (Ω) and the resistor to remain as a 20 Ω resistor. The source of $10u(t)$ becomes $\frac{10}{s}$. The current flowing through the circuit is $I(s)$. Since all impedances in the frequency domain are presented in Ωs, the voltage drop across each element is the product of the impedance's current $I(s)$. For instance, the voltage drop across the inductor is $10e - 3sI(s)$, and the voltage drop across the resistor is $20I(s)$ (Fig. 7.9).

**Fig. 7.9**  Laplace transform of the circuit (Example 7.45)

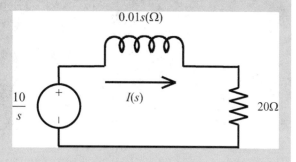

(continued)

**Example 7.45** (continued)

Writing a KVL starting from the negative terminal of the source and circling clockwise results in

$$-\frac{10}{s} + 10e - 3sI(s) + 20I(s) = 0$$

Solving for $I(s)$ results in

$$I(s) = \frac{\frac{10}{s}}{10e - 3s + 20}$$

Simplifying the equation by factoring the $10e - 3$ out to make the denominator a monic polynomial results in

$$I(s) = \frac{10}{0.01s\left(s + \frac{20}{0.01}\right)} = \frac{1000}{s(s + 2000)}$$

Taking Laplace inverse by partial fraction results in

$$I(s) = \frac{A}{s} + \frac{B}{(s + 2000)}$$

$$A = s\frac{1000}{s(s + 2000)}\Big|_{s=0} = \frac{1}{2}$$

$$B = (s + 2000)\frac{1000}{s(s + 2000)}\Big|_{s=-2000} = -\frac{1}{2}$$

Therefore, the function can be written as

$$I(s) = \frac{0.5}{s} - \frac{0.5}{(s + 2000)}$$

The time domain, by taking Laplace inverse and using the templates introduced earlier, can be obtained as

$$i(t) = 0.5u(t) - 0.5e^{-2000t}u(t)A$$

**Example 7.46 Considering the circuit of the previous example (7.41), find the voltage across the 20 Ω resistor and the voltage across the 10 mH inductor both in frequency and time domains.**

*Solution.* The voltage of 20 Ω in the frequency domain can be obtained as

$$V_{20\Omega} = 20I(s) = 20\left(\frac{0.5}{s} - \frac{0.5}{(s+2000)}\right) = \frac{10}{s} - \frac{10}{(s+2000)}$$

Moreover, the voltage in the time domain is

$$v_{20\Omega}(t) = 10u(t) - 10e^{-2000t}u(t)V$$

The voltage across the 10 mH inductor is obtained in the Laplace domain as (remember the ohmic value of this inductor in the Laplace domain is $10e - 3s = 0.01s$ (Ω)):

$$V_{10\,mH}(s) = 0.01sI(s) = 0.01s\frac{1000}{s(s+2000)} = \frac{10}{(s+2000)}$$

In the time domain, this current is

$$v_{10\,mH}(t) = 10e^{-2000t}u(t)V$$

**Example 7.47 Considering the circuit in Fig. 7.10, find the current of each branch $I_1$ and $I_2$ in Laplace and the time domain.**

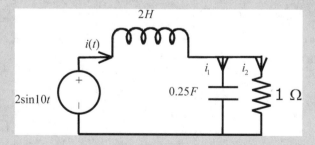

**Fig. 7.10** Circuit of Example 7.47 in the time domain

*Solution.* Circuit element and source equivalent in Laplace result in the following circuit (Fig. 7.11):

(continued)

**Example 7.47** (continued)

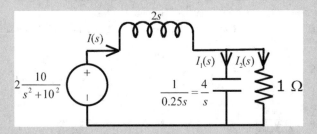

**Fig. 7.11** Laplace transform of the circuit in Fig. 7.10

The current $I(s)$ can be obtained by dividing the voltage by the total impedance observed at the source terminals. Therefore,

$$I(s) = \frac{\dfrac{20}{s^2 + 100}}{2s + \left(1 \left\| \dfrac{4}{s}\right.\right)} = \frac{\dfrac{20}{s^2 + 100}}{2s + \dfrac{4}{1 + \dfrac{4}{s}}} = \frac{\dfrac{20}{s^2 + 100}}{2s + \dfrac{4}{s+4}} = \frac{\dfrac{20}{s^2 + 100}}{\dfrac{2s(s+4)+4}{s+4}}$$

$$= \frac{20(s+4)}{(s^2 + 100)(2s^2 + 8s + 4)}$$

$$I(s) = \frac{10(s+4)}{(s^2 + 100)(s^2 + 4s + 2)}$$

This fraction can be split into two partial fractions as follows:

$$I(s) = \frac{10(s+4)}{(s^2 + 100)(s^2 + 4s + 2)} = \frac{As + B}{s^2 + 100} + \frac{Cs + D}{s^2 + 4s + 2}$$

where $A = -0.1017$, $B = 0.0071$, $C = 0.1017$, and $D = 0.399$.
Therefore,

$$I(s) = \frac{-0.1017s + 0.0071}{s^2 + 100} + \frac{0.1017s + 0.399}{(s^2 + 4s + 2) = (s+2)^2 - 2}$$

Splitting the fractions results into the known templates for sin, cos, sinh, and cosh as follows:

$$I(s) = \frac{-0.1017s}{s^2 + 100} + \frac{0.0071}{s^2 + 100} + 0.1017\frac{s + 3.932}{(s+2)^2 - 2}$$

(continued)

**Example 7.47** (continued)

The coefficient of the last fraction, 3.932, can be written as 2 + 1.932 to show the same frequency shift by 2 rad/s as the term $(s + 2)^2$ demonstrates.

$$I(s) = \frac{-0.1017s}{s^2 + 100} + \frac{0.0071}{s^2 + 100} + 0.1017\frac{(s + 2) + 1.932}{(s + 2)^2 - 2}$$

$$I(s) = \frac{-0.1017s}{s^2 + 100} + \frac{0.0071}{s^2 + 100} + 0.1017\left(\frac{(s + 2)}{(s + 2)^2 - 2} + \frac{1.932}{(s + 2)^2 - 2}\right)$$

Therefore, the time domain expression of the current can be obtained as follows (note that the frequency shift becomes an exponential in the time domain):

$$i(t) = \left(-0.1017\cos 10t + \frac{0.0071}{10}\sin 10t\right.$$
$$\left. +0.1017\left(e^{-2t}\cos h\sqrt{2}t - \frac{1.932}{\sqrt{2}}e^{-2t}\sin h\sqrt{2}t\right)\right)u(t)A$$

The current division between the branches of $\frac{4}{s}$ Ω the capacitor and the 1 Ω resistor is as follows:

$$I_1(s) = \frac{1}{\frac{4}{s} + 1}I(s) = \frac{s}{s + 4}I(s) = \frac{s}{s + 4}\frac{10(s + 4)}{(s^2 + 100)(s^2 + 4s + 2)}$$

$$= \frac{10s}{(s^2 + 100)(s^2 + 4s + 2)} = \frac{As + B}{s^2 + 100} + \frac{Cs + D}{s^2 + 4s + 2}$$

$$A = -0.0875, B = 0.3570, C = 0.0875, D = -0.0071.$$

Therefore,

$$I_1(s) = \frac{-0.875s}{s^2 + 100} + \frac{0.3570}{s^2 + 100} + 0.0875\left(\frac{s + 0.0811}{(s + 2)^2 - 2}\right)$$

$$I_1(s) = \frac{-0.875s}{s^2 + 100} + \frac{0.3570}{s^2 + 100} + 0.0875\left(\frac{s + 2 - 1.9189}{(s + 2)^2 - 2}\right)$$

$$I_1(s) = \frac{-0.875s}{s^2 + 100} + \frac{0.3570}{s^2 + 100} + 0.0875\left(\frac{s + 2}{(s + 2)^2 - 2} - \frac{1.9189}{(s + 2)^2 - 2}\right)$$

(continued)

**Example 7.47** (continued)

$$i_1(t) = \left( -0.875 \cos 10t + \frac{0.3570}{10} \sin 10t \right.$$

$$+0.0875 \left( e^{-2t} \cos h\sqrt{2}t - \frac{1.9189}{\sqrt{2}} e^{-2t} \sin h\sqrt{2}t \right) \bigg) u(t)A$$

The current passing $1\,\Omega$ resistor is obtained by current division as follows:

$$I_2(s) = \frac{\dfrac{4}{s}}{\dfrac{4}{s}+1} I(s) = \frac{4}{s+4} I(s) = \frac{4}{s+4} \frac{10(s+4)}{(s^2+100)(s^2+4s+2)}$$

$$= \frac{40}{(s^2+100)(s^2+4s+2)} = \frac{As+B}{s^2+100} + \frac{Cs+D}{s^2+4s+2}$$

$$A = -0.0143, B = -0.3499, C = 0.0143, D = 0.4070.$$

Therefore,

$$I_2(s) = \frac{-0.0143s}{s^2+100} + \frac{-0.3499}{s^2+100} + 0.0143 \left( \frac{s+28.4615}{(s+2)^2-2} \right)$$

$$I_2(s) = \frac{-0.0143s}{s^2+100} + \frac{-0.3499}{s^2+100} + 0.0143 \left( \frac{s+2}{(s+2)^2-2} + \frac{26.4615}{(s+2)^2-2} \right)$$

$$i_2(t) = \left( -0.0143 \cos 10t - \frac{0.3499}{10} \sin 10t \right.$$

$$+0.0143 \left( e^{-2t} \cos h\sqrt{2}t - \frac{26.4615}{\sqrt{2}} e^{-2t} \sin h\sqrt{2}t \right) \bigg) u(t)A$$

**Example 7.48** Considering the circuit in the previous example (7.47), shown as follows, find the voltage across the $1\,\Omega$ resistor (Fig. 7.12).

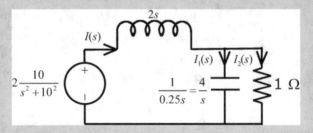

**Fig. 7.12** Laplace transform of the circuit of Example 7.48

(continued)

**Example 7.48** (continued)

*Solution.* The current $I(s)$ passes through the parallel of $1\,\Omega$ and $\frac{4}{s}$ impedances. Therefore, the voltage becomes

$$V_o(s) = I(s)\left(1\left\|\frac{4}{s}\right.\right)$$

Replacing the results of the value in

$$V_o(s) = \frac{10(s+4)}{(s^2+100)(s^2+4s+2)}\left(1\left\|\frac{4}{s}\right.\right)$$

$$V_o(s) = \frac{10(s+4)}{(s^2+100)(s^2+4s+2)}\frac{4}{(s+4)} = \frac{40}{(s^2+100)(s^2+4s+2)}$$

$$= \frac{As+B}{s^2+100} + \frac{Cs+D}{s^2+4s+2}$$

$$A = -0.0143, B = -0.3499, C = 0.0143, D = 0.4070.$$

Therefore,

$$V_o(s) = \frac{-0.0143s}{s^2+100} + \frac{-0.3499}{s^2+100} + 0.0143\left(\frac{s+28.4615}{(s+2)^2-2}\right)$$

$$V_o(s) = \frac{-0.0143s}{s^2+100} + \frac{-0.3499}{s^2+100} + 0.0143\left(\frac{s+2}{(s+2)^2-2} + \frac{26.4615}{(s+2)^2-2}\right)$$

$$v_o(t) = \left(-0.0143\cos 10t - \frac{0.3499}{10}\sin 10t\right.$$

$$+0.0143\left(e^{-2t}\cos h\sqrt{2}t - \frac{26.4615}{\sqrt{2}}e^{-2t}\sin h\sqrt{2}t\right)\right)u(t)\,V$$

**Example 7.49 Considering the circuit shown in Fig. 7.13, find the current in each loop, $I_1(s)$ and $I_2(s)$, as a function of the input voltage $V_{in}(s)$.**
*Solution.* The circuit in Laplace form has two loops as follows:

- KVL ① $-V_{in}(s) + RI_1(s) + sL_1(I_1(s) - I_2(s)) = 0$
- KVL ② $\frac{1}{sC}I_2(s) + sL_2I_2(s) + sL_1(I_2(s) - I_1(s)) = 0$

(continued)

**Example 7.49** (continued)

**Fig. 7.13** Circuit of
Example 7.49

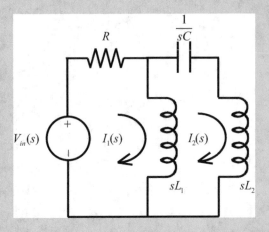

Solving for $I_1(s)$, $I_2(s)$ results in

$$(R + sL_1)I_1(s) - sL_1I_2(s) = V_{\text{in}}(s)$$

$$\left(\frac{1}{sC} + sL_1 + sL_2\right)I_2(s) - sL_1I_1(s) = 0$$

Finding $I_1(s)$ from the second equation results in $I_1(s) = \frac{1}{sL_1} \times$
$\left(\frac{1}{sC} + sL_1 + sL_2\right)I_2(s)$, and replacing it in the first equation yields

$$(R + sL_1)\left(\frac{1}{sL_1}\left(\frac{1}{sC} + sL_1 + sL_2\right)I_2(s)\right) - sL_1I_2(s) = V_{\text{in}}(s)$$

Solving for $I_2(s)$ results in

$$I_2(s) = \frac{1}{\left((R + sL_1)\left(\frac{1}{sL_1}\left(\frac{1}{sC} + sL_1 + sL_2\right)\right) - sL_1\right)} V_{\text{in}}(s)$$

Simplifying

$$I_2(s) = \frac{CL_1s^2}{CL_1L_2s^3 + RC(L_1 + L_2)s^2 + L_1s + R} V_{\text{in}}(s)$$

(continued)

**Example 7.49** (continued)

This function represents the current in the second loop as a function of the input voltage source. Considering $I_1(s) = \frac{1}{sL_1}\left(\frac{1}{sC} + sL_1 + sL_2\right)I_2(s)$ and replacing this from the second loop current, $I_1(s)$ becomes

$$I_1(s) = \frac{1}{sL_1}\left(\frac{1}{sC} + sL_1 + sL_2\right)$$

$$\times \frac{1}{\left((R + sL_1)\left(\frac{1}{sL_1}\left(\frac{1}{sC} + sL_1 + sL_2\right)\right) - sL_1\right)} V_{in}(s)$$

Simplifying current $I_1(s)$ in terms of voltage $V_{in}(s)$ yields

$$I_1(s) = \frac{C(L_1 + L_2)s^2 + 1}{CL_1L_2s^3 + RC(L_1 + L_2)s^2 + L_1s + R} V_{in}(s)$$

**Example 7.50** Consider the circuit shown in Fig. 7.14. Find the voltage of nodes in terms of the input currents using the Laplace transform.

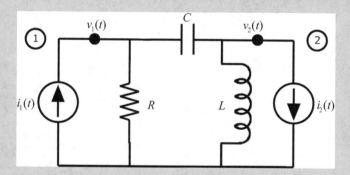

**Fig. 7.14** Circuit of Example 7.50 in the time domain

*Solution.* Taking the Laplace transform of the circuit, the circuit shown in Fig. 7.15 is obtained

(continued)

**Example 7.50** (continued)

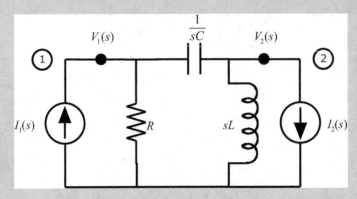

**Fig. 7.15**  Laplace transform of the circuit (Example 7.45)

For simplicity in writing KCL equations (in this example), the Laplace operator $s$ is omitted from the functions.

- KCL ① $-I_1 + \frac{V_1}{R} + sC(V_1 - V_2) = 0$
- KCL ② $I_2 + \frac{V_2}{sL} + sC(V_2 - V_1) = 0$

Simplifying results in

$$\left(\frac{1}{R} + sC\right)V_1 - sCV_2 = I_1$$

$$-sCV_1 + \left(\frac{1}{sL} + sC\right)V_2 = -I_2$$

$$V_1 = \frac{R(CL(I_1 - I_2) + I_1)}{CLs^2 + RCs + 1} = \frac{R\left(1 + \frac{1}{LC}\right)}{s^2 + \frac{R}{L}s + \frac{1}{LC}}I_1 - \frac{R}{s^2 + \frac{R}{L}s + \frac{1}{LC}}I_2$$

$$V_2 = -\frac{Ls(I_2 - CRsI_1 + CRsI_2)}{CLs^2 + RCs + 1} = \frac{Rs^2}{s^2 + \frac{R}{L}s + \frac{1}{LC}}I_1 - \frac{RLs\left(\frac{1}{RC} + s\right)}{s^2 + \frac{R}{L}s + \frac{1}{LC}}I_2$$

As these equations show, the voltages of $V_1$ and $V_2$ are functions of both input currents $I_1$ and $I_2$.

## Problems

7.1. Find the Laplace transform of the following functions:

(a) $f(t) = 10u(t)$
(b) $f(t) = -20\delta(t)$
(c) $f(t) = -5tu(t)$
(d) $f(t) = 10e - 30tu(t)$
(e) $f(t) = 3te - 5tu(t)$
(f) $f(t) = 20 \sin 75tu(t)$
(g) $f(t) = \sin 2\omega tu(t)$
(h) $f(t) = \cos 2\omega tu(t)$
(i) $f(t) = 10 \sin (5t + 30)u(t)$
(j) $f(t) = 110\sqrt{2} \cos 377tu(t)$
(k) $f(t) = 12 \sin h20tu(t)$
(l) $f(t) = 10 \cos h2\pi tu(t)$

7.2. Find the Laplace of the following function operations:

(a) $f(t) = 2u(t) + 5tu(t)$
(b) $f(t) = 10 \sin 20t + 5 \cos 20t$
(c) $f(t) = 5 \sin 3t - \cos 15t$
(d) $f(t) = 3e - 10t \sin 15t$
(e) $f(t) = 75e - 100t \cos 314t$
(f) $f(t) = 3e - 10t \sin h15t$
(g) $f(t) = 75e - 100t \cos h314t$
(h) $f(t) = 13tu(t - 5)$
(i) $f(t) = 10te - 3tu(t - 4)$
(j) $f(t) = 5t3u(t)$

7.3. Find the Laplace inverse of the following functions:

(a) $F(s) = 1 + s$
(b) $F(s) = \frac{1}{s+5}$
(c) $F(s) = \frac{s}{s+5}$
(d) $F(s) = \frac{s-1}{s+5}$
(e) $F(s) = \frac{1}{s^2+25}$
(f) $F(s) = \frac{s+1}{s^2+25}$
(g) $F(s) = \frac{s^2-1}{s^2+25}$
(h) $F(s) = \frac{1}{s^2-25}$
(i) $F(s) = \frac{s+1}{s^2-25}$
(j) $F(s) = \frac{1}{s^2+7s+12}$
(k) $F(s) = \frac{s+1}{s^2+7s+12}$

(l) $F(s) = \frac{s^2+5s+6}{s^2+7s+12}$

(m) $F(s) = \frac{1}{s^2+6s+100}$

(n) $F(s) = \frac{s+1}{s^2+6s+100}$

(o) $F(s) = \frac{s^2+1}{s^2+6s+100}$

(p) $F(s) = \frac{s^2+s+1}{s^2+6s+100}$

(q) $F(s) = \frac{1}{s(s^2+100)}$

(r) $F(s) = \frac{1}{s(s+3)(s+7)}$

(s) $F(s) = \frac{(s+1)(s+2)}{s(s+3)(s+7)}$

(t) $F(s) = \frac{1}{s^3(s+3)^2(s+7)}$

(u) $F(s) = \frac{1}{s}e^{-3s}$

(v) $F(s) = \frac{1}{s^2}\left(e^{-3s}+e^{-5s}\right)$

(w) $F(s) = e^{-10s}\frac{s}{s+5}$

(x) $F(s) = \frac{e^{-s}}{s^2+25}$

(y) $F(s) = \frac{(s+1)e^{2s}}{s^2+25}$

(z) $F(s) = \frac{20}{(s+3\pi)^5}$

7.4. Prove that

(a) $L\{\cos hat\cos at\} = \frac{s^3}{s^4+4a^4}$

(b) $L\{t^2\cos\omega t\} = \frac{2s(s^2-3\omega^2)}{(s^2+\omega^2)^3}$

7.5. Find the convolution of $h(t) = f(t) * g(t)$, wherein $f(t)$ and $g(t)$ are as follows:

(a) $f(t) = u(t) - u(t-5)$, $g(t) = 3u(t-4) - 3u(t-5)$.

(b) $f(t) = u(t) - u(t-5)$, $g(t) = 3tu(t) - 3(t-5)u(t-5) - 3u(t-5)$.

(c) $f(t) = 5u(t) - u(t-1) - u(t-2) - 3u(t-3)$, $g(t) = 3tu(t) - 6(t-1)u(t-1) + 3(t-2)u(t-2)$.

7.6. Do the following convolution integrals using Laplace and direct method:

(a) $f_1(t) * f_2(t)$

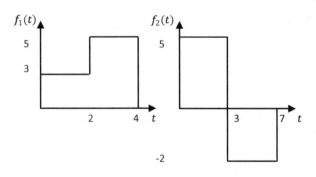

(b) $f_3(t) * f_4(t)$

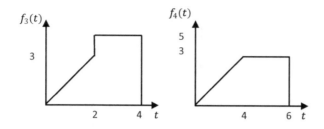

(c) $f_5(t) * f_6(t)$

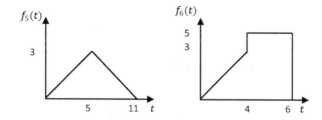

7.7. Solve the following differential equations using Laplace transform:

| (a) | $\dot{y} + 9y = 0$ | $y(0) = 5$ |
|---|---|---|
| (b) | $\ddot{y} + 9y = \delta(t)$ | $y(0) = 0, y'(0) = 2$ |
| (c) | $\ddot{y} + 2\dot{y} = 10u(t)$ | $y(0) = 1, y'(0) = -1$ |
| (d) | $\ddot{y} + 2\dot{y} = 10tu(t)$ | $y(0) = 1, y'(0) = -1$ |
| (e) | $\ddot{y} - 2\dot{y} - 3y = \dot{\delta}(t)$ | $y(0) = 1, y'(0) = 7$ |
| (f) | $4\ddot{y} + y = 0$ | $y(0) = 1, y'(0) = -2$ |
| (g) | $4\ddot{y} + y = \sin(t)$ | $y(0) = 0, y'(0) = -1$ |
| (h) | $\ddot{y} + 2\dot{y} + 5y = 0$ | $y(0) = 2, y'(0) = -4$ |
| (i) | $\ddot{y} + 2y = u(t) - u(t-1)$ | $y(0) = 0, y'(0) = 0$ |
| (j) | $\ddot{y} + 3\dot{y} + 2y = \delta(t-a)$ | $y(0) = 0, y'(0) = 0, a > 0.$ |

7.8. Using Laplace transform, find $i_1(t)$, $i_2(t)$, and $v(t)$.

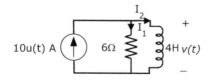

7.9. Using Laplace transform, find $v_o(t)$ where $V_{in} = u(t)$.

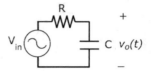

7.10. Using Laplace transform, find $v_o(t)$.

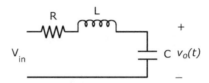

7.11. Using Laplace transform, find $I_1$ and $V_o(t)$.

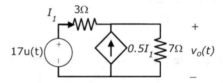

7.12. Using Laplace transform, find $i(t)$ and $v(t)$.

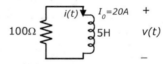

7.13. Using Laplace transform, find $i(t)$ and $v(t)$.

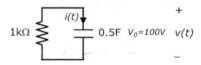

7.14. Using Laplace transform, find $v_1(t)$, $v_2(t)$, and $v_3(t)$.

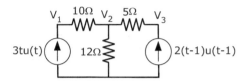

7.15. Using Laplace transform, find $v_1(t)$, $v_2(t)$, $i_1(t)$, and $i_2(t)$.

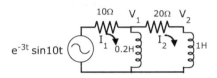

7.16. Using Laplace transform, find $v_1(t)$, $v_2(t)$, $i_1(t)$, and $i_2(t)$.

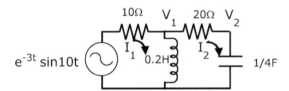

7.17. Using Laplace transform, find $i_1(t)$, $i_2(t)$, and $i_3(t)$.

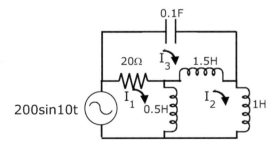

7.18. Using Laplace transform, find $v_1(t)$ and $v_o(t)$.

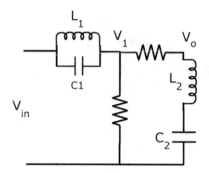

# Chapter 8
# Transfer Functions

## Definition of Transfer Function

A linear physical system with one or multiple sets of input and output can be represented by mathematical functions that relate any outputs to any inputs. These functions are unique and are defined based on the systems governing equations. The transfer function of a system is defined as the Laplace transform of the output response over the Laplace transform of the input excitation. Transfer functions are defined for any desired set of input and output functions that may relate the input and output together. Considering the Laplace transform of the input function as $X(s)$ and the output as $Y(s)$, the transfer function $H(s)$can be defined as

$$H(s) \triangleq \frac{Y(s)}{X(s)}$$

In the time domain, the transfer function $h(t)$ is defined through the convolution product (*) as follows:

$$y(t) = x(t) * h(t)$$

The Laplace transform of the convolution product is obtained as

$$L\{x(t) * h(t)\} = X(s)H(s)$$

This chapter identifies the transfer function of linear circuits in both the frequency and time domains. Figure 8.1 demonstrates the system response when the desired input is applied to the system's transfer function.

A. Izadian, *Fundamentals of Modern Electric Circuit Analysis and Filter Synthesis*,
https://doi.org/10.1007/978-3-031-21908-5_8

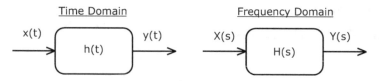

**Fig. 8.1** Relation of the input, output, and transfer function in both time and frequency domains. In time domain, convolution integral obtains the output for a given input signal $x(t)$ by $y(t) = x(t) * h(t)$. In the frequency domain, a product of the system transfer function $H(s)$ by the given input signal $X(s)$ determines the output by $Y(s) = X(s)H(s)$

**Example 8.1 A circuit has a response function of $y(t) = 2e - t \sin (10\ t)$ when the input function of $x(t) = 3u(t)$ is applied. Find the transfer function in the frequency domain.**

*Solution.* According to the definition, the transfer function is a division of the Laplace transforms of the output response over the Laplace transform of the input signal. Hence, the transfer function is obtained as

$$H(s) = \frac{L\{y(t)\}}{L\{x(t)\}} = \frac{L\{2e^{-t}\sin(10t)\}}{L\{3u(t)\}} = \frac{\frac{20}{(s+1)^2+100}}{\frac{3}{s}} = \frac{\frac{20}{3}s}{(s+1)^2+100}$$

**Example 8.2 The transfer function of a system is given as $H(s) = \frac{1}{s+5}$. If the system is excited by input $x(t) = 5u(t)$, find the output.**

*Solution.* In the frequency domain, the transfer function definition finds the output in the Laplace domain to be

$$H(s) \triangleq \frac{Y(s)}{X(s)} \Rightarrow Y(s) = X(s)H(s)$$

The Laplace of input is obtained by

$$X(s) = L\{5u(t)\}$$

Hence, the output becomes

$$Y(s) = \frac{5}{s} \frac{1}{s+5} = \frac{5}{s(s+5)}$$

In the time domain, the output can be obtained by

(continued)

**Example 8.2** (continued)

$$y(t) = 5u(t) * e^{-5t}u(t).$$

In this chapter, a technique for obtaining this convolution product is introduced.

## Multi-input-Multi-output Systems

In physical systems, an output may be influenced by several inputs. The transfer function of this system is the linear summation of all transfer functions excited by various inputs that contribute to the desired output. For instance, if inputs $x_1(t)$ and $x_2(t)$ directly influence the output $y(t)$, respectively, through transfer functions $h_1(t)$ and $h_2(t)$, the output is therefore obtained as

$$y(t) = x_1(t) * h_1(t) + x_2(t) * h_2(t)$$

In the Laplace domain, the output is obtained as

$$Y(s) = X_1(s)H_1(s) + X_2(s)H_2(s)$$

Figure 8.2 demonstrates this system's configuration.

**Note 8.1** In MIMO systems, individual transfer functions are obtained by measuring the output signal while applying an input signal to one input port and zeroing out the rest of the inputs.

**Note 8.2** In the MIMO system with two inputs, transfer function $H_1(s)$ is obtained when a signal is applied to $X_1(s)$, and $X_2(s)$ is kept at zero.

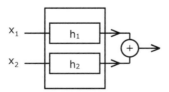

**Fig. 8.2** Two systems in parallel receive two separate inputs. The overall system output is an algebraic summation of the separately excited system outputs

$$H_1(s) = \frac{Y(s)}{X_1(s)}\Big|_{X_2(s)=0}$$

Similarly,

$$H_2(s) = \frac{Y(s)}{X_2(s)}\Big|_{X_1(s)=0}$$

## Obtaining Transfer Function of Electric Circuits

Transfer functions of electric circuits demonstrate a mathematical representation of the circuit behavior. A transfer function defined for an input and output set is unique and depends only on the circuit characteristics, not the input or output signals. However, circuit parameters such as voltage and current can be used as input and output signals to obtain the transfer functions. For instance, $V_{in}$ can be considered an input signal in the following circuit, and $I$ and $V_{out}$ are considered output signals. It is a desired signal consideration and has to match the reality of the circuit operation. In this circuit, a source is the actual driver of the current and the voltage measured across the inductor. Therefore, it makes sense to consider it as an input. The results are the current flowing in the circuit (Fig. 8.3) and the voltage appearing at the inductor.

Based on this logical approach, the following transfer functions can be defined:

$$H_1 \triangleq \frac{I}{V_{in}}$$

$$H_2 \triangleq \frac{V_{out}}{V_{in}}$$

Now the question is how to find these transfer functions. One of the best approaches is to solve the circuit in the Laplace domain for the desired output defined in the transfer function. For instance, to find the transfer function $H_1$, the circuit can be solved to obtain $I$, and to obtain the transfer function $H_2$, the circuit can be solved for $V_{out}$.

**Fig. 8.3** RL circuit. The voltage source forces the input function, and the output is either the measured voltage across the inductor or the current in the circuit

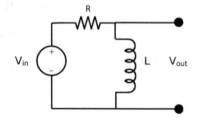

When the desired output parameter is identified, it must make changes and use replacements to express the desired output in the circuit input signal. Then the ratio of the output over input can be easily obtained.

**Example 8.3 In the given circuit of Fig. 8.3, find the transfer functions defined as $H_1 \triangleq \frac{I}{V_{in}}$ or $H_2 \triangleq \frac{V_{out}}{V_{in}}$.**

*Solution.* To find $H_1$, as explained, the circuit is solved for the current $I$ while the input source is applied.

The circuit current is obtained using a KVL as follows:

$$- V_{in} + RI + sLI = 0$$

The current becomes

$$I = \frac{V_{in}}{R + sL}$$

The output $I$ is in terms of the input voltage $V_{in}$. Therefore, no further replacement is needed. A ratio of the output $I$ over the input $V_{in}$ can be obtained as

$$H_1(s) = \frac{I}{V_{in}} = \frac{1}{R + sL} = \frac{\frac{1}{L}}{s + \frac{R}{L}}$$

To find $H_2(s)$, the circuit is solved for the desired output $V_{out}$ as follows:

$$V_{out} = I \times sL$$

The current is not the type of input the transfer function was defined for. Rather, the input has to be the input voltage. Therefore, a replacement of $I$ with its equivalent $\frac{V_{in}}{R+sL}$ seems necessary. This yields

$$V_{out} = sL \frac{V_{in}}{R + sL}$$

The ratio of the output over input is obtained as

$$H_2(s) = \frac{V_{out}}{V_{in}} = \frac{sL}{R + sL} = \frac{s}{s + \frac{R}{L}}$$

**Example 8.4 Show the equivalent of the circuit transfer functions obtained in Example 8.3.**
*Solution.* See Fig. 8.4.

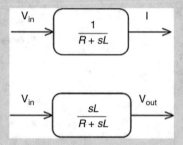

**Fig. 8.4** Equivalent system transfer functions to obtain the current or the voltage when the input is the voltage

**Example 8.5 In the circuit of Fig. 8.5, find the transfer function $H(s) \triangleq \frac{I}{I_s}$.**
*Solution.* The transfer function's output refers to the current through the inductor. In the frequency domain, the current of the inductor can be found as

$$I = \frac{5}{100e - 3s + 5}I_s = \frac{50}{s + 50}I_s$$

Therefore, the division of the inductor current over the source current can be found as

$$\frac{I}{I_s} = \frac{50}{s + 50}$$

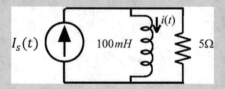

**Fig. 8.5** Circuit of Example 8.5

**Example 8.6 In the circuit of Fig. 8.6, find the transfer function** $H(s) \triangleq \frac{I}{V_s}$.

*Solution.* The current through the inductor can be obtained by dividing the voltage drop across the $7\Omega \| 200e - 3\ s$ by the impedance of the inductor. The voltage drop is

$$V_o = \frac{(0.2s \mid \mid 7)}{(0.2s \mid \mid 7) + 10} V_s = \frac{\frac{0.2s \times 7}{0.2s+7}}{\frac{0.2s \times 7}{0.2s+7} + 10} V_s = \frac{1.4s}{1.4s + 2s + 70} V_s$$

$$= \frac{1.4s}{3.4s + 70} V_s = \frac{\frac{1.4}{3.4}s}{s + \frac{70}{3.4}} V_s = \frac{0.412s}{s + 28.89} V_s$$

$$I = \frac{V_o}{0.2s} = \frac{\frac{0.412s}{s+28.89}}{0.2s} V_s = \frac{2.06}{s + 28.89} V_s$$

Therefore,

$$H(s) = \frac{I}{V_s} = \frac{2.06}{s + 28.89}$$

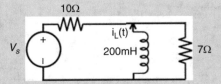

**Fig. 8.6** Circuit of Example 8.6

**Example 8.7 In the circuit of Fig. 8.7, find the transfer function** $H(s) \triangleq \frac{V}{V_s}$.

*Solution.* The voltage across the capacitor can be found through a voltage division between the parallel of $200k\Omega$ and $\frac{1}{100e - 6\ s}$ and the $50k\Omega$. The voltage is

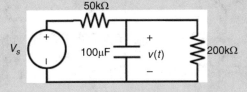

**Fig. 8.7** Circuit of Example 8.7

(continued)

**Example 8.7** (continued)

$$V = \frac{200k \left\| \frac{1}{100e-6\ s} \right.}{200k \left\| \frac{1}{100e-6\ s} \right. + 50k} V_s$$

$$V = \frac{\frac{200k \times \frac{10k}{s}}{200k + \frac{10k}{s}}}{\frac{200k \times \frac{10k}{s}}{200k + \frac{10k}{s}} + 50k} V_s = \frac{\frac{20 \times \frac{1k}{s}}{20 + \frac{1}{s}}}{\frac{20 \times \frac{1k}{s}}{20 + \frac{1}{s}} + 50k} V_s = \frac{\frac{20k}{s}}{\frac{20k}{s} + 100k + \frac{50k}{s}} V_s$$

$$V = \frac{20k}{70k + 100ks} V_s = \frac{2}{7 + 10s} V_s$$

Therefore,

$$H(s) = \frac{V}{V_s} = \frac{2}{7 + 10s} = \frac{0.2}{s + 0.7}$$

**Example 8.8 Find the transfer function of the circuit shown in Fig. 8.8.**
The circuit has two loops with current flowing $I_1$ and $I_2$ and the source of $V_{in}$.
The transfer functions can be defined as $H_1(s) = \frac{I_1(s)}{V_{in}(s)}$ and $H_2(s) = \frac{I_2(s)}{V_{in}(s)}$. To
get the transfer functions, a good starting point would be to find $I_1$ and $I_2$. The
results are shown as follows:

$$I_1(s) = \frac{C(L_1 + L_2)s^2 + 1}{CL_1L_2s^3 + RC(L_1 + L_2)s^2 + L_1s + R} V_{in}(s)$$

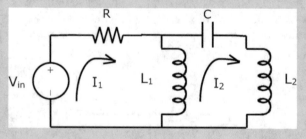

**Fig. 8.8** Circuit of Example 8.8

(continued)

**Example 8.8** (continued)

$$I_2(s) = \frac{CL_1s^2}{CL_1L_2s^3 + RC(L_1 + L_2)s^2 + L_1s + R} V_{in}(s)$$

The remaining step is obtaining the output ratio over input signals defined in transfer functions. Therefore,

$$H_1(s) = \frac{I_1(s)}{V_{in}(s)} = \frac{C(L_1 + L_2)s^2 + 1}{CL_1L_2s^3 + RC(L_1 + L_2)s^2 + L_1s + R}$$

$$H_2(s) = \frac{I_2(s)}{V_{in}(s)} = \frac{CL_1s^2}{CL_1L_2s^3 + RC(L_1 + L_2)s^2 + L_1s + R}$$

**Example 8.9 Considering the circuit shown in Fig. 8.9, find the transfer functions defined as** $H_1(s) \triangleq \frac{V_1(s)}{I_1(s)}$, $H_2(s) \triangleq \frac{V_1(s)}{I_2(s)}$, $H_3(s) \triangleq \frac{V_2(s)}{I_1(s)}$, **and** $H_4(s) \triangleq \frac{V_2(s)}{I_2(s)}$.

*Solution.* The circuit has two inputs, $I_1$ and $I_2$, and two outputs, $V_1$ and $V_2$, were defined. Therefore, four transfer functions can be defined (Fig. 8.9).

Considering the transfer functions defined, one can see that the problem is asking for voltages $V_1$ and $V_2$.

The solutions are as follows:

$$V_1 = \frac{R(CL(I_1 - I_2) + I_1)}{CLs^2 + RCs + 1} = \frac{R\left(1 + \frac{1}{LC}\right)}{s^2 + \frac{R}{L}s + \frac{1}{LC}} I_1 - \frac{R}{s^2 + \frac{R}{L}s + \frac{1}{LC}} I_2$$

$$V_2 = -\frac{Ls(I_2 - CRsI_1 + CRsI_2)}{CLs^2 + RCs + 1} = \frac{Rs^2}{s^2 + \frac{R}{L}s + \frac{1}{LC}} I_1 - \frac{RLs\left(\frac{1}{RC} + s\right)}{s^2 + \frac{R}{L}s + \frac{1}{LC}} I_2$$

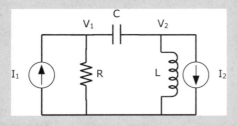

**Fig. 8.9** Circuit of Example 8.9

(continued)

**Example 8.9** (continued)

As the solutions demonstrate, each voltage depends on the input current sources. For instance, $V_1$ is a function of both $I_1$ and $I_2$. To obtain the transfer function $H_1(s) \triangleq \frac{V_1(s)}{I_1(s)}$, the second current source value should be zero, $I_2 = 0$, or

$$H_1(s) \triangleq \frac{V_1(s)}{I_1(s)}\bigg|_{I_2(s)=0}, \qquad H_2(s) \triangleq \frac{V_1(s)}{I_2(s)}\bigg|_{I_1(s)=0}. \quad \text{Hence,}$$

$$H_1(s) = \frac{R\left(1 + \frac{1}{LC}\right)}{s^2 + \frac{R}{L}s + \frac{1}{LC}}, \qquad H_2(s) = -\frac{R}{s^2 + \frac{R}{L}s + \frac{1}{LC}}$$

$$H_3(s) \triangleq \frac{V_2(s)}{I_1(s)}\bigg|_{I_2(s)=0}, \qquad H_4(s) \triangleq \frac{V_2(s)}{I_2(s)}\bigg|_{I_1(s)=0}. \quad \text{Hence,}$$

$$H_3(s) = \frac{Rs^2}{s^2 + \frac{R}{L}s + \frac{1}{LC}}, \qquad H_4(s) = -\frac{RLs\left(\frac{1}{RC} + s\right)}{s^2 + \frac{R}{L}s + \frac{1}{LC}}$$

## Transfer Function Operations

Consider a large circuit with several branches, nodes, and sources that generate several responses. The circuit can be broken into manageable sections to find the transfer function. Each section's transfer function can be obtained. The overall transfer function is obtained through some operations defined for parallel, series, feedback, and feedforward connection of transfer functions. In this section, these operations are introduced and analyzed.

**Series connection**  Consider two or more transfer functions connected in series. The output of one transfer function is fed to the input of the other, as shown in Fig. 8.10.

The overall transfer function is the product of all transfer functions:

$$\text{TF}(s) = \prod_{i=1}^{n} \text{TF}_i(s)$$

**Fig. 8.10**  Tandem connection or series connection of transfer functions

**Example 8.10 Consider a multi-loop circuit as given in Fig. 8.11. Find the circuit's transfer function** $H(s) = \frac{V_o}{V_{in}}$.

*Solution.* The circuit can be split into three sections which are connected in series. The transfer function of each section is obtained to find the overall transfer function. It can be observed that

$$\frac{V_o}{V_{in}} = \frac{V_o}{V_2} \times \frac{V_2}{V_1} \times \frac{V_1}{V_{in}}$$

Each of these transfer functions can be obtained as follows:

$$\frac{V_1}{V_{in}} = \frac{\frac{1}{sC_1}}{R_1 + \frac{1}{sC_1}} = \frac{1}{1 + R_1 C_1 s} = \frac{\frac{1}{R_1 C_1}}{s + \frac{1}{R_1 C_1}}$$

$$\frac{V_2}{V_1} = \frac{R_2}{R_2 + sL_1} = \frac{\frac{R_2}{L_1}}{s + \frac{R_2}{L_1}}$$

$$\frac{V_o}{V_2} = \frac{\frac{1}{sC_2}}{sL_2 + \frac{1}{sC_2}} = \frac{1}{1 + L_2 C_2 s^2} = \frac{\frac{1}{L_2 C_2}}{s^2 + \frac{1}{L_2 C_2}}$$

Replacing into the overall transfer equation,

$$\frac{V_o}{V_{in}} = \frac{\frac{1}{R_1 C_1}}{s + \frac{1}{R_1 C_1}} \frac{\frac{R_2}{L_1}}{s + \frac{R_2}{L_1}} \frac{\frac{1}{L_2 C_2}}{s^2 + \frac{1}{L_2 C_2}} = \frac{\frac{R_2}{R_1 C_1 C_2 \, L_1 L_2}}{\left(s + \frac{1}{R_1 C_1}\right)\left(s + \frac{R_2}{L_1}\right)\left(s^2 + \frac{1}{L_2 C_2}\right)}$$

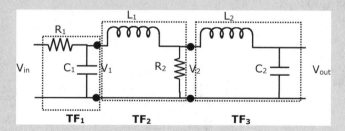

**Fig. 8.11** A circuit is broken into three tandem-connected subsystems. The transfer function of the overall system is the product of the individual transfer functions of subsystems

## Parallel Connection

If two or more transfer functions are connected in parallel, their operation will depend on the final operation designed by the circuit. For instance, consider the following transfer functions (Fig. 8.12).

The equivalent transfer function becomes the summation of two systems as

$$TF = TF_1 + TF_2$$

## Feedback Connection

The connection of transfer functions as feedback suggests measuring the output compared with the desired reference to generate an error signal. This signal is used to correct the behavior of the system. In feedback systems, the output signals may be added or subtracted from the reference signal, which generates positive or negative feedback.

Consider the transfer function of plant $G$ and the feedback signal transducer $F$, as shown in Fig. 8.13.

Signal $E$ is obtained as $E = R - CF$. The output of plant $G$ is labeled as signal $C = EG$. Replacing error into this equation yields

$$C = (R - CF)G$$

The transfer function is defined as the ratio of output over input transfer functions $\frac{C}{R}$ obtained as follows:

**Fig. 8.12** Parallel connection of two systems receiving the same input signal

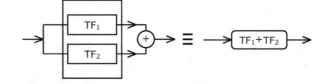

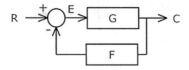

**Fig. 8.13** Closed-loop connection of system $G$ through the system of $F$. Since the feedback signal is subtracted from the reference $R$, the system is called "a negative feedback closed-loop system"

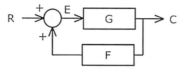

**Fig. 8.14** Closed-loop connection of system $G$ through the system of $F$. Since the feedback signal is added to the reference $R$, the system is called "a positive feedback closed-loop system"

$$C(1 + FG) = RG$$

$$\frac{C}{R} = \frac{G}{1 + FG}$$

The transfer function of negative feedback in a single loop system is obtained by the ratio of feedforward gain $G$ over 1 plus the loop gain $FG$. The feedforward gain is the gain observed in a forward path from the input to the output, and the loop gain is the product of all transfer functions existing in the loop circulating once.

In a positive feedback system, as shown in Fig. 8.14.

The error signal becomes $E = R + CF$. Considering the output signal $C = EG$ and the replacement from the error yields

$$C = (R + CF)G$$

The transfer function of a positive feedback system becomes

$$\frac{C}{R} = \frac{G}{1 - FG}$$

Considering the feedback system shown in Fig. 8.14, the transfer function defined as $H(s) = \frac{E(s)}{R(s)}$ can be obtained through the feedforward (FF) over 1+loopgain $(1 + LG)$ as follows:

$$H(s) = \frac{E(s)}{R(s)} = \frac{FF}{1 + LG} = \frac{1}{1 + FG}$$

**Example 8.11 The system diagram of a circuit is shown in Fig. 8.15. Find the transfer function of the closed-loop system.** *Solution.* The closed-loop system's transfer function $\frac{C}{R}$ is obtained by identifying the feedforward path and the loop gain.

The feedforward path is the direct connection from input $R$ to output $C$ identified $\frac{K}{s+5}$, and the loop gain is identified as the product of the system and the feedback $\frac{K}{s+5} \frac{2}{s+1}$. Therefore, the closed-loop transfer function becomes

(continued)

**Example 8.11** (continued)

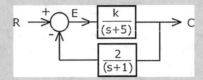

**Fig. 8.15**   Closed-loop system of Example 8.11

$$\frac{C}{R} = \frac{\frac{K}{s+5}}{1 + \frac{K}{s+5}\frac{2}{s+1}} = \frac{K(s+1)}{s^2 + 6s + 5 + 2K}$$

## Feedback and Change of Order of Circuit

The order of a circuit is determined as the higher power of $s$ in the denominator of its transfer function. The feedback system may change the order of a circuit too. As explained in the previous example, the plant $\frac{K}{s+5}$ is a first-order circuit. However, the closed-loop system $\frac{C}{R} = \frac{K(s+1)}{s^2 + 6s + 5 + 2K}$ is a second-order system.

**Example 8.12 Find the transfer function of the system** $G(s) = \frac{k}{s+a}$**with negative unity feedback.** *Solution.* The closed-loop transfer function is obtained by

$$H(s) = \frac{FF}{1 + LG} = \frac{\frac{k}{s+a}}{1 + \frac{k}{s+a}} = \frac{k}{s + (a + k)}$$

As this example explains, the gain of the system dynamics can change the location of the poles in the system from $s = -a$ to $s = -k - a$.

**Example 8.13 Using the negative unity feedback closed-loop system, find the system's open loop transfer function in which the closed-loop transfer function becomes** $H(s) = \frac{k}{s+a}$**.**
*Solution.* Consider $G(s)$ as the open-loop transfer function.

(continued)

**Example 8.13** (continued)

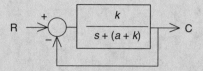

$$H(s) = \frac{G(s)}{1 + G(s)} \equiv \frac{k}{s + a}$$

Therefore,

$$sG(s) + aG(s) = k + kG(s)$$

Solving for $G(s)$ yields

$$G(s) = \frac{k}{s + (a + k)}$$

**Example 8.14** Considering the open-loop system as a simple integrator $G(s) = \frac{1}{s}$, find the feedback gain to form the closed-loop transfer function of $H(s) = \frac{1}{s+a}$.

*Solution.* The closed-loop transfer function of the system is

$$H(s) = \frac{\frac{1}{s}}{1 + \frac{k}{s}} \equiv \frac{1}{s + a}$$

Therefore,

$$H(s) = \frac{1}{s + k} = \frac{1}{s + a}$$

Hence, $k = a$.

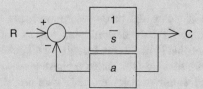

**Example 8.15 Consider a circuit with input $x(t)$ and output $y(t)$, presented in its differential equation as follows:**

$$\ddot{y} + 2\dot{y} + y = \ddot{x} + 6\dot{x} + x$$

*Find the transfer function of the system $G(s) = \frac{Y(s)}{X(s)}$.*

*Solution.* The transfer function is the ratio of the Laplace of output over the Laplace of input signals. In this example, the Laplace transform of the differential equation is identified as

$$L\{\ddot{y} + 2\dot{y} + y\} = L\{\ddot{x} + 6\dot{x} + x\}$$

$$(s^2 + 2s + 1)Y(s) = (s^2 + 6s + 1)X(s)$$

The transfer function becomes

$$G(s) = \frac{Y(s)}{X(s)} = \frac{s^2 + 6s + 1}{s^2 + 2s + 1}$$

## Poles and Zeros

Consider a transfer function defined as the ratio of two polynomials: the numerator and the denominator. These polynomials depending on their order might have several roots. Any root of the numerator polynomial makes the transfer function zero. Hence, the roots of the numerator polynomial are defined as zeros of the transfer function. Any root of the denominator polynomial makes the transfer function $\infty$. The roots of the denominator polynomial are called poles of the transfer functions. Poles and zeros are measured in the frequency domain and have the unit of rad/sec. These frequencies/roots may become real numbers or complex conjugate numbers. The real and imaginary parts indicate various parts of a response, as explained in Chap. 4 on the response of second-order systems.

## Phase Plane

The type and location of poles and zeros determine many system characteristics, including the system response to the desired input, and its stability, among others. A complex conjugate plane that indicates the location of all poles with a cross $\times$ and the location of all zeros with a circle $\bigcirc$ is called a phase plane. As the system gains or parameters change, the location of poles and zeros changes, showing a trajectory that

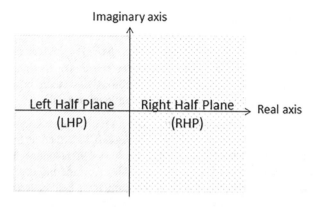

**Fig. 8.16** A phase plane. The real and imaginary axes are shown. The plane is divided into left and right half planes that show entirely different system characteristics. This plane shows the location of poles and zeros, and their trajectories should the system parameters change (due to controller effects or the system changes). Poles and zeros in RHP possess a positive real part, and those in LHP possess a negative real part

indicates system characteristics. This topic will be discussed in detail in the rest of this chapter. The phase plane has one real and one imaginary axis showing the poles and zeros' real and imaginary parts. This distinction is important as it indicates the limits of stability and the effect of controller gains.

The real value axis can be divided into positive (right-hand) and negative (left-hand) sides. This divides the phase plane into right half plane (RHP) and left half plane (LHP) regions (Fig. 8.16).

## Limit of Stability

The phase plane is one indicator that the system is stable and whether a change occurs in the system remains stable. This indicator depends on the time domain response of a system.

For instance, consider a system response as $y(t) = 2e^{+3t}u(t)$. This exponential value increases as time increases. This does not reach a steady value, and if this is the energy of a system, it shows a growing level of energy in the system and can virtually reach an infinite value that is out of control. This system is unstable because its internal energy does not remain bounded. It remains stable if the internal energy decreases or remains constant. For instance, consider the same system with a negative exponential argument as $y(t) = 2e^{-3t}u(t)$. As time increases, the system loses energy and reaches a stable operation, the origin.

Translating these systems into the frequency domain through Laplace transform reveals that the unstable system $Y(s) = \frac{2}{s-3}$ had a pole in the RHP (positive real part pole) and the stable system $Y(s) = \frac{2}{s+3}$ had a pole in LHP (negative real part pole).

*It can be concluded that stable systems have no poles in RHP, and they also should not have repeated poles at the origin.*

To understand this analogy, consider a system as $G(s) = \frac{1}{s^2}$ which has two poles at the origin. The Laplace inverse of this system is $g(t) = t\, u(t)$. As time increases, the value of $g(t)$ increases and reaches infinite. Therefore, since there is no limit for this output (or energy) increment, the system is considered unstable.

**The phase of the system**   The system may have a phase depending on the location of poles and zeros. The phase of a transfer function is identified as the numerator's phase minus the denominator's phase. For a stable system response, as explained in detail, the system phase should be less than $-180°$. Considering that this phase is a subtraction of two numbers, the phase of the numerator helps decrease the phase, making it more stable and showing a better/faster response. Zeros in RHP show a positive phase as their phase is calculated by $\tan^{-1} \frac{Im\{root\}}{Re\{root\}}$. This takes the system to move away from its best phase angle and closer to the limit $-180°$. Therefore, the systems with zeros in RHP are called non-minimum phase systems. Their response to control action in the time domain shows an initial decrease and moves away from the reference before moving toward it. An initial dip causes much trouble for the control system. Examples of these systems can be found in power electronics boost converters or the water level controls in the drum of power plant boilers.

---

**Example 8.16 Find the poles and zeros of the following transfer functions.**

1. $G(s) = \frac{s+1}{s^2+1}$.

This system has a first-order numerator, resulting in one zero as the root of $s + 1 = 0$. Therefore, the zero of the transfer function is $s = -1$ rad/s. The denominator is a second-order system with two poles as the root of $s^2 + 1 = 0$. This yields two complex conjugate roots of $s = \pm j1$ rad/s.

The phase plane of these roots can be shown in Fig. 8.17

The system is stable as there are no poles in RHP.

2. $G(s) = \frac{s^2+4}{s^2+2s+5}$.

The system has two zeros and two poles. The zeros are obtained from the roots of $s^2 + 4 = 0$ as $s = \pm j^2$ rad/s. The poles are obtained by the roots of $s^2 - 2s + 5 = 0$ as $s = +1 \pm j^2$ rad/s. The system is unstable as the poles have all positive real parts (Fig. 8.18).

3. $G(s) = \frac{s^2-4}{s^2+2s+5}$

The system has two zeros at $s^2 - 4 = 0$ as $s = \pm 2$ rad/s. One of the zeros is located in RHP, making the system a non-minimum phase. The poles are obtained by the roots of $s^2 + 2s + 5 = 0$ as $s = -1 \pm j^2$ rad/s. As the poles have all negative real parts, the system is stable. Therefore, the system is stable but non-minimum phase (Fig. 8.19).

---

(continued)

**Example 8.16** (continued)

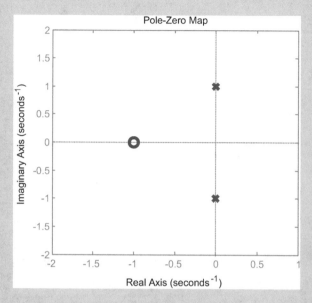

**Fig. 8.17** The figure of poles and zeros of $G(s) = \frac{s+1}{s^2+1}$

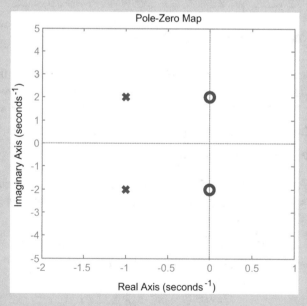

**Fig. 8.18** Poles and zeros of $G(s) = \frac{s^2+4}{s^2+2s+5}$

(continued)

**Example 8.16** (continued)

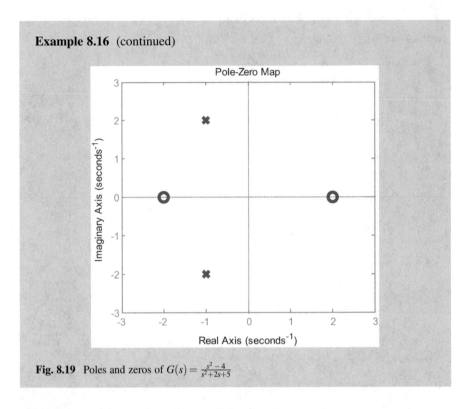

**Fig. 8.19** Poles and zeros of $G(s) = \frac{s^2-4}{s^2+2s+5}$

## Initial Value and Final Value Theorems

A partial time-domain response of a circuit can be approximated by its initial and final values. The system response limit is obtained as the time reaches zero or infinite to obtain these values. Consider $y(t)$ as the system response. Therefore,

- Initial value becomes $y(0) = \lim_{t \to 0} y(t)$
- Final value becomes $y(\infty) = \lim_{t \to \infty} y(t)$

If all values are given in the frequency domain, one way to obtain the initial and final values will be through the Laplace inverse transforms of the response and then evaluate the system's initial and final values in the time domain.

An alternative approach will be using initial and final value theorems, which directly utilize the frequency domain system responses. According to the initial value theorem, the initial amount of a function is obtained by

$$y(0) = \lim_{t \to 0} y(t) \equiv \lim_{s \to \infty} sY(s)$$

**Note 8.3** The left-hand side of this theorem is in the time domain, and the right-hand side is in the frequency domain. Once the time reaches zero, the frequency must reach infinite.

**Note 8.4**

There is a factor $s$ imposed in the frequency domain.

The final value, *which is also called the steady-state response*, is accordingly defined as

$$y(\infty) = y_{ss} = \lim_{t \to \infty} y(t) \equiv \lim_{s \to 0} sY(s)$$

**Note 8.5** The left-hand side of this theorem is in the time domain, and the right-hand side is in the frequency domain. Once the time reaches infinite, the frequency must reach zero.

**Note 8.6**

There is a factor $s$ imposed in the frequency domain.

**Example 8.17 Find the initial and final value of the following system.**

$$Y(s) = \frac{s^2 + 3s + 0.5}{s(s+1)^2}$$

*Solution.* The initial value is found $y(t \to 0) = \lim_{s \to \infty} s \, \frac{s^2+3s+0.5}{s(s+1)^2} = 1.$

*Hint.* When a polynomial of order $n$ reaches infinite, it is equivalent to its highest-order component, e.g., when $s \to \infty$, $s^2 + 3s + 1 \equiv s^2$.

The final value is found as $y(t \to \infty) = \lim_{s \to 0} s \, \frac{s^2+3s+0.5}{s(s+1)^2} = \frac{1}{2}.$

This means that the signal $y(t)$ started at value 1 and reached value 1/2, as shown in Fig. 8.20.

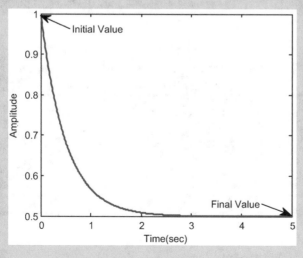

**Fig. 8.20** Time representation of $y(t)$ also shows $y(0) = 1$ and $y(\infty) = 0.5$

**Example 8.18 Find the initial value and final value of the system**
$Y(s) = \frac{s-10}{s^2+3s+1}$.

*Solution.*        The        initial        value        is        obtained

as $y(t \to 0) = \lim_{s \to \infty} s \; \frac{s-10}{s^2+3s+1} = \lim_{s \to \infty} \frac{s^2-10s}{s^2+3s+1} = \lim_{s \to \infty} \frac{s^2}{s^2} = 1$

The final value is obtained as $y(t \to \infty) = \lim_{s \to 0} s \; \frac{s-10}{s^2+3s+1} = \lim_{s \to 0} 0 \; \frac{-10}{1} = 0$.

The signal $y(t)$ starts from 1 and reaches 1 at a very large time. Figure 8.21 shows the sketch of the response.

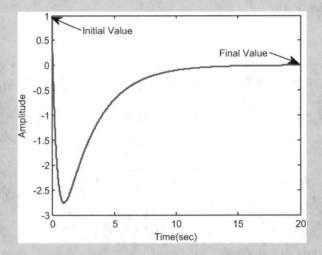

**Fig. 8.21** Time representation of $y(t)$ also shows $y(0) = 1$ and $y(\infty) = 0$

**Example 8.19 The process $G(s)$ is controlled by a unity feedback system. Find the closed-loop system steady-state response to a unit step input.**
*Solution.* The closed-loop system transfer function with input $x(t) = u(t)$ and output of $y(t)$ as the transfer function of

$$H(s) = \frac{G(s)}{1 + G(s)} = \frac{Y(s)}{X(s)}$$

Therefore, the output of the system can be obtained as

(continued)

**Example 8.19** (continued)

$$Y(s) = X(s) \frac{G(s)}{1 + G(s)}$$

The steady-state response of the system is its final value as follows:

$$y_{ss} = \lim_{s \to 0} sY(s) = \lim_{s \to 0} sX(s) \frac{G(s)}{1 + G(s)}$$

Replacing the input with a step function yields

$$y_{ss} = \lim_{s \to 0} s \frac{1}{s} \frac{G(s)}{1 + G(s)}$$

$$y_{ss} = \lim_{s \to 0} \frac{G(s)}{1 + G(s)}$$

**Example 8.20 Closed-loop poles of a system are $s = -2 \pm j5$, and the zero is located at $s = -1$. Find the steady-state step response of the system. Find its steady state error.**

*Solution.* The closed-loop poles and zero results in the system transfer function as

Poles: $s = -2 \pm j5 \rightarrow s + 2 = \pm j5 \rightarrow s^2 + 4s + 4 = -25 \rightarrow$
Poles: $s^2 + 4s + 29$

$$H(s) = \frac{s + 1}{s^2 + 4s + 29}$$

$$H(s) = \frac{Y(s)}{X(s)}$$

$$y_{ss} = \lim_{s \to 0} sY(s) = \lim_{s \to 0} sX(s)H(s) = \lim_{s \to 0} s \frac{1}{s} \frac{s + 1}{s^2 + 4s + 29} = \frac{0 + 1}{0^2 + 4 \times 0 + 29}$$

$$= \frac{1}{29}$$

The steady-state error of this system is

$$e_{ss} = 1 - y_{ss} = 1 - \frac{1}{29} = \frac{28}{29}$$

**Example 8.21 In the previous example, the zero location has moved to $s = -20$. Find the steady-state step response of the system. Find its steady-state error.**

*Solution.* The closed-loop poles and zero results in the system transfer function as

Poles: $s = -2 \pm j5 \rightarrow s + 2 = \pm j5 \rightarrow s^2 + 4s + 4 = -25 \rightarrow$
Poles: $s^2 + 4s + 29$

$$H(s) = \frac{s + 20}{s^2 + 4s + 29}$$

$$H(s) = \frac{Y(s)}{X(s)}$$

$$y_{ss} = \lim_{s \to 0} sY(s) = \lim_{s \to 0} sX(s)H(s) = \lim_{s \to 0} s\frac{1}{s}\frac{s + 20}{s^2 + 4s + 29} = \frac{0 + 20}{0^2 + 4 \times 0 + 29}$$

$$= \frac{20}{29}$$

The steady-state error of this system is

$$e_{ss} = 1 - y_{ss} = 1 - \frac{20}{29} = \frac{9}{29}$$

**Example 8.22 Find steady-state response of the closed-loop system of $H(s) = \frac{7}{s+5}$ to a unit step function.**

*Solution.*

$$H(s) = \frac{Y(s)}{X(s)}$$

$$Y(s) = X(s)H(s) = \frac{1}{s}\frac{7}{s + 5}$$

$$y_{ss} = \lim_{s \to 0} sY(s) = \lim_{s \to 0} s\frac{1}{s}\frac{7}{s + 5} = \lim_{s \to 0}\frac{7}{s + 5} = \frac{7}{5}$$

**Example 8.23 Find the steady-state error of a unity feedback closed-loop system with transfer function** $H(s) = \frac{Y(s)}{X(s)}$.

*Solution.* Error in closed-loop systems is defined as (note that all Laplace operators are omitted for simplicity, $X \equiv X(s)$, $Y \equiv Y(s)$, $E \equiv E(s)$)

$$E = X - Y$$

Therefore,

$$\frac{E}{X} = \frac{X}{X} - \frac{Y}{X} = 1 - \frac{Y}{X}$$

Hence,

$$E = X\left(1 - \frac{Y}{X}\right)$$

Therefore,

$$e_{ss} = \lim_{s \to 0} sE = \lim_{s \to 0} sX\left(1 - \frac{Y}{X}\right)$$

$$e_{ss} = \lim_{s \to 0} sX(1 - H(s))$$

**Example 8.24 Transfer-function of a second-order system is** $H(s) = \frac{s+1}{s^2+7s+1}$. **Find the steady-state step response of the system.**
*Solution.*

$$H(s) = \frac{Y(s)}{X(s)} \rightarrow Y(s) = X(s)H(s)$$

$$y_{ss} = \lim_{s \to 0} sY(s) = \lim_{s \to 0} sX(s)H(s) = \lim_{s \to 0} s\frac{1}{s}\frac{s+1}{s^2+7s+1}$$

$$= \lim_{s \to 0} \frac{0+1}{0^2 + 7 \times 0 + 1} = 1$$

The steady-state error in this case is

(continued)

**Example 8.24** (continued)

$$e_{ss} = 1 - y_{ss} = 1 - 1 = 0$$

The control system performs the control with zero tracking error. It may take some time, but $e_{ss} = 0$ shows that the control system eventually reaches zero.

## Order and Type of a System

A system may be presented by its governing differential equation. A standard differential equation form has two polynomials representing the input and output dynamics. The order of the system is the highest degree of differential in the output equation. Consider the following differential equation where $y$ is the output and $u$ is the input:

$$y^{(n)}(t) + a_1 y^{(n-1)}(t) + \ldots + a_n y(t) = b_0 x^{(m)}(t) + b_1 x^{(m-1)}(t) + \ldots + b_m x(t)$$

The polynomial is a function of $y$: output is the output equation, and the polynomial involves $x$: input is the input equation.

Output equation: $y^{(n)}(t) + a_1 y^{(n-1)}(t) + \ldots + a_n y(t)$
Input equation: $b_0 x^{(m)}(t) + b_1 x^{(m-1)}(t) + \ldots + b_m x(t)$

*Order of a system*: The order is the highest order of the differential in the output equation or $n$. This system will have $n$ poles.

The system transfer function can be obtained as the ratio of the Laplace of the output over the Laplace of the input as follows:

$$H(s) = \frac{b_0 s^m + b_1 s^{m-1} + \ldots + b_m}{s^n + a_1 s^{n-1} + \ldots + a_n}$$

*Type of a system*: The type of the system $H(s)$ is the number of its poles at the origin.

### First-Order Systems

According to our definition, the first-order system is a system with one pole. However, many transfer functions have one pole. Here, two of these systems are introduced. One is used in modeling system processes and one in a model of filters. However, these systems can be converted together and may have similar behavior.

*The first-order system to model processes has an ultimate amplitude of k and a time constant τ as follows*:

$$H(s) = \frac{k}{\tau s + 1}$$

This system has a pole at $s = -\frac{1}{\tau}$. The time constant $(t = \tau)$ refers to the time that it takes for its response to reach ~ 63% of the ultimate value, shown as follows:

$$y(t = \tau) = 1 - e^{-\frac{t}{\tau}} = 1 - e^{-1} = 1 - \frac{1}{2.718} = 0.63 \to 63\%$$

**Example 8.25 Find the step response to the system** $H(s) = \frac{10}{0.5s+1}$.
*Solution.* The system is in standard form for a process that reaches an amplitude of 10 with a time constant of 0.5 sec. The time constant is obtained by finding 63% of the amplitude 10 or 6.3. The time corresponding to this amplitude is ~ 0.5 sec. In this system, 0.497 sec is shown in the following figure.

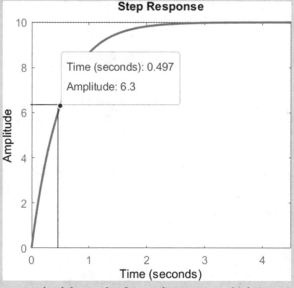

*The other standard form of a first-order system, which is mainly used in analyzing filters, can be expressed as follows*:

(continued)

**Example 8.25** (continued)

$$H(s) = \frac{k}{s+k}$$

This system has an ultimate DC gain of 1 or 100% and reaches 70.7% of the ultimate value at the frequency of $s = k\left(\frac{rad}{sec}\right)$.

**Example 8.26 Find the order and type of the following system.**

$$H(s) = \frac{5}{s+5}$$

*Solution.* This system has a pole at

$$s + 5 = 0$$

$$s = -5$$

*It is a first-order system. Since this system has no poles at the origin, the system is type 0.*

**Example 8.27 Find the order and type of the following system.**
*Solution.* Consider the system

$$H(s) = \frac{s+0.5}{s-4}$$

The pole of the system is located at

$$s - 4 = 0$$

$$s = +4$$

The system has a zero at

$$s + 0.5 = 0$$

(continued)

**Example 8.27** (continued)

$$s = -0.5$$

It is a first-order system. Since the system has no poles at the origin, the system is type 0.

**Example 8.28 Find the order and type of the following system.**

$$H(s) = \frac{10}{s}$$

*Solution.* This system has a pole at

$$s = 0$$

It is a first-order system. Since the pole is located at the origin, the system is type 1.

## Second-Order Systems

The highest-order differential of the output equation of a second-order system is 2. Therefore, the denominator of second-order systems is a quadratic equation resulting in two poles.

**Standard form** Second-order systems have a standard form as follows:

$$H(s) = \frac{\omega_n^2}{s^2 + 2\zeta\omega_n s + \omega_n^2}$$

where $\zeta$ is the damping factor of the system and $\omega_n$ is the system's natural or resonant frequency.

The location of poles in a second-order system on the phase plane depends on the amount of damping factor and the damping frequency.

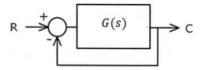

Consider a unity negative feedback system with an open-loop transfer function of $G(s)$. The closed-loop transfer function is

$$H(s) = \frac{G}{1+G} = \frac{\omega_n^2}{s^2 + 2\zeta\omega_n s + \omega_n^2}$$

Therefore,

$$G\left(s^2 + 2\zeta\omega_n s + \omega_n^2\right) = \omega_n^2 + G\omega_n^2$$

Solving for $G$ results in

$$G(s) = \frac{\omega_n^2}{s^2 + 2\zeta\omega_n s}$$

$$G(s) = \frac{\omega_n^2}{s(s + 2\zeta\omega_n)}$$

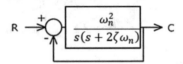

Consider a unity negative feedback system with an open-loop transfer function of $G(s)$. The closed-loop transfer function is

$$H(s) = \frac{G}{1+G} = \frac{\omega_n^2}{s^2 + 2\zeta\omega_n s + \omega_n^2}$$

Therefore,

$$G\left(s^2 + 2\zeta\omega_n s + \omega_n^2\right) = \omega_n^2 + G\omega_n^2$$

Solving for $G$ results in:

$$G(s) = \frac{\omega_n^2}{s^2 + 2\zeta\omega_n s}$$

$$G(s) = \frac{\omega_n^2}{s(s + 2\zeta\omega_n)}$$

The step response according to the location of poles and the damping conditions are discussed as follows:

In general, the location of poles in a second-order system is

$$s_{1,2} = -\zeta\omega_n \pm \omega_n\sqrt{\zeta^2 - 1}$$

## Case 1. Oscillatory

The system is oscillatory when the damping is zero ($\zeta = 0$):

$$\text{if} : \zeta = 0, \text{ then} : s_{1,2} = \pm j\omega_n$$

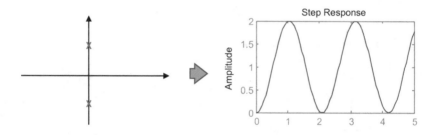

## Case 2. Underdamped

The system is underdamped when the damping is $0 < \zeta < 1$:

$$\text{if} : 0 < \zeta < 1, \text{then} : s_{1,2} = -\zeta\omega_n \pm j\omega_n\sqrt{1 - \zeta^2}$$

considering

$$\omega_d = \omega_n\sqrt{1 - \zeta^2}$$

where $\omega_d$ is the damping frequency.
Therefore,

$$s_{1,2} = -\zeta\omega_n \pm j\omega_d$$

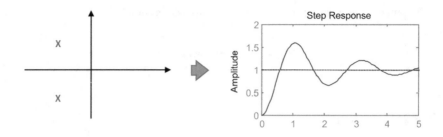

## Case 3. Critically Damped

The system is critically damped when the damping is one $\zeta = 1$.

$$\text{if} : \zeta = 1, \text{then} : s_{1,2} = -\omega_n$$

In this case, the system response has no oscillation.

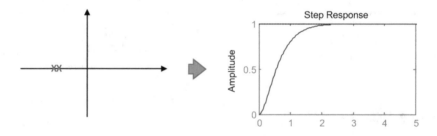

## Case 4. Overdamped

The system is overdamped when the damping factor is larger than 1, $\zeta > 1$.

$$s_{1,2} = -\zeta\omega_n \pm \omega_n\sqrt{\zeta^2 - 1}$$

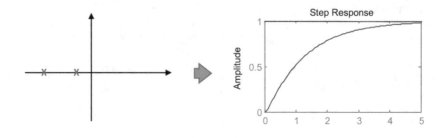

**Other forms** A nonstandard second-order system still has two poles but may have zeros. A transfer function of a second-order system with a zero may exist as follows:

$$H(s) = \frac{2\zeta\omega_n s}{s^2 + 2\zeta\omega_n s + \omega_n^2}$$

$$H(s) = \frac{s^2 + \omega_n^2}{s^2 + 2\zeta\omega_n s + \omega_n^2}$$

**Example 8.29** In a second-order system, the damping factor is $\zeta = 0.7$, and the resonant frequency is $\omega_n = 450\,\frac{rad}{sec}$. Find the damping frequency and the transfer function of the system.

*Solution.* The damping frequency is

$$m,mm\ mn\ m\omega_d = \omega_n\sqrt{1 - \zeta^2} = 450\sqrt{1 - 0.7^2} = 321.36\,\frac{rad}{sec}$$

The transfer function of the system is

$$H(s) = \frac{\omega_n^2}{s^2 + 2\zeta\omega_n s + \omega_n^2} = \frac{450^2}{s^2 + 2 \times 0.7 \times 450 s + 450^2} = \frac{450^2}{s^2 + 630 s + 450^2}$$

**Example 8.30** A system has a transfer function of $G(s) = \frac{10}{s^2 + \sqrt{3}s - 7}$. Find the damping factor, and the natural frequency of the system is a unity gain closed-loop structure.

*Solution.* The closed-loop transfer function of the system is

$$H(s) = \frac{G}{1 + G} = \frac{\frac{10}{s^2 + \sqrt{3}s - 7}}{1 + \frac{10}{s^2 + \sqrt{3}s - 7}} = \frac{10}{s^2 + \sqrt{3}s + 3}$$

As the system suggests, by comparing the denominator with the standard system, the system parameters are

$$s^2 + 2\zeta\omega_n s + \omega_n^2 = s^2 + \sqrt{3}s + 3$$

(continued)

**Example 8.30** (continued)

$$\therefore 2\zeta\omega_n = \sqrt{3}\,\&\,\omega_n^2 = 3$$

$$\omega_n = \sqrt{3}\,\left(\frac{\text{rad}}{\text{sec}}\right),\zeta = \frac{\sqrt{3}}{2\times\sqrt{3}} = 0.5$$

## *Analysis of Step Response of Second-Order System*

In a second-order underdamped system, the response generates an overshoot and settles overtime to reach the reference.

The step response of an underdamped second-order system shown in Fig. 8.22 can be predicted by knowing the damping factor $\zeta$ and the natural frequency $\omega_n$.

The step response of an underdamped system generates an overshoot and some form of oscillation that takes time to settle. The overshoot, the time of the first peak, and the time to settle within $\pm 5\%$ and $\pm 2\%$ variation are specific to the system characteristics, namely, its damping factor and the damping frequency. They can be defined as follows:

- *Rise time* ($t_r$): The time it takes for the output to reach from 10% to 90% of the reference or 100%. It can be calculated as

$$t_r = \frac{\pi - \cos^{-1}\zeta}{\omega_d}$$

- *Delay time* ($t_d$): The time it takes for the response to reach 50% of the reference. It can be calculated as

**Fig. 8.22** Step response of a second-order underdamped system

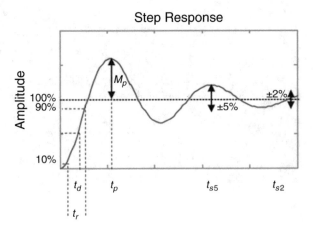

$$t_d = \frac{1 - 0.7\zeta}{\omega_d}$$

- *Peak time* $(t_p)$: The time at which the first overshoot peak appears. It can be calculated as

$$t_p = \frac{\pi}{\omega_d}$$

- *Maximum overshoot* $(M_p)$: The amount from which the peak of the overshoot exceeds the reference line or 100%. It can be calculated as

$$M_p = e^{\frac{-\pi\zeta}{\sqrt{1-\zeta^2}}}$$

- *Settling time % $(t_{s5})$*: The time it takes for the response oscillations to reach within $\pm 5\%$ of the reference. It can be calculated as

$$t_{s5} = \frac{3}{\zeta\omega_n}$$

- *Settling time % $(t_{s2})$*: The time it takes for the response oscillations to reach within $\pm 2\%$ of the reference. It takes more time for the response to settle to $\pm 2\%$ than $\pm 5\%$. It can be calculated as

$$t_{s2} = \frac{4}{\zeta\omega_n}$$

**Example 8.31** A second-order system has damping factor $\zeta = 0.6$, and the natural frequency of $\omega_n = 100$ rad/sec. Find the transfer function of the system. Determine the step response overshoot and peak time.

*Solution.* The standard second-order system is expressed as follows:

$$H(s) = \frac{100^2}{s^2 + 2 \times 0.6 \times 100 \ s + 100^2}$$

$$M_p = e^{\frac{-\pi\zeta}{\sqrt{1-\zeta^2}}} = e^{\frac{-\pi 0.6}{\sqrt{1-0.6^2}}} = 0.094$$

(continued)

**Example 8.31** (continued)

This means there exists a 9.4% overshoot. The peak value becomes $1 + 0.094 = 1.094$:

$$t_p = \frac{\pi}{\omega_d}$$

$$\omega_d = \omega_n \sqrt{1 - \zeta^2} = 100 \sqrt{1 - 0.6^2} = 80 \; \frac{rad}{sec}$$

$$t_p = \frac{\pi}{80} = 0.0393 \; sec$$

**Example 8.32 The step response of a system is shown as follows. Find the transfer function of the system (Fig. 8.23).**
*Solution*. The system step response reaches the maximum value of 1.16. This means a 0.16 Maximum Peak or

$$M_p = 0.16$$

This results in

$$M_p = e^{\frac{-\pi\zeta}{\sqrt{1-\zeta^2}}} = 0.16$$

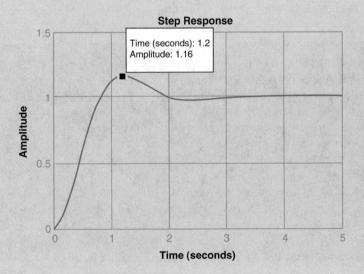

**Fig. 8.23** Step response of Example 8.32

(continued)

**Example 8.32** (continued)

Taking the natural log of both sides results in

$$Ln\left(e^{\frac{-\pi\zeta}{\sqrt{1-\zeta^2}}}\right) = Ln\ 0.16$$

$$\frac{-\pi\zeta}{\sqrt{1-\zeta^2}} = -1.83$$

$$\frac{\pi^2\zeta^2}{1-\zeta^2} = 1.83^2$$

$$1-\zeta^2 = \frac{\pi^2}{1.83^2}\zeta^2$$

$$1 = (2.944+1)\zeta^2$$

$$\zeta = 0.5$$

From the response, the peak time is recorded as 1.2 s. Therefore,

$$t_p = \frac{\pi}{\omega_d} = 1.2$$

$$\frac{\pi}{\omega_d} = \frac{\pi}{\omega_n\sqrt{1-\zeta^2}} = 1.2$$

Replacing the damping value found earlier results in

$$\frac{\pi}{\omega_n\sqrt{1-0.5^2}} = 1.2$$

$$\omega_n = \frac{\pi}{1.2\times\sqrt{1-0.5^2}} = 3\frac{rad}{s}$$

Knowing the damping $\zeta = 0.5$ and the natural frequency as $\omega_n = 3$, a standard transfer function becomes

$$H(s) = \frac{\omega_n^2}{s^2 + 2\zeta\omega_n s + \omega_n^2}$$

$$H(s) = \frac{3^2}{s^2 + 2\times0.5\times3\ s + 3^2}$$

$$H(s) = \frac{9}{s^2 + 3\ s + 9}$$

**Example 8.33** **A second-order system has a transfer function of** $H(s) = \frac{900}{s^2+6s+900}$. **Find the damping factor, natural frequency, damping frequency, amount of overshoot, time of the first peak, the settling time for $\pm 5\%$, and settling time for $\pm 2\%$.**

*Solution.* The transfer function shows the natural frequency of

$$\omega_n = \sqrt{900} = 30$$

$$2 \times \zeta\omega_n = 6 \rightarrow \zeta = \frac{6}{2 \times 30} = 0.1$$

The damping frequency $\omega_d = \omega_n \sqrt{1 - \zeta^2} = 30\sqrt{1 - 0.1^2} = 29.84$.

$$M_p = e^{\frac{-\pi\zeta}{\sqrt{1-\zeta^2}}} = e^{\frac{-\pi \times 0.1}{\sqrt{1-0.1^2}}} = 0.72$$

This means that the system reaches $1 + 0.72 = 1.72$ at the peak.

$$t_p = \frac{\pi}{\omega_d} = \frac{\pi}{29.84} = 0.105 \text{ sec}$$

$$t_{s5} = \frac{3}{\zeta\omega_n} = \frac{3}{0.1 \times 30} = 1 \text{ sec}$$

$$t_{s2} = \frac{4}{\zeta\omega_n} = \frac{4}{0.1 \times 30} = 1.3 \text{ sec}$$

## The Effect of Controller on Type-Zero Systems

Consider a plant (system) with one non-zero pole and either one or no zeros. The type of a system refers to the number of poles located at the origin. The system can be expressed as $G(s) = \frac{k}{s+a}$, where $k$ is the gain and $a$ is the pole, decay factor, or the inverse of the time constant. This system can be tuned and influenced by another block named a controller. The controller in a closed-loop system is shown in Fig. 8.24.

The closed-loop system has the following transfer function: $\frac{C}{R} = \frac{FF}{1+LG} =$

$\frac{G_c\frac{k}{s+a}}{1+G_c\frac{k}{s+a}} = \frac{kG_c}{s+(a+kG_c)}$. The transfer function poles can be placed at any desired location

**Fig. 8.24** The figure of the closed-loop controller

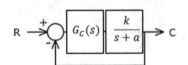

on the phase plane to obtain the desired system responses. The gain $k$ not only tunes the system gain but also influences the location of poles.

The controller can be considered as follows:

(a) *Simple gain controller*, proportional, $G_c = k_p$. The closed-loop system becomes

$$\frac{C}{R} = \frac{kk_p}{s + (a + kk_p)}$$

This results in a straightforward pole placement and controller through both gains of $k$ and $k_c$.

The system error shown in Fig. 8.24 can be calculated as $E = R - C$. A division by $R$ shows the error signal as

$$\frac{E}{R} = 1 - \frac{C}{R}$$

Replacing from the transfer function results in

$$\frac{E}{R} = 1 - \frac{kk_p}{s + (a + kk_p)} = \frac{s + a}{s + (a + kk_p)}$$

The error depends on the input $R$ to the system.

1. Consider a unit step $r = u(t)$ or $R = \frac{1}{s}$

The error becomes

$$E = R \frac{s + a}{s + (a + kk_p)} = \frac{1}{s} \frac{s + a}{s + (a + kk_p)}$$

$$e_{ss} = \lim_{s \to 0} s \frac{1}{s} \frac{s + a}{s + (a + kk_p)} = \frac{a}{(a + kk_p)}$$

It is a limited error and reaches zero only if a very high gain $k_p$ is applied. High-gain controllers are particularly not desirable as they tend to amplify noise too and are expensive to implement in hardware.

2. Consider a ramp input as $r = tu(t)$ or $R = \frac{1}{s^2}$

The steady-state error becomes

$$e_{ss} = \lim_{s \to 0} s \frac{1}{s^2} \frac{s + a}{s + (a + kk_p)} = \frac{a}{(a + kk_p)} = \infty$$

The proportional gain controller in a type zero system cannot limit the system error under ramp excitation.

(b) *Effect of a proportional-integral (PI) controller on type-zero systems.* The control system is now considered to be $G_c = k_p + \frac{k_i}{s}$. The transfer function of the closed system considering this controller becomes

$$\frac{C}{R} = \frac{k\left(k_p + \frac{k_i}{s}\right)}{s + \left(a + k\left(k_p + \frac{k_i}{s}\right)\right)}$$

(c) The error is always

$$E = R\left(1 - \frac{C}{R}\right)$$

1. Consider a unit step $r = u(t)$ or $R = \frac{1}{s}$

The error becomes

$$E = \frac{1}{s}\left(1 - \frac{k\left(k_p + \frac{k_i}{s}\right)}{s + \left(a + k\left(k_p + \frac{k_i}{s}\right)\right)}\right)$$

The steady-state error is calculated as

$$e_{ss} = \lim_{s \to 0} s\frac{1}{s}\left(1 - \frac{k\left(k_p + \frac{k_i}{s}\right)}{s + \left(a + k\left(k_p + \frac{k_i}{s}\right)\right)}\right) = 1 - \frac{k\left(k_p + \frac{k_i}{s}\right)}{s + \left(a + k\left(k_p + \frac{k_i}{s}\right)\right)} = 0$$

**Note 8.7** It is observed that the integral part of the controller can force the system to reach a zero error condition. Therefore, it can be concluded that the PI controller has zero control error in type-zero systems with a unit step input.

2. Consider a ramp input as $r = tu(t)$ or $R = \frac{1}{s^2}$.

The steady-state error becomes

$$e_{ss} = \lim_{s \to 0} s \frac{1}{s^2} \left( 1 - \frac{k\left(k_p + \frac{k_i}{s}\right)}{s + \left(a + k\left(k_p + \frac{k_i}{s}\right)\right)} \right)$$

$$= \lim_{s \to 0} \frac{1}{s} \left( 1 - \frac{k\left(k_p + \frac{k_i}{s}\right)}{s + \left(a + k\left(k_p + \frac{k_i}{s}\right)\right)} \right) = -\frac{2k_p}{k_i}$$

**Note 8.8** The system tracks the ramp input with a non-zero steady-state error.

3. Consider a hyperbolic input as $r = t^n u(t)$, $n \geq 2$, or $R = \frac{n!}{s^{n+1}}$.

The steady-state error becomes

$$e_{ss} = \lim_{s \to 0} s \frac{1}{s^{n+1}} \left( 1 - \frac{k\left(k_p + \frac{k_i}{s}\right)}{s + \left(a + k\left(k_p + \frac{k_i}{s}\right)\right)} \right)$$

$$= \lim_{s \to 0} \frac{1}{s^n} \left( 1 - \frac{k\left(k_p + \frac{k_i}{s}\right)}{s + \left(a + k\left(k_p + \frac{k_i}{s}\right)\right)} \right) = \infty$$

**Note 8.9** This means the system cannot track the inputs with higher order than $tu(t)$.

## Tracking Error Considering the Type and the Input as Reference Waveform

A summary of what is explained in the steady-state error calculations while the type of the combined controller and the plant are considered against the input reference is summarized in the following table:

As a hint: The Laplace of input reference determines the input type.

- If the type of the controller and the system combined match the input type, the error is limited.
- If the controller and system combined type is less than the input type, tracking is impossible.
- If the type of the controller and the system combined are higher than the input type, the error is zero.

| System input | Tracking error | | |
| --- | --- | --- | --- |
| | System type 0 | System type 1 | System type 2 |
| Unit step | Limited error | Zero error | Zero error |
| Ramp | Cannot track. Error $= \infty$ | Limited error | Zero error |
| Parabolic | Cannot track. Error $= \infty$ | Cannot track. Error $= \infty$ | Limited error |

**Example 8.34 A system with transfer-function** $G(s) = \frac{1}{s+3}$ **is being controlled by a proportional controller** $G_c(s) = k$. **Find the steady-state error to a step input. Find the steady-state error to a ramp input.**
*Solution.* The controller and the system combined have a transfer-function of

$$G(s)G_c(s) = k\frac{1}{s+3} = \frac{k}{s+3}$$

This system is type zero. However, the step input $x(t)$ in the frequency domain is expressed as a type 1 system, $X(s) = \frac{1}{s}$. Since the type of the system is one less than the type of input, there is a limited error in the steady-state response. No matter how much the gain increases, the error is not zero. This can be mathematically proven as follows:

$$e_{ss} = \lim_{s \to 0} sX(s)(1 - H(s)) = \lim_{s \to 0} s\frac{1}{s}\left(1 - \frac{\frac{k}{s+3}}{1 + \frac{k}{s+3}}\right) = \lim_{s \to 0}\left(1 - \frac{\frac{k}{0+3}}{1 + \frac{k}{0+3}}\right)$$

$$e_{ss} = 1 - \frac{\frac{k}{3}}{1 + \frac{k}{3}} = \frac{1}{1 + \frac{k}{3}} = \frac{3}{k + 3}$$

At $k = 10$, the steady-state error is $e_{ss} = \frac{3}{10+3} = 0.23$ or 23%.
At $k = 100$, the steady-state error is $e_{ss} = \frac{3}{100+3} = 0.029$ or 2.9%.
High-gain controllers are not desirable as they may also amplify the noise in the system.

For the ramp input $x(t) = tu(t)$, the Laplace transform suggests $X(s) = \frac{1}{s^2}$ which is a type 2 system. In this case, the controller cannot follow the input, and the error reaches infinite. This can be mathematically proven as follows:

$$e_{ss} = \lim_{s \to 0} sX(s)(1 - H(s)) = \lim_{s \to 0} s\frac{1}{s^2}\left(1 - \frac{\frac{k}{s+3}}{1 + \frac{k}{s+3}}\right) = \lim_{s \to 0}\frac{1}{0}\left(1 - \frac{\frac{k}{0+3}}{1 + \frac{k}{0+3}}\right)$$

$$e_{ss} = +\infty$$

**Example 8.35** A system has an open-loop transfer function of $G(s) = \frac{3}{s+5}$. Design a controller which can track the step input with zero error.

*Solution.* The input to this system is step function $x(t) = u(t)$ which is a type 1 system $X(s) = \frac{1}{s}$. The system is, however, type zero. To control this system with zero tracking error, the controller and the system must be at least the same type as the input. Therefore, the type of the controller should be at least 1, for which when it is combined with the system, they provide a type 1. An example of a type 1 system is an integrator controller of $G_c(s) = \frac{k_i}{s}$. Therefore, an integrator can reach a zero tracking error.

**Example 8.36** A system has an open-loop transfer function of $G(s) = \frac{3}{s(s+5)}$.

Design a controller which can track the step input with zero error.

*Solution.* The input to this system is step function $x(t) = u(t)$ which is a type 1 system $X(s) = \frac{1}{s}$. The system is already type 1. To control this system with zero tracking error, the controller and the system must be at least the same type as the input. Therefore, the type of the controller can be 0, for which, when combined with the system, they provide a type 1. An example of a type 0 system is a proportional controller of $G_c(s) = k_p$. Therefore, a proportional controller can reach a zero tracking error.

**Example 8.37** A system has an open-loop transfer function of $G(s) = \frac{3}{s+5}$. Design a controller which can track the ramp input with zero error.

*Solution.* The input to this system is step function $x(t) = tu(t)$ which is a type 2 system $X(s) = \frac{1}{s^2}$. The system is, however, type zero. To control this system with zero tracking error, the controller and the system must be at least the same type as the input. Therefore, the type of the controller should be at least 2, for which when it is combined with the system, they provide a type 2. An example of a type 2 system is a double integrator controller of $G_c(s) = \frac{k_i}{s^2}$. Therefore, a double integrator can reach a zero tracking error.

**Example 8.38 A system has an open-loop transfer function of** $G(s) = \frac{3}{s(s+5)}$.

**Design a controller which can track the ramp input with zero error.**

*Solution.* The input to this system is step function $x(t) = tu(t)$ which is a type 2 system $X(s) = \frac{1}{s^2}$. The system is already type 1. To control this system with zero tracking error, the controller and the system must be at least the same type as the input. Therefore, the type of the controller can be 1,for which, when combined with the system, they provide a type 2. An example of a type 1 system is an integrator controller of $G_c(s) = \frac{k_i}{s}$. Therefore, an integral controller can reach a zero tracking error.

## Convolution Integral

Consider a system presented by a function in time as $h(t)$. Once an input signal $x(t)$ is applied to this system function, an output signal $y(t)$ will be generated. Almost all systems operate based on this principle, and there is a need to identify the output signal. This analysis aims to determine the system's stability, the type, and the shape of the output. The input signal is often identified as a series of recorded data points, which might make the analysis difficult. The convolution integral is one technique to obtain the output (Fig. 8.25).

In the time domain, the output is obtained as follows:

$$y(t) = h(t) * x(t) = x(t) * h(t)$$

where the "*" represents the convolution.

**Example 8.39 The system $h(t) = 2tu(t)$ receives an input signal of $x(t) = 3u(t)$ and explains how the input and output are related.** *Solution.* The output is obtained through convolution integral as follows:

$$y(t) = x(t) * h(t)$$
$$y(t) = 2tu(t) * 3u(t)$$

**Fig. 8.25**   Time-domain representation of a system $h(t)$, input $x(t)$, and the output $y(t)$

**Properties of convolution integral**
Consider three signals $x(t)$, $h(t)$ and $y(t)$:

1. $x(t) * h(t) = h(t) * x(t)$
2. $x(t) * (h(t) * z(t)) = (x(t) * h(t)) * z(t)$
3. $x(t) * (h(t) + z(t)) = (x(t) * h(t)) + (x(t) + z(t))$

It is of utmost importance to describe the integral convolution operation. This operation can be defined as an integral product of the two given signals in partial time steps as follows:

$$y(t) = x(t) * h(t) = \int_{-\infty}^{+\infty} x(\lambda)h(t - \lambda)d\lambda = \int_{-\infty}^{+\infty} x(t - \lambda)h(\lambda)d\lambda$$

Let us consider that the signals are defined for positive time, i.e., $t > 0$, and that the systems are "causal," i.e., the more number of poles than zeros, and then the integral can be modified to

$$y(t) = x(t) * h(t) = \int_{0}^{t} x(\lambda)h(t - \lambda)d\lambda = \int_{0}^{t} x(t - \lambda)h(\lambda)d\lambda$$

To evaluate the integral, the functions $x(\lambda)$ and $h(t - \lambda)$ are rolled over each other, and their functions will be evaluated in a specific time shift. The following steps are needed:

1. Mirror one of the functions, either $x(t)$ *or* $h(t)$, concerning the vertical axis. This results in $x(-\lambda)$ *or* $h(-\lambda)$.
2. Shift the mirrored signal by $t$ seconds to obtain $x(t - \lambda)$ *or* $h(t - \lambda)$. This time shift allows the mirrored signal to roll over the other signal.
3. Specific time partitions are obtained by observing the changes in the product of the two signals $x(t - \lambda)h(\lambda)$ *or* $x(\lambda)h(t - \lambda)$, resulting in the time shift.
4. Evaluate the integral in the specified times until two signals have no overlap or the results remain similar (no new condition is generated).

**Example 8.40 A system $h(t)$ is a pulse of amplitude 2 stretched from time 1 to 3 s. Find the output response due to an input signal $x(t) = u(t)$.**
*Solution.* It is possible to mirror either one of the x or h signals in this solution. Let us mirror $x(t)$ with respect to the vertical axis. As the horizontal axis changes from $t \rightarrow \lambda$, the function $x(-\lambda) = u(-\lambda)$.
  A shift by $t$ seconds results in $x(t - \lambda) = u(t - \lambda)$.
  Figure 8.26 shows the position of two signals when the time shift does not reach the $h(\lambda)$. Hence, for any time shift less than $t < 1$, the two signals generate 0 as product (Fig. 8.27).

(continued)

**Example 8.40** (continued)

**Fig. 8.26** Functions of $h(t)$
and $x(t)$ defined in Example
8.40

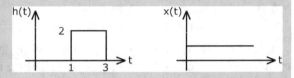

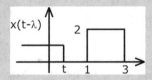

**Fig. 8.27** The position of two functions when their product is zero. This means the two functions do not have any common area. This occurs all the time when $t < 1$

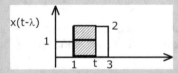

**Fig. 8.28** Collision of two functions when their product is still a function of $t$. This variable can vary from $1 < t < 3$

$$t < 1, \quad \int_0^t x(t - \lambda)h(\lambda)d\lambda = 0$$

Figure 8.28 shows the position of two signals when a time shift generates a collision.

$$1 < t < 3,$$

The two signals have output from $1 < \lambda < t$. Therefore, the $\int_0^t x(t - \lambda)h(\lambda)d\lambda$ becomes

$$\int_1^t x(t - \lambda)h(\lambda)d\lambda = \int_1^t 1 \times 2 \ d\lambda = 2\lambda\Big|_1^t = 2(t - 1)$$

(continued)

**Example 8.40** (continued)

$t > 3$; the product of two signals will change to a fixed value (0) (Fig. 8.29)

$$\int_3^t x(t-\lambda)h(\lambda)d\lambda = \int_3^t 0\,d\lambda = 0$$

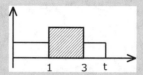

**Fig. 8.29** Collision of two functions when their product is no longer a function of time. Note that for all times $t > 3$, the common area covered by both signals is fixed

**Example 8.41 Find the convolution of Fig. 8.29 two signals as shown in Fig. 8.30.**

*Solution.* Obtaining $x(-\lambda)$ and shifting by $t$ seconds results in $x(t-\lambda)$. Rolling this signal over $h(t)$ in segments results in

If $t < 1$, the two signals generate zero product, as shown in Fig. 8.30. Therefore,

$$\int_0^t x(t-\lambda)h(\lambda)d\lambda = 0$$

If $1 < t < 2$, the product of two signals generates a value, as shown in Fig. 8.31.

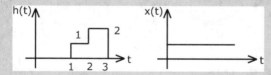

**Fig. 8.30** Signals used in Example 8.41

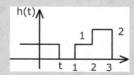

**Fig. 8.31** The position of two functions when their product is zero. This means the two functions do not have any common area. This occurs all the time when $t < 1$

(continued)

**Example 8.41** (continued)

The integral is evaluated in the range of $1 < \lambda < t$ as follows:

$$\int_1^t x(t-\lambda)h(\lambda)d\lambda = \int_1^t 1 \times 1 d\lambda = t - 1$$

As the time shift continues over 2 sec, the $h(\lambda)$ value changes. Therefore, a new limit of integral is needed.

If $2 < t < 3$, the convolution results in $1 < \lambda < 2$ & $2 < \lambda < t$. Figure 8.32 shows these functions.

The convolution integral becomes

$$\int_1^2 x(t-\lambda)h(\lambda)d\lambda + \int_2^t x(t-\lambda)h(\lambda)d\lambda \Big|_2^t =$$
$$\int_1^2 1 \times 1 d\lambda + \int_2^t 1 \times 2 d\lambda = (2-1) + 2(t-2) = 2t - 3$$

If $t > 3$, the value of the functions will not change, but the integral value will not depend on $t$.

Figure 8.33 shows the signals.

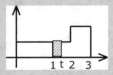

**Fig. 8.32** As the function $h(t)$ has two steps, when the function $x(t)$ moves forward, the area it covers by colliding with the $h(t)$ changes depending on the position of $t$. If $1 < t < 2$, the area covered is as shown, depending on variable $t$

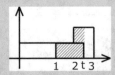

**Fig. 8.33** When $t$ passes point 2 but are still below ($2 < t < 3$), part of the collision area becomes fixed and not a function of $t$, and part of it will still depend on $t$. For this reason, the area between 1 and 3 should be split into two sections

(continued)

**Example 8.41** (continued)

The convolution integral becomes (Fig. 8.34)

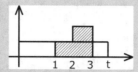

**Fig. 8.34** When $t > 3$, the Collision of two functions is no longer a function of variable $t$

$$\int_1^2 x(t-\lambda)h(\lambda)d\lambda + \int_2^3 x(t-\lambda)h(\lambda)d\lambda =$$

$$\int_1^2 1 \times 1 d\lambda + \int_2^3 1 \times 2 d\lambda = (2-1) + 2(3-2) = 3$$

Therefore,

$$x(t) * h(t) = \begin{cases} 0 & t < 1 \\ 2t - 2 & 1 < t < 2 \\ 2t - 3 & 2 < t < 3 \\ 3 & t > 3 \end{cases}$$

## Laplace of convolution integral

Laplace of convolution of two signals in the time domain is equivalent to the product of their Laplace transforms in the frequency domain, as follows:

$$L\{x(t) * h(t)\} = X(s)H(s)$$

**Example 8.42 Redo Example 8.40 using Laplace transform.**
Considering the waveforms of $h(t)$ and $x(t)$, their Laplace transforms becomes

$$h(t) = 2u(t-1) - 2u(t-3) \rightarrow H(s) = \frac{2}{s}e^{-s} - \frac{2}{s}e^{-3s}$$

$$x(t) = u(t) \rightarrow H(s) = \frac{1}{s}$$

(continued)

**Example 8.42** (continued)

Therefore, the convolution integral becomes

$$x(t) * h(t) = L^{-1}\left\{\left(\frac{2}{s}e^{-s} - \frac{2}{s}e^{-3s}\right)\left(\frac{1}{s}\right)\right\} = L^{-1}\left\{\left(\frac{2}{s^2}e^{-s} - \frac{2}{s^2}e^{-3s}\right)\right\}$$
$$= 2(t-1)u(t-1) - 2(t-3)u(t-3)$$

**Example 8.43 Redo Example 8.41 using Laplace transform.**
In this example, the signals are $x(t) = u(t)$ and $h(t) = u(t-1) + u(t-2) - 2u(t-3)$. Therefore, their Laplace transform becomes

$$X(s) = \frac{1}{s}, H(s) = \frac{1}{s}\left(e^{-s} + e^{-2s} - 2e^{-3s}\right)$$

$$x(t) * h(t) = L^{-1}\left\{\left(\frac{1}{s}\right)\left(\frac{1}{s}\left(e^{-s} + e^{-2s} - 2e^{-3s}\right)\right)\right\}$$
$$= L^{-1}\left\{\left(\frac{1}{s^2}\left(e^{-s} + e^{-2s} - 2e^{-3s}\right)\right)\right\}$$
$$= (t-1)u(t-1) + (t-2)u(t-2) - 2(t-3)u(t-3).$$

**Example 8.44 Find the convolution of the functions shown in Fig. 8.35.**
Functions $x(t)$ and $h(t)$ are given. It is simpler to mirror and shift $h(t)$ concerning the vertical axis. Then the function $h(t-\lambda)$ is rolled over the function $x(\lambda)$.

**Fig. 8.35** Functions of
Example 8.44

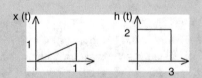

For $t < 0$, the two functions generate no product. Therefore (Fig. 8.36),

$$y(t) = \int_0^t x(\lambda)h(t-\lambda)d\lambda = 0$$

(continued)

**Example 8.44** (continued)

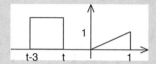

**Fig. 8.36** The function $h(t)$ has been mirrored and shifted. The product of two functions is zero before they collide for the time $t < 0$

For $0 < t < 1$, the two functions generate (Fig. 8.37).

$$y(t) = \int_0^t x(\lambda)h(t-\lambda)d\lambda = \int_0^t \lambda \times 2d\lambda = t^2$$

**Fig. 8.37** The product of two signals is a function of $t$ for the time $0 < t < 1$

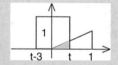

For $1 < t < 3$, the two functions generate a new product as (Fig. 8.38)

$$y(t) = \int_0^t x(\lambda)h(t-\lambda)d\lambda = \int_0^1 \lambda \times 2d\lambda = 1$$

**Fig. 8.38** The product of two signals is no longer a function of $t$ when the function $h(t)$ has moved such that $1 < t < 3$

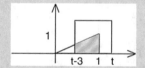

For $3 < t < 4$, the two signals have a new product as (Fig. 8.39):

$$y(t) = \int_0^t x(\lambda)h(t-\lambda)d\lambda = \int_{t-3}^1 \lambda \times 2d\lambda = 1 - (t-3)^2$$

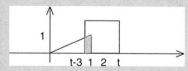

**Fig. 8.39** The overlap part of function $h(t)$ still makes the product of two functions dependent on time $t$ when $3 < t < 4$. Since the length of function $h(t)$ is 3, for the start point $t$, the endpoint becomes $(t-3)$. Therefore, the product of these functions becomes non-zero, starting from the point $(t-3)$ to point 1, at which the function $x(t)$ ends

(continued)

**Example 8.44**   (continued)

For $t > 4$, the two signals have no product, and the output becomes zero, as shown in Fig. 8.40.

**Fig. 8.40**  The two functions have no non-zero product when $t > 4$

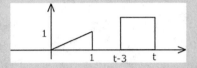

$$y(t) = \int_0^t x(\lambda)h(t-\lambda)d\lambda = 0$$

The function becomes $f(t) = \begin{cases} 0 & t < 0 \\ t^2 & 0 < t < 1 \\ 1 & 1 < t < 3 \\ 1 - (t-3)^2 & 3 < t < 4 \\ 0 & t > 4. \end{cases}$

**Example 8.45 Redo the previous example using Laplace transform.**
*Solution.* In this example mathematical representation of the signals become $x(t) = tu(t) - (t - 1)u(t - 1) - u(t - 1)$ and $h(t) = 2u(t) - 2u(t - 3)$. Therefore, their Laplace transforms becomes

$$X(s) = \frac{1}{s^2} - \frac{1}{s^2}e^{-s} - \frac{1}{s}e^{-s}, \quad H(s) = \frac{2}{s} - \frac{2}{s}e^{-3s}$$

The convolution becomes

$$y(t) = x(t) * h(t) = L^{-1}\left\{ \left(\frac{1}{s^2} - \frac{1}{s^2}e^{-s} - \frac{1}{s}e^{-s}\right)\left(\frac{2}{s} - \frac{2}{s}e^{-3s}\right) \right\}$$

$$= L^{-1}\left\{ \frac{2}{s^2}\left(\frac{1}{s} - \frac{1}{s}e^{-s} - e^{-s}\right)(1 - e^{-3s}) \right\}$$

$$= L^{-1}\left\{ 2\left(\frac{1}{s^3} - \frac{e^{-3s}}{s^3} - \frac{1}{s^3}e^{-s} - \frac{1}{s^3}e^{-4s} - \frac{1}{s^2}e^{-s} + \frac{1}{s^2}e^{-4s}\right) \right\}$$

$$2\left(\frac{1}{2}t^2u(t) - \frac{1}{2}(t-3)^2u(t-3) - \frac{1}{2}(t-1)^2u(t-1)\right.$$

$$\left. - \frac{1}{2}(t-4)^2u(t-4) - (t-1)u(t-1) + (t-4)u(t-4)\right)$$

## State Space Analysis

Analysis of electric circuits leads to differential equations representing voltage and current variations. The order of these differential equations depends on the order of the circuit. Considering an $n$th-order system, the differential equation results in one variable, either the current or the voltage of components. Extra mathematical operations may be needed to obtain the rest of the variables. However, the voltages of capacitors and the current of inductors can be obtained individually. The variables that show the energy content of the system (voltage of capacitors and current of inductors) are called state variables. State-space equations are another form of system dynamics representation, showing the dynamics of individual state variables. Therefore, the variation of all system dynamics can be observed simultaneously.

The general form of state-space representation is

$$\dot{x}(t) = Ax(t) + Bu(t)$$
$$y(t) = Cx(t) + Du(t)$$

where $x(t)$ is the vector of state variables at the same size as the order of the system $n \times 1$, $u(t)$ is the vector of input variables of size $m \times 1$, $y(t)$ is the vector of output variables of size $p \times 1$, $A_{n \times n}$ is the system matrix, $B_{n \times m}$ is the input matrix, $C_{p \times n}$ is the output matrix, and $D_{p \times m}$ is the feedforward matrix. The system input, state, and output signals are as follows:

$$x(t) = \begin{bmatrix} x_1(t) \\ x_2(t) \\ \vdots \\ x_n(t) \end{bmatrix}, u(t) = \begin{bmatrix} u_1(t) \\ u_2(t) \\ \vdots \\ u_m(t) \end{bmatrix}, y(t) = \begin{bmatrix} y_1(t) \\ y_2(t) \\ \vdots \\ y_p(t) \end{bmatrix}.$$

A variable is a state variable if, with a given initial value and with the dynamics, the future values of the variable can be predicted.

A state variable shows the internal energy variation in the system. For instance, the energy stored in a capacitor is shown in its voltage, and the energy stored in an inductor is shown by its current. Therefore, studying the state variables provides insight into the internal energy stored in the circuit.

According to the passivity notion in system stability analysis, a stable system dissipates more energy than it receives. Therefore, the net energy variation in a stable system is either zero or negative.

$$\text{If} : \Delta E_{\text{internal}} \leq 0 \rightarrow \text{Then} : \text{System is Stable}$$

## Obtaining State-Space Equations from Differential Equations

State-space representation of a system is not unique. This means that a system can be presented by various state-space equations, i.e., matrices $A$, $B$, $C$, and $D$ may have various forms. Consider an $n$th order differential equation as follows:

$$y^{(n)}(t) + a_1 y^{(n-1)}(t) + \ldots + a_n y(t) = bu(t)$$

To obtain a canonical form state-space representation of this system, the following steps are needed:

Step 1: Consider

$$y(t) = x_1(t)$$

This yields

$$x_1^{(n)}(t) + a_1 x_1^{(n-1)}(t) + \ldots + a_n x_1(t) = bu(t)$$

Step 2: Consider the rest of the state variables except the last one as follows:

$$\dot{x}_1(t) = x_2(t)$$
$$\dot{x}_2(t) = x_3(t)$$
$$\vdots$$
$$\dot{x}_{n-1}(t) = x_n(t)$$

Replacing these assumptions into the differential equation obtains $\dot{x}_n(t)$ as follows:

$$\dot{x}_n(t) = -a_1 x_{n-1}(t) - \ldots - a_n x_1(t) + bu(t)$$

$$
\begin{bmatrix} \dot{x}_1(t) \\ \vdots \\ \dot{x}_n(t) \end{bmatrix} = \begin{bmatrix} 0 & 1 & 0 \\ \vdots & \vdots & \vdots \\ -a_1 & \cdots & -a_n \end{bmatrix} \begin{bmatrix} x_1(t) \\ \vdots \\ x_n(t) \end{bmatrix} + \begin{bmatrix} 0 \\ \vdots \\ b \end{bmatrix} u(t)
$$

$$
y(t) = [1 \cdots 0] \begin{bmatrix} x_1(t) \\ \vdots \\ x_n(t) \end{bmatrix}
$$

**Example 8.46 Find the state-space representation of the system $\ddot{y} + 3\dot{y} + 2y = 1.5u$ with input $u(t)$ and output $y(t)$.**

*Solution.* This is a second-order system, as the output equation is a second-order differential equation. Therefore there are two state variables to show the internal energy of this system, namely, $x_1(t)$, $x_2(t)$.

The following steps are a general approach to finding a state-space model of the system (canonical form).

$$y = x_1$$

One of the state equations is already found as $\dot{x}_1 = x_2$. This is a standard consideration to obtain the canonical form of state-space representation. Replacing all the assumed $y = x_1$ and $\dot{x}_1 = x_2$ yields

$$\ddot{x}_1 + 3\dot{x}_1 + 2x_1 = 1.5u$$

$$\dot{x}_2 + 3x_2 + 2x_1 = 1.5u$$

The second state equation can be found by soling for $\dot{x}_2$ as follows:

$$\dot{x}_2 = -3x_2 - 2x_1 + 1.5u$$

Therefore,

$$\dot{x}_1 = x_2$$

$$\dot{x}_2 = -3x_2 - 2x_1 + 1.5u$$

$$y = x_1$$

In the matrix form, the state space equations are

$$\begin{pmatrix} \dot{x}_1 \\ \dot{x}_2 \end{pmatrix} = \begin{pmatrix} 0 & 1 \\ -2 & -3 \end{pmatrix} \begin{pmatrix} x_1 \\ x_2 \end{pmatrix} + \begin{pmatrix} 0 \\ 1.5 \end{pmatrix} u$$

$$y = \begin{bmatrix} 1 & 0 \end{bmatrix} \begin{pmatrix} x_1 \\ x_2 \end{pmatrix} + 0u$$

**Example 8.47 Find the state space representation of the following differential equation.**

$$\ldots y + 5\ddot{y} + 10\dot{y} + 7y = u$$

*Solution.*
Consider

$$y = x_1$$

This yields

$$\ldots x_1 + 5\ddot{x}_1 + 10\,\dot{x}_1 + 7x_1 = u$$

This is a third-order differential equation, which means there are dynamics for three state variables, namely, $x_1$, $x_2$, and $x_3$. Therefore, there is a need to assume two of these dynamics:

$$\dot{x}_1 = x_2$$

$$\dot{x}_2 = x_3$$

Replacing into differential equation results in

$$\dot{x}_3 + 5x_3 + 10x_2 + 7x_1 = u$$

Solving for $\dot{x}_3$ yields

$$\dot{x}_3 = -7x_1 - 10x_2 - 5x_3 + u$$

In matrix form, the system is represented as

$$\begin{bmatrix} \dot{x}_1 \\ \dot{x}_2 \\ \dot{x}_3 \end{bmatrix} = \begin{bmatrix} 0 & 1 & 0 \\ 0 & 0 & 1 \\ -7 & -10 & -5 \end{bmatrix} \begin{bmatrix} x_1 \\ x_2 \\ x_3 \end{bmatrix} + \begin{bmatrix} 0 \\ 0 \\ 1 \end{bmatrix} u$$

$$y = [1 \ 0 \ 0] \begin{bmatrix} x_1 \\ x_2 \\ x_3 \end{bmatrix}$$

## Obtaining a Block Diagram of a State-Space Equation

Block diagram representation of a state-space equation is possible with the notion that each set of the equation is a first-order differential equation. Therefore, the dynamics of each state variable can be obtained by an integrator $\int$ or in the Laplace domain by $\frac{1}{s}$, as shown in Fig. 8.41.

The total number of integrators must equal the number of state variables. It is assumed that the integrators generate a state variable at their outputs, e.g., $x$. Therefore, their input is the derivative of the state variable, e.g., $\dot{x}$. A given $\dot{x}$ dynamic must be configured using the integrators' state variables, inputs, and the required gains. This is explained in an example as follows.

**Example 8.48 Draw a block diagram representation of the following state-space model.**

$$\begin{cases} \dot{x}_1 = 2x_1 + x_2 - 5x_3 - 2u \\ \dot{x}_2 = -5x_1 - x_2 - x_3 + u \\ \dot{x}_3 = x_1 + 9x_2 - x_3 \end{cases}$$

*Solution.* The system has three state variables that suggest the existence of three integrators. Figure 8.41 shows these integrators.

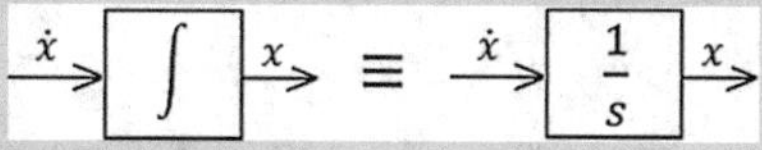

$$\dot{x} \rightarrow \boxed{\int} \xrightarrow{x} \equiv \dot{x} \rightarrow \boxed{\dfrac{1}{s}} \xrightarrow{x}$$

**Fig. 8.41** Integrators in the time domain and frequency domain. Considering the output of integrators as state variables, the input to these units expresses the formation of state-space equations

As Fig. 8.42 shows, input to the integrators is the derivative of the state variable. Therefore, these dynamics can be set up. Figure 8.43 shows the setup for the first dynamic $\dot{x}_1 = 2x_1 + x_2 - 5x_3 - 2u$.

**Fig. 8.42** The system with three state variables has to have three integrators, each of which generates one of the state space equations

(continued)

**Example 8.48** (continued)

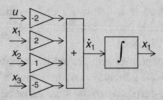

**Fig. 8.43** The formation of the first state-space equation is shown. The input to the equations is state variables or the inputs with proper gains

Now, the signals $x_1, x_2, x_3$, and $u$ can be connected to the output of the right integrators. The same procedure is utilized to build the other state variable dynamics. Figure 8.44 shows the block diagram of the entire system.

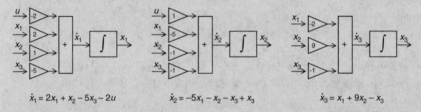

$$\dot{x}_1 = 2x_1 + x_2 - 5x_3 - 2u \qquad \dot{x}_2 = -5x_1 - x_2 - x_3 + x_3 \qquad \dot{x}_3 = x_1 + 9x_2 - x_3$$

**Fig. 8.44** A Block diagram of three state-space equations is formed according to the equations

$$\dot{x}_1 = 2x_1 + x_2 - 5x_3 - 2u; \dot{x}_2 = -5x_1 - x_2 - x_3 + u \ \dot{x}_3 = x_1 + 9x_2 - x_3$$

**Example 8.49 Draw a block diagram representation of the following state-space model.**

$$\begin{cases} \dot{x}_1 = & -x_1 + x_2 \\ \dot{x}_2 = & 2x_1 - x_2 + 5u \end{cases}$$

*Solution*. Since there are two state variables, the system dynamics has two integrators that result in $x_1$ and $x_2$. The input to these integrators is $\dot{x}_1$ and $\dot{x}_2$. Therefore, the diagrams can be implemented as shown in Fig. 8.45.

(continued)

**Example 8.49** (continued)

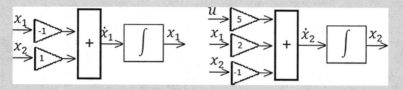

**Fig. 8.45** Block diagram of the differential equation as $\dot{x}_1 = -x_1 + x_2$, $\dot{x}_2 = 2x_1 - x_2 + 5u$

## Obtaining State-Space of Differential Equations that Involve Differential of the Input Signals

Consider a causal differential equation that involves $n$th order differential of the output and $m$th order differential of the output signal $(n < m)$ as follows:

$$y^{(n)}(t) + a_1 y^{(n-1)}(t) + \ldots + a_n y(t) = b_m u^{(m)}(t) + b_{m-1} u^{(m-1)}(t) + \ldots + b_0 u(t)$$

There are several techniques to obtain the state space equation of this system. The technique introduced in this chapter is based on block diagrams. To do that, this equation is solved for $y$ by taking $n$ integrals as follows:

$$\underbrace{\iiint \cdots \int}_{n \text{ Integrals}} \left( y^{(n)}(t) + a_1 y^{(n-1)}(t) + \cdots + a_n y(t) \right) = \underbrace{\iiint \cdots \int}_{n \text{ Integrals}} \left( b_m u^{(m)}(t) + b_{m-1} u^{(m-1)}(t) + \cdots + b_0 u(t) \right)$$

$$y + a_1 \int y + a_2 \iint y + \cdots + a_n \underbrace{\iiint \cdots \int}_{n \text{ Integrals}} y = b_m \underbrace{\iiint \cdots \int}_{n - m \text{ Integrals}} u + \cdots + b_0 \underbrace{\iiint \cdots \int}_{n \text{ Integrals}} u$$

Solving for $y$, the block diagram is formed as follows:

$$y = -a_1 \int y - a_2 \iint y - \ldots - a_n \iiint \cdots \int y + b_m \iiint \cdots \int u + \ldots$$
$$+ b_0 \iiint \cdots \int u$$

To obtain $y$, the differential equation suggests taking $n$ integrals. The output of the $n$th integrator is $y$.

**Example 8.50 Find the state-space representation of the following differential equation.**

$$\ldots y + 5\ddot{y} + 6\dot{y} + 11y = 2\ddot{u} + \dot{u} + u$$

*Solution.* This is a third-order differential equation, and since the second-order input equation $2\ddot{u} + \dot{u} + u$ has less order than the third-order output differential equation; the system is also causal. There is a need to take three times integral from both sides of the equation to obtain $y$. Solving for $y$ yields

$$y = -5\int y - 6\iint y - 11\iiint y + 2\int u + \iint u + \iiint u$$

$$\text{①}\qquad\text{②}\qquad\quad\text{③}\qquad\text{④}\quad\text{⑤}\quad\text{⑥}$$

To build $y$ through integrals of $y$ and $u$, there is a need to feedback on $y$ considering the gains. Three integrators exist in this differential equation. In the next step, consider three integrator blocks with an adder in the input of each block, and the output of the far-right integrator block is $y$.

Now, consider the differential equation for each of the right-hand side components of the equation (*) is built. To build ①, the feedback is taken from $y$, and considering gain $-5$, it is added to the summation of the right integral block. When the system runs numerically, this loop forms the $-5\int y$ component. The input signal ④ is added to the far-right integrator by gain 2 to be integrated once by reaching signal $y$. Segment ⑤ is added to the second integrator, and segment ⑥ is added to the left integrator. Figure 8.46 shows block diagram implementation.

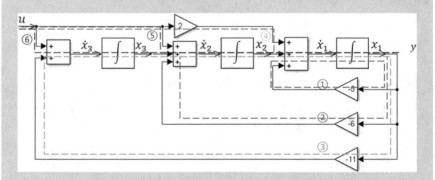

**Fig. 8.46** Block diagram implementation of the system in example 8.24

## Obtaining Transfer Function from State-Space Representation

Finding the state space representation of a system might have several approaches. These techniques' representations might look different, but they all show the same dynamics. The state-space representations are not unique. However, the transfer function of a system is unique. There is a relation between the state space and the transfer function of a system expressed as follows:

Consider a state-space system as

$$\dot{x}(t) = Ax(t) + Bu(t)$$
$$y(t) = Cx(t) + Du(t)$$

Taking the Laplace of this system results in

$$sX(s) - X(0) = AX(s) + BU(s)$$
$$Y(s) = CX(s) + DU(s)$$

The transfer function is obtained at the zero initial conditions. Therefore, $X(0) = 0$.

This results in

$$sX(s) - AX(s) = BU(s)$$

Solving for $X(s)$ yields

$$(sI - A)X(s) = BU(s)$$

And

$$X(s) = (sI - A)^{-1}BU(s)$$

Replacing this into the output equation results in

$$Y(s) = C(sI - A)^{-1}BU(s) + DU(s)$$
$$Y(s) = \left(C(sI - A)^{-1}B + D\right)U(s)$$

The system transfer function, which is unique over various combinations of matrices $A$, $B$, $C$, and $D$, can be obtained and presented as

$$\frac{Y(s)}{U(s)} = C(sI - A)^{-1}B + D$$

**Example 8.51 A state-space representation of a dynamical system is given as**

$$\dot{X} = \begin{bmatrix} 1 & -1 \\ -3 & -5 \end{bmatrix} X + \begin{bmatrix} 0 \\ 1 \end{bmatrix} u$$

$$y = \begin{bmatrix} 1 & 0 \end{bmatrix} X$$

*Find the transfer function of the system.*
*Solution.* The transfer function is defined as

$$\frac{Y(s)}{U(s)} = C(sI - A)^{-1}B + D$$

From the system, the following matrices can be obtained:

$$A = \begin{bmatrix} 1 & -1 \\ -3 & -5 \end{bmatrix}, B = \begin{bmatrix} 0 \\ 1 \end{bmatrix}, C = \begin{bmatrix} 1 & 0 \end{bmatrix}$$

$$D = 0 \text{ and } I_{2 \times 2} = \begin{bmatrix} 1 & 0 \\ 0 & 1 \end{bmatrix}$$

Therefore, the transfer function is

$$\frac{Y(s)}{U(s)} = \begin{bmatrix} 1 & 0 \end{bmatrix} \left( s \begin{bmatrix} 1 & 0 \\ 0 & 1 \end{bmatrix} - \begin{bmatrix} 1 & -1 \\ -3 & -5 \end{bmatrix} \right)^{-1} \begin{bmatrix} 0 \\ 1 \end{bmatrix} + 0$$

Simplifying results in

$$\frac{Y(s)}{U(s)} = \begin{bmatrix} 1 & 0 \end{bmatrix} \begin{bmatrix} s-1 & 1 \\ 3 & s+5 \end{bmatrix}^{-1} \begin{bmatrix} 0 \\ 1 \end{bmatrix}$$

$$\frac{Y(s)}{U(s)} = \begin{bmatrix} 1 & 0 \end{bmatrix} \frac{1}{\det \begin{bmatrix} s-1 & 1 \\ 3 & s+5 \end{bmatrix}} \begin{bmatrix} s+5 & -1 \\ -3 & s+1 \end{bmatrix} \begin{bmatrix} 0 \\ 1 \end{bmatrix}$$

(continued)

**Example 8.51** (continued)

$$\frac{Y(s)}{U(s)} = \frac{[1 \quad 0]\begin{bmatrix} s+5 & -1 \\ -3 & s+1 \end{bmatrix}\begin{bmatrix} 0 \\ 1 \end{bmatrix}}{(s-1)(s+5)-1\times 3}$$

$$\frac{Y(s)}{U(s)} = \frac{[1\times(s+5)+0\times-3 \quad 1\times-1+0\times(s+1)]\begin{bmatrix} 0 \\ 1 \end{bmatrix}}{s^2+4s-8}$$

$$\frac{Y(s)}{U(s)} = \frac{[s+5-1]\begin{bmatrix} 0 \\ 1 \end{bmatrix}}{s^2+4s-8}$$

$$\frac{Y(s)}{U(s)} = \frac{[(s+5)\times 0 - 1\times 1]}{s^2+4s-8}$$

$$\frac{Y(s)}{U(s)} = \frac{-1}{s^2+4s-8}$$

**Example 8.52 A state-space representation of a dynamical system is given as**

$$\dot{X} = \begin{bmatrix} 0 & 1 \\ -3 & -5 \end{bmatrix} X + \begin{bmatrix} 0 \\ 1 \end{bmatrix} u$$

$$y = [1 \quad 0]X$$

*Find the system's transfer function and poles and zeros of the system.*
**Solution.** The transfer function is defined as

$$\frac{Y(s)}{U(s)} = C(sI-A)^{-1}B+D$$

From the system, the following matrices can be obtained

$$A = \begin{bmatrix} 0 & 1 \\ -3 & -5 \end{bmatrix}, B = \begin{bmatrix} 0 \\ 1 \end{bmatrix}, C = [1 \quad 0]$$

(continued)

**Example 8.52** (continued)

$$D = 0 \text{ and } I_{2 \times 2} = \begin{bmatrix} 1 & 0 \\ 0 & 1 \end{bmatrix}$$

Therefore, the transfer function is

$$\frac{Y(s)}{U(s)} = \begin{bmatrix} 1 & 0 \end{bmatrix} \left( s \begin{bmatrix} 1 & 0 \\ 0 & 1 \end{bmatrix} - \begin{bmatrix} 0 & 1 \\ -3 & -5 \end{bmatrix} \right)^{-1} \begin{bmatrix} 0 \\ 1 \end{bmatrix} + 0$$

Simplifying results in

$$\frac{Y(s)}{U(s)} = \begin{bmatrix} 1 & 0 \end{bmatrix} \begin{bmatrix} s & -1 \\ 3 & s+5 \end{bmatrix}^{-1} \begin{bmatrix} 0 \\ 1 \end{bmatrix}$$

$$\frac{Y(s)}{U(s)} = \begin{bmatrix} 1 & 0 \end{bmatrix} \frac{1}{\det \begin{bmatrix} s & -1 \\ 3 & s+5 \end{bmatrix}} \begin{bmatrix} s+5 & 1 \\ -3 & s \end{bmatrix} \begin{bmatrix} 0 \\ 1 \end{bmatrix}$$

$$\frac{Y(s)}{U(s)} = \frac{\begin{bmatrix} 1 & 0 \end{bmatrix} \begin{bmatrix} s+5 & 1 \\ -3 & s \end{bmatrix} \begin{bmatrix} 0 \\ 1 \end{bmatrix}}{s(s+5) - 1 \times -3}$$

$$\frac{Y(s)}{U(s)} = \frac{\begin{bmatrix} 1 \times (s+5) + 0 \times -3 & 1 \times 1 + 0 \times s \end{bmatrix} \begin{bmatrix} 0 \\ 1 \end{bmatrix}}{s^2 + 5s + 3}$$

$$\frac{Y(s)}{U(s)} = \frac{\begin{bmatrix} s+5 & 1 \end{bmatrix} \begin{bmatrix} 0 \\ 1 \end{bmatrix}}{s^2 + 5s + 3}$$

$$\frac{Y(s)}{U(s)} = \frac{[(s+5) \times 0 - 1 \times 1]}{s^2 + 5s + 3}$$

$$\frac{Y(s)}{U(s)} = \frac{-1}{s^2 + 5s + 3}$$

There are no zeros in the transfer function.

There are two poles at $s^2 + 5s + 3 = 0$. The poles are $s = -2.5 \pm \frac{\sqrt{13}}{2}$.

**Example 8.53 A state-space representation of a dynamical system is given as**

$$\dot{X} = \begin{bmatrix} 0 & 1 \\ -1 & -1 \end{bmatrix} X + \begin{bmatrix} 0 \\ 1 \end{bmatrix} u$$

$$y = [1 \quad 0] X$$

*Find the system's transfer function and poles and zeros of the system.*
*Solution.* The transfer function is defined as

$$\frac{Y(s)}{U(s)} = C(sI - A)^{-1} B + D$$

From the system, the following matrices can be obtained

$$A = \begin{bmatrix} 0 & 1 \\ -1 & -1 \end{bmatrix}, B = \begin{bmatrix} 1.5 \\ 1 \end{bmatrix}, C = [1 \quad 0]$$

$$D = 0 \text{ and } I_{2 \times 2} = \begin{bmatrix} 1 & 0 \\ 0 & 1 \end{bmatrix}$$

Therefore, the transfer function is

$$\frac{Y(s)}{U(s)} = [1 \quad 0] \left( s \begin{bmatrix} 1 & 0 \\ 0 & 1 \end{bmatrix} - \begin{bmatrix} 0 & 1 \\ -1 & -1 \end{bmatrix} \right)^{-1} \begin{bmatrix} 1.5 \\ 1 \end{bmatrix} + 0$$

Simplifying results in

$$\frac{Y(s)}{U(s)} = [1 \quad 0] \begin{bmatrix} s & -1 \\ 1 & s+1 \end{bmatrix}^{-1} \begin{bmatrix} 1.5 \\ 1 \end{bmatrix}$$

$$\frac{Y(s)}{U(s)} = [1 \quad 0] \frac{1}{\det \begin{bmatrix} s & -1 \\ 1 & s+1 \end{bmatrix}} \begin{bmatrix} s+1 & 1 \\ -1 & s \end{bmatrix} \begin{bmatrix} 1.5 \\ 1 \end{bmatrix}$$

$$\frac{Y(s)}{U(s)} = \frac{[1 \quad 0] \begin{bmatrix} s+1 & 1 \\ -1 & s \end{bmatrix} \begin{bmatrix} 1.5 \\ 1 \end{bmatrix}}{s(s+1) - 1 \times -1}$$

$$\frac{Y(s)}{U(s)} = \frac{[1 \times (s+1) + 0 \times -1 \quad 1 \times 1 + 0 \times s] \begin{bmatrix} 1.5 \\ 1 \end{bmatrix}}{s^2 + s + 1}$$

(continued)

**Example 8.53** (continued)

$$\frac{Y(s)}{U(s)} = \frac{[s+1 \quad -1]\begin{bmatrix} 1.5 \\ 1 \end{bmatrix}}{s^2 + s + 1}$$

$$\frac{Y(s)}{U(s)} = \frac{[(s+1) \times 1.5 - 1 \times 1]}{s^2 + s + 1}$$

$$\frac{Y(s)}{U(s)} = \frac{1.5s + 0.5}{s^2 + s + 1} = 1.5 \frac{s + 0.3}{s^2 + s + 1}$$

This system has one zero and two poles. The zero is calculated from the numerator polynomial as

$$s + 0.3 = 0 \rightarrow s = -0.3$$

The poles are obtained from the denominator polynomial as

$$s^2 + s + 1 = 0$$

$$(s + 0.5)^2 + 0.75 = 0$$

$$s = -0.5 \pm j\sqrt{0.75}$$

**Example 8.54 Find the poles of a process with a system matrix of $A = \begin{bmatrix} -1 & 2 \\ -1 & -3 \end{bmatrix}$.**

*Solution.* The poles are the roots of the transfer function's denominator. The denominator is the determinant of the matrix $(sI - A)$. Therefore,

Poles are located at

$$\det(sI - A) = \det\left( s\begin{pmatrix} 1 & 0 \\ 0 & 1 \end{pmatrix} - \begin{pmatrix} -1 & 2 \\ -1 & -3 \end{pmatrix} \right)$$

$$= \det\begin{pmatrix} s+1 & -2 \\ +1 & s+3 \end{pmatrix} = (s+1)(s+3) - 1 \times -2 = s^2 + 4s + 5$$

$$s^2 + 4s + 5 = 0 \rightarrow (s+2)^2 + 1 = 0 \rightarrow s + 2 = \pm j \rightarrow s = -2 \pm j$$

**Example 8.55** The poles of a system are $s = -1, s = -2$. Determine how many possible system matrices can be found that result in these poles.

*Solution.* The poles can determine a transfer function denominator as $(s + 1)$ $(s + 2) = s^2 + 3s + 1$. However, a general second-order system matrix can be presented as $A = \begin{bmatrix} a & b \\ c & d \end{bmatrix}$. This system results in poles as

$$\det(sI - A) = \det\left(s\begin{pmatrix} 1 & 0 \\ 0 & 1 \end{pmatrix} - \begin{pmatrix} a & b \\ c & d \end{pmatrix}\right) = \det\begin{pmatrix} s-a & -b \\ -c & s-d \end{pmatrix}$$

$$= (s - a)(s - d) + bc$$

$$\therefore \quad s^2 - (a + d)s - ad + bc \equiv s^2 + 3s + 1$$

$$-(a + d) = 3, \quad bc - ad = 1$$

This forms two equations and four unknown system. This may have many solutions which confirm that the state-space representations are not unique.

## Bode Diagram

Transfer functions are defined in the Laplace domain using operation $s$. As the Laplace operator is a function frequency, the change of operating frequencies influences the transfer function. As with all complex functions, the transfer function shows amplitude and phase that are respected to any operating frequency. The amplitude and phase change profile with respect to the operating frequency is also called the "Bode" plot or Bode diagram.

The simplest way of obtaining the Bode diagram is to evaluate the amplitude and phase at each frequency for a given range. The frequency range starts from DC ($s = j0$) and is extended as long as there is variation in the phase and amplitude. However, this is computationally expensive. There are indicators in the transfer function that determine the amplitude and phase diagrams, and this section identifies these indicators and determines the Bode diagram accordingly.

## Transfer Function Amplitude and Phase

The effective range of frequencies needed to be checked for amplitude, and the phase might extend from zero to extremely high frequencies. To obtain a readable chart, the axis of frequencies is shown as logarithmic values.

The amplitude of the transfer function might also be a very large number. The amplitude unit is presented in decibels to avoid operation with large numbers.

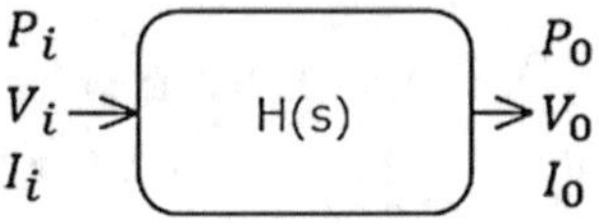

**Fig. 8.47** Input-output
signal types to a system $H(s)$

$$
\begin{array}{ll}
P_i & P_0 \\
V_i \rightarrow \boxed{\quad H(s) \quad} \Rightarrow V_0 \\
I_i & I_0
\end{array}
$$

**Decibel** Consider the transfer functions that determine the power ratio (output over input) as $H(s) = \frac{P_0}{P_i}$, the ratio of voltages $H(s) = \frac{V_0}{V_i}$, or currents $H(s) = \frac{I_0}{I_i}$. This is shown in Fig. 8.47.

Transfer function gain in decibel is defined as

$$
|H|\mathrm{dB} = 10 \log \frac{P_0}{P_i}
$$

Considering that the power is proportional to the square of voltage or current as $P = RI^2 = \frac{V^2}{R}$, the transfer functions amplitude becomes

$$
|H|\mathrm{dB} = 10 \log \frac{P_0}{P_i} = 10 \log \frac{\frac{V_o^2}{R}}{\frac{V_i^2}{R}}
$$

$$
= 10 \log \frac{V_o^2}{V_i^2}
$$

$$
= 20 \log \frac{V_0}{V_i}
$$

Similarly, the current gain of an amplifier in dB is obtained as follows:

$$
|H|\mathrm{dB} = 10 \log \frac{P_0}{P_i} = 10 \log \frac{RI_o^2}{RI_i^2}
$$

$$
= 10 \log \frac{I_o^2}{I_i^2}
$$

$$
= 20 \log \frac{I_0}{I_i}
$$

---

**Example 8.56 The output power delivery of an amplifier is 10 W. Find the amplifier gain in dB if the input signal was 0.1 W.**

*Solution.* The ratio of $\frac{P_0}{P_i}$ is $\frac{10}{0.1}$. Therefore, the amplifier gain in dB is

$$
|H|\mathrm{dB} = 10 \log \frac{P_0}{P_i} = 10 \log \frac{10}{0.1} = 20 (\mathrm{dB})
$$

**Example 8.57 Calculate the gain of a circuit when the output voltage drops by 70.7% at a certain frequency.**

*Solution.* Since voltage gain is used, the gain in dB is obtained by

$$|H|\text{dB} = 20\log\frac{V_0}{V_i} = 20\log 0.707 = -3 \ \text{dB}$$

**Example 8.58 Find the power gain reduction of a circuit at the frequency in which the voltage gain is −3dB.**

*Solution.* When the voltage gain is $-3 \ dB$, the voltage gain $A_V = \frac{V_0}{V_i}$ becomes

$$-3 = 20\log\frac{V_0}{V_i}$$

$$\frac{V_0}{V_i} = 10^{\frac{-3}{20}} = 0.707$$

$$\frac{P_0}{P_i} = \left(\frac{V_0}{V_i}\right)^2 = 0.5$$

*It can be concluded that, at a certain frequency, when the voltage gain drops to 70.7%, the power gain drops to 50%. This frequency is also called half-power point frequency.*

## Bode Plot of A Transfer Function

A transfer function is a combination of several components, and there is a template for each element of a transfer function. Therefore, the transfer function must be broken into these elements and combined with frequency responses in the first step.

**Element 1**

*Gain or DC gain.* $H(s) = K$. To obtain the gain of a transfer function, the poles and zeros must be written as

$$H(s) = K \frac{\prod\limits_{i=1}^{m} \left(1 + \frac{s}{z_m}\right)}{\prod\limits_{j=1}^{n} \left(1 + \frac{s}{p_j}\right)}$$

The amplitude of this transfer function at a certain frequency $s = j\omega$ is obtained as

$$|H(j\omega)| = K \frac{\prod\limits_{i=1}^{m} \sqrt{1 + \left(\frac{\omega}{z_m}\right)^2}}{\prod\limits_{j=1}^{n} \sqrt{1 + \left(\frac{\omega}{p_j}\right)^2}}$$

The DC gain in dB is

$$20 \log K$$

The phase for this element is zero. Therefore, the Bode plot is obtained for a simple gain element, as shown in Fig. 8.48.

**Fig. 8.48** Amplitude and phase diagram of $H(s) = K$

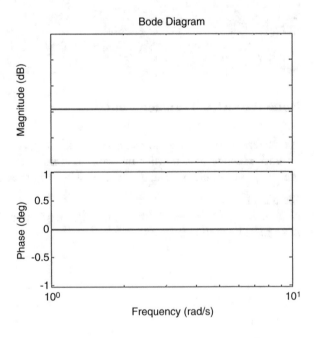

**Example 8.59 Find the Bode plot of the following function.**

$$H(s) = 30$$

*Solution.* The transfer function using $s = j\omega$ results in

$$H(j\omega) = 30$$

The amplitude of this function is

$$|H| = 20\log 30 = 29.54 \ \text{dB}$$

The phase of this function is constant at $0^\circ$. Figure 8.49 shows the Bode diagram.

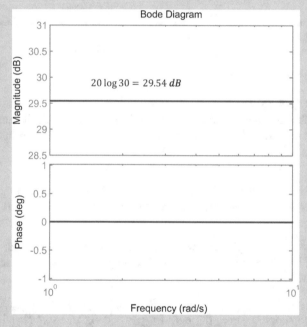

**Fig. 8.49** Amplitude and phase diagram of $H(s) = 30$

**Fig. 8.50** Amplitude and
phase diagram of $H(s) = s$

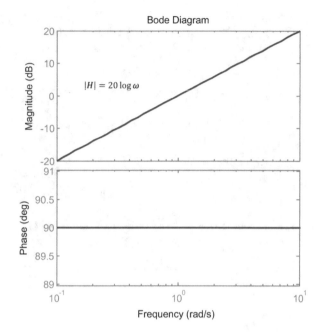

**Element 2**

*Zero at the origin. $H(s) = s$.*

   This element has the transfer function as $H(j\omega) = j\omega$, with the amplitude of I
$H\,I\, = 20 \log \omega$ (dB) and the phase of $\angle H = 90°$. The Bode plot of this element is
shown in Fig. 8.50.

**Note 8.10**

As the graph shows, the amplitude increased 20 dB in each decade of the frequency.
When the frequency is increased tenfold, the gain is changed by 20 dB.

**Note 8.11**

The amplitude is a straight line with a positive slope of 20 dB/dec, passing the
frequency of 1 rad/sec or $10^0$ in Fig. 8.50. The amplitude is negative for frequencies
less than 1 rad/sec.

---

**Example 8.60 Find the Bode plot of the following transfer function:**

$$H(s) = 2s$$

   *Solution.* This function has two elements: the gain of 2 ①, and one is a zero
at the origin, s②. This suggests two plots for the amplitude |$H$| in Fig. 8.51.

(continued)

**Example 8.60** (continued)

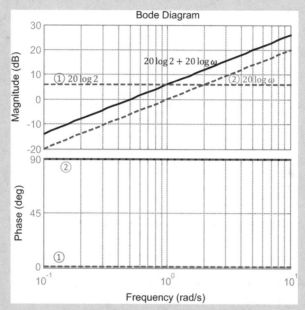

**Fig. 8.51** Amplitude and phase diagram of H(s) = 2 ① and H(s) = s ②, separately and combined

### Element 3
*A repeated zero at the origin. $H(s) = s^N$, N times.*

The repeat of zero $N$ times at the origin increases the slope by the same factor $20N$db/dec, and the phase is shifted to $N(90)$.

### Element 4
*A pole at the origin. $H(s) = \frac{1}{s}$*

In this case, the amplitude is inversely proportional to the frequency's increase $|H| = 20 \log \frac{1}{\omega}$. The phase is independent of frequency at $-90°$. The Bode plot is shown in Fig. 8.52.

### Note 8.12
The slope of this amplitude is $-20$ dB per decade. The amplitude drops 20 dB when the frequency increases from 1 to 10 rad/s.

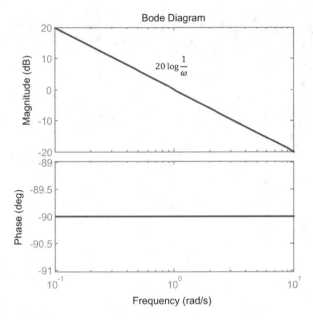

**Fig. 8.52** Amplitude and phase diagram of $H(s) = \frac{1}{s}$

**Note 8.13**
The amplitude decreases from near zero to 1 rad/s onward.

**Note 8.14**
The phase is constant and does not change with frequency.

**Example 8.61 Find the Bode plot of the function:** $H(s) = \frac{2}{s}$.
*Solution.* The function has two elements, element 2 shown by ① and element $\frac{1}{s}$ by ②. This results in two amplitudes as $20 \log 2$ and $20 \log \frac{1}{\omega}$. A separate plot of these two elements is shown in Figs. 8.53 and 8.54.

The combined elements have the following frequency response (Fig. 8.54):

(continued)

**Example 8.61** (continued)

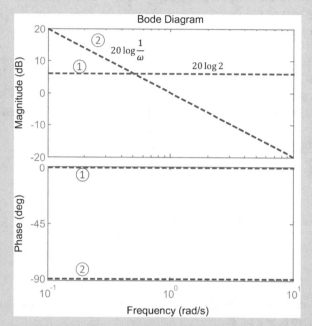

**Fig. 8.53** Amplitude and phase diagram of $H(s) = \frac{1}{s}$ and $H(s) = 2$, separately

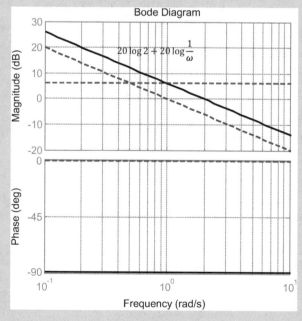

**Fig. 8.54** Amplitude and phase diagram of $H(s) = \frac{1}{s}$ and $H(s) = 2$, combined

**Element 5**

*A repeated pole at the origin N times.* $H(s) = \frac{1}{s^N}$.

The Bode plot of this function has a slope $-20N$ dB/dec, and the phase of this system is constant $-N(90°)$.

**Element 6**

*A unity gain transfer function with a zero at z.* $H(s) = \left(1 + \frac{s}{z}\right)$. The amplitude of this transfer function is evaluated for frequencies higher than $z$ rad/sec. The phase of this transfer function needs to be evaluated in a decade below and a decade above the $z$ rad/sec.

Considering $s = j\omega$, the transfer function becomes

$$H(j\omega) = \left(1 + \frac{j\omega}{z}\right)$$

The amplitude of the transfer function in $dB$ becomes

$$|H| = 20\log\sqrt{1 + \left(\frac{\omega}{z}\right)^2}$$

This function shows $\approx 0$ dB gain for all frequencies from $\omega = \frac{1}{10}z$ up to the location of zero $\omega = z$ rad/sec:

$$\left|H\left(\omega = \frac{1}{10}z\right)\right| = 20\log\sqrt{1 + \left(\frac{\frac{1}{10}z}{z}\right)^2} = 0 \text{ dB}$$

At the location of zero, the gain becomes

$$|H(\omega = z)| = 20\log\sqrt{1 + \left(\frac{z}{z}\right)^2} = 3 \text{ dB}$$

The gain at ten times the zero $\omega = 10z$ becomes

$$|H(\omega = 10z)| = 20\log\sqrt{1 + \left(\frac{10z}{z}\right)^2} = 20 \text{ dB}$$

It can be concluded that the gain is increased from the location of zero by a slope of 20 dB/dec.

To analyze the phase of this function, consider $\omega = \frac{1}{10}z$ $\omega = z$, and $\omega = 10z$. The phase can be calculated as follows:

$$\angle H = \tan^{-1} \frac{\frac{\omega}{z}}{1}$$

At a decade below the location of zero $\omega = \frac{1}{10}z$, the phase is

$$\angle H\left(\omega = \frac{1}{10}z\right) = \tan^{-1} \frac{\frac{\frac{1}{10}z}{z}}{1} = 5.7°$$

An asymptote of the phase variation for smaller frequencies can be considered approximately zero. At the location of zero $\omega = z$, the phase reaches

$$\angle H(\omega = z) = \tan^{-1} \frac{\frac{z}{z}}{1} = 45°$$

At a decade higher than the location of zero $\omega = 10z$, the phase of transfer function becomes

$$\angle H(\omega = 10z) = \tan^{-1} \frac{\frac{10z}{z}}{1} = 85°$$

As an asymptote, the phase saturates to $90°$ when the frequency increases very high.

Therefore, the Bode plot of the function is shown in Fig. 8.55.

**Fig. 8.55** Amplitude and phase diagram of $H(s) = \left(1 + \frac{s}{z}\right)$

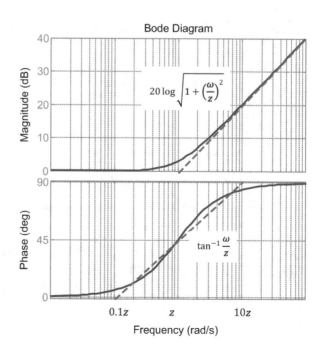

**Example 8.62 Find the Bode diagram of the following function:**

$$H(s) = (s+7)$$

*Solution.* This function has a zero at $s = -7$ or $z = 7$. To convert this function to the form of $\left(1 + \frac{s}{z}\right)$, 7 is factored out as follows:

$$H(s) = 7\left(1 + \frac{s}{7}\right)$$

Therefore, this function has an amplitude of 20 dB. The Bode plot of each element is $(20\log 7)$①$\&20\log\sqrt{1 + \left(\frac{\omega}{7}\right)^2}$② obtained as shown in Fig. 8.56.

As observed, the transfer function's amplitude at zero $s = -7$ rad/sec starts at an increment of 20 dB/dec. The effective range of phase starts a decade below at $\frac{1}{10}7$ rad/ sec and ends a decade higher at $10 \times 7$ rad/sec. Combining the bode plot of these two elements is shown in Figs. 8.56 and 8.57.

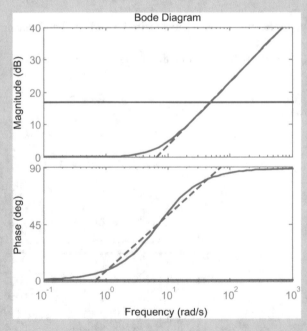

**Fig. 8.56** Amplitude and phase diagram of $H(s) = 7$ and $H(s) = \left(1 + \frac{s}{7}\right)$, separately

(continued)

**Example 8.62** (continued)

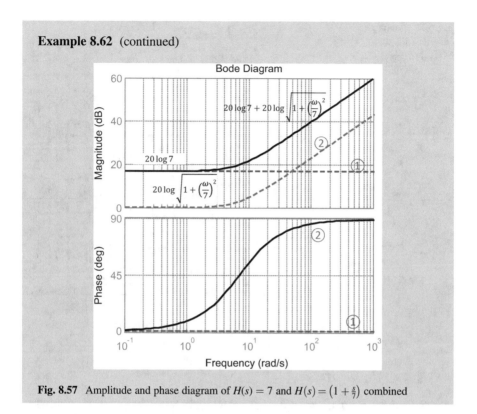

**Fig. 8.57** Amplitude and phase diagram of $H(s) = 7$ and $H(s) = \left(1 + \frac{s}{7}\right)$ combined

### Element 7

*A repeated zero at $s = z$.* $H(s) = \left(1 + \frac{s}{z}\right)^{N}$. The repeated zero at $z$ increases the slope by factor $N$ to become 20 $N$ dB/dec, and the phase ultimately reaches $N(90°)$, suggesting an increased phase slope of $N(45)$ deg /dec.

### Element 8

*A unity gain transfer function with a pole at $s = p$.* $H(s) = \frac{1}{\left(1 + \frac{s}{p}\right)}$.

The amplitude of this transfer function can be obtained as

$$|H| = 20 \log \frac{1}{\sqrt{1 + \left(\frac{\omega}{p}\right)^2}}$$

The phase of this transfer function can be obtained as

**Fig. 8.58** Amplitude and phase diagram of $H(s) = \frac{1}{\left(1+\frac{s}{p}\right)}$

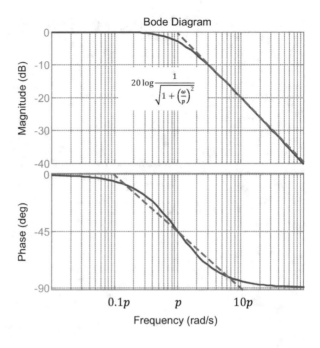

$$\angle H = 0 - \tan^{-1} \frac{\omega}{p}$$

The evaluation points are a decade below and above the pole's location. The amplitude is zero until the pole location starts to drop with a slope of $-20$ dB/dec and the phase starts to drop a decade below the pole location at $1/10\,p$ and reaches $-90°$ a decade above the pole at $10p$. At the pole location, the phase is $-45°$, suggesting a $-45$ deg/dec slope of the phase diagram. This can result in asymptotes, and the Bode diagram is shown in Fig. 8.58:

---

**Example 8.63 Find the Bode diagram of the following function:** $H(s) = \frac{1}{1+\frac{s}{7}}$.

*Solution.* The transfer function has a pole at $\omega = 7$. Therefore, the phase is zero for frequencies below 7 rad/sec and decreases by a $-20$ dB/dec slope. This means that the amplitude reaches $-20$ dB at a frequency of $10 \times 7 = 70$ rad/sec. The phase starts to decrease a decade below the pole location $\frac{1}{10}7 = 0.7$ rad/s and reaches $-90°$ a decade higher at 70 rad/sec. Figure 8.59 shows the Bode diagram asymptotes and its accurate graph. The phase reaches $-45°$ at the location of the pole. Therefore, the phase drops by a $-45$ deg/dec slope and saturates to $-90°$.

(continued)

**Example 8.63** (continued)

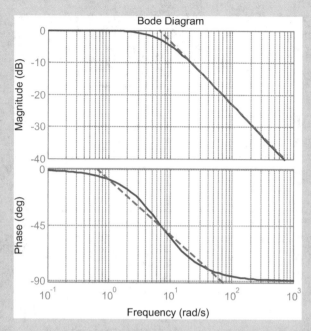

**Fig. 8.59** Amplitude and phase diagram of $H(s) = \frac{1}{1+\frac{s}{7}}$

### Element 9

*Repeated pole at p. $H(s) = \frac{1}{\left(1+\frac{s}{p}\right)^N}$*. The amplitude of this function starts to drop at the location of pole $p$ and drops with slope $-N(20)$ dB/dec. The phase starts to drop with a slope of $-N45$ deg/dec at a decade below the pole and saturates with $-N(90°)$ at a decade above the pole.

**Example 8.64** Find the Bode plot of the following function $H(s) = \frac{(s+50)}{(s+1)(s+10)}$.

*Solution.* The first step in obtaining the Bode plot of a transfer function is to convert it to the form with gain factored out as follows:

$$H(s) = \frac{(s+50)}{(s+1)(s+10)} = \frac{50\left(1+\frac{s}{50}\right)}{10(s+1)\left(1+\frac{s}{10}\right)} = \frac{5\left(1+\frac{s}{50}\right)}{(s+1)\left(1+\frac{s}{10}\right)}$$

There are four elements: gain 5 ①, zero at 50 of $\left(1+\frac{s}{50}\right)$ ②, poles at 1 and 10 through $\frac{1}{s+1}$ ③ and $\frac{1}{1+\frac{s}{10}}$ ④, respectively.

(continued)

**Example 8.64** (continued)

Therefore, four amplitude and four phase signals are combined to get the transfer function response. The Bode plot elements combined are shown in Fig. 8.60.

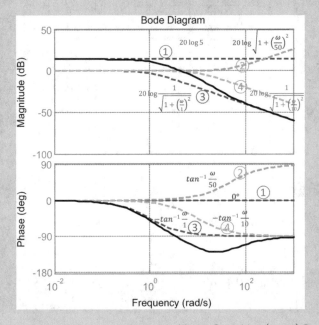

**Fig. 8.60** Amplitude and phase diagram of $H(s) = 5$ ①, $H(s) = \left(1 + \frac{s}{50}\right)$②, $H(s) = \frac{1}{(s+1)}$③, $H(s) = \frac{1}{\left(1+\frac{s}{10}\right)}$④, separately and combined

---

**Example 8.65 Find Bode plot of $H(s) = \frac{4}{(s+5)^3}$.**

*Solution.* The transfer function has an amplitude of

$$H(s) = \frac{4}{5^3\left(1 + \frac{s}{5}\right)^3} = \frac{4}{125}\,\frac{1}{\left(1 + \frac{s}{5}\right)^3}$$

The Bode plot of each of these elements $\frac{4}{125}$ ① and $\frac{1}{\left(1+\frac{s}{5}\right)^3}$ ② is shown in Fig. 8.61. As the Figure shows, the triple poles at 5 rad/sec show a slope of $-60$ dB/dec, and the phase reaches $-3 \times 90° = -270°$ a decade above 5 or 50 rad/sec. The slope of phase drop is $-45 \times 3 = -135$ deg/dec.

(continued)

**Example 8.65**  (continued)

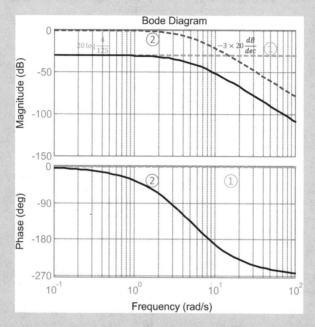

**Fig. 8.61**  Amplitude and phase diagram of $H(s) = \frac{4}{125}$ ①, $H(s) = \frac{1}{\left(1+\frac{s}{5}\right)^3}$ ②, separately and combined

**Example 8.66 Find Bode plot of $H(s) = \frac{s}{s+10}$.**
*Solution.* After converting the transfer function to the known template (factoring out the gain), it is known that there are three elements of gain $\frac{1}{10}$ ①, a zero at the origin $s$ ②, and a pole at 10, $\frac{1}{\left(1+\frac{s}{10}\right)}$ ③.

The Bode plot of these three elements is shown in Fig. 8.62.

(continued)

**Example 8.66** (continued)

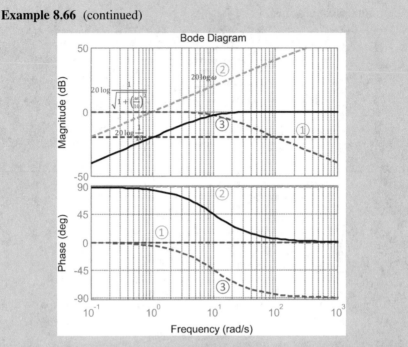

**Fig. 8.62** Amplitude and phase diagram of $H(s) = \frac{1}{10}$①, $H(s) = 2$②, $H(s) = \left(1 + \frac{s}{10}\right)$③. separately and combined

## Element 10

*A unity-gain complex-conjugate zero $s_{1,2} = -\zeta\omega_n \pm j\omega_n\sqrt{1-\zeta^2}$.*

$$H(s) = \left(1 + \frac{2\zeta}{\omega_n}s + \frac{1}{\omega_n^2}s^2\right)$$

In a complex conjugate roots condition, the damping factor $0 < \zeta < 1$. Therefore, an overshoot is observed in the Bode diagram's amplitude. The amplitude is 0 dB until the reach of resonant frequency $\omega_n$. At this point, the amplitude is increased by 40 dB/dec.

The phase starts to increase a decade below the resonant frequency $\omega_n$ with a slope of 90 deg/dec. At the resonant frequency, the phase reaches $90°$, and a decade above the resonant, it is saturated to $180°$.

Figure 8.63 shows the actual and asymptotes of the Bode plot.

Effect of damping factor: As the damping factor becomes less than 1, an overshoot is observed in the system's time response. This overshoot increases in damping factors near zero when the system becomes oscillatory. This effect is shown in Fig. 8.64. As the damping factor drops, the amplitude change at the resonant frequency becomes larger, and the rate of phase change becomes faster.

**Fig. 8.63** Amplitude and phase diagram of $H(s) = \left(1 + \frac{2\zeta}{\omega_n}s + \frac{1}{\omega_n^2}s^2\right)$

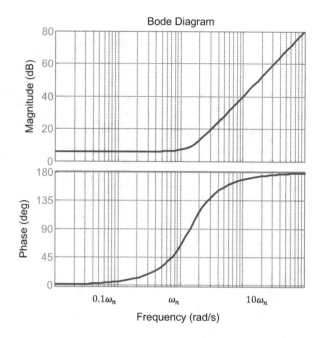

**Fig. 8.64** Amplitude and phase diagram of $H(s) = \left(1 + \frac{2\zeta}{\omega_n}s + \frac{1}{\omega_n^2}s^2\right)$

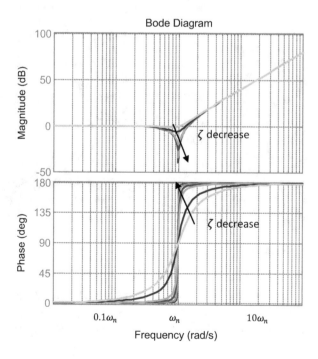

**Example 8.67** Find Bode plot of the following function: $H(s) = (1 + 0.02\, s + 0.01\, s^2)$.

*Solution.* Compared to the template $H(s) = \left(1 + \frac{2\zeta}{\omega_n}s + \frac{1}{\omega_n^2}s^2\right)$, the parameters are found to be $\zeta = 0.1$ and $\omega_n = 10$. Therefore, a sketch of the Bode diagram for the amplitude starts at 10 rad/sec with an increase of slope + 40 dB/dec.

The phase change starts from $0°$ at $\frac{1}{10} \times 10 = 1$ rad/ sec and ends at $180°$ at $10 \times 10 = 100$ rad/sec. The slope of change is +90 deg/dec. Figure 8.65 shows the Bode diagram.

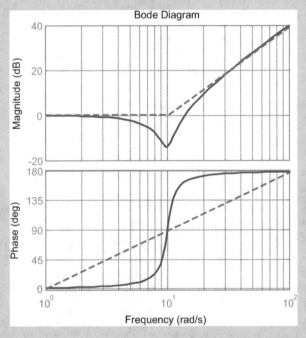

**Fig. 8.65** Amplitude and phase diagram of $H(s) = (1 + 0.02\, s + 0.01\, s^2)$. This has a damping factor $\zeta = 0.1$ and a resonant frequency of $\omega_n = 10$

**Element 11**

*Repeated component N times.* $H(s) = \left(1 + \frac{2\zeta}{\omega_n}s + \frac{1}{\omega_n^2}s^2\right)^N$.

In this case, the resonant frequency and the damping factor do not change. However, the amplitude increases sharper with a slope of $N \times 40$ dB/dec, and the phase reaches the final value of $N \times 180°$ with a slope of $N \times 90$ deg/dec.

**Example 8.68 Find the Bode plot of the following transfer function.**
$H(s) = (1 + 0.02\ s + 0.01\ s^2)^2.$

*Solution.* Compared to the template $H(s) = \left(1 + \frac{2\zeta}{\omega_n}s + \frac{1}{\omega_n^2}s^2\right)$, the parameters
are found to be $\zeta = 0.1$ and $\omega_n = 10$. Therefore, a sketch of the Bode diagram
for the amplitude starts at 10 rad/sec with an increase of slope $+ 2 \times 40$ dB/dec.

The phase change starts from $0°$ at $\frac{1}{10} \times 10 = 1$ rad/s and ends at $2 \times 180°$ at
$10 \times 10 = 100$ rad/sec. The slope of change is $+2 \times 90$ deg/dec. Figure 8.66
shows the Bode diagram.

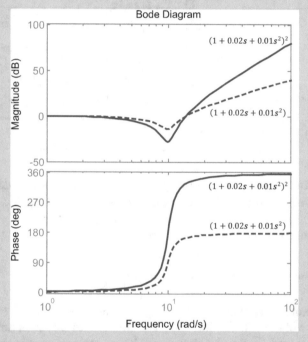

**Fig. 8.66** Amplitude and phase diagram of $H(s) = (1 + 0.02\ s + 0.01\ s^2)$ and
$H(s) = (1 + 0.02\ s + 0.01\ s^2)^2$

**Element 12**

$$H(s) = \frac{1}{1 + \frac{2\zeta}{\omega_n}s + \frac{1}{\omega_n^2}s^2}.$$

This transfer function has two poles $s_{1,2} = -\zeta\omega_n \pm j\omega_n\sqrt{1-\zeta^2}$. The amplitude
is zero until the reach of resonant frequency $\omega_n$; it starts to drop with the slope of $-$

**Fig. 8.67** Amplitude and phase diagram of $H(s) = \frac{1}{1+\frac{2\zeta}{\omega_n}s+\frac{1}{\omega_n^2}s^2}$

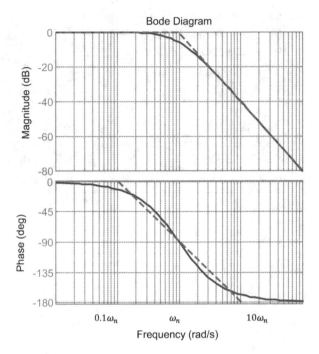

40 dB/dec. The phase also starts to drop from zero beginning at a decade below the resonant frequency $\frac{1}{10}\omega_n$, continues to drop with a slope of $-90$ deg/dec, and reaches a saturated value of $-180°$ a decade above the resonant frequency $10\omega_n$. The amplitude and phase are shown in Fig. 8.67.

**Element 13**

*Repeated complex conjugate poles* $H(s) = \dfrac{1}{\left(1+\frac{2\zeta}{\omega_n}s+\frac{1}{\omega_n^2}s^2\right)^N}$.

The amplitude is zero until the reach of resonant frequency $\omega_n$; it starts to drop with the slope of $-N \times 40$ dB/dec. The phase also starts to drop from zero beginning at a decade below the resonant frequency, $\frac{1}{10}\omega_n$; continues to drop with slope $-N \times 90$ deg/dec; and reaches a saturated value of $-N \times 180°$ a decade above the resonant frequency $10\omega_n$.

The phase drop sharpness also depends on the damping, and as the damping factor is decreased, the slope of the phase changes, and the overshoot of the amplitude increase. This is shown in Fig. 8.68.

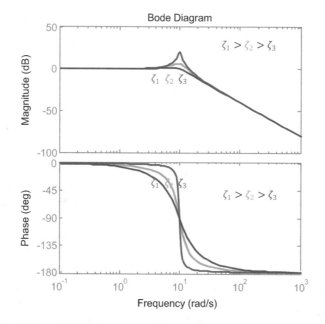

**Fig. 8.68** Amplitude and phase diagram of $H(s) = \frac{1}{1+\frac{2\zeta}{\omega_n}s+\frac{1}{\omega_n^2}s^2}$ when damping decreases

**Example 8.69 Find the Bode diagram of the following transfer function.**

$$H(s) = \frac{1}{(1+0.02s+0.01s^2)^2}$$

*Solution*. Since this is a unity gain transfer function of a second-order unit, the entry gain (or the gain at low frequencies close to DC) is 0 dB. The amplitude drops by a slope of $-40 \times 2 = -80$ dB/dec at 10 rad/sec resonant frequency. The phase diagram starts at a decade below the resonant 1 rad/sec from 0° and reaches $-180 \times 2 = -360°$ by the frequency a decade above the resonant, i.e., 100 rad/sec. Figure 8.69 shows the amplitude and phase variations. The transfer function shows a damping factor of $\zeta = 0.1$. Therefore, there is an overshoot observed in the time response. This overshoot is also shown in the figure by exceeding the approximate responses (dotted lines).

(continued)

**Example 8.69**  (continued)

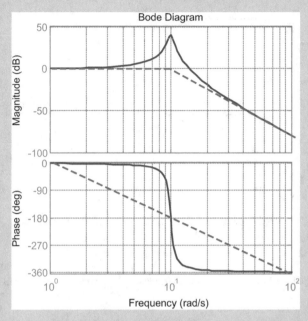

**Fig. 8.69**  Amplitude and phase diagram of $H(s) = \frac{1}{\left(1+0.02s+0.01s^2\right)^2}$

**Example 8.70 Find the Bode diagram** $H(s) = \frac{(s+5)}{(s+7)(s+10)^2}$.

*Solution.* First, the transfer function gain is obtained by factoring the gains to form the element templates.

$$H(s) = \frac{(s+5)}{(s+7)(s+10)^2} = \frac{5}{7 \times 10^2} \frac{\left(1+\frac{s}{5}\right)}{\left(1+\frac{s}{7}\right)\left(1+\frac{s}{10}\right)^2}$$

Therefore, the transfer function has a gain of $\frac{5}{700}$ ① and three elements of $\left(1+\frac{s}{5}\right)$ ②, $\left(1+\frac{s}{7}\right)$ ③, and $\left(1+\frac{s}{10}\right)^2$ ④.

Individual Bode diagram of each element can be obtained as shown in Fig. 8.70:

(continued)

**Example 8.70** (continued)

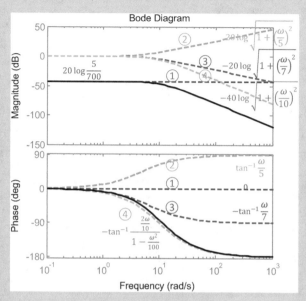

**Fig. 8.70** Amplitude and phase diagram of $H(s) = \frac{5}{700}$ ①, $H(s) = \left(1 + \frac{s}{5}\right)$ ②, $H(s) = \left(1 + \frac{s}{7}\right)$ ③, and $H(s) = \left(1 + \frac{s}{10}\right)^2$ ④ separately and combined

**Example 8.71 Find the Bode diagram of** $H(s) = \frac{100 \ (s+10)}{(s+5)(s+20)}$.

*Solution.* To obtain the DC gain, frequencies are factors out, and the template is identified as

$$H(s) = \frac{100(s+10)}{(s+5)(s+20)} = \frac{100 \times 10}{5 \times 20} \frac{\left(1 + \frac{s}{10}\right)}{\left(1 + \frac{s}{5}\right)\left(1 + \frac{s}{20}\right)}$$

The function has four elements: a DC gain of $\frac{100 \times 10}{5 \times 20}$ ①, zero of $\left(1 + \frac{s}{10}\right)$ ②, two-pole elements of $\left(1 + \frac{s}{5}\right)$ ③, and $\left(1 + \frac{s}{20}\right)$ ④.

A plot of these four elements and the overall summation of these elements is shown in Fig. 8.71.

(continued)

**Example 8.71** (continued)

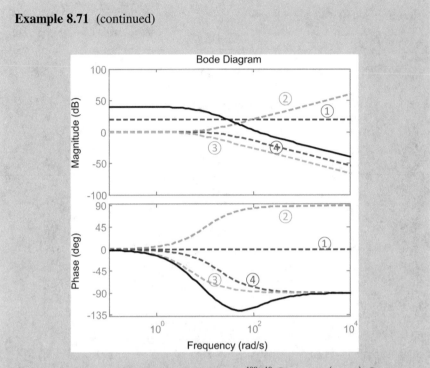

**Fig. 8.71** Amplitude and phase diagram of $H(s) = \frac{100 \times 10}{5 \times 20}$ ①, $H(s) = \left(1 + \frac{s}{10}\right)$ ②, $H(s) = \left(1 + \frac{s}{5}\right)$ ③, and $H(s) = \left(1 + \frac{s}{20}\right)$ ④, separately and combined

## Problems

Find the transfer function of the following circuits as they are defined for each case:

8.1. Find $H(s) = \frac{V_o}{V_{in}}$.

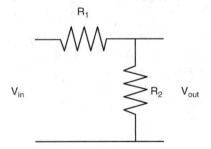

8.2. Find $H(s) = \frac{V_o}{V_{in}}$.

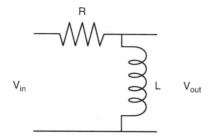

8.3. Find $H(s) = \frac{V_o}{V_{in}}$.

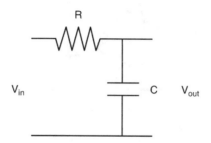

8.4. Find $H(s) = \frac{V_o}{V_{in}}$.

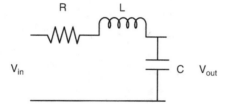

8.5. Find $H(s) = \frac{V_o}{V_{in}}$.

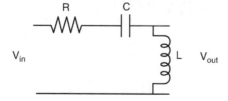

8.6. Find $H(s) = \frac{V_o}{V_{in}}$.

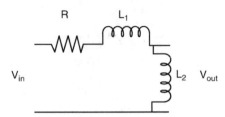

8.7. Find $H(s) = \frac{V_o}{V_{in}}$.

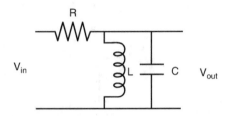

8.8. Find $H(s) = \frac{V_o}{V_{in}}$.

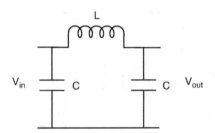

8.9. Find $H(s) = \frac{V_o}{V_{in}}$.

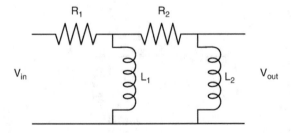

8.10. Find $H(s) = \frac{V_o}{V_{in}}$.

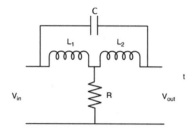

8.11. Find the transfer function $H(s) = \frac{V_0}{V_{in.}}$.

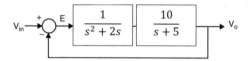

8.12. Find the transfer function $H(s) = \frac{V_0}{V_{in}}$. Find $K$ to map the pole of the system in LHP.

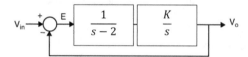

8.13. Find the transfer function $H(s) = \frac{V_0}{V_{in}}$.

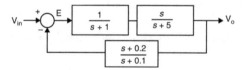

8.14. Find the transfer function $H(s) = \frac{V_0}{V_{in}}$.

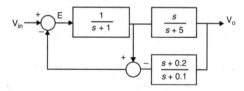

8.15. Find poles and zeroes of the open loop and closed systems described in Problems 8.11–8.14.

8.16. Find the transfer function of the following systems, where $x$ is the input and $y$ is the output:

$$\dot{y} + 5y = 5x$$
$$\ddot{y} + 9y = \dot{x} + x$$
$$\ddot{y} + 2\dot{y} = \ddot{x} + 2\dot{x} + 5x$$
$$\ldots y + 7\ddot{y} + 2\dot{y} = \dot{x}$$
$$\ddot{y} - 2\dot{y} - 3y = \ddot{x} + 2x$$

8.17. Find poles and zeroes of the transfer functions described in Problem 8.16a–e.

8.18. A dynamic system with a differential equation that relates the output $y$ to the input $x$ is

$$\ddot{y} + 2\dot{y} = \ddot{x} + 2\dot{x} + 5x$$

Find the transfer function of a unity-negative feedback closed-loop system defined as $H(s) = \frac{Y(s)}{X(s)}$.

8.19. In the systems of problem 8.18:

(a) Find the poles and zeros of the transfer functions.
(b) Determine whether the poles are in the right half plane or the left half plane.
(c) Determine the stability of the systems.

8.20. The systems of problem 8.16 are located in a unity-negative feedback closed-loop system.

(a) Find the transfer function of the closed-loop system.
(b) Find the poles and zeros of the closed-loop system.
(c) Compare the poles obtained in problem 8.17 with the ones from the closed loop.
(d) Determine whether the systems have changed their stability.

8.21. Consider a gain $K$ in the closed-loop negative feedback system (Problem 8.16). Find the $K$ or a limit for $K$ such that the poles of the closed-loop systems obtained in Problem 18.15 become stable.

8.22. Find the initial and final values in the transfer functions of Problem 8.15.

8.23. Find the initial and final values in the transfer functions of Problem 8.16.

8.24. Find the error value in the negative feedback closed-loop systems of Problem 8.16 when:

(a) A unit step function is applied in the input.
(b) A ramp function is applied in the input.
(c) $f(t) = t^2$ is applied in the input.

8.25. The system $h(t)$ is defined in each case (below) and is excited by the given input $x(t)$. Find the output signals $y(t)$ using convolution integral, and verify your results by its Laplace approach.

(a) $h(t) = 2e^{-10t}$, $x(t) = u(t)$
(b) $h(t) = 2u(t - 5) - 2u(t - 10)$, $x(t) = u(t)$
(c) $h(t) = 2u(t - 5) - 2u(t - 10)x(t) = tu(t) - 2(t - 2)u(t - 2) + (t - 3)u(t - 3)$
(d) $h(t) = 2u(t - 5) - 2u(t - 10)$, $x(t) = u(t) + u(t - 1) - 2u(t - 2)$

8.26. Find the state space representation of the circuits shown in Problems 8.1–8.10.
8.27. Find the state space representation of the following differential equations, where $x$ is the input and $y$ is the output. Put the models in matrix form $\dot{X} = AX + Bu$, $y = CX + Du$:

(a) $\dot{y} + 5y = 5u$
(b) $\ddot{y} - 2\dot{y} - 3y = 2u$
(c) $\ddot{y} + 9y = \dot{u} + u$
(d) $\ddot{y} + 2\dot{y} = \ddot{u} + 2\dot{u} + 5u$
(e) $\ldots y + 7\ddot{y} + 2\dot{y} = \dot{u}$
(f) $\ddot{y} - 2\dot{y} - 3y = \ddot{u} + 2u$

8.28. State space models found in Problem 8.27a–f are used in a unity feedback system. Find the state space model of the closed-loop system.
8.29. Show the state space representation of Problem 8.27 in block diagram format.
8.30. Determine the $M_p$, $t_p$, $t_r$, $t_{s5}$, $t_{s2}$, $t_d$, and sketch the step response of the following systems:

(a) $H(s) = \frac{100}{s^2 + 10s + 100}$
(b) $H(s) = \frac{1}{s^2 + 0.2s + 1}$
(c) $H(s) = \frac{900}{s^2 + 12s + 900}$

8.31. Find the transfer function of the following systems:

(a) $\dot{X} = \begin{bmatrix} 0 & 1 \\ -11 & -6 \end{bmatrix} X + \begin{bmatrix} 0 \\ 5 \end{bmatrix} u$
$y = \begin{bmatrix} 1 & 0 \end{bmatrix} X$

(b) $\dot{X} = \begin{bmatrix} 1 & -1 \\ -1 & 1 \end{bmatrix} X + \begin{bmatrix} 2 \\ 1 \end{bmatrix} u$
$y = \begin{bmatrix} 1 & 0 \end{bmatrix} X$

(c)
$$\dot{X} = \begin{bmatrix} 1 & -1 \\ -3 & -5 \end{bmatrix} X + \begin{bmatrix} 0 \\ 1 \end{bmatrix} u$$
$$y = [1 \ 0]X$$

(d)
$$\dot{X} = \begin{bmatrix} 0 & 1 & 0 \\ 0 & 0 & 1 \\ -6 & -11 & -5 \end{bmatrix} X + \begin{bmatrix} 0 \\ 1 \\ -4 \end{bmatrix} u$$
$$y = [1 \ 0 \ 0]X$$

(e)
$$\dot{X} = \begin{bmatrix} -1 & 1 & 0 \\ 0 & -1 & 1 \\ -6 & -11 & -5 \end{bmatrix} X + \begin{bmatrix} 0 \\ 0 \\ -4 \end{bmatrix} u$$
$$y = [1 \ 2 \ -1]X$$

8.32. Find the Bode diagram of the following transfer functions:

(a) $H(s) = 2 s$

(b) $H(s) = \frac{2}{s}$

(c) $H(s) = s + 1$

(d) $H(s) = \frac{1}{s+1}$

(e) $H(s) = s + 100$

(f) $H(s) = \frac{1}{s+100}$

(g) $H(s) = \frac{s}{s+100}$

(h) $H(s) = \frac{s+3}{s+5}$

(i) $H(s) = \frac{1}{s(s+20)}$

(j) $H(s) = \frac{1}{s^2(s+20)}$

(k) $H(s) = \frac{50}{s+0.1}$

(l) $H(s) = \frac{100(s+40)}{(s+15)(s+25)}$

(m) $H(s) = \frac{100(s+40)}{(s+15)(s+25)^2}$

(n) $H(s) = \frac{(s+4)}{(s+5)^3(s+10)}$

(o) $H(s) = \frac{100}{(s^2+15s+100)}$

(p) $H(s) = \frac{100(s+25)}{(s^2+15s+100)}$

(q) $H(s) = \frac{100}{(s^2+15s+100)(s+25)}$

(r) $H(s) = \frac{100}{(s^2+15s+100)^2(s+25)}$

(s) $H(s) = (s^2 + 10s + 100)^2$

# Chapter 9
# Passive Filters

## Introduction

Frequency-selective circuits or "filters" are circuits that either pass or attenuate signals at a specific frequency or a range of desired frequencies. These filters can be implemented in hardware to be programmed in software that runs on a processor to accomplish similar tasks. For instance, consider old cassette players, or video players, in which several sets of filters are required to make the quality of sound or video expected from high-quality sets, whereas considering a modern MP3 player, or a digital satellite receiver, in which the same or even better quality of signal processing is expected. However, the sets have no actual hardware to accomplish the signal filtration. In this chapter, both hardware circuits and methods to implement the filters in software are discussed.

## Passive and Active Filters

When a signal is passed through a filter, its amplitude may drop a small percentage because the circuit elements are not ideal, and the signal may lose some power. Therefore, the filters are passive (with no amplification) or active (with amplification). This chapter discusses passive filters and Chap. 11 discusses active filters.

### *Category of Passive Filter Circuits*

To determine the type of filter, the ratio of the output voltage to the input voltage of a filter or its gain is analyzed over a range of frequencies at the input. The gain variation with respect to the frequency is called frequency response. Bode diagrams

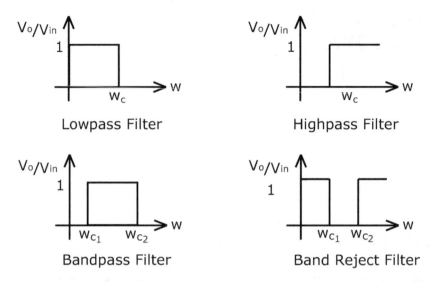

**Fig. 9.1** Frequency response of ideal filters, their cutoff frequencies, passing band, and rejecting band. The filter passes a range of frequencies if, at those frequencies, the output receives a majority of the input signal or ideally 100% of it $\left(\frac{V_o}{V_{in}} = 1\right)$. The filter rejects a range of frequencies if, at those frequencies, the output is attenuated, and the signal's power is reduced significantly. Ideally, at the rejection band, the output is zero $\left(\frac{V_o}{V_{in}} = 0\right)$. Accordingly, the response of various filters can be recognized

also show the frequency response of a circuit. Several frequencies directly contribute to the characteristics of a filter. Some of these frequencies are discussed here. All filters have a range of frequencies at which they fully pass, or they fully stop. The frequencies at which a filter switches from passing to no pass or vice versa are called cutoff frequencies and are shown by $\omega_c$. These frequencies are also called corner frequencies. Based on the frequency response of a certain circuit, there are four major types of filters. The first is the one to pass all frequencies below a certain frequency or low-pass filters; the second is a filter to pass all signals above a certain frequency or high-pass filter, third is the filter to pass all signals in the desired band of frequency or band-pass filter, and fourth is a filter to reject all signals in a band of frequency or band-reject filter. Figure 9.1 shows the ideal operation of these filters. It is desired that a filter drops the output voltage to zero at the no pass range and passes 100% of the signal in the pass range.

As shown, high-pass and low-pass filters have one cutoff frequency each. However, band-pass and band-reject filters have two cutoff frequencies. The band of frequencies determines the bandwidth of the filter. This is all the range of frequencies at which the filter passes the signals (signals at these frequencies pass through the filter). This bandwidth is obtained as the difference between two cutoff frequencies.

It is required to revisit the impedance of inductors and capacitors to understand the operation of a filter better.

The impedance of an inductor $L(H)$ with respect to a signal at frequency $\omega$ is $Z = jX_L = j\omega L$ ($\Omega$). It shows that if the signal's frequency is increased, the impedance of the inductor also increases. At low frequencies, the inductor shows a very low impedance and passes the low-frequency signals, and at high frequencies, the impedance of the inductor is increased to high values that can block the signals from passing. Of course, the use of term frequency is relative to the operation of the filter, and the cutoff frequency is determined based on the inductance and other circuit parameters.

The impedance of a capacitor $C(F)$ is inversely proportional to the frequency of a signal $Z = -jX_C = \frac{-j}{C\omega}$. It means that a capacitor blocks low-frequency signals and passes high-frequency signals because; at low frequencies, the capacitor demonstrates a significant ohmic impedance that can block the signals; at high frequencies, the ohmic impedance of the capacitor shows a short circuit that allows signals to pass.

Based on these facts, an inductor is shown with a low-frequency (LF) arrow $\overset{\rightarrow LF}{L}$, and a capacitor with a high-frequency (HF) arrow on the top $\overset{\rightarrow HF}{C}$.

There are several techniques to determine the cutoff frequency of a filter. In this section, sinusoidal signal analysis is used, and in the next section, the Laplace technique is used to analyze the filters.

**Example 9.1** In an **RL** circuit, the input frequency is variable. Find the output voltage at the given frequencies when $V_{in} = 10$ at the frequency $\omega = 1,\ 10,\ 100,\ 1000,\ 2000,$ and $10000$ $\left(\frac{rad}{s}\right)$ (Fig. 9.2).

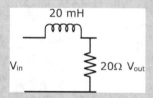

**Fig. 9.2** Circuit of Example 9.1

*Solution.* The circuit shows a voltage divider between the 20 $\Omega$ resistor and the $j\omega L$ inductor. The output voltage is

$$V_o = \frac{20}{20 + j0.02\omega} V_{in}$$

The amplitude of the output voltage becomes

$$|V_o| = \frac{20}{\sqrt{20^2 + (0.02\omega)^2}} 10$$

At $\omega = 1$

(continued)

**Example 9.1** (continued)

$$|V_o| = \frac{20}{\sqrt{20^2 + (0.02)^2}} 10 \cong 10$$

At $\omega = 10$

$$|V_o| = \frac{20}{\sqrt{20^2 + (0.02 \times 10)^2}} 10 \cong 10$$

At $\omega = 100$

$$|V_o| = \frac{20}{\sqrt{20^2 + (0.02 \times 100)^2}} 10 = 9.95$$

At $\omega = 1000$

$$|V_o| = \frac{20}{\sqrt{20^2 + (0.02 \times 1000)^2}} 10 = 7.07$$

At $\omega = 2000$

$$|V_o| = \frac{20}{\sqrt{20^2 + (0.02 \times 2000)^2}} 10 = 4.47$$

At $\omega = 10,000$

$$|V_o| = \frac{20}{\sqrt{20^2 + (0.02 \times 10,000)^2}} 10 \cong 1$$

As observed, the output voltage decreases as the source frequency increases. The voltage is high and almost passes to the output without much change. At the frequency $\omega = 1000$ rad/s, the output becomes 70.7% of the input, known as cutoff frequency (in actual filters. Ideally, the signal would become zero). The output voltages above this frequency drop rapidly to very low values at high frequencies.

## *Filter Gains*

At a certain frequency, filters can change the power of a signal with respect to the input power of the same signal. This gain can be positive, meaning the output power has increased, and it can be negative, meaning that the output power has decreased. The gain of power $G_P$ measured in decibel (dB) is as follows:

$$G_P = 10 \log \frac{P_o}{P_{in}} \ (dB)$$

where $P_o$ is the output power and $P_{in}$ is the input power. Considering the input voltage and output voltage signals at the same frequency, the power of these signals is proportional to the square of their voltages, or

$$P_o \propto V_o^2$$

$$P_{in} \propto V_{in}^2$$

Hence, the voltage gain of the filter can be expressed as

$$G_V = 10 \log \frac{P_o}{P_{in}} = 10 \log \frac{V_o^2}{V_{in}^2} = 10 \log \left( \frac{V_o}{V_{in}} \right)^2 = 20 \log \frac{V_o}{V_{in}} (dB)$$

$$G_V = 20 \log \frac{V_o}{V_{in}}$$

Measuring the gains based on the current measurement results in the following calculations:

$$P_o \propto I_o^2$$

$$P_{in} \propto I_{in}^2$$

Hence, the current gain of the filter can be expressed as

$$G_I = 10 \log \frac{P_{in}}{P_o} = 10 \log \frac{I_o^2}{I_{in}^2} = 10 \log \left( \frac{I_o}{I_{in}} \right)^2 = 20 \log \frac{I_o}{I_{in}} (dB)$$

$$G_I = 20 \log \frac{I_o}{I_{in}}$$

**Example 9.2 A 100 kHz signal at a voltage of 16 mV passes a filter and is measured at 8 mV at the output. Find the gain of the filter.**
*Solution.* The gain in dB is obtained as follows:

$$G_V = 20 \log \frac{V_o}{V_{\text{in}}}$$

$$G_V = 20 \log \frac{8 \text{ m}}{16 \text{ m}} = -6 \text{ dB}$$

**Example 9.3 Find the power gain of a +1 dB filter.**
*Solution.*

$$G_P = 10 \log \frac{P_o}{P_{\text{in}}} = +1$$

$$\frac{P_o}{P_{\text{in}}} = 10^{\frac{1}{10}} = 1.26$$

It means that the output power has increased by 26%.

**Example 9.4 Find the power gain of a +2 dB filter.**
*Solution.*

$$G_P = 10 \log \frac{P_o}{P_{\text{in}}} = +2$$

$$\frac{P_o}{P_{\text{in}}} = 10^{\frac{2}{10}} = 1.58$$

It means that the output power has increased by 58%.

**Example 9.5 Find the power gain of a 3 dB filter.**
*Solution.*

$$G_P = 10 \log \frac{P_o}{P_{in}} = +3$$

$$\frac{P_o}{P_{in}} = 10^{\frac{3}{10}} = 2$$

It means that the output power has increased by 100%.

**Example 9.6 Find the power gain of a −1 dB filter.**
*Solution.*

$$G_P = 10 \log \frac{P_o}{P_{in}} = -1$$

$$\frac{P_o}{P_{in}} = 10^{\frac{-1}{10}} = 0.79$$

It means that the output power has dropped by 21%.

**Example 9.7 Find the power gain of a −2 dB filter.**
*Solution.*

$$G_P = 10 \log \frac{P_o}{P_{in}} = -2$$

$$\frac{P_o}{P_{in}} = 10^{\frac{-2}{10}} = 0.63$$

It means that the output power has dropped by 37%.

**Example 9.8 Find the power gain of a −3 dB filter.**
*Solution.*

$$G_P = 10 \log \frac{P_o}{P_{in}} = -3$$

$$\frac{P_o}{P_{in}} = 10^{\frac{-3}{10}} = 0.5$$

It means that the output power has dropped by 50%.

## Cutoff and Half-Power Point Frequencies

A summary of Examples 9.1, 9.2, 9.3, 9.4, 9.5, 9.6, and 9.7 is given in the following table:

| Power gain (dB) | Power ratio | % of output power change |
|---|---|---|
| +1 | 1.26 | 26% increase |
| +2 | 1.58 | 58% increase |
| +3 | 2 | 100% increase |
| −1 | 0.79 | 21% decrease |
| −2 | 0.63 | 37% decrease |
| −3 | 0.5 | 50% decrease |

At the gain of −3 dB, the signal's power is cut in half. Therefore, the frequency at which the power of a signal is dropped in half is called "the half-power point" frequency.

This is the frequency at which filters also show effective action of filtration. This frequency was also called the "cutoff" frequency.

The voltage and current gain at the cutoff frequency become

$$-3 = 20 \log \frac{V_o}{V_{in}}$$

$$\frac{V_o}{V_{in}} = 10^{\frac{-3}{20}} = 0.707 \ \text{ or } \ \frac{1}{\sqrt{2}}$$

If a signal is tuned at the cutoff frequency, its output voltage becomes 0.707 of the signal's input voltage.

# Low-Pass Filter

A LPF is designed to pass all frequencies below the cutoff frequency, and it ideally has an amplitude of 100% (of input appears in the output). However, in reality, the amplitude drops before the cutoff frequency. In this case, the cutoff frequency is identified when the transfer function gain drops 3 dB, or the output signal amplitude reaches 70.7% of the input signal.

Inductors and capacitors can be used to realize a low-pass filter. These circuit topologies may differ, but their operating characteristics are the same. They may be tuned to attenuate signals above a certain cutoff frequency.

The following procedure is repeated for all types of filters:

- Find the transfer function of the circuit.
- Find the cutoff frequencies using the transfer function.
- Find the cutoff frequencies using the $\frac{1}{\sqrt{2}} H_{max}$ rule.
- Find the bandwidth and resonant frequencies.
- Determine the frequency response in the form of a Bode diagram (dB) and an absolute amplitude value.
- Double check the Bode diagram by evaluating extreme frequency response $(\omega \rightarrow 0 \ \& \ \omega \rightarrow \infty)$ or $(s \rightarrow 0 \ \& \ s \rightarrow \infty)$ and the resonant frequency.

## *First-Order* **RL** *Low-Pass Filter*

To pass low-frequency signals to the output of a filter, an inductor may be used to bridge the input to output ports. This circuit element parameter is tuned to obtain the desired cutoff frequency beyond the cutoff frequency; higher frequency signals generate high impedance and are automatically filtered out (Fig. 9.3).

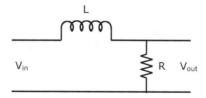

**Fig. 9.3** An *RL* circuit forms a low-pass filter. Inductors pass low frequencies, and if connected between the source and the output, they provide a path to pass the low frequencies to the output, forming a low-pass filter. A combination of resistance and inductance values determines the cutoff frequencies

**Identifying the Cutoff Frequency from the Ratio of the Output Over Input**

At the cutoff frequency, the amplitude of the transfer function becomes 70.7% of the maximum or $\frac{1}{\sqrt{2}} H_{max}$. $H(j\omega)$ is the filter's transfer function or the ratio of the output voltage over the input voltage. Therefore,

$$|H(j\omega)|_{\omega=\omega_c} = \frac{1}{\sqrt{2}} H_{max}$$

Maximum output can reach 1. Therefore,

$$|H(j\omega)|_{\omega=\omega_c} = \frac{V_o(j\omega)}{V_{in}(j\omega)} = \left|\frac{\frac{R}{L}}{j\omega_c + \frac{R}{L}}\right| = \frac{1}{\sqrt{2}}$$

$$\frac{\frac{R}{L}}{\sqrt{\omega_c^2 + \left(\frac{R}{L}\right)^2}} = \frac{1}{\sqrt{2}}$$

Solving for $\omega_c$ results in

$$\frac{\left(\frac{R}{L}\right)^2}{\omega_c^2 + \left(\frac{R}{L}\right)^2} = \frac{1}{2}$$

$$\omega_c = \pm \frac{R}{L}$$

where the positive frequency is acceptable. Therefore,

$$\omega_c = \frac{R}{L} \rightarrow f_c = \frac{R}{2\pi L}$$

**Note 9.1** The cutoff frequency of a first-order filter is the inverse of its time constant. In this $RL$ circuit, the time constant is $\tau = \frac{L}{R}$; therefore, the cutoff frequency is $\omega_c = \frac{1}{\tau}$ or $\omega_c = \frac{1}{\frac{L}{R}} = \frac{R}{L}$.

**Using Laplace Transform to Find the Cutoff Frequency**

The output voltage is obtained by a voltage division between $R$ in the output and $L$ in the bridge to the output. The transfer function is as follows:

$$\frac{V_o}{V_{in}} = \frac{R}{R + sL} = \frac{\frac{R}{L}}{s + \frac{R}{L}}$$

This transfer function has no zeros and one pole at $s + \frac{R}{L} = 0$ or

$$s = -\frac{R}{L}$$

The Bode diagram's amplitude drops at the pole's location by 3 dB. This qualifies the pole $\left(\frac{R}{L}\right)$ to be the cutoff frequency. Therefore,

$$j\omega = -\frac{R}{L}$$

$$\omega_c = \frac{R}{L}\left(\frac{rad}{s}\right)$$

To obtain an approximate frequency response of the circuit, the amplitude can be checked in two extreme frequencies of $\omega = 0$ and $\omega = \infty$ as follows:

$$\left|\frac{V_o}{V_{in}}(j0)\right| = \lim_{s \to 0}\left|\frac{\frac{R}{L}}{s + \frac{R}{L}}\right| = 1$$

This means 100% of the input signal appears in the output.

$$\left|\frac{V_o}{V_{in}}(j\infty)\right| = \lim_{s \to \infty}\left|\frac{\frac{R}{L}}{s + \frac{R}{L}}\right| = 0$$

The Bode plot of the transfer function $\quad H(s) = \frac{\frac{R}{L}}{s + \frac{R}{L}}$ is obtained as Fig. 9.4. There is just one pole element $\frac{R}{L}$ with unity gain.

Utilizing the Bode diagram to identify the cutoff frequency requires identifying the frequency at which a $-3$ dB drop from the maximum amplitude occurs. Drawing a straight horizontal line of $-3$ dB cuts the amplitude and identifies the frequency $\omega_c = \frac{R}{L}$.

The circuit phase changes from $0°$ to $-90°$, and at the pole's location (cutoff frequency), the phase reaches $-45°$.

The cutoff frequency can also be obtained from the output ratio over input when it reaches 70.7%.

**Fig. 9.4** Frequency response of a non-ideal first-order low-pass filter. For the frequencies above the cutoff frequency, the power of signals drops below an effective level. Hence, they are called rejected. Magnitude in this frequency response is shown in terms of decibel. At the cutoff frequency, a −3 dB (drop) is observed at the magnitude of the signals. Note that the phase shift reaches a −45° drop by the cutoff frequency and is saturated to −90° at frequencies a decade higher than $\omega_c$ (Fig. 9.5)

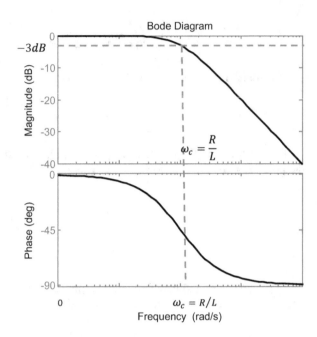

**Fig. 9.5** The low-pass filter frequency response is formed with an *RL* circuit. The magnitude of the signal is shown in absolute values. This shows the ratio of the voltages observed at the output. A 0.707 or 70.7% ratio determines the cutoff frequency. Note that the phase shift reaches a −45° drop by the cutoff frequency and is saturated to −90° at frequencies a decade higher than $\omega_c$

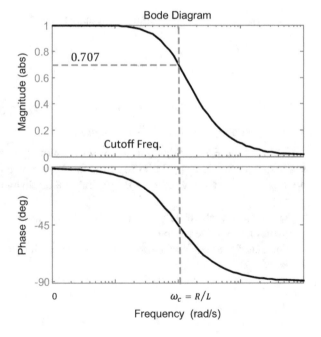

**Example 9.9** Design an *RL* LPF to cut frequencies higher than 1000 rad/s. Consider a 20-mH inductor (Fig. 9.6).

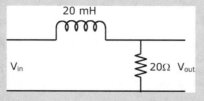

**Fig. 9.6** Circuit of Example 9.9

*Solution.* An *RL* LPF circuit has the cutoff frequency $\omega_c = \frac{R}{L}$ which is desired to be $1k$ rad/s.

$$\frac{R}{20e - 3} = 1000$$

This results in

$$R = 20 \ \Omega$$

The transfer function of the circuit is

$$H(s) = \frac{1000}{s + 1000}.$$

The frequency response of this transfer function is shown in Fig. 9.7.

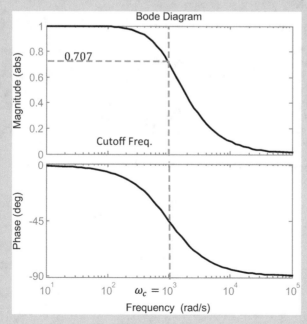

**Fig. 9.7** Frequency response of the circuit Example 9.9

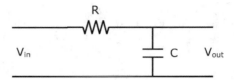

**Fig. 9.8** An *RC* circuit to form a low-pass filter. Capacitors pass high frequencies. If they are connected to direct all high frequencies to the ground, the output equals a short circuit (or very low impedance) at high frequencies. Hence, the voltage measured at the output is nearly zero at high frequencies. A combination of resistance and capacitance determines the cutoff frequency

## *First-Order RC Low-Pass Filter*

As explained earlier, capacitors conduct high frequencies. If a capacitor is connected such that the frequencies it passes are connected to the ground, then only low frequencies reach the output. Figure 9.8 shows an *RC* circuit that forms a low-pass filter.

### Identifying the Cutoff Frequency from the Ratio of the Output Over Input

At the cutoff frequency, the amplitude of the transfer function becomes 70.7% of the maximum $\frac{1}{\sqrt{2}} H_{max}$. Therefore,

$$|H(j\omega)|_{\omega=\omega_c} = \frac{1}{\sqrt{2}} H_{max}$$

Maximum output can reach 1. Therefore,

$$|H(j\omega)|_{\omega=\omega_c} = \frac{V_o(j\omega)}{V_{in}(j\omega)} = \left|\frac{\frac{1}{RC}}{j\omega_c + \frac{1}{RC}}\right| = \frac{1}{\sqrt{2}}$$

$$\frac{\frac{1}{RC}}{\sqrt{\omega_c{}^2 + \left(\frac{1}{RC}\right)^2}} = \frac{1}{\sqrt{2}}$$

Solving for $\omega_c$ results in

$$\frac{\left(\frac{1}{RC}\right)^2}{\omega_c{}^2 + \left(\frac{1}{RC}\right)^2} = \frac{1}{2}$$

$$\omega_c = \pm\frac{1}{RC}$$

which positive and negative frequencies show the same frequency amplitude. Therefore,

$$\omega_c = \frac{1}{RC} \& f_c = \frac{1}{2\pi RC}$$

**Note 9.2** The cutoff frequency of a first-order filter is the inverse of its time constant. In this $RC$ circuit, the time constant is $\tau = RC$; therefore, the cutoff frequency is $\omega_c = \frac{1}{\tau}$ or $\omega_c = \frac{1}{RC}$.

**Using Laplace Transform to Find the Cutoff Frequency**

The transfer function is as follows:

$$\frac{V_o}{V_{in}} = \frac{\frac{1}{sC}}{R + \frac{1}{sC}} = \frac{1}{RCs + 1} = \frac{\frac{1}{RC}}{s + \frac{1}{RC}}$$

This transfer function has no zeros and one pole at $s + \frac{1}{RC} = 0$ or

$$s = -\frac{1}{RC}$$

The amplitude of the Bode diagram drops at the pole's location by 3 dB. This qualifies the pole $\left(\frac{1}{RC}\right)$ to be the cutoff frequency. Therefore,

$$j\omega = -\frac{1}{RC}$$

$$\omega_c = \frac{1}{RC} \left(\frac{rad}{s}\right)$$

To obtain an approximate frequency response of the circuit, the amplitude can be checked in two extreme frequencies of $\omega = 0$ and $\omega = \infty$ as follows:

$$\left|\frac{V_o}{V_{in}}(j0)\right| = \lim_{s \to 0} \left|\frac{\frac{1}{RC}}{s + \frac{1}{RC}}\right| = 1$$

This means 100% of the input signal appears in the output.

$$\left|\frac{V_o}{V_{in}}(j\infty)\right| = \lim_{s \to \infty} \left|\frac{\frac{1}{RC}}{s + \frac{1}{RC}}\right| = 0$$

Bode plot of the transfer function $H(s) = \frac{\frac{1}{RC}}{s + \frac{1}{RC}}$ is obtained as Fig. 9.8. There is just one element of the pole $\frac{1}{RC}$ with unity gain (Figs. 9.9 and 9.10).

**Fig. 9.9** The frequency response of a low-pass filter built by an *RL* or an *RC* circuit is similar. The parametric expression of the cutoff frequency might be different, and two different circuits might have identical frequency responses. The magnitude at this frequency response is shown in terms of decibel. At the cutoff frequency, a $-3$ dB (drop) is observed at the magnitude of the signals. Note that the phase shift reaches a $-45°$ drop by the cutoff frequency and is saturated to $-90°$ at frequencies a decade higher than $\omega_c$

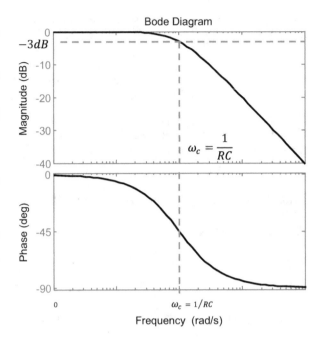

**Fig. 9.10** The low-pass filter frequency response is formed with an *RC* circuit. The magnitude of the signal is shown in absolute values. This shows the ratio of the voltages observed at the output. A 0.707 or 70.7% ratio determines the cutoff frequency. Note that the phase shift reaches a $-45°$ drop by the cutoff frequency and is saturated to $-90°$ at frequencies a decade higher than $\omega_c$

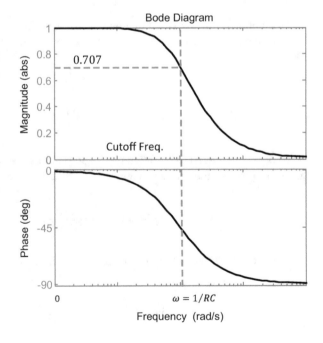

Utilizing the Bode diagram to identify the cutoff frequency requires identifying the frequency at which a $-3$ dB drop from the maximum amplitude occurs. Drawing a straight line of $-3$ dB cuts the amplitude and identifies the frequency $\omega = \frac{1}{RC}$.

The circuit phase changes from $0°$ to $-90°$, and at the pole's location (cutoff frequency), the phase reaches $-45°$.

The cutoff frequency can also be obtained from the output ratio over input when it reaches 70.7%.

**Example 9.10 Design an *RC* LPF to cut frequencies higher than 1000 rad/s. Consider a 20-μF capacitor (Fig. 9.11).**

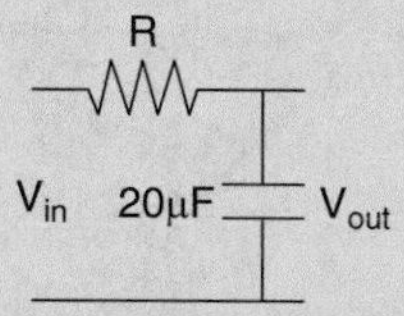

**Fig. 9.11** Circuit for Example 9.10

*Solution.* An *RC* LPF circuit has the cutoff frequency $\omega_c = \frac{1}{RC}$ which is desired to be $1k$ rad/s.

$$\frac{1}{R20e-6} = 1000$$

This results in

$$R = 50 \ \Omega$$

The transfer function of the circuit is

$$H(s) = \frac{1000}{s+1000}.$$

(continued)

**Example 9.10** (continued)

The frequency response of the filter is shown in Fig. 9.12.

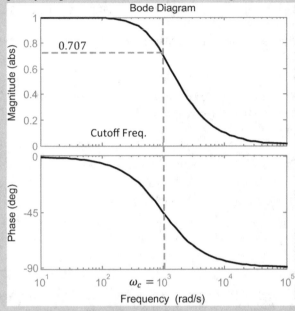

**Fig. 9.12** Frequency response of the circuit for Example 9.10

**Note 9.3** First-order low-pass filters with a cutoff frequency of $\omega_c$ have a transfer function template of

$$H(s) = \frac{\omega_c}{s + \omega_c}.$$

**Example 9.11 Design a LPF to filter out frequencies higher than 200 Hz.**
*Solution.* The template for a LPF is

$$H(s) = \frac{\omega_c}{s + \omega_c}$$

$$\omega_c = 2\pi f_c = 2\pi \times 200 = 1256.6 \ \frac{\text{rad}}{\text{s}}$$

Therefore, the filter becomes

$$H(s) = \frac{1256.6}{s + 1256.6}$$

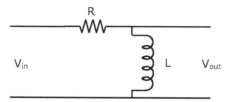

**Fig. 9.13** An *RL* circuit forms a high-pass filter. Inductors pass low frequencies. If they are connected to direct all low frequencies to the ground, the output equals a short circuit (or very low impedance) at low frequencies. Hence, the voltage measured at the output is nearly zero at low frequencies. A combination of the resistance and inductance determines the cutoff frequency

## *First-Order High-Pass Filter*

A HPF is a circuit that eliminates frequencies below the cutoff frequency and allows passing signals of a higher frequency than the cutoff frequency. This characteristic can be obtained using an inductor- or a capacitor-based circuit.

### *RL* **HPF**

A first-order *RL* HPF is shown in Fig. 9.13.

### **Identifying the Cutoff Frequency from the Ratio of the Output Over Input**

Obtaining cutoff frequency using the $\frac{1}{\sqrt{2}} H_{max}$ rule:

$$|H(j\omega)|_{\omega=\omega_c} = \frac{1}{\sqrt{2}} H_{max}$$

Maximum output can reach 1. Therefore,

$$|H(j\omega)|_{\omega=\omega_c} = \frac{V_o(j\omega)}{V_{in}(j\omega)} = \left|\frac{j\omega_c}{j\omega_c + \frac{R}{L}}\right| = \frac{1}{\sqrt{2}}$$

$$\frac{\omega_c}{\sqrt{\omega_c^2 + \left(\frac{R}{L}\right)^2}} = \frac{1}{\sqrt{2}}$$

Solving for $\omega_c$ results in

$$\frac{\omega_c{}^2}{\omega_c{}^2 + \left(\frac{R}{L}\right)^2} = \frac{1}{2}$$

$$\omega_c = \pm \frac{R}{L}$$

where positive and negative frequencies show the same frequency amplitude. Therefore,

$$\omega_c = \frac{R}{L}$$

### Using Laplace Transform to Find the Cutoff Frequency

The inductor passes low frequencies to the ground and allows a path for high-frequency signals to pass to the output. The circuit transfer function is obtained as

$$H(s) = \frac{sL}{R + sL} = \frac{s}{s + \frac{R}{L}}$$

The circuit has one zero at the origin $s = 0$, and one pole of the circuit is $s + \frac{R}{L} = 0$ or $s = -\frac{R}{L}$. The cutoff frequency is, therefore,

$$\omega_c = \frac{R}{L}\left(\frac{\text{rad}}{\text{s}}\right)$$

Replacing this frequency in the transfer function results in

$$H(s) = \frac{s}{s + \omega_c}$$

Evaluating the transfer function on two extreme frequencies results in

$$\left|\frac{V_o}{V_{in}}(j0)\right| = \lim_{s \to 0}\left|\frac{s}{s + \frac{R}{L}}\right| = 0$$

$$\left|\frac{V_o}{V_{in}}(j\infty)\right| = \lim_{s \to \infty}\left|\frac{s}{s + \frac{R}{L}}\right| = 1$$

The circuit blocks all low frequencies and passes 100% of the high-frequency signals to the output.

The Bode plot of the transfer function $H(s) = \frac{s}{s + \frac{R}{L}}$ is obtained as Fig. 9.13. There is just one pole element $\frac{R}{L}$ with unity gain (Fig. 9.14). Figure 9.15 shows the frequency response with normalized amplitude.

**Fig. 9.14** The frequency response of a high-pass filter by an *RL* circuit. The magnitude at this frequency response is shown in terms of decibel. At the cutoff frequency, a $-3$ dB (drop) is observed at the magnitude of the signals. Note that the phase shift starting at $+90°$ reaches $45°$ by the cutoff frequency and is saturated to $0°$ at frequencies a decade higher than $\omega_c$

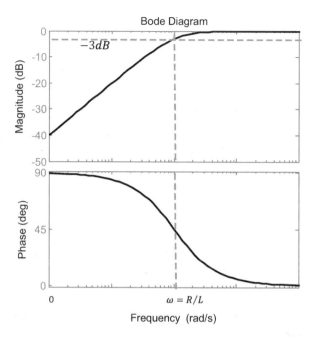

**Fig. 9.15** The frequency response of a high-pass filter is formed with an *RL* circuit. The magnitude of the signal is shown in absolute values. This shows the ratio of the voltages observed at the output. A 0.707 or 70.7% ratio determines the cutoff frequency. Note that the phase shift starting at $+90°$ reaches $45°$ by the cutoff frequency and is saturated to $0°$ at frequencies a decade higher than $\omega_c$

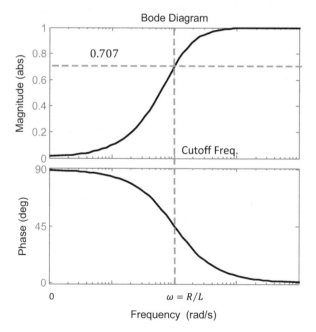

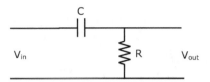

**Fig. 9.16** An *RC* circuit to form a high-pass filter. Capacitors pass high frequencies. The output equals the source voltage at high frequencies if connected directly between the source and the load. Hence, the voltage measured at the output is ideally the source voltage. A combination of resistance and capacitance determines the cutoff frequency

## *First-Order RC High-Pass Filter*

Capacitors pass high-frequency signals and block low frequencies. Using a capacitor to bridge the input to the output of a resistive load provides a path for high-frequency signals to pass and low-frequency signals to be rejected. Figure 9.16 shows an *RC* circuit that behaves as a high-pass filter.

### Identifying the Cutoff Frequency from the Ratio of the Output Over Input

Obtaining cutoff frequency using the $\frac{1}{\sqrt{2}} H_{\max}$ rule:

$$|H(j\omega)|_{\omega=\omega_c} = \frac{1}{\sqrt{2}} H_{\max}$$

Maximum output can reach 1. Therefore,

$$|H(j\omega)|_{\omega=\omega_c} = \frac{V_o(j\omega)}{V_{in}(j\omega)} = \left|\frac{j\omega_c}{j\omega_c + \frac{1}{RC}}\right| = \frac{1}{\sqrt{2}}$$

$$\frac{\omega_c}{\sqrt{\omega_c^2 + \left(\frac{1}{RC}\right)^2}} = \frac{1}{\sqrt{2}}$$

Solving for $\omega_c$ results in

$$\frac{\omega_c^2}{\omega_c^2 + \left(\frac{1}{RC}\right)^2} = \frac{1}{2}$$

$$\omega_c = \pm \frac{1}{RC}$$

where the positive frequency is acceptable. Therefore,

$$\omega_c = \frac{1}{RC}$$

## Using Laplace Transform to Find the Cutoff Frequency

The circuit transfer function is obtained as

$$H(s) = \frac{R}{R + \frac{1}{sC}} = \frac{RCs}{1 + RCs} = \frac{s}{s + \frac{1}{RC}}$$

The circuit has one zero at the origin $s = 0$, and one pole of the circuit is $s + \frac{1}{RC} = 0$, or $s = -\frac{1}{RC}$. The cutoff frequency is, therefore,

$$\omega_c = \frac{1}{RC} \left(\frac{\text{rad}}{\text{s}}\right)$$

Replacing this frequency in the transfer function results in

$$H(s) = \frac{s}{s + \omega_c}$$

Evaluating the transfer function on two extreme frequencies results in

$$\left|\frac{V_o}{V_{in}}(j0)\right| = \lim_{s \to 0}\left|\frac{s}{s + \frac{1}{RC}}\right| = 0$$

$$\left|\frac{V_o}{V_{in}}(j\infty)\right| = \lim_{s \to \infty}\left|\frac{s}{s + \frac{1}{RC}}\right| = 1$$

The circuit blocks all low frequencies and passes 100% of the high-frequency signals to the output.

The Bode plot of the transfer function $H(s) = \frac{s}{s + \frac{1}{RC}}$ is obtained as Fig. 9.17. There is just one pole element $\frac{1}{RC}$ with unity gain (Fig. 9.18).

**Note 9.4** A first-order HPF with a cutoff frequency of $\omega_c$ has a transfer function of

$$H(s) = \frac{s}{s + \omega_c}.$$

**Fig. 9.17** The frequency response of a high-pass filter by an *RC* circuit. Magnitude at this frequency response is shown in terms of absolute values. At the cutoff frequency, a −3 dB (drop) is observed at the magnitude of the signals. Note that the phase shift starting at +90° reaches 45° by the cutoff frequency and is saturated to 0° at frequencies a decade higher than $\omega_c$

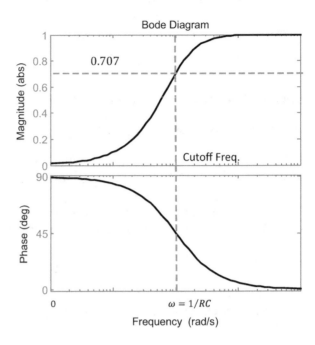

**Fig. 9.18** The frequency response of a high-pass filter built by an *RL* or an *RC* circuit is similar. The parametric expression of the cutoff frequency might be different, and two different circuits might have identical frequency responses. The frequency response of a high-pass filter by an *RC* circuit is shown. The magnitude at this frequency response is shown in terms of decibel. At the cutoff frequency, a −3 dB (drop) is observed at the magnitude of the signals. Note that the phase shift starting at +90° reaches 45° by the cutoff frequency and is saturated to 0° at frequencies a decade higher than $\omega_c$

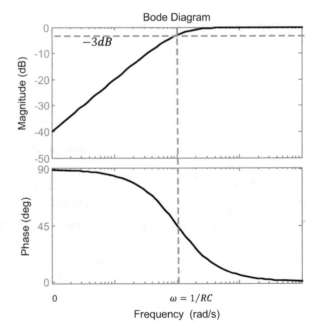

**Example 9.12 Design a HPF to pass frequencies higher than 1500 Hz.**
*Solution.* Considering a first-order HPF template, the filter has a cutoff
frequency of

$$\omega_c = 2\pi f_c = 2\pi \times 1500 = 9429.7 \ \frac{\text{rad}}{\text{s}}$$

Therefore, the transfer function becomes

$$H(s) = \frac{s}{s + 9429.7}$$

This circuit can be realized using *RL* or *RC* elements. Each of these
frequencies results in a specific cutoff frequency expression as follows:

The cutoff frequency of an *RL* circuit is the inverse of the time constant
$\omega_c = \frac{R}{L} = 9424.7$. Considering $L = 1$ mH results in $R = 9.424 \ \Omega$.

In an *RC* circuit, the cutoff frequency becomes $\omega_c = \frac{1}{RC} = 9424.7$. Considering $C = 1 \ \mu$F results in $R = 106.10 \ \Omega$.

## Analysis of *LC* Circuits

### *Parallel* **LC** *Circuit*

Consider a parallel connection of inductor $L$ and capacitor $C$ shown in Fig. 9.19.

**Identifying the Cutoff Frequency from the Ratio of the Output Over**
**Input**

The impedance of this parallel unit is obtained as follows:

$$Z = \frac{1}{j\omega C} \left\| j\omega L = \frac{\frac{1}{j\omega C} j\omega L}{\frac{1}{j\omega C} + j\omega L} = \frac{\frac{1}{C} j\omega}{-\omega^2 + \frac{1}{LC}} \right.$$

This circuit has a resonant frequency of $\frac{1}{\sqrt{LC}}$ at which it acts as an open circuit
with impedance reaching infinite:

**Fig. 9.19** A parallel *LC*
circuit to be used as a
building block forming
other types of filters

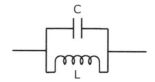

$$\lim_{\omega \to \frac{1}{\sqrt{LC}}} \left( \frac{1}{j\omega C} \middle\| j\omega L \right) = \infty$$

This parallel unit reaches zero impedance at two extreme frequencies as follows:

$$\lim_{\omega \to 0} \left( \frac{1}{j\omega C} \middle\| j\omega L \right) = 0 \, \& \lim_{\omega \to \infty} \left( \frac{1}{j\omega C} \middle\| j\omega L \right) = 0$$

**Parallel *LC* Circuit Using Laplace**

The impedance of this parallel unit is obtained as follows:

$$Z = \frac{1}{sC} \middle\| sL = \frac{\frac{1}{sC} sL}{\frac{1}{sC} + sL} = \frac{\frac{1}{C} s}{s^2 + \frac{1}{LC}}$$

This circuit has a resonant frequency $\frac{1}{\sqrt{LC}}$ and a pole at the same frequency. Therefore, at the resonant frequency, this unit will act as an open circuit with impedance reaching infinite:

$$\lim_{s \to \frac{1}{\sqrt{LC}}} \left( \frac{1}{sC} \middle\| sL \right) = \infty$$

This parallel unit reaches zero impedance at two extreme frequencies as follows:

$$\lim_{s \to 0} \left( \frac{1}{sC} \middle\| sL \right) = 0 \, \& \lim_{s \to \infty} \left( \frac{1}{sC} \middle\| sL \right) = 0$$

This unit can be utilized in circuits to bridge the input to output or connect across the output port. Either case, it should be used with this characteristic in mind that it becomes an open circuit at the resonant frequency.

## *Series LC Circuit*

Consider a series connection of inductor $L$ and capacitor $C$ shown in Fig. 9.20.

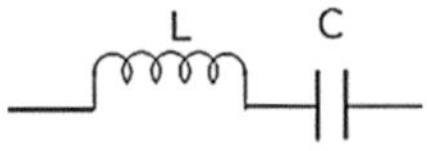

**Fig. 9.20**  A series *LC* circuit to be used as a building block forming other types of filters

**Identifying the Cutoff Frequency from the Ratio of the Output Over Input**

The impedance of this series unit is obtained as follows:

$$Z = \frac{1}{j\omega C} + j\omega L = \frac{1 - LC\omega^2}{j\omega C} = \frac{\frac{1}{LC} - \omega^2}{j\omega \frac{1}{L}}$$

This parallel unit reaches zero impedance at two extreme frequencies as follows:

$$\lim_{\omega \to 0}\left(\frac{1}{j\omega C}\middle\| j\omega L\right) = \infty \,\&\, \lim_{\omega \to \infty}\left(\frac{1}{j\omega C}\middle\| j\omega L\right) = \infty$$

This circuit has a resonant frequency of $\frac{1}{\sqrt{LC}}$ at which it acts as a short circuit with impedance reaching zero:

$$\lim_{\omega \to \frac{1}{\sqrt{LC}}}\left(\frac{1}{j\omega C} + j\omega L\right) = 0$$

**Series *LC* Circuit Using Laplace**

The impedance of this series unit is obtained as follows:

$$Z = \frac{1}{sC} + sL = \frac{1 + LCs^2}{sC} = \frac{\frac{1}{LC} + s^2}{\frac{1}{L}s}$$

This parallel unit reaches zero impedance at two extreme frequencies as follows:

$$\lim_{s \to 0}\left(\frac{1}{sC}\middle\| sL\right) = \infty \,\&\, \lim_{s \to \infty}\left(\frac{1}{sC}\middle\| sL\right) = \infty$$

This circuit has a resonant frequency $\frac{1}{\sqrt{LC}}$ and a zero at the same frequency. Therefore, this unit acts as a short circuit at the resonant frequency with an impedance reaching zero.

$$\lim_{s \to \frac{1}{\sqrt{LC}}}\left(\frac{1}{sC} + sL\right) = 0$$

This unit can be utilized in circuits to bridge the input to output or connect across the output port. Either case, it should be used with this characteristic in mind that it becomes a short circuit at the resonant frequency.

## *Summary of LC Series and Parallel Circuits*

| Circuit | $\omega \to 0$ | $\omega = \omega_0$ | $\omega \to \infty$ |
|---|---|---|---|
| C / L | $Z = 0\,\Omega$ <br> Short | $Z = \infty\,\Omega$ <br> Open | $Z = 0\,\Omega$ <br> Short |
| L / C | $Z = \infty\,\Omega$ <br> Open | $Z = 0\,\Omega$ <br> Short | $Z = \infty\,\Omega$ <br> Open |

## Band-Pass Filters

A band-pass filter is designed to pass signals of frequencies in a specific range (bandwidth) and block the rest. A BPF reaches its highest amplitude at a frequency in the bandwidth BW known as resonant frequency $\omega_0$. There are two cutoff frequencies, $\omega_{c1}$ and $\omega_{c2}$, to form the passband and determine the bandwidth as

$$BW = \omega_{c2} - \omega_{c1}.$$

Therefore, a BPF has a minimum second-order circuit structure. Figure 9.21 shows the frequency response of a band-pass filter:

**Fig. 9.21** Typical frequency response of a band-pass filter. The cutoff frequencies and the resonant frequency are shown. The frequency response reaches its maximum amplitude and phase $45^\circ$ at the resonant frequency

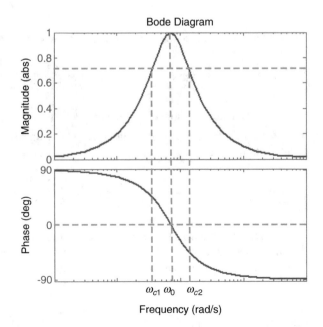

**Fig. 9.22** Magnitude and
phase variation as a function
of frequency and the
damping factor $\zeta$ of a BPF.
As the damping factor is
increased, the bandwidth of
the filter is increased, and
the rate of the phase shift
transition at $45°$ point is
decreased

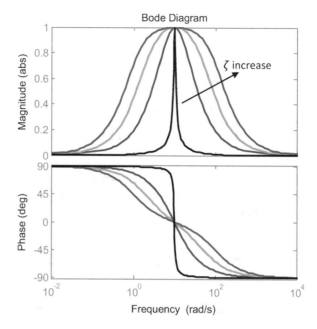

As an accurate measure, the resonant frequency $\omega_0$ is a geometrical average of
$\omega_{c1}$ and $\omega_{c2}$ as

$$\omega_0 = \sqrt{\omega_{c1}\omega_{c2}}$$

An approximate considers the resonant and arithmetic average in the middle of
bandwidth. Therefore, each cutoff frequency is approximately located in $\frac{BW}{2}$ from the
resonant $\omega_0$. Therefore,

$$\omega_{c1,2} = \omega_0 \mp \frac{BW}{2}$$

Several circuits behave similarly to a band-pass filter. However, a second-order
circuit with resonant frequency requires an inductor and a capacitor. Two circuits are
proposed considering the parallel and series connection of these elements and their
operation at the resonant frequency.

The frequency response of BPF is shown in Fig. 9.22.

## *BPF Circuit 1: Using LC Series*

Series connection of *LC* shows a short circuit at the resonant frequency. Given the
required bandwidth, this filter can pass a desired range of frequencies. To obtain this

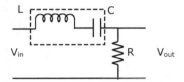

**Fig. 9.23** Using the series $LC$ block to build a BPF. The series $LC$ circuit shows a short circuit at the resonant, passing input signals to the output. The impedance increases as the input signal frequencies move away from the resonant, resulting in attenuation in the output voltage. This results in a band-pass filter

characteristic, a series $LC$ circuit is utilized to connect the input to the output, where the output is the voltage drop observed at resistance in the terminals. Figure 9.23 shows the circuit.

### Identifying the Cutoff Frequency from the Ratio of the Output Over Input

Cutoff frequencies can also be obtained utilizing the circuit parameters as follows:

$$|H(j\omega)|_{\omega=\omega_c} = \frac{1}{\sqrt{2}} \; H_{max}$$

The maximum output of the filter occurs at the resonant frequency $\omega_0$, obtained as follows:

$$H_{max} = |H(j\omega)|_{\omega=\omega_0} = \left| \frac{R}{R + j\omega L + \frac{1}{j\omega C}} \right| = 1$$

Therefore,

$$|H(j\omega)|_{\omega=\omega_c} = \left| \frac{RC\omega}{\sqrt{\left( RC\omega \right)^2 + \left( LC\omega^2 - 1 \right)^2}} \right| = \frac{1}{\sqrt{2}}$$

$$\frac{RC\omega_c}{\sqrt{\left( RC\omega_c \right)^2 + \left( LC\omega_c^2 - 1 \right)^2}} = \frac{1}{\sqrt{2}}$$

Solving for $\omega_c$ results in

$$\omega_{c1,2} = \mp \frac{R}{2L} + \sqrt{\left(\frac{R}{2L}\right)^2 + \frac{1}{LC}}$$

Replacing the BW and resonant frequency,

$$\omega_{c1,2} = \mp \frac{BW}{2} + \sqrt{\left(\frac{BW}{2}\right)^2 + \omega_0{}^2}$$

Considering $Q = \frac{\omega_0}{BW}$,

$$\omega_{c1,2} = \omega_0 \left( \mp \frac{1}{2Q} + \sqrt{1 + \left(\frac{1}{2Q}\right)^2} \right)$$

Utilizing these cutoff frequencies, the bandwidth can be obtained as follows:

$$BW = \omega_{c2} - \omega_{c1} = \frac{R}{L}$$

$$Q = \frac{\omega_0}{BW} = \frac{\frac{1}{\sqrt{LC}}}{\frac{R}{L}} = \sqrt{\frac{L}{CR^2}}$$

## Using Laplace Transform to Find the Cutoff Frequency

The circuit has a transfer function defined as

$$\frac{V_o}{V_{in}} = \frac{R}{R + sL + \frac{1}{sC}} = \frac{RCs}{RCs + LCs^2 + 1} = \frac{\frac{R}{L}s}{s^2 + \frac{R}{L}s + \frac{1}{LC}}$$

As the transfer function shows, there is one zero at the origin, resulting in no pass of low frequencies. There are two poles observed as $\omega_{c1}$ and $\omega_{c2}$. At these poles' locations, the transfer function's amplitude drops by $-3$ dB from the peak $H_{max}$ observed at the resonant $\omega_0$.

Compared to the denominator of a standard second-order system $H(s) = \frac{k}{s^2 + 2\zeta\omega_0 s + \omega_0{}^2}$, it is obtained that

$$\omega_0{}^2 = \frac{1}{LC}$$

$$2\zeta\omega_0 = \frac{R}{L} = BW$$

The quality of a filter depends on the shape and sharpness of the filtration. Known as $Q$, the quality factor is obtained as follows:

$$Q = \frac{\omega_0}{\text{BW}} = \frac{1}{2\zeta}$$

Therefore, the transfer function of a second-order BPF can be written as a template of

$$H(s) = \frac{V_o}{V_{in}} = \frac{\text{BW}s}{s^2 + \text{BW}s + \omega_0{}^2}$$

The location of cutoff frequencies is obtained, as approximate, as follows:

$$\omega_{c1,2} \approx \omega_0 \mp \frac{\text{BW}}{2}$$

The exact location of the poles is obtained as follows:

- If $0 < \zeta < 1$, the poles of the filter are located at $s_{1,2} = -\zeta\omega_0 \pm j\omega_d$. For very small damping factors, $\zeta \ll 1$, $\omega_d \approx \omega_0$. Therefore, poles and cutoff frequencies match $s_{1,2} \approx \omega_{c1,2}$. Note that $\omega_d$ is the damping frequency $\omega_d = \omega_0\sqrt{1 - \zeta^2}$.
- If $\zeta > 1$, the filter's poles are located at $\qquad s_{1,2} = -\zeta\omega_0 \pm \omega_0\sqrt{\zeta^2 - 1}$ or $s_{1,2} = -\frac{\text{BW}}{2} \pm \omega_d$.

*Design a BPF with the resonant frequency* 100 rad/s *and bandwidth of* 10 rad/s. *Solution.* BW $= 10$ and $\omega_0 = 100$; therefore,

$$H(s) = \frac{V_o}{V_{in}} = \frac{10s}{s^2 + 10s + 100^2} = \frac{10s}{s^2 + 10s + 10000}$$

---

**Example 9.13**  The damping of a BPF is designed to be $\zeta = 0.01$. What is the $Q$ of the circuit?
   *Solution.*

$$Q = \frac{1}{2\zeta} = \frac{1}{2 \times 0.01} = 50$$

*Find the transfer function of a BPF with cutoff frequencies of* $\omega_{c1} = 200$ rad/s *and* $\omega_{c2} = 1500$ rad/s.

(continued)

**Example 9.13** (continued)
*Solution.* The bandwidth can be obtained as

$$BW = \omega_{c2} - \omega_{c1} = 1500 - 200 = 1300 \ \frac{rad}{s}$$

The resonant frequency is obtained as

$$\omega_0 = \sqrt{\omega_{c1}\omega_{c2}} = \sqrt{200 \times 1500} = 547.7 \ \frac{rad}{s}$$

Knowing these two factors, the transfer function is obtained as

$$H(s) = \frac{V_o}{V_{in}} = \frac{BWs}{s^2 + BWs + \omega_0{}^2} = \frac{1300s}{s^2 + 1300s + 547.7^2}$$

## *BPF Circuit 2: Using LC Parallel*

The parallel connection of *LC* shows an open circuit at the resonant frequency. Given the required bandwidth, this filter can block a desired range of frequencies from being connected to the ground. A parallel *LC* circuit is utilized to connect across the output to obtain this characteristic. A resistor bridges the input to the output to prevent short circuit connection of signals at the resonant. Figure 9.24 shows the circuit.

### Identifying the Cutoff Frequency from the Ratio of the Output Over Input

Cutoff frequencies can also be obtained utilizing the circuit parameters as follows:

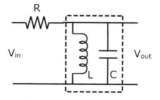

**Fig. 9.24** Using a parallel *LC* block to build a BPF. The parallel *LC* unit exhibits an open circuit at the resonant frequency and a dropping impedance at other frequencies. Therefore, a band-pass behavior is obtained. The overall behavior is shown as a band-pass filter

$$|H(j\omega)|_{\omega = \omega_c} = \frac{1}{\sqrt{2}} \, H_{\text{max}}$$

The maximum output of the filter occurs at the resonant frequency $\omega_0$, obtained as follows:

$$H_{\text{max}} = |H(j\omega)|_{\omega = \omega_0} = \left| \frac{\frac{1}{j\omega C} \left\| j\omega L \right.}{R + \frac{1}{j\omega C} \left\| j\omega L \right.} \right| = 1$$

Therefore,

$$|H(j\omega)|_{\omega = \omega_c} = \left| \frac{\frac{\frac{1}{C} j\omega}{-\omega^2 + \frac{1}{LC}}}{R + \frac{\frac{1}{C} j\omega}{-\omega^2 + \frac{1}{LC}}} \right| = \frac{1}{\sqrt{2}}$$

$$\frac{\frac{\frac{1}{C} \omega_c}{-\omega_c^2 + \frac{1}{LC}}}{\sqrt{R^2 + \left( \frac{\frac{1}{C} \omega_c}{-\omega_c^2 + \frac{1}{LC}} \right)^2}} = \frac{1}{\sqrt{2}}$$

Solving for $\omega_c$ results in

$$\omega_{c1,2} = \mp \frac{1}{2RC} + \sqrt{\left( \frac{1}{2RC} \right)^2 + \frac{1}{LC}}$$

Replacing from the circuit values,

$$\omega_{c1,2} = \mp \frac{\text{BW}}{2} + \sqrt{\left( \frac{\text{BW}}{2} \right)^2 + \omega_0^2}$$

Considering $Q = \frac{\omega_0}{\text{BW}}$,

$$\omega_{c1,2} = \omega_0 \left( \mp \frac{1}{2Q} + \sqrt{1 + \left( \frac{1}{2Q} \right)^2} \right)$$

Utilizing these cutoff frequencies, the bandwidth can be obtained as follows:

$$\text{BW} = \omega_{c2} - \omega_{c1} = \frac{1}{RC}$$

$$Q = \frac{\omega_0}{\mathrm{BW}} = \frac{\frac{1}{\sqrt{LC}}}{\frac{1}{RC}} = \sqrt{\frac{CR^2}{L}}$$

## Using Laplace Transform to Find the Cutoff Frequency

The circuit has a transfer function defined as

$$\frac{V_o}{V_{in}} = \frac{sL\left\|\frac{1}{sC}\right.}{R + \left(sL\left\|\frac{1}{sC}\right.\right)} = \frac{\frac{sL}{LCs^2+1}}{R + \frac{sL}{LCs^2+1}} = \frac{sL}{RLCs^2 + Ls + R} = \frac{\frac{1}{RC}s}{s^2 + \frac{1}{RC}s + \frac{1}{LC}}$$

As the transfer function shows, there is one zero at the origin, resulting in no low-frequency signals passing. There are two poles observed as $\omega_{c1}$ and $\omega_{c2}$. At these poles' locations, the transfer function's amplitude drops by $-3$ dB from the peak $H_{max}$ observed at the resonant $\omega_0$.

Compared to the denominator of a standard second-order system $H(s) = \frac{k}{s^2+2\zeta\omega_0 s+\omega_0^2}$, it is obtained that

$$\omega_0^2 = \frac{1}{LC}$$

$$2\zeta\omega_0 = \frac{1}{RC} = \mathrm{BW}$$

The quality of a filter depends on the shape and sharpness of the filtration. Known as $Q$, the quality factor is obtained as follows:

$$Q = \frac{\omega_0}{\mathrm{BW}} = \frac{1}{2\zeta}$$

Therefore, the transfer function of a second-order BPF can be written as a template of

$$H(s) = \frac{V_o}{V_{in}} = \frac{\mathrm{BW}s}{s^2 + \mathrm{BW}s + \omega_0^2}$$

The location of cutoff frequencies is obtained, as approximate, as follows:

$$\omega_{c1,2} \approx \omega_0 \mp \frac{\mathrm{BW}}{2}$$

Cutoff frequencies can also be obtained utilizing the circuit parameters as follows:

$$|H(j\omega)|_{\omega=\omega_c} = \frac{1}{\sqrt{2}} H_{\text{max}}$$

Repeating the same procedure as in the $LC$ series results in

$$\omega_{c1,2} = \mp \frac{\text{BW}}{2} + \sqrt{\left(\frac{\text{BW}}{2}\right)^2 + \omega_0{}^2}$$

Replacing from the circuit values,

$$\omega_{c1,2} = \mp \frac{1}{2RC} + \sqrt{\left(\frac{1}{2RC}\right)^2 + \frac{1}{LC}}$$

Considering $Q = \frac{\omega_0}{\text{BW}}$,

$$\omega_{c1,2} = \omega_0 \left( \mp \frac{1}{2Q} + \sqrt{1 + \left(\frac{1}{2Q}\right)^2} \right)$$

Utilizing these cutoff frequencies, the bandwidth can be obtained as follows:

$$\text{BW} = \omega_{c2} - \omega_{c1} = \frac{1}{RC}$$

$$Q = \frac{\omega_0}{\text{BW}} = \frac{\frac{1}{\sqrt{LC}}}{\frac{1}{RC}} = \sqrt{\frac{CR^2}{L}}$$

## Band-Reject Filters

A band-reject filter is a circuit that passes signals outside a specific range and blocks the signals in the range, also known as bandwidths. The frequency response of a band-reject filter is shown in Fig. 9.25.

Considering the characteristics observed from the series and parallel $LC$ circuits, a BRF behavior can be obtained from these circuits as follows:

### BRF Circuit 1: LC Series

Consider that a series $LC$ circuit becomes a short circuit at the resonant frequency; placing this circuit parallel to the output results in a BRF. A resistor bridges the input to the output and prevents a short circuit current at the resonant frequency signals. The filter is shown in Fig. 9.26.

**Fig. 9.25** Magnitude and phase variation as a function of frequency and the damping factor $\zeta$ of a BRF. As the damping factor is increased, the bandwidth of the filter is increased, and the rate of the phase shift transition at $45°$ point is decreased

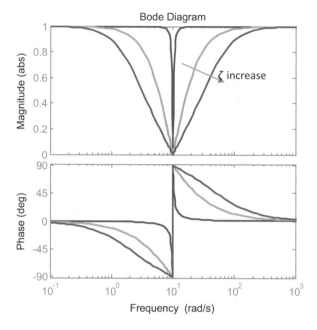

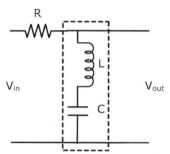

**Fig. 9.26** *LC* series block is utilized to build a BRF. The series *LC* element exhibits a short circuit at the resonant frequency that imposes zero volts at the output. The impedance is increased once the input signal frequency moves away from the resonant frequency, showing a band-reject response

## Identifying the Cutoff Frequency from the Ratio of the Output Over Input

Cutoff frequencies can also be obtained utilizing the circuit parameters as follows:

$$|H(j\omega)|_{\omega=\omega_c} = \frac{1}{\sqrt{2}} H_{max}$$

The ratio of the output voltage over the input voltage is obtained through a voltage divider as follows:

$$|H(j\omega)| = \frac{V_o(j\omega)}{V_{in}(j\omega)} = \left|\frac{\left(j\omega L + \frac{1}{j\omega C}\right)}{R + j\omega L + \frac{1}{j\omega C}}\right| = \left|\frac{1 - LC\omega^2}{1 - LC\omega^2 + j\omega RC}\right|$$

The maximum output of the filter occurs at DC, and the resonant frequency is obtained as follows:

$$H_{max} = |H(j\omega)|_{\omega=0} = \left|\frac{1 - LC\omega^2}{1 - LC\omega^2 + j\omega RC}\right| = 1$$

The cutoff frequencies are obtained as follows:

$$|H(j\omega)|_{\omega=\omega_c} = \left|\frac{1 - LC\omega_c^2}{1 - LC\omega_c^2 + j\omega_c RC}\right| = \frac{1}{\sqrt{2}}$$

$$\frac{1 - LC\omega_c^2}{\sqrt{\left(1 - LC\omega_c^2\right)^2 + (\omega_c RC)^2}} = \frac{1}{\sqrt{2}}$$

Solving for $\omega_c$ results in

$$\omega_{c1,2} = \mp\frac{R}{2L} + \sqrt{\left(\frac{R}{2L}\right)^2 + \frac{1}{LC}}$$

Replacing the circuit values, the cutoff frequencies are expressed as follows:

$$\omega_{c1,2} = \mp\frac{BW}{2} + \sqrt{\left(\frac{BW}{2}\right)^2 + \omega_0^2}$$

Considering $Q = \frac{\omega_0}{BW}$ yields

$$\omega_{c1,2} = \omega_0\left(\mp\frac{1}{2Q} + \sqrt{1 + \left(\frac{1}{2Q}\right)^2}\right)$$

Utilizing these cutoff frequencies, the bandwidth can be obtained as follows:

$$BW = \omega_{c2} - \omega_{c1} = \frac{R}{L}$$

$$Q = \frac{\omega_0}{BW} = \frac{\frac{1}{\sqrt{LC}}}{\frac{R}{L}} = \sqrt{\frac{L}{CR^2}}$$

## Using Laplace Transform to Find the Cutoff Frequencies

The circuit has a transfer function defined as

$$\frac{V_o}{V_{in}} = \frac{sL + \frac{1}{sC}}{R + sL + \frac{1}{sC}} = \frac{LCs^2 + 1}{RCs + LCs^2 + 1} = \frac{s^2 + \frac{1}{LC}}{s^2 + \frac{R}{L}s + \frac{1}{LC}}$$

As the transfer function shows, two zeros produce a short circuit output of signals at the resonant frequency. There are two poles observed which relate to $\omega_{c1}$ and $\omega_{c2}$. At these cutoff frequencies' locations, the transfer function's amplitude drops by $-3$ dB from the peak $H_{max}$.

Compared to the denominator of a standard second-order system $H(s) = \frac{k}{s^2 + 2\zeta\omega_0 s + \omega_0^2}$, it is obtained that

$$\omega_0^2 = \frac{1}{LC}$$

$$2\zeta\omega_0 = \frac{R}{L} = BW$$

The quality of a filter depends on the shape and sharpness of the filtration. Known as $Q$, the quality factor is obtained as follows:

$$Q = \frac{\omega_0}{BW} = \frac{1}{2\zeta}$$

Therefore, the transfer function of a second-order BPF can be written as follows:

$$H(s) = \frac{V_o}{V_{in}} = \frac{s^2 + \omega_0^2}{s^2 + BWs + \omega_0^2}$$

The location of cutoff frequencies is obtained, as approximate, as follows:

$$\omega_{c1,2} \approx \omega_0 \mp \frac{BW}{2}$$

The exact locations of poles are obtained as follows:

- *Underdamped: $0 < \zeta < 1$*

  - *The poles of the filter are located at $s_{1,2} = -\zeta\omega_0 \pm j\omega_d$.*
  - *If $\zeta "1$, then $\omega_d \approx \omega_0$. Poles and cutoff frequencies match $s_{1,2} \approx \omega_{c1,2}$.*
  - *$\omega_d$ is the damping frequency $\omega_d = \omega_0\sqrt{1 - \zeta^2}$.*

- *Overdamped: $\zeta > 1$*

  - *The poles of the filter are $s_{1,2} = -\zeta\omega_0 \pm \omega_0\sqrt{\zeta^2 - 1}$ or $s_{1,2} = -\frac{BW}{2} \pm \omega_d$.*

*Design a BRF with a resonant frequency of* 100 rad/s *and bandwidth of* 10 rad/s.
*Solution.* BW $= 10$ *and* $\omega_0 = 100$; therefore,

$$H(s) = \frac{V_o}{V_{in}} = \frac{s^2 + 100^2}{s^2 + 10s + 100^2} = \frac{s^2 + 100^2}{s^2 + 10s + 10000}$$

**Example 9.14 The damping of a BRF is designed to be $\zeta = 0.2$. What is the $Q$ of the circuit?**
*Solution.*

$$Q = \frac{1}{2\zeta} = \frac{1}{2 \times 0.2} = 2.5$$

*Find the transfer function of a BRF with cutoff frequencies of* $\omega_{c1} = 200$ rad/s, $\omega_{c2} = 1500$ rad/s.
*Solution.* The bandwidth can be obtained as

$$BW = \omega_{c2} - \omega_{c1} = 1500 - 200 = 1300 \ \frac{rad}{s}$$

The resonant frequency is obtained as

$$\omega_0 = \sqrt{\omega_{c1}\omega_{c2}} = \sqrt{200 \times 1500} = 574.7 \ \frac{rad}{s}$$

Knowing these two factors, the transfer function is obtained as

$$H(s) = \frac{V_o}{V_{in}} = \frac{s^2 + \omega_0^2}{s^2 + BWs + \omega_0^2} = \frac{s^2 + 574.7^2}{s^2 + 1300s + 574.7^2}$$

## *BRF Circuit 2: Using LC Parallel*

The parallel connection of *LC* demonstrates an open circuit at the resonant frequency. This filter blocks the desired range of frequencies in the bandwidth. It can also be used in the bridge connection from input to output. Figure 9.27 shows the circuit.

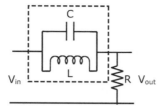

**Fig. 9.27** Using a parallel $LC$ block to build a BRF. The parallel $LC$ unit exhibits an open circuit at the resonant frequency and a dropping impedance at other frequencies. Therefore, a band-reject behavior is obtained

### Identifying the Cutoff Frequency from the Ratio of the Output Over Input

Cutoff frequencies can also be obtained utilizing the circuit parameters.

The ratio of the output over input voltages is obtained through a voltage divider as follows:

$$|H(j\omega)| = \frac{V_o}{V_{in}} = |\frac{R}{R + \frac{1}{j\omega C}||j\omega L}|$$

$$|H(j\omega)| = |\frac{R}{R + \frac{\frac{1}{C}j\omega}{-\omega^2 + \frac{1}{LC}}}|$$

$$|H(j\omega)| = |\frac{1 - LC\omega^2}{1 - LC\omega^2 + j\omega\frac{1}{RC}}|$$

$$|H(j\omega)|_{\omega = \omega_c} = \frac{1}{\sqrt{2}} H_{max}$$

Repeating the same procedure as in $LC$ parallel and considering the maximum amplitude, the cutoff frequencies can be obtained.

$$H_{max} = |H(j\omega)|_{\omega = 0} = |\frac{1 - LC\omega^2}{1 - LC\omega^2 + j\omega\frac{1}{RC}}| = 1$$

This results in

$$|H(j\omega)|_{\omega = \omega_c} = |\frac{1 - LC\omega_c^2}{1 - LC\omega_c^2 + j\omega_c\frac{1}{RC}}| = \frac{1}{\sqrt{2}}$$

$$\left|\frac{1 - LC\omega_c^2}{\sqrt{\left(1 - LC\omega_c^2\right)^2 + \left(\omega_c \frac{1}{RC}\right)^2}}\right| = \frac{1}{\sqrt{2}}$$

$$\omega_{c1,2} = \mp \frac{1}{2RC} + \sqrt{\left(\frac{1}{2RC}\right)^2 + \frac{1}{LC}}$$

Replacing the circuit equivalent values results in

$$\omega_{c1,2} = \mp \frac{BW}{2} + \sqrt{\left(\frac{BW}{2}\right)^2 + \omega_0^2}$$

Considering $Q = \frac{\omega_0}{BW}$ yields,

$$\omega_{c1,2} = \omega_0 \left( \mp \frac{1}{2Q} + \sqrt{1 + \left(\frac{1}{2Q}\right)^2} \right)$$

The bandwidth can be obtained as follows:

$$BW = \omega_{c2} - \omega_{c1} = \frac{1}{RC}$$

$$Q = \frac{\omega_0}{BW} = \frac{\frac{1}{\sqrt{LC}}}{\frac{1}{RC}} = \sqrt{\frac{CR^2}{L}}$$

**Using Laplace Transform to Find the Cutoff Frequency**

The circuit has a transfer function defined as

$$\frac{V_o}{V_{in}} = \frac{R}{R + \left(sL \| \frac{1}{sC}\right)} = \frac{R}{R + \frac{sL}{LCs^2 + 1}} = \frac{R\left(LCs^2 + 1\right)}{RLCs^2 + Ls + R} = \frac{RLCs^2 + R}{RLCs^2 + Ls + R}$$

$$\frac{V_o}{V_{in}} = \frac{s^2 + \frac{1}{LC}}{s^2 + \frac{1}{RC}s + \frac{1}{LC}}$$

As the transfer function shows, there are two complex conjugate zeros at the resonant frequency. This guarantees that the output reaches zero at the resonant frequency.

There are two poles observed which are related to $\omega_{c1}$ and $\omega_{c2}$. At these cutoff frequencies' locations, the transfer function's amplitude drops by $-3$ dB from the peak $H_{\max}$. Compared to the denominator of a standard second-order system $H(s) = \frac{k}{s^2 + 2\zeta\omega_0 s + \omega_0^2}$, it is obtained that

$$\omega_0^2 = \frac{1}{LC}$$

$$2\zeta\omega_0 = \frac{1}{RC} = \text{BW}$$

The quality of a filter depends on the shape and sharpness of the filtration. Known as $Q$, the quality factor is obtained as follows:

$$Q = \frac{\omega_0}{\text{BW}} = \frac{1}{2\zeta}$$

Therefore, the transfer function of a second-order BRF can be written as follows:

$$H(s) = \frac{V_o}{V_{\text{in}}} = \frac{s^2 + \omega_0^2}{s^2 + \text{BW}s + \omega_0^2}$$

The location of cutoff frequencies is obtained, as approximate, as follows:

$$\omega_{c1,2} \approx \omega_0 \mp \frac{\text{BW}}{2}$$

## Summary of Filters in Laplace

Transfer functions of filters discussed in this chapter are examples of a larger body of circuits that can perform frequency selectivity and realize filters. These transfer functions and their characteristics are summarized in this section. The following table shows the filters and their location of poles and zeros.

| Filter type | A transfer function | Circuit variations | Pole-zero map |
|---|---|---|---|
| Low-pass filter | $H(s) = \frac{\omega_c}{s + \omega_c}$ | RL. $\omega_c = \frac{R}{L}$ <br> RC. $\omega_c = \frac{1}{RC}$ | A pole on the real axis |

(continued)

| Filter type | A transfer function | Circuit variations | Pole-zero map |
|---|---|---|---|
| High-pass filter | $H(s) = \frac{s}{s+\omega_c}$ | $RL.\ \omega_c = \frac{R}{L}$  $RC.\ \omega_c = \frac{1}{RC}$ | A pole on the real axis. $s = -\omega_c$  A zero at the origin. $s = 0$ |
| Band-pass filter | $H(s) = \frac{BWs}{s^2+BWs+\omega_0^2}$ | **LC: Series**  $BW = \frac{R}{L}$  $Q = \sqrt{\frac{L}{CR^2}}$  **LC: Parallel**  $BW = \frac{1}{RC}$  $Q = \sqrt{\frac{CR^2}{L}}$ | $\zeta > 1$  Two poles on the real axis. $s_{1,2} = -\frac{BW}{2} \pm \omega_d$  A zero at the origin. $s = 0$  $0 < \zeta < 1$  Two comp. conj. poles. If $\zeta \gg 1 \rightarrow s_{1,2} \approx \omega_{c_{1,2}}$  A zero at the origin |
| Band-reject filter | $H(s) = \frac{s^2+\omega_0^2}{s^2+BWs+\omega_0^2}$ | **LC: Series**  $BW = \frac{R}{L}$  $Q = \sqrt{\frac{L}{CR^2}}$  **LC: Parallel**  $BW = \frac{1}{RC}$  $Q = \sqrt{\frac{CR^2}{L}}$ | $\zeta > 1$  $+j\omega_0$  $-j\omega_0$  Two poles on the real axis. $s_{1,2} = -\frac{BW}{2} \pm \omega_d$  Two comp. conj. zeros on the imaginary axis. $s = \pm j\omega_0$  $0 < \zeta < 1$  $+j\omega_0$  $-j\omega_0$  Two comp. conj. poles. If $\zeta \gg 1 \rightarrow s_{1,2} \approx \omega_{c_{1,2}}$  Two comp. conj. zeros on the imaginary axis. $s = \pm j\omega_0$ |

# Higher-Order Filters

The frequency response of an ideal filter is different from what has been observed from an actual circuit. Inaccuracies are mostly introduced around the cutoff frequency as a simple structure first-order filter might not be able to reduce enough the power of a selected signal. This means the bandwidth of a second-order BPF or BRF might not be as accurate as desired.

In previous sections, it also showed that BPF and BRF benefited from the existence of a resonant frequency observed from a second-order system. However, in the simplest form, the LPF and HPF were first-order, utilizing only one energy-storing element. This may result in an average performance from the transfer functions and the circuits. Increasing the order of a filter may improve its frequency response making it closer to an ideal performance with sharp drops at the frequencies close to the cutoff frequencies. There are multiple higher-order filters existing that have specific frequency responses. Since the second-order RLC circuits were analyzed for the BPF and BRF applications, transfer functions are obtained using the same circuit to build LPF and HPF.

## *Second-Order Low-Pass Filer*

A low-pass filter is shown in Fig. 9.28. This is the same RLC series circuit used to make BPF and BRF. At extremely low frequencies, the capacitor shows an open circuit behavior, and the inductor shows a short circuit behavior. Therefore, the input signals reach the output through the resistor, and this will cause low frequencies to pass. The inductor is an open circuit at high frequencies, and the capacitor is a short circuit. Both of which eliminate high frequencies from reaching out the output. Therefore, the circuit is LPF. The system's transfer function is defined as the output voltage ratio to the input voltage, revealing more details about this filter's operation.

This circuit has a transfer function defined as the ratio of the output voltage over the input voltage. It can be found as follows:

$$H(s) \triangleq \frac{V_{out}}{V_{in}}$$

**Fig. 9.28** Second-order low-pass filter

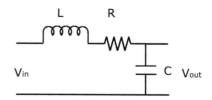

$$H(s) = \frac{\frac{1}{sC}}{R + sL + \frac{1}{sC}} = \frac{\frac{1}{LC}}{s^2 + \frac{R}{L}s + \frac{1}{LC}}$$

To find the amplitude and the phase, $s = j\omega$. Therefore,

$$H(j\omega) = \frac{\frac{1}{LC}}{-\omega^2 + \frac{R}{L}j\omega + \frac{1}{LC}} = \frac{\frac{1}{LC}}{\left(\frac{1}{LC} - \omega^2\right) + \frac{R}{L}j\omega}$$

The cutoff frequency is the frequency at which the amplitude of the transfer function reaches $\frac{1}{\sqrt{2}}H_{max}$. Hence,

$$|H_{max}|_{\omega = \omega_c} = \left|\frac{\frac{1}{LC}}{\left(\frac{1}{LC} - \omega_c^2\right) + \frac{R}{L}j\omega_c}\right| = \frac{1}{\sqrt{2}}$$

$$\frac{\frac{1}{LC}}{\sqrt{\left(\frac{1}{LC} - \omega_c^2\right)^2 + \left(\frac{R}{L}\omega_c\right)^2}} = \frac{1}{\sqrt{2}}$$

$$\left(\frac{1}{LC} - \omega_c^2\right)^2 + \left(\frac{R}{L}\omega_c\right)^2 = \frac{2}{L^2C^2}$$

$$\omega_c^4 + \left(\frac{R^2}{L^2} - \frac{2}{LC}\right)\omega_c^2 - \frac{1}{L^2C^2} = 0$$

$$\omega_c^2 = \frac{-\left(\frac{R^2}{L^2} - \frac{2}{LC}\right) \pm \sqrt{\left(\frac{R^2}{L^2} - \frac{2}{LC}\right)^2 + 4\frac{1}{L^2C^2}}}{2}$$

Therefore, the cutoff frequency is the positive real value of $\omega_c$ as follows:

$$\omega_c = \pm\sqrt{\frac{-\left(\frac{R^2}{L^2} - \frac{2}{LC}\right) \pm \sqrt{\left(\frac{R^2}{L^2} - \frac{2}{LC}\right)^2 + 4\frac{1}{L^2C^2}}}{2}}$$

Considering $\alpha = \frac{R}{L}$ & $\omega_0^2 = \frac{1}{LC}$, yields:

$$\omega_c = \pm\sqrt{\frac{-\left(\alpha^2 - 2\omega_0^2\right) \pm \sqrt{\left(\alpha^2 - 2\omega_0^2\right)^2 + 4\omega_0^2}}{2}}$$

**Example 9.15 Find the cutoff frequency of a second-order LPF realized through an RLC series where $R = 10\ \Omega$, $L = 1$ mH, and $C = 200\ \mu$F shown in Fig. 9.28.**

*Solution.*

$$\omega_c = \pm \sqrt{\frac{-\left(\frac{R^2}{L^2} - \frac{2}{LC}\right) \pm \sqrt{\left(\frac{R^2}{L^2} - \frac{2}{LC}\right)^2 + 4\frac{1}{L^2C^2}}}{2}}$$

Evaluation results in four roots of $\omega_c = \pm j9501.4$; $\omega_c = \pm 526.23$. The cutoff frequency is $\omega_c = 526.23\ \frac{rad}{s}$. The filter's frequency response, shown in Fig. 9.29, confirms the LPF behavior and cutoff frequency.

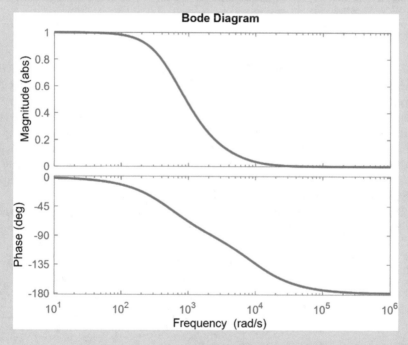

**Fig. 9.29** The frequency response of the LPF $H(s) = \frac{\frac{1}{LC}}{s^2 + \frac{R}{L}s + \frac{1}{LC}} = \frac{5e6}{s^2 + 1e4s + 5e6}$ with a cutoff frequency of $\omega_c = 526\ \frac{rad}{s}$

**Example 9.16 Find the cutoff frequency of a second-order LPF realized through an RLC series where $R = 10\,\Omega, L = 1\text{mH},$ and $C = 200\,\mu\text{F}$ shown in Fig. 9.28.**

*Solution.*

$$\omega_c = \pm\sqrt{\dfrac{-\left(\frac{R^2}{L^2} - \frac{2}{LC}\right) \pm \sqrt{\left(\frac{R^2}{L^2} - \frac{2}{LC}\right)^2 + 4\frac{1}{L^2C^2}}}{2}}$$

Evaluation results in four roots of $\omega_c = \pm j9501.4$; $\omega_c = \pm 526.23$. The cutoff frequency is $\omega_c = 526.23\,\frac{\text{rad}}{\text{s}}$. The filter's frequency response, shown in Fig. 9.29, confirms the LPF behavior and cutoff frequency.

## Second-Order High-Pass Filer

A HPF can be obtained from an RLC series when the output is measured across the inductor. The circuit schematic is shown in Fig. 9.30. At extremely low frequencies, the capacitor shows an open circuit behavior, and the inductor shows a short circuit behavior. Therefore, the input signals are blocked from the output, and this will cause the measured output to reach zero at low frequencies. The inductor is an open circuit at high frequencies, and the capacitor is a short circuit. Both provide a path for the high frequencies to reach the output without attenuation. Therefore, the circuit is HPF. The system's transfer function is defined as the output voltage ratio to the input voltage, revealing more details about this filter's operation.

This circuit has a transfer function defined as the ratio of the output voltage over the input voltage. It can be found as follows:

$$H(s) \triangleq \frac{V_{\text{out}}}{V_{\text{in}}}$$

$$H(s) = \frac{sL}{R + sL + \frac{1}{sC}} = \frac{LCs^2}{RCs + LCs^2 + 1} = \frac{s^2}{s^2 + \frac{R}{L}s + \frac{1}{LC}}$$

To find the amplitude and the phase, $s = j\omega$. Therefore,

**Fig. 9.30**  Second-order high-pass filter

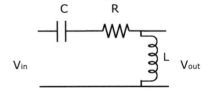

$$H(j\omega) = \frac{-\omega^2}{-\omega^2 + \frac{R}{L}j\omega + \frac{1}{LC}} = \frac{-\omega^2}{\left(\frac{1}{LC} - \omega^2\right) + \frac{R}{L}j\omega}$$

The cutoff frequency is the frequency at which the amplitude of the transfer function reaches $\frac{1}{\sqrt{2}}H_{max}$. Hence,

$$|H_{max}|_{\omega=\omega_c} = \left|\frac{\omega_c^2}{\left(\frac{1}{LC} - \omega_c^2\right) + \frac{R}{L}j\omega_c}\right| = \frac{1}{\sqrt{2}}$$

$$\frac{\omega_c^2}{\sqrt{\left(\frac{1}{LC} - \omega_c^2\right)^2 + \left(\frac{R}{L}\omega_c\right)^2}} = \frac{1}{\sqrt{2}}$$

$$\left(\frac{1}{LC} - \omega_c^2\right)^2 + \left(\frac{R}{L}\omega_c\right)^2 = 2\omega_c^4$$

$$\omega_c^4 - \left(\frac{R^2}{L^2} - \frac{2}{LC}\right)\omega_c^2 - \frac{1}{L^2C^2} = 0$$

$$\omega_c^2 = \frac{+\left(\frac{R^2}{L^2} - \frac{2}{LC}\right) \pm \sqrt{\left(\frac{R^2}{L^2} - \frac{2}{LC}\right)^2 + 4\frac{1}{L^2C^2}}}{2}$$

Therefore, the cutoff frequency is the positive real value of $\omega_c$ as follows:

$$\omega_c = \pm\sqrt{\frac{\left(\frac{R^2}{L^2} - \frac{2}{LC}\right) \pm \sqrt{\left(\frac{R^2}{L^2} - \frac{2}{LC}\right)^2 + 4\frac{1}{L^2C^2}}}{2}}$$

Considering $\alpha = \frac{R}{L}$ & $\omega_0^2 = \frac{1}{LC}$, yields:

$$\omega_c = \pm\sqrt{\frac{\left(\alpha^2 - 2\omega_0^2\right) \pm \sqrt{\left(\alpha^2 - 2\omega_0^2\right)^2 + 4\omega_0^2}}{2}}$$

**Example 9.17 Find the cutoff frequency of a second-order HPF realized through an RLC series where $R = 10\,\Omega, L = 1\,\text{mH}$, and $C = 200\,\mu\text{F}$ shown in Fig. 9.28.**

*Solution.*

$$\omega_c = \pm \sqrt{\frac{\left(\frac{R^2}{L^2} - \frac{2}{LC}\right) \pm \sqrt{\left(\frac{R^2}{L^2} - \frac{2}{LC}\right)^2 + 4\frac{1}{L^2C^2}}}{2}}$$

Evaluation results in four roots of $\omega_c = \pm\,j5262.4$; $\omega_c = \pm\,9501.4$. The cutoff frequency is $\omega_c = 9501.4\,\frac{\text{rad}}{\text{s}}$. The filter's frequency response, shown in Fig. 9.31, confirms the HPF behavior and cutoff frequency.

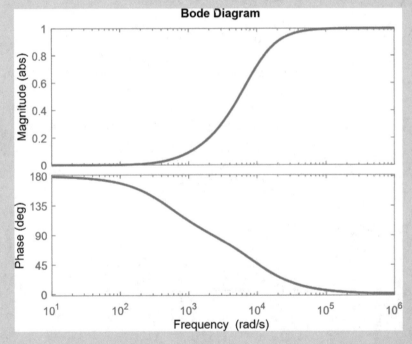

**Fig. 9.31** The frequency response of the LPF $H(s) = \frac{s^2}{s^2 + \frac{R}{L}s + \frac{1}{LC}} = \frac{s^2}{s^2 + 1e4s + 5e6}$ with a cutoff frequency of $\omega_c = 9501.4\,\frac{\text{rad}}{\text{s}}$

**Example 9.18 Find the cutoff frequency of a second-order HPF realized through an RLC series where $R = 10\,\Omega$, $L = 1$ mH, and $C = 200\,\mu$F shown in Fig. 9.28.**

*Solution.*

$$\omega_c = \pm\sqrt{\frac{\left(\frac{R^2}{L^2} - \frac{2}{LC}\right) \pm \sqrt{\left(\frac{R^2}{L^2} - \frac{2}{LC}\right)^2 + 4\frac{1}{L^2C^2}}}{2}}$$

Evaluation results in four roots of $\omega_c = \pm j526.4$; $\omega_c = \pm 9501.4$. The cutoff frequency is $\omega_c = 9501.4\,\frac{\text{rad}}{\text{s}}$. The filter's frequency response, shown in Fig. 9.31, confirms the HPF behavior and cutoff frequency.

## Higher-Order Filter by Repeated Circuits

Another method to obtain higher-order filters is to repeat the existing filters multiple times. This results in closer to the ideal system response. However, this action changes the filter characteristics, shifts the cutoff frequencies, and shrinks or expands the bandwidth discussed in this section.

### Repeated LPF

Figure 9.32 shows a repeated low-pass filter, $n$ times.
Considering the transfer function of a low-pass filter as

$$H(s) = \frac{\omega_c}{s + \omega_c}$$

The overall $n$th-order filter results in

$$H(s) = \frac{\omega_c{}^n}{(s + \omega_c)^n}$$

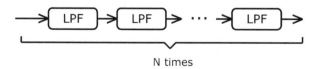

N times

**Fig. 9.32** Tandem connection of $n$ identical low-pass filters. The cutoff frequency of the overall filter is shifted from the cutoff frequency of an individual filter

The effect of this repeated filter is a shift in cutoff frequency, which is obtained as follows:

At $\omega = \omega_{cn}$

$$-3 \text{ dB} = 20 \log \left| \frac{\omega_c{}^n}{(s+\omega_c)^n} \right|$$

$$-3 \text{ dB} = n20 \log \left| \frac{\omega_c}{\sqrt{\omega_{cn}{}^2 + \omega_c{}^2}} \right|$$

$$-\frac{3}{n} \text{dB} = 20 \log \left| \frac{\omega_c}{\sqrt{\omega_{cn}{}^2 + \omega_c{}^2}} \right|$$

$$10^{-\frac{3}{20n}} = \frac{\omega_c}{\sqrt{\omega_{cn}{}^2 + \omega_c{}^2}}$$

The cutoff frequency of $n$th-order LPF is obtained as follows:

$$\omega_{cn} = \omega_c \sqrt{10^{\left(\frac{3}{10n}\right)} - 1}$$

The cutoff frequency of a first-order filter, if repeated $n$ times, is shifted by factor $\sqrt{10^{\left(\frac{3}{10n}\right)} - 1}$. The trade-off is a higher-quality filter with a sharper drop of power at frequencies outside of the desired range.

---

**Example 9.19 A low-pass filter has a cutoff frequency of 1000 rad/s. Find the cutoff frequency of this filter if it is repeated two and three times in cascade.**

*Solution.* For two times repeat of this filter, $n = 2$ (Fig. 9.33 shows the shift in frequencies):

$$\omega_{c2} = \omega_c \sqrt{10^{\left(\frac{3}{10n}\right)} - 1} = 1000 \sqrt{10^{\left(\frac{3}{10 \times 2}\right)} - 1} = 1000 \times 0.6422 = 642.2 \ \frac{\text{rad}}{\text{s}}$$

For $n = 3$,

$$\omega_{cn} = \omega_c \sqrt{10^{\left(\frac{3}{10n}\right)} - 1} = 1000 \sqrt{10^{\left(\frac{3}{10 \times 3}\right)} - 1} = 1000 \times 0.5088 = 508.8 \ \frac{\text{rad}}{\text{s}}$$

(continued)

**Example 9.19** (continued)

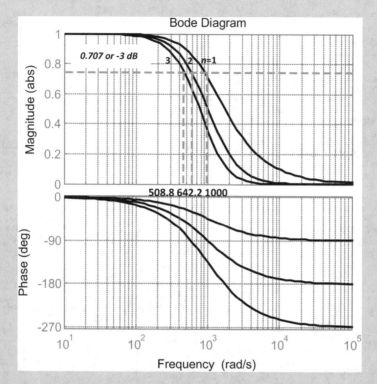

**Fig. 9.33** The shift of the cutoff frequency in the case of two tandem connections and three tandem connections of low-pass filters. The cutoff frequency of one filter was 1000 rad/s, and using two of these filters in tandem reduces the cutoff frequency to 642.2 rad/s, and three of the same filters in tandem reduce the cutoff frequency 508.8 rad/s. The phase shifts also increase as the order of circuits in higher tandem filters increases. A single pole (in one filter) reaches $-90°$, two poles in a second-order filter drop the phase to $-180°$, and three poles in a third-order filter drop the phase to $-270°$

**Example 9.20** It is desired to design a third-order low-pass filter with the cutoff frequency $\omega_c = 1000$ rad/s. Find an appropriate first-order LPF to be repeated three times.

*Solution.* The cutoff frequency of the third-order system is given, and the first-order frequency is needed. Hence,

$$\omega_{c3} = 1000.$$

(continued)

**Example 9.20** (continued)

$$n = 3.$$

Therefore,

$$1000 = \omega_c \sqrt{10^{\left(\frac{3}{10 \times 3}\right)} - 1}$$

$$\omega_c = \frac{1000}{0.508} = 1965.4 \ \frac{\text{rad}}{\text{s}}$$

**Example 9.21 Find the cutoff frequency of the following filter.**

$$H(s) = \frac{2000^4}{(s + 2000)^4}$$

*Solution.* The transfer function shows a four-time repeat of a low-pass filter with $\omega_c = 2000$. Therefore,

$$\omega_{c4} = \omega_c \sqrt{10^{\left(\frac{3}{10 \times 4}\right)} - 1}$$

$$\omega_{c4} = 868.3 \ \frac{\text{rad}}{\text{s}}$$

**Repeated HPF**

Considering a cascade connection of HPFs, as shown in Fig. 9.34, the transfer function of the overall system is obtained as follows (Fig. 9.34):

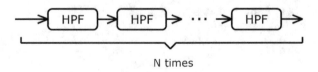

N times

**Fig. 9.34** Tandem connection of $N$ identical high-pass filters. The cutoff frequency of the overall filter is shifted from the cutoff frequency of an individual filter

$$H(s) = \frac{s}{s + \omega_c}$$

The repeated transfer function is obtained as

$$H(s) = \frac{s^n}{(s + \omega_c)^n}$$

The cutoff frequency of the overall system can be obtained at the location of 3 dB drop gain as follows:

$$-3 \ \text{dB} = 20 \log \left| \frac{s^n}{(s + \omega_c)^n} \right|$$

$$-3 \ \text{dB} = n20 \log \left| \frac{\omega_{cn}}{\sqrt{\omega_{cn}^2 + \omega_c^2}} \right|$$

$$-\frac{3}{n} \text{dB} = 20 \log \left| \frac{\omega_{cn}}{\sqrt{\omega_{cn}^2 + \omega_c^2}} \right|$$

$$10^{-\frac{3}{20n}} = \frac{\omega_{cn}}{\sqrt{\omega_{cn}^2 + \omega_c^2}}$$

The cutoff frequency of $n$th-order HPF is obtained as follows:

$$\omega_{cn} = \frac{\omega_c}{\sqrt{10^{\left(\frac{3}{10n}\right)} - 1}}$$

**Example 9.22 A high-pass filter has a cutoff frequency of 1000 rad/s. Find the filter's cutoff frequency if repeated twice and three times in a cascade (Fig. 9.35).**
*Solution.* For two times repeat of this filter, $n = 2$:

$$\omega_{c2} = \frac{\omega_c}{\sqrt{10^{\left(\frac{3}{10n}\right)} - 1}} = \frac{1000}{\sqrt{10^{\left(\frac{3}{10 \times 2}\right)} - 1}} = \frac{1000}{0.6422} = 1557.1 \ \frac{\text{rad}}{\text{s}}$$

For $n = 3$,

$$\omega_{cn} = \frac{\omega_c}{\sqrt{10^{\left(\frac{3}{10n}\right)} - 1}} = \frac{1000}{\sqrt{10^{\left(\frac{3}{10 \times 3}\right)} - 1}} = \frac{1000}{0.5088} = 1965.4 \ \frac{\text{rad}}{\text{s}}$$

(continued)

**Example 9.22** (continued)

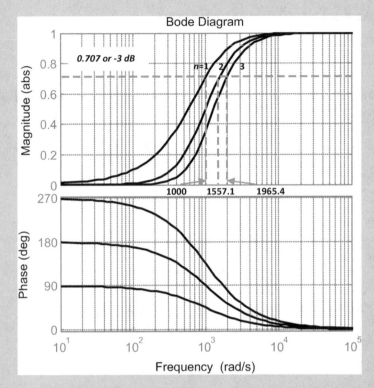

**Fig. 9.35** The shift of the cutoff frequency in the case of two tandem connections and three tandem connections of high-pass filters. The cutoff frequency of one filter was 1000 rad/s and using two of these filters in tandem increased the cutoff frequency to 1557.1 rad/s. Moreover, three of the same filters in tandem increased the cutoff frequency to 1965.4 rad/s. The phase shifts also increase as the order of circuits in higher tandem filters increases. A single pole (in one filter) started from $+90°$, two poles in a second-order filter started the phase from $+180°$, and three poles in a third-order filter started the phase from $+270°$

**Repeated BPF**

If a band-pass filter is repeated $n$ times, each cutoff frequency is shifted to narrow the filter's bandwidth. It results in a higher-quality filter. The low cutoff frequency is shifted as a high-pass filter, and the high cutoff frequency is shifted as a low-pass filter (shown in Fig. 9.36).

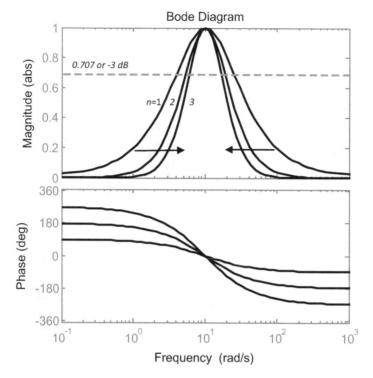

**Fig. 9.36** The behavior of repeated BPF can be observed in both the low-pass and high-pass sides. As the cutoff frequency of low-pass filters is decreased and that of the high-pass filters is increased, it results in a narrower bandwidth of the BPF

## Repeated BRF

If a band-reject filter is repeated $n$ times, the cutoff frequency of the high-pass filter part is shifted toward higher frequencies, and the cutoff frequency of the low-pass filter part is shifted to lower frequencies. This results in wider bandwidth of the filter, i.e., a larger range of frequencies are blocked. Figure 9.37 shows the frequency response of a repeated band-reject filter (Fig. 9.37).

## Butterworth Filters

Some filters can be designed to perform superior at the desired order. Butterworth is a filter that can be designed as an $n$th-order transfer function. These filters' performance becomes closer to an ideal filter without the significant shift in their cutoff frequency. This is achieved through a design table for the cutoff frequency $\omega_c = 1$; the transfer function is shifted to the desired cutoff frequency.

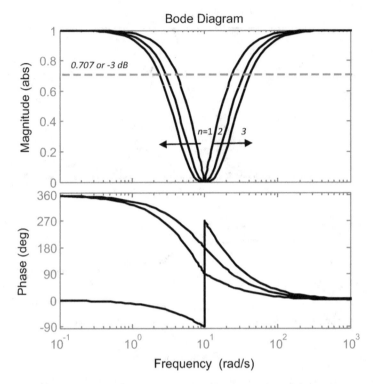

**Fig. 9.37** The behavior of repeated BPF can be observed in both the low-pass and high-pass sides. As the cutoff frequency of low-pass filters is decreased and that of the high-pass filters is increased, it results in a wider bandwidth of the BRF

## *Butterworth Low-Pass Filter*

The transfer function of a Butterworth LPF of order $n$ has no zeros but $n$ poles at the predefined normalized locations to improve the filtration performance, i.e., the frequency response. The transfer function is expressed as

$$H_n(s) = \frac{1}{B_n(s)}$$

where $B_n(s)$ is polynomial of order $n$. The polynomial can be analytically obtained. However, a table can also be utilized to simplify the design of these filters (all the polynomials are repeated for a specific order). The table determines the denominator's polynomial at $\omega_c = 1\,\frac{\text{rad}}{\text{s}}$.

## Butterworth Denominator Polynomials

| n | Denominator $B_n(s)$ |
|---|---|
| 1 | $(s + 1)$ |
| 2 | $(s^2 + 1.414214s + 1)$ |
| 3 | $(s + 1)(s^2 + s + 1)$ |
| 4 | $(s^2 + 0.765367s + 1)(s^2 + 1.847759s + 1)$ |
| 5 | $(s + 1)(s^2 + 0.618034s + 1)(s^2 + 1.618034s + 1)$ |
| 6 | $(s^2 + 0.517638s + 1)(s^2 + 1.414214s + 1)(s^2 + 1.931852s + 1)$ |
| 7 | $(s + 1)(s^2 + 0.445042s + 1)(s^2 + 1.246980s + 1)(s^2 + 1.801398s + 1)$ |
| 8 | $(s^2 + 0.390181s + 1)(s^2 + 1.111140s + 1)(s^2 + 1.662939s + 1)(s^2 + 1.961571s + 1)$ |
| 9 | $(s + 1)(s^2 + 0.347296s + 1)(s^2 + s + 1)(s^2 + 1.532089s + 1)(s^2 + 1.879385s + 1)$ |
| 10 | $(s^2 + 0.312869s + 1)(s^2 + 0.907981s + 1)(s^2 + 1.414214s + 1)(s^2 + 1.782013s + 1)$ $(s^2 + 1.975377s + 1)$ |

**Example 9.23 Design standard first-order Butterworth LPF.**
*Solution.* Since the cutoff frequency is $\omega_c = 1$, there is no need to shift the frequency. Therefore, the polynomial from the table for $n = 1$ works.

$$H_1(s) = \frac{1}{s + 1}$$

This is consistent with what was obtained earlier in this chapter as LPF at $\omega_c = 1$. To scale to the desired cutoff frequency, a shift of $s \to \frac{s}{\omega_c}$ needs to be applied.

*Design a first-order Butterworth LPF with $\omega_c = 100$.*
*Solution.* The standard ($\omega_c = 1$) first-order LPF Butterworth filter is expressed as

$$H_1(s) = \frac{1}{s + 1}$$

The shift of the cutoff frequency to $\omega_c = 100, s \to \frac{s}{100}$. Therefore;

$$H_1\left(\frac{s}{100}\right) = \frac{1}{\frac{s}{100} + 1} = \frac{100}{s + 100}$$

*Design a third-order Butterworth LPF at $\omega_c = 100$. Compare the filtering performance with the first-order Butterworth LPF at $\omega_c = 100$.*
*Solution.* According to the table, $B_3(s) = (s + 1)(s^2 + s + 1)$. Therefore, the filter at $\omega_c = 1$ is:

(continued)

**Example 9.23** (continued)
$$H_3(s) = \frac{1}{(s+1)(s^2+s+1)}$$

The scaling of $s \rightarrow \frac{s}{100}$ is needed to obtain the desired cutoff frequency as follows:

$$H_3(s) = \frac{1}{\left(\frac{s}{100}+1\right)\left(\left(\frac{s}{100}\right)^2+\frac{s}{100}+1\right)} = \frac{100^3}{(s+100)\left(s^2+100s+100^2\right)}$$

Figure 9.38 shows the frequency response of the first- and third-order Butterworth filters. Both have the same cutoff frequency $\omega_c = 100$, but they perform differently at frequencies close to the cutoff frequency. As the order of the filter increases, its performance gets closer to an ideal filter.

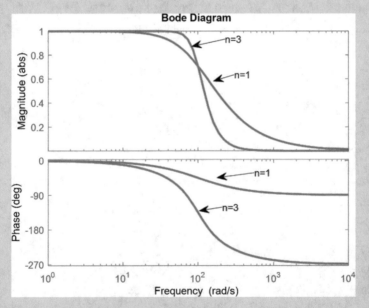

**Fig. 9.38** Comparison of a first-order and third-order Butterworth filter tuned at $\omega_c = 100$

## Butterworth High-Pass Filter

The transfer function of a Butterworth HPF of order $n$ has $n$ zeros at the origin and $n$ poles at the predefined normalized locations to improve the filtration performance, i.e., the frequency response. The transfer function is expressed as

$$H_n(s) = \frac{s^n}{B_n(s)}$$

where $B_n(s)$ is polynomial of order $n$ defined the same as for the Butterworth HPF. The polynomial can be analytically obtained. However, a table can also be utilized to simplify the design of these filters (all the polynomials are repeated for a specific order). The table determines the denominator's polynomial at $\omega_c = 1 \frac{rad}{s}$.

**Example 9.24 Design standard first-order Butterworth HPF.**
*Solution.* Since the cutoff frequency is $\omega_c = 1$, there is no need to shift the frequency. Therefore, the polynomial from the table for $n = 1$ works.

$$H_1(s) = \frac{s}{s+1}$$

This is consistent with what was obtained earlier in this chapter as HPF at $\omega_c = 1$. To scale to the desired cutoff frequency, a shift of $s \rightarrow \frac{s}{\omega_c}$ needs to be applied.

*Design a first-order Butterworth HPF with $\omega_c = 100$.*
*Solution.* The standard ($\omega_c = 1$) first-order HPF Butterworth filter is expressed as

$$H_1(s) = \frac{s}{s+1}$$

The shift of the cutoff frequency to $\omega_c = 100, s \rightarrow \frac{s}{100}$. Therefore,

$$H_1\left(\frac{s}{100}\right) = \frac{\frac{s}{100}}{\frac{s}{100}+1} = \frac{s}{s+100}$$

*Design a third-order Butterworth HPF at $\omega_c = 100$. Compare the filtering performance with the first-order Butterworth LPF at $\omega_c = 100$.*
*Solution.* According to the table, $B_3(s) = (s + 1)(s^2 + s + 1)$. Therefore, the filter at $\omega_c = 1$ is:

$$H_3(s) = \frac{s^3}{(s + 1)(s^2 + s + 1)}$$

The scaling of $s \rightarrow \frac{s}{100}$ is needed to obtain the desired cutoff frequency as follows:

(continued)

**Example 9.24** (continued)

$$H_3(s) = \frac{\left(\frac{s}{100}\right)^3}{\left(\frac{s}{100} + 1\right)\left(\left(\frac{s}{100}\right)^2 + \frac{s}{100} + 1\right)} = \frac{s^3}{(s + 100)(s^2 + 100s + 100^2)}$$

Figure 9.39 compares the effect of higher-order filters designed to improve the frequency response performance by becoming closer to the ideal filter.

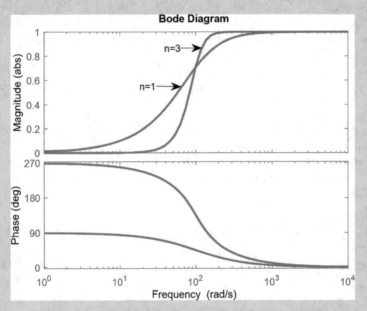

**Fig. 9.39** Comparison of a first-order and third-order Butterworth filter tuned at $\omega_c = 100$

## Problems

### Low-Pass Filter

9.1. A low-pass filter has a cutoff frequency of $f_c = 3000$ Hz.

    (a) Design an *RC* circuit to realize this filter. Consider $C = 1$ µF. Draw the circuit and analyze its operation.

    (b) Design an *RL* circuit to realize this filter. Consider $L = 150$ mH. Draw the circuit and analyze its operation.

9.2. A low-pass filter has a cutoff frequency of $f_c = 60$ Hz.

    (a) Design an $RC$ circuit to realize this filter. Consider $C = 10$ μF. Draw the circuit and analyze its operation.

    (b) Design an $RL$ circuit to realize this filter. Consider $L = 10$ mH. Draw the circuit and analyze its operation.

9.3. In an $RL$ LPF, the circuit parameters are $R = 2\ \Omega$ and $L = 0.05$ mH. Find the cutoff frequency of the filter. Determine the output voltage amplitude and its phase when a $V_{in} = 10$ V input is applied and only the input frequency is adjusted at the following frequencies one by one $\omega = 4000, 40,000, 400,000 \left(\frac{rad}{s}\right)$.

9.4. In an $RC$ LPF, the circuit parameters are $R = 2\ \Omega$ and $C = 10$ μF. Find the cutoff frequency of the filter. Determine the output voltage amplitude and phase when a $V_{in} = 10$ V input is tuned at the cutoff frequency, a decade above and a decade below the cutoff.

## Low-Pass Filter Using Laplace

9.5. Design a unity-gain LPF with a cutoff frequency of $\omega_c = 15k$ rad/s. Find its transfer function.

9.6. Design a unity gain LPF to cut the frequencies higher than the heartbeat of 1 Hz. The frequency of 1 Hz should fully pass to the output. Find the transfer function of the filter.

9.7. Determine the following systems' filter type and cutoff frequency.

    (a) $G(s) = \frac{200}{s+200}$

    (b) $G(s) = \frac{1}{1e - 3s+1}$

    (c) $G(s) = \frac{400}{2e - 4s+2}$

9.8. In Problem 9.7, find the poles and zeros of the transfer functions, and identify the correlation between the poles, cutoff frequency, and filter types.

9.9. Determine and sketch the frequency response of the transfer functions in Problem 9.7.

## High-Pass Filter

9.10. A high-pass filter has a cutoff frequency of $f_c = 3000$ Hz.

    (a) Design an $RC$ circuit to realize this filter. Consider $C = 1$ μF. Draw the circuit and analyze its operation.

    (b) Design an $RL$ circuit to realize this filer. Consider $L = 150$ mH. Draw the circuit and analyze its operation.

9.11. A high-pass filter has a cutoff frequency of $f_c = 60$ Hz.

(a) Design an $RC$ circuit to realize this filter. Consider $C = 10$ μF. Draw the circuit and analyze its operation.
(b) Design an $RL$ circuit to realize this filer. Consider $L = 10$ mH. Draw the circuit and analyze its operation.

9.12. In a $RL$ HPF, the circuit parameters are $R = 2$ Ω and $L = 0.05$ mH. Find the cutoff frequency of the filter. Determine the output voltage amplitude and phase when a $V_{in} = 10$ V input is applied and only the input frequency is adjusted at the following frequencies one by one $\omega = 4000, 40,000, 400,000$ $\left(\frac{rad}{s}\right)$.

9.13. In a $RC$ HPF, the circuit parameters are $R = 2$ Ω and $C = 10$ μF. Find the cutoff frequency of the filter. Determine the output voltage amplitude and phase when a $V_{in} = 10$ V input is tuned at the cutoff frequency, a decade above and a decade below the cutoff.

## High-Pass Filter Using Laplace

9.14. Design a unity gain HPF with a cutoff frequency of $\omega_c = 15k$ rad/s. Find its transfer function.

9.15. Design a unity gain HPF to cut the frequencies lower than 60 Hz. The frequency of 60 Hz should fully pass to the output. Find the transfer function of the filter.

9.16. Determine the filter type of the following systems and determine their cutoff frequency.

(a) $G(s) = \frac{200s}{s+200}$
(b) $G(s) = \frac{s}{1e-3s+1}$
(c) $G(s) = \frac{400s}{2e-4s+2}$

9.17. In Problem 9.16, find the poles and zeros of the transfer functions, and identify the correlation between the poles, cutoff frequency, and filter types.

9.18. Determine and sketch the frequency response of the transfer functions in Problem 9.17.

## Series and Parallel LC Circuits

9.19. Tune an $LC$ series circuit at 60 Hz. Find a reasonable value for $L$ and $C$. Analyze the operation of this circuit as a bridge between the input and output (of a filter). Analyze the operation of this circuit in parallel to the output (of a filter).

9.20. Tune an *LC* parallel circuit at 60 Hz. Find a reasonable value for *L* and *C*. Analyze the operation of this circuit as a bridge between the input and output (of a filter). Analyze the operation of this circuit in parallel to the output (of a filter).

## Band-Pass Filters

9.21. Design a BPF to show a bandwidth of 100 Hz and a resonant frequency of 5000 Hz. Find the transfer function using an *LC* series and an *LC* parallel in two cases. Use existing range components.

9.22. Design a BPF to show a bandwidth of 10 Hz and a resonant frequency of 5000 Hz. Find the transfer function using an *LC* series and a parallel in two cases. Use existing range components.

9.23. Design a BPF to show a bandwidth of 1 Hz and a resonant frequency of 5000 Hz. Find the transfer function using an *LC* series and a parallel in two cases. Use existing range components.

9.24. Design a BPF to show a bandwidth of 500 Hz and a lower cutoff frequency of 10,000 Hz. Find the transfer function using an *LC* series and a parallel in two cases. Use existing range components.

9.25. Design a BPF to show a bandwidth of 500 Hz and a resonant frequency of 40,000 Hz. Find the transfer function using an *LC* series and a parallel in two cases. Use existing range components.

9.26. The damping of a BPF is designed to be $\zeta = 0.5$. What is the $Q$ of the circuit? Design the filter if the resonant frequency is 100 Hz. Sketch the frequency response.

9.27. The damping of a BPF is designed to be $\zeta = 0.1$. What is the $Q$ of the circuit? Design the filter if the resonant frequency is 100 Hz. Sketch the frequency response.

9.28. The damping of a BPF is designed to be $\zeta = 0.05$. What is the $Q$ of the circuit? Design the filter if the resonant frequency is 100 Hz. Sketch the frequency response.

## Band-Pass Filters Using Laplace

9.29. Find the transfer function of a BPF with a bandwidth of 100 Hz and resonant frequency of 5000 Hz. Find the transfer function using an *LC* series and a parallel in two cases. Use existing range components. Determine the poles and zeros of the circuit and estimate the cutoff frequency from the values of the poles.

9.30. Find the transfer function of a BPF with a bandwidth of 10 Hz and a resonant frequency of 5000 Hz. Find the transfer function using an *LC* series and a parallel in two cases. Use existing range components. Determine the poles and zeros of the circuit and estimate the cutoff frequency from the values of the poles.

9.31. Find the transfer function of a BPF with a bandwidth of 1 Hz and resonant frequency of 5000 Hz. Find the transfer function using an $LC$ series and a parallel in two cases. Use existing range components. Determine the poles and zeros of the circuit and estimate the cutoff frequency from the values of the poles.

9.32. Find the transfer function of a BPF with a bandwidth of 0.5 Hz and a resonant frequency of 60 Hz. Find the transfer function using an $LC$ series and a parallel in two cases. Use existing range components. Determine the poles and zeros of the circuit and estimate the cutoff frequency from the values of the poles.

9.33. Determine the filter type, cutoff frequency, resonant frequency, quality factor, damping factor, and bandwidth of the following transfer functions:

(a) $G(s) = \frac{20s}{s^2+20s+10000}$

(b) $G(s) = \frac{200s}{s^2+200s+9000}$

(c) $G(s) = \frac{0.1s}{s^2+0.1s+(120\pi)^2}$

(d) $G(s) = \frac{0.01s}{s^2+0.01s+(2\pi)^2}$

9.34. Realize the transfer functions of Problem 9.32 in the $LC$ series.

9.35. Realize the transfer functions of Problem 9.32 in $LC$ parallel.

9.36. The damping of a BPF is designed to be $\zeta = 0.5$. What is the $Q$ of the circuit? Find the transfer function if the resonant frequency is 100 Hz. Sketch the frequency response.

9.37. The damping of a BPF is designed to be $\zeta = 0.1$. What is the $Q$ of the circuit? Find the transfer function if the resonant frequency is 100 Hz. Sketch the frequency response.

9.38. The damping of a BPF is designed to be $\zeta = 0.05$. What is the $Q$ of the circuit? Find the transfer function if the resonant frequency is 100 Hz. Sketch the frequency response.

9.39. Find the transfer function of a BPF with cutoff frequencies of $f_{c1} = 100$ Hz, $f_{c2} = 500$ Hz.

9.40. Find the transfer function of a BPF with cutoff frequencies of $\omega_{c1} = 58$ Hz $f_{c_2} = 60$ Hz.

9.41. Find the transfer function of a BPF with cutoff frequencies of $\omega_{c1} = 1000$ rad/ s and $\omega_{c2} = 5000$ rad/s.

## Band-Reject Filters

9.42. Design a BRF to show a bandwidth of 100 Hz and a resonant frequency of 400 Hz. Find the transfer function using an $LC$ series and a parallel in two cases. Use existing range components.

9.43. Design a BRF to show a bandwidth of 10 Hz and a resonant frequency of 400 Hz. Find the transfer function using an *LC* series and a parallel in two cases. Use existing range components.

9.44. Design a BRF to show a bandwidth of 1 Hz and a resonant frequency of 400 Hz. Find the transfer function using an *LC* series and a parallel in two cases. Use existing range components.

9.45. Design a BRF to show a bandwidth of 0.5 Hz and a resonant frequency of 60 Hz. Find the transfer function using an *LC* series and a parallel in two cases. Use existing range components.

9.46. The damping of a BRF is designed to be $\zeta = 0.5$. What is the $Q$ of the circuit? Design the filter if the resonant frequency is 100 Hz. Sketch the frequency response.

9.47. The damping of a BRF is designed to be $\zeta = 0.1$. What is the $Q$ of the circuit? Design the filter if the resonant frequency is 100 Hz. Sketch the frequency response.

9.48. The damping of a BRF is designed to be $\zeta = 0.05$. What is the $Q$ of the circuit? Design the filter if the resonant frequency is 100 Hz. Sketch the frequency response.

## Band-Pass Filters Using Laplace

9.49. Find the transfer function of a BRF with a bandwidth of 100 Hz and resonant frequency of 5000 Hz. Find the transfer function using an *LC* series and a parallel in two cases. Use existing range components. Determine the poles and zeros of the circuit and estimate the cutoff frequency from the values of the poles.

9.50. Find the transfer function of a BRF with a bandwidth of 10 Hz and a resonant frequency of 5000 Hz. Find the transfer function using an *LC* series and a parallel in two cases. Use existing range components. Determine the poles and zeros of the circuit and estimate the cutoff frequency from the values of the poles.

9.51. Find the transfer function of a BRF with a bandwidth of 1 Hz and resonant frequency of 5000 Hz. Find the transfer function using an *LC* series and a parallel in two cases. Use existing range components. Determine the poles and zeros of the circuit and estimate the cutoff frequency from the values of the poles.

9.52. Find the transfer function of a BRF with a bandwidth of 0.5 Hz and a resonant frequency of 60 Hz. Find the transfer function using an *LC* series and a parallel in two cases. Use existing range components. Determine the poles and zeros of the circuit and estimate the cutoff frequency from the values of the poles.

9.53. Determine the filter type, cutoff frequency, resonant frequency, quality factor, damping factor, and bandwidth of the following transfer functions:

(a) $G(s) = \frac{s^2+10000}{s^2+20s+10000}$

(b) $G(s) = \frac{s^2+9000}{s^2+200s+9000}$

(c) $G(s) = \frac{s^2+(120\pi)^2}{s^2+0.1s+(120\pi)^2}$

(d) $G(s) = \frac{s^2+(2\pi)^2}{s^2+0.01s+(2\pi)^2}$

9.54. Realize the transfer functions of Problem 9.52 in the $LC$ series.

9.55. Realize the transfer functions of Problem 9.52 in $LC$ parallel.

9.56. The damping of a BRF is designed to be $\zeta = 0.5$. What is the $Q$ of the circuit? Find the transfer function if the resonant frequency is 100 Hz. Sketch the frequency response.

9.57. The damping of a BRF is designed to be $\zeta = 0.1$. What is the $Q$ of the circuit? Find the transfer function if the resonant frequency is 100 Hz. Sketch the frequency response.

9.58. The damping of a BRF is designed to be $\zeta = 0.05$. What is the $Q$ of the circuit? Find the transfer function if the resonant frequency is 100 Hz. Sketch the frequency response.

9.59. Find the transfer function of a BRF with cutoff frequencies of $f_{c1} = 100$ Hz, $f_{c2} = 500$ Hz.

9.60. Find the transfer function of a BRF with cutoff frequencies of $\omega_{c1} = 58$ Hz, $f_{c2} = 60$ Hz.

9.61. Find the Transfer Function of a BRF with cutoff Frequencies of $\omega_{c1} = 1000$ rad/s and $\omega_{c2} = 5000$ rad/s.

## Overall Filtration Process

9.62. Determine whether the signals $V_1 - V_7$ in group (a) passes through each filter (b). For instance, will $V_1$pass $H_1(s)$, $H_2(s)$, etc. ?

(a) Input signals

    (i) $V_1 = 5$

    (ii) $V_3 = 5 \sin 10\,t$

    (iii) $V_4 = 5 \sin 100\,t$

    (iv) $V_5 = 5 \sin 1000\,t$

    (v) $V_6 = 5 \sin 5000\,t$

    (vi) $V_7 = 5 \sin 10{,}000\,t$

(b) Filters

    (i) $H_1(s) = \frac{1000}{s+1000}$

    (ii) $H_2(s) = \frac{s}{s+1000}$

(iii) $H_3(s) = \frac{4000s}{s^2+4000s+3000^2}$

(iv) $H_4(s) = \frac{s^2+3000^2}{s^2+2000s+3000^2}$

## *Butterworth Filters*

9.63. Design a first-order Butterworth LPF with a cutoff frequency of $\omega_c = 1000 \frac{rad}{s}$.

9.64. Design a second-order Butterworth LPF with a cutoff frequency of $\omega_c = 1000 \frac{rad}{s}$.

9.65. Design a third-order Butterworth LPF with a cutoff frequency of $\omega_c = 1000 \frac{rad}{s}$.

9.66. Design a fourth-order Butterworth LPF with a cutoff frequency of $\omega_c = 1000 \frac{rad}{s}$.

9.67. Design a first-order Butterworth HPF with a cutoff frequency of $\omega_c = 1000 \frac{rad}{s}$.

9.68. Design a second-order Butterworth HPF with a cutoff frequency of $\omega_c = 1000 \frac{rad}{s}$.

9.69. Design a third-order Butterworth HPF with a cutoff frequency of $\omega_c = 1000 \frac{rad}{s}$.

9.70. Design a fourth-order Butterworth HPF with a cutoff frequency of $\omega_c = 1000 \frac{rad}{s}$.

## *Higher-Order Filter*

9.71. A LPF has a cutoff frequency of $\omega_c = 100$ rad/s. The filter is repeated three times. Find the new cutoff frequency. Sketch an *RL* and an *RC* filter separately to realize the third-order filter. Sketch the frequency response of the first order and third order.

9.72. A third-order LPF has to have a cutoff frequency of $\omega_c = 100$ rad/s. Find the cutoff frequency of the first-order circuit. Sketch an *RL* and an *RC* filter separately to realize the third-order filter. Sketch the frequency response of the first order and third order.

9.73. A HPF has a cutoff frequency of $\omega_c = 100$ rad/s. The filter is repeated three times. Find the new cutoff frequency. Sketch an *RL* and an *RC* filter separately to realize the third-order filter. Sketch the frequency response of the first order and third order.

9.74. A third-order HPF has to have a cutoff frequency of $\omega_c = 100$ rad/s. Find the cutoff frequency of the first-order circuit. Sketch an *RL* and an *RC* filter separately to realize the third-order filter. Sketch the frequency response of the first order and third order.

## Higher-Order Filter Using Laplace

9.75. Determine the type, order, cutoff frequency, bandwidth, and quality factor of the following filters (whichever applies):

(a) $G(s) = \dfrac{100^3}{(s+100)^3}$

(b) $G(s) = \dfrac{s^4}{(s+10000)^4}$

(c) $G(s) = \dfrac{400s^2}{(s^2+20s+10000)^2}$

(d) $G(s) = \dfrac{\left(s^2+(1200\pi)^2\right)^3}{\left(s^2+20\pi s+(1200\pi)^2\right)^3}$

# Chapter 10
# Operational Amplifiers

Years before introducing microcontrollers and digital computers, control system operations, industrial computations, and even simulation of dynamic systems were made possible using analog computers. The heart of an industrial analog computer is a device called an operational amplifier or "Opamp." These amplifiers contain many transistors to accomplish theoretically infinite gain to either input.

In a simple form, an Opamp is packaged in an eight-port device, as shown in Fig. 10.1.

As the figure shows, the Opamp has major ports as follows:

- Two input ports

    – Inverting
    – Noninverting

- One output port
- Positive power supply
- Negative power supply

These ports show the Opamp; ideally, power supplies are also eliminated from schematics. Figure 10.2 shows the ideal Opamp.

## Ideal Opamp

Ideally, Opamps are considered devices that have the following:

1. Infinite gain. The Opamp has no limit on the voltage it can generate in the output, and the supply voltages can be infinitely large.
2. Infinitely a fast response. It means a large (close to infinite) slew rate.
3. Infinite bandwidth means that the gain of Opamp does not drop at infinitely large frequencies.

© The Author(s), under exclusive license to Springer Nature Switzerland AG 2023           555
A. Izadian, *Fundamentals of Modern Electric Circuit Analysis and Filter Synthesis*,
https://doi.org/10.1007/978-3-031-21908-5_10

**Fig. 10.1**  Pin layout of a typical eight-pin operational amplifier. This chip contains only one Opamp. Some chips contain two or more Opamps

**Dual-In-Line Package**

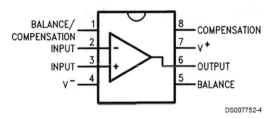

DS007752-4

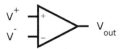

**Fig. 10.2**  Schematic of an ideal Opamp without showing the power ports. In this type of application, it is considered that the Opamp can take unlimited voltage levels and has no delay in response or what is known as a slew rate

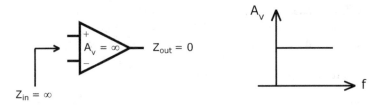

**Fig. 10.3**  The frequency response of an ideal Opamp shows high gain across the frequency spectrum. However, the gain of the actual Opamp drops as the operating frequency reaches the physical limitations of the device. Hence, this resulting a high gain over an unlimited range of frequencies

4. Infinite input impedance. It means that the input current to each input port (inverting and noninverting) is zero. These Opamps are voltage-controlled devices.
5. Zero output impedance. The output voltage is not dependent on the load impedance (Fig. 10.3).

Operation of an ideal Opamp. Considering infinitely fast response devices with unlimited voltage levels, the voltage observed at the output port of an Opamp becomes the value of its + supply if the + port voltage becomes slightly higher than the voltage applied at the − port.

For instance, $v^+ = 4$ V and $v^- = 3.99$ V result in $+V_{cc}$ in the output. $v^+ = 1.5$ V and $v^- = 1.6$ V result in $-V_{cc}$ in the output. As the Opamp is ideal, the output voltage switches to $+V_{cc}$ or $-V_{cc}$ is infinitely fast (Fig. 10.4).

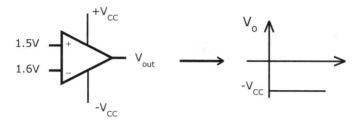

**Fig. 10.4** Application as a comparator. When inverting and noninverting pins have slightly different voltages, the output voltage switches to one of the supply voltages connected to the pin with a higher voltage

## Slew Rate

The speed of an Opamp (actually measured in $\frac{V}{\mu s}$ or $\frac{V}{ms}$) to follow a reference signal with amplitude $V$ at frequency $f$ (Hz) is measured as slew rate, which can be found as

$$\text{Slew Rate} = 2\pi f V \; \left(\frac{V}{s}\right)$$

$$\text{Slew Rate} = \frac{2\pi f V}{1e3} \; \left(\frac{V}{ms}\right)$$

$$\text{Slew Rate} = \frac{2\pi f V}{1e6} \; \left(\frac{V}{\mu s}\right)$$

This means that the variations in the reference signal can be amplified in the output of the Opamp; this requires that the Opamp's slew rate be equal to or higher than $2\pi f V$. For instance, a 150 mV signal at a frequency of 10 kHz requires an Opamp with a minimum slew rate of

**Fig. 10.5** The inverting and noninverting ports are virtually short circuits

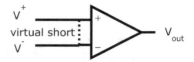

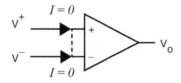

**Fig. 10.6** The inverting and noninverting ports are virtually short circuits. However, the high-input impedance of the ports makes the current entering each port equal to zero

$$\text{Slew Rate} = \frac{(2\pi 10e3 \times 150e - 3)}{1000}$$

$$\text{Slew Rate} = 9.42 \left(\frac{\text{V}}{\text{ms}}\right)$$

## Opamp in Circuits

Two rules simplify the analysis of an Opamp in a circuit.

**Rule 1** There is a *virtual short circuit* between the two input ports, $v^+$ and $v^-$. Figure 10.5 shows this simplification notation.

**Rule 2** Despite a short circuit between the ports of an Opamp, no current flows to either port. Figure 10.6 shows this simplification notation.

**Example 10.1 Find the output voltage of the following circuit (Fig. 10.7).**
*Solution.* Since the noninverting port is connected to the ground, it has zero potential or $v^+ = 0$. Applying rule 1, the voltage of the inverting port $v^-$ due to the virtual short circuit between the inverting and noninverting ports taking up the voltage of the noninverting port. Therefore, virtually, $v^+ = v^- = 0$.

Imposing a zero volt potential at the inverting port, the current $I_{in} = \frac{V_{in} - 0}{Z_{in}}$.
According to rule 2, there is no current entering the ports of Opamp. Therefore, a KCL at the node ① indicates that the current must flow through the feedback impedance as follows:

(continued)

**Example 10.1** (continued)

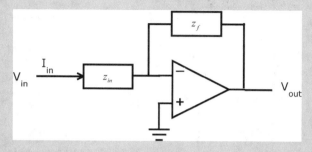

**Fig. 10.7** Figure of Example 10.1. Inverting amplifier

$$I_{\text{in}} = I_f$$

$$I_f = \frac{0 - V_o}{Z_f}$$

Therefore,

$$I_{\text{in}} = I_f \rightarrow \frac{V_{\text{in}} - 0}{Z_{\text{in}}} = \frac{0 - V_o}{Z_f}$$

The output voltage in terms of input voltage can be obtained as

$$\frac{V_{\text{in}} - 0}{Z_{\text{in}}} = \frac{0 - V_o}{Z_f}$$

$$\frac{V_o}{V_{\text{in}}} = \frac{-Z_f}{Z_{\text{in}}}$$

This shows two important applications of an Opamp:

1. The input voltage can be amplified by the ratio of feedback impedance over the input impedance $\frac{Z_f}{Z_{\text{in}}}$ as long as it is not reached the supply voltage limits
2. The output voltage phase is 180° apart from the input voltage indicated by the (−) sign. For this reason, this circuit is also called inverting amplifier.

**Example 10.2 Noninverting amplifier. Find the output voltage of the following circuit (Fig. 10.8).**

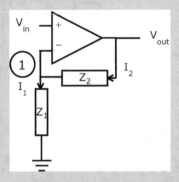

**Fig. 10.8** Figure of the circuit in Example 10.2, a noninverting amplifier

*Solution.* The noninverting port $v^+$ is connected to the input voltage $V_{in}$. Therefore, according to rule 1, the inverting port voltage is virtually $v^- = V_{in}$. This voltage forces current to ground through impedance $Z_1$ as

$$I_1 = \frac{V_{in}}{Z_1}$$

According to rule 2, there is no current passing to the port. Therefore, a KCL at node ① is written as

$$I_1 = I_2 = \frac{V_o - v^-}{Z_2} = \frac{V_o - V_{in}}{Z_2}$$

$$\frac{V_{in}}{Z_1} = \frac{V_o - V_{in}}{Z_2}$$

The ratio of output voltage over input voltage is obtained as

$$\frac{V_o}{V_{in}} = 1 + \frac{Z_2}{Z_1}$$

As the equation shows, the output voltage depends on the ratio of feedback impedance over the ground impedance $1 + \frac{Z_2}{Z_1}$, which is in phase with the input voltage.

**Example 10.3 Considering an ideal Opamp and unlimited voltage supply, find the output voltage in the following circuit if the input is excited by a $V_{in} = 2 \sin(10\,t)$ V (Fig. 10.9).**

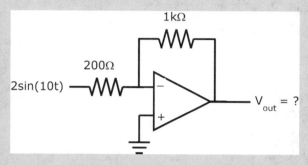

**Fig. 10.9** Figure of the circuit in Example 10.3, an inverting amplifier

*Solution.* The transfer function of an inverting amplifier is given by the ratio of the feedback over the input impedances. Therefore,

$$\frac{V_o}{V_{in}} = -\frac{Z_f}{Z_{in}} = -\frac{1000}{200} = -5$$

The output voltage becomes

$$V_o = -5V_{in} = -5 \times 2 \sin 10t = -10 \sin 10t \ \text{V}$$

**Example 10.4 Considering a supply voltage limit at $\pm Vcc = \pm 8$ V. Find the output voltage from the circuit discussed in the previous example.**
*Solution.* The circuit's output voltage, without any limit in supply voltage amplitude, was expected to reach a $\pm 10$ V peak. However, as the voltage is limited to $\pm 4$ V, the peak voltage is clamped at 8 V. The result is a sinusoidal clamped at a 4 V peak.

$$V_o = \begin{cases} -4 & V_o < -4 \\ -10 \sin 10t & -4 < V_o < 4 \\ +4 & V_o > 4 \end{cases}$$

(continued)

**Example 10.4** (continued)

Figure 10.10 shows the output voltage of the amplifier. The figure shows that a large gain does not mean that the output voltage is amplified to a very large voltage.

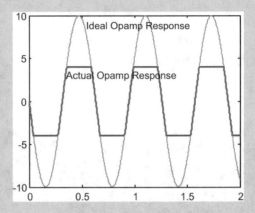

**Fig. 10.10** Clamped output in a saturated amplifier

**Example 10.5 Consider the circuit shown in Fig. 10.11. Find the limit of input voltage amplitude before the output voltage is saturated (Fig. 10.11).**

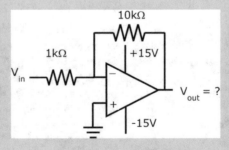

**Fig. 10.11** The circuit in Example 10.5

(continued)

**Example 10.5** (continued)

*Solution.* The figure shows an inverting amplifier with the gain of

$$\frac{V_o}{V_{in}} = \frac{-Z_f}{Z_{in}} = -\frac{10 \text{ k}}{1 \text{ k}} = -10$$

The input voltage is therefore amplified tenfold. The power supply limit is shown as $\pm 15$ V. Any amplified voltage outside this band is clamped. To reach an output of 15 V with a gain of 10, the input voltage must not be above $V_{in} = \frac{15}{10} = 1.5$ V.

**Example 10.6 Find the output voltage of the circuit shown in Fig. 10.12 when a unit step is applied to the input (Fig. 10.12).**

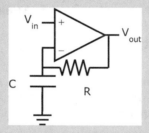

**Fig. 10.12** Figure of the circuit in Example 10.6

*Solution.* The figure shows a noninverting amplifier, in which the transfer function is obtained as

$$\frac{V_o}{V_{in}} = 1 + \frac{Z_2}{Z_1}$$

From the circuit,

$$Z_2 = R$$

$$Z_1 = \frac{1}{sC}$$

Therefore,

(continued)

**Example 10.6** (continued)

$$\frac{V_o}{V_{in}} = 1 + RCs$$

Applying a unit step function

$$V_{in} = \frac{1}{s}$$

results in

$$V_o = \frac{1}{s}(1 + RCs)$$

Taking Laplace inverse obtains

$$v_o(t) = u(t) + RC\delta(t)$$

## Mathematical Operations

So far, Opamps have been used to scale a signal as an amplifier. Opamps can also perform other mathematical operations such as addition, subtraction, differentiation, integration, unity follower, and comparator.

### Adder

Operational amplifiers can add multiple signals with individually controlled gains. Consider an inverting amplifier connected to the inverting port with multiple input impedance sources by various voltages. This circuit is shown in Fig. 10.13.

The voltage at node ① is virtually ground because the noninverting port $v^+$ is connected to the ground. This causes currents to flow from the voltages to the node and from the node to the output. A summation of these currents flow to the feedback resistor as follows:

$$I_1 = \frac{V_1 - 0}{R_1}$$

**Fig. 10.13** Adder circuit
with three inputs and
inverting

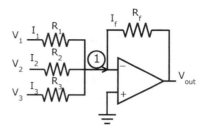

$$I_2 = \frac{V_2 - 0}{R_2}$$

$$I_3 = \frac{V_3 - 0}{R_3}$$

KCL at node ① shows

$$I_1 + I_2 + I_3 = I_f$$

$$\frac{V_1}{R_1} + \frac{V_2}{R_2} + \frac{V_3}{R_3} = \frac{0 - V_o}{R_f}$$

Therefore,

$$V_o = -\frac{R_f}{R_1}V_1 - \frac{R_f}{R_2}V_2 - \frac{R_f}{R_3}V_3$$

If the input resistors are equal,

$$R_1 = R_2 = R_3 = R$$

Then,

$$V_o = -\frac{R_f}{R}(V_1 + V_2 + V_3)$$

**Example 10.7 Find the circuit's output voltage shown in Fig. 10.14, once
without the supply voltage limit and once considering the given supply
voltage limits. Consider $V_1 = 10$ V, $V_2 = -18$ V, and $V_3 = 20$ V.**

(continued)

**Example 10.7** (continued)

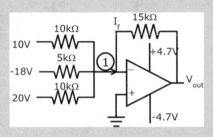

**Fig. 10.14** The circuit in Example 10.7

*Solution.* The output voltage is obtained as

$$V_o = -\frac{R_f}{R_1}V_1 - \frac{R_f}{R_2}V_2 - \frac{R_f}{R_3}V_3$$

$$V_o = -\frac{15\ \text{k}}{10\ \text{k}}10 - \frac{15\ \text{k}}{5\ \text{k}}(-18) - \frac{15\ \text{k}}{10\ \text{k}}20 = +9\ \text{V}$$

When there is no saturation, the output voltage reaches +9 V. However, the supply voltage limit of 4.7 V limits the calculated output voltage of 9 V to 4.7 V. Therefore, the output does not exceed 4.7 V.

**Example 10.8** In an adder circuit, if all input resistors are equal, $R_1 = R_2 = R_3 = R$, and the feedback resistor is $R_f = \frac{1}{3}R$, find the output voltage.

*Solution.* The output voltage is calculated according to

$$V_o = -\frac{R_f}{R_1}V_1 - \frac{R_f}{R_2}V_2 - \frac{R_f}{R_3}V_3$$

Replacing the values yields

$$V_o = -\frac{1}{3}(V_1 + V_2 + V_3)$$

**Example 10.9 Design a four-bit digital to analog converter.**
*Solution.* A four-bit binary number is known as $b_3b_2b_1b_0$. The number in base-10 equals

$$2^0b_0 + 2^1b_1 + 2^2b_2 + 2^3b_3$$

Considering the bits $b_0$–$b_3$ as various inputs to the DAC, the gains that need to be obtained for each of these inputs are $2^0$, $2^1$, $2^2$, $2^3$, respectively. Therefore, the circuit is obtained as follows (Fig. 10.15):

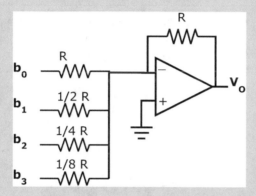

**Fig. 10.15** A four-bit digital to analog converter

## Subtraction

Consider the circuit shown in Fig. 10.16. Two input voltages are applied to the noninverting port and a negative feedback amplifier through the inverting port. This results in the output being a gained subtract of the two voltages.

Since there are two outputs, the superposition can be applied. For this purpose, let us consider input voltages one by one while the other voltage is turned off, i.e., zero.

**Fig. 10.16** Subtracting circuit

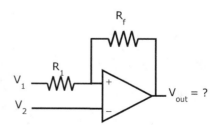

Effect of $V_1$ while $V_2 = 0$. The circuit becomes an inverting amplifier with the output to be

$$\frac{V_{o1}}{V_1} = -\frac{R_f}{R_1}$$

Now, consider the effect of $V_2$ while $V_1 = 0$. The circuit becomes a noninverting amplifier with the gain calculated as

$$\frac{V_{o2}}{V_2} = 1 + \frac{R_f}{R_1}$$

Therefore, the output voltage under the effect of two signals is the summation of

$$V_o = \frac{V_{o1}}{V_1} + \frac{V_{o2}}{V_2}$$

$$V_o = -\frac{R_f}{R_1} V_1 + \left(1 + \frac{R_f}{R_1}\right) V_2$$

If $\frac{R_f}{R_1} \ll 1$

$$V_o \approx -\frac{R_f}{R_1} V_1 + \left(\frac{R_f}{R_1}\right) V_2$$

$$V_o = \frac{R_f}{R_1} (V_2 - V_1)$$

**Example 10.10 The measured signal must be subtracted from the reference signal in a vehicle cruise control system. Using an Opamp, design a circuit to accomplish this operation. (In this example, the speed is measured by an equivalent voltage.)**

*Solution.* Two signals are being subtracted. Figure 10.17 shows the circuit diagram.

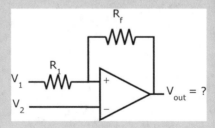

**Fig. 10.17** The circuit used in Example 10.10

**Fig. 10.18** An integrating circuit

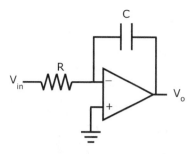

## Integrator

The inverting amplifier circuit was discussed earlier. The gain of the amplifier was obtained as the ratio of the feedback impedance over the input impedance. If the feedback impedance is replaced with an integrating element, such as a capacitor with impedance $\frac{1}{sC}$, the resultant transfer function becomes an integrator (Fig. 10.18).

$$\frac{V_0}{V_{in}} = - \frac{\frac{1}{sC}}{R} = - \left(\frac{1}{RC}\right) \frac{1}{s}$$

$$v_0(t) = - \left(\frac{1}{RC}\right) \int v_{in}(t) \ dt$$

In this transfer function $\frac{1}{RC}$ is the gain and $\frac{1}{s}$ is the integrating agent. This transfer function takes the integral of the input signal and scales the integrated signal.

**Example 10.11 Find the output voltage of the following circuit considering a sinusoidal input as $v_{in} = 0.1 \sin 377 \, t$ (Fig. 10.19).**

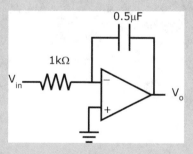

**Fig. 10.19** The circuit in Example 10.11

(continued)

**Example 10.11** (continued)

$$v_0(t) = -\left(\frac{1}{RC}\right) \int v_{in}(t) \ dt$$

$$v_0 = \left(\frac{-1}{1e3 \times 0.5e - 6}\right) \int 0.1 \sin 377t dt$$

$$v_0 = -2000 \times 0.1 \int \sin 377t dt$$

$$v_0 = \frac{200}{377} \cos 377t \ V$$

$$v_0 = 0.53 \cos 377t \ V$$

**Example 10.12 Find the limit of gain in which the integrator circuit introduced in the previous example remains unsaturated. The supply voltage amplitude is ±4.7 V.**

*Solution.* The output voltage of the integrator when the sinusoidal voltage of frequency $\omega$ is applied (in the time domain) is obtained as

$$v_0 = -\left(\frac{1}{RC}\right) \int v \sin \omega t = \frac{v}{RC\omega} \cos \omega t$$

$$\frac{v}{RC\omega} = 4.7$$

Given $v = 0.1$ V and $\omega = 377$ rad/s.

$$\frac{0.1}{RC \times 377} = 4.7$$

$$RC = 56.43e - 6$$

Considering $C = 1$ μF results in $R = 56.43 \ \Omega$.

## Differentiator

Considering the gain of inverting amplifier, as the ratio of feedback to input impedance, and taking input impedance as $\frac{1}{sC}$ yield a Laplace operator in the transfer function numerator, making the system a differentiator. The differentiator circuit is shown in Fig. 10.20.

The transfer function is obtained as follows:

**Fig. 10.20** A differentiator
circuit

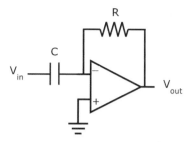

$$\frac{V_0}{V_{\text{in}}} = -\frac{R}{\frac{1}{sC}} = -RCs$$

Taking Laplace inverse of the function results in the output voltage as

$$v_0(t) = -RC\frac{dv_{\text{in}}(t)}{dt}$$

**Example 10.13** Considering $v_{\text{in}}(t) = 5e - 3\sin 4000\pi t$, $R = 1$ M$\Omega$,
and $C = 1$ µF, in a differentiator circuit, find the output voltage.
*Solution.*

$$v_0(t) = -RC\frac{dv_{\text{in}}(t)}{dt}$$

$$v_0(t) = -1e6 \times 1e - 6\frac{d5e - 3\sin 4000\pi t}{dt}$$

$$v_0(t) = -20\pi\cos 4000\pi t \ \text{V}$$

## *Comparator*

This circuit compares two signals and sets the output as the supply voltage source
associated with the highest of the two. For instance, the output is $+V_{\text{cc}}$ if the voltage
applied to the noninverting port is higher than the voltage applied to the inverting
port. The output becomes $-V_{\text{cc}}$ if the voltage applied to the inverting port is higher
than the voltage applied to the noninverting port. This is shown as follows
(Fig. 10.21):
   If

**Fig. 10.21** A comparator
circuit

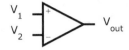

$$v^+ > v^- \Rightarrow v_0 = + V_{cc}$$
$$v^+ < v^- \Rightarrow v_0 = - V_{cc}$$

This is due to the highest gain applied to the difference of

$$v^+ - v^-$$

Infinite gain is saturated at the level of $\pm V_{cc}$ to result in the pulse of appropriate polarity. The speed of reaching the output to the highest level or "slew rate" is also very high, making the comparator operate almost instantaneously.

## Pulse Width Modulation (PWM)

Knowing the operation of comparator circuits, if a DC voltage is compared with sawtooth signals, the result is a pulse width modulation signal. The circuit is shown in Fig. 10.22.

The figure shows that the sawtooth is compared with a DC level voltage. Since the DC voltage is connected to the noninverting port, when the DC voltage is higher than the value of the sawtooth, the output voltage becomes the value of $V_{cc}$. As the sawtooth becomes larger than the DC, the output becomes zero. This results in a pulse. To modulate the pulse width, the DC voltage level can change, and increasing the level of DC extends the time of the $V_{cc}$ and therefore extends the pulse width. Lowering the DC value also shortens the pulse width, making the train of pulses a controllable PWM.

It should be noted that the maximum value of DC voltage should not increase the amplitude of pulses; otherwise, the output voltage becomes a continuous step function. In this condition, no pulse is created.

## Unit Follower

Buffer layers or isolation layers are utilized to isolate different stages of an amplifier because of their power needs. Opamps can be used as a buffer stage to isolate the input layer from the output, while the output follows the input signal at a higher power rating. Figure 10.23 shows a unit follower circuit.

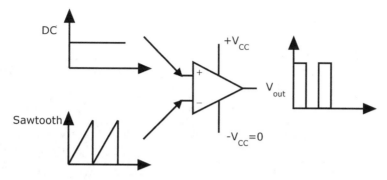

**Fig. 10.22** Inputs of a comparator to generate controlled width pulses

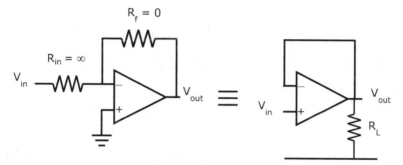

**Fig. 10.23** A unit follower circuit

In this circuit, the noninverting input is virtually connected to the output layer, while it is isolated through the infinite input impedance of the Opamp. The amount of power provided at the output depends on the power rating of the Opamp.

In this circuit, the noninverting amplifier has zero ohm feedback and an infinite ohm impedance to the ground. Therefore,

$$\frac{V_0}{V_{in}} = 1 + \frac{R_f}{R_1} = 1 + \frac{0}{\infty} = 1$$

$$V_0 = V_{in}$$

$$Z_{in} = \infty$$

## Function Builder

As explained earlier, operational amplifiers were used to build functions in analog computers. What was explained earlier in this chapter was the formation of simple algebraic equations, differentiators, and simple integrators. However, these functions can vary from simple algebraic operations to complicated differential equations.

**Example 10.14 Using operational amplifiers, build a circuit to realize the following differential equation.**

$$\dot{y} + 2y = \sin t$$

*Solution.* The system requires an integrator and a summation. To convert the equation to an integral equation, there is a need to take the integral of both sides as follows:

$$\int (\dot{y} + 2y = \sin t)$$

$$y + 2\int y = \int \sin t$$

Solving for $y$ results in

$$y = -2\int y + \int \sin t$$

As one of the signals is negative and one is positive, there are multiple ways to implement the function. It is recommended that the signal $\sin t$ is passed through an inverting unity gain amplifier and then summed and integrated, as shown in Fig. 10.24.

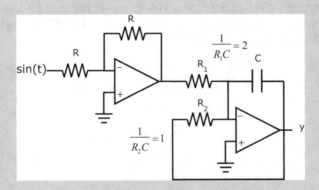

**Fig. 10.24** Circuit realization for Example 10.14

(continued)

**Example 10.14** (continued)

The gains of the inverting integrator are the coefficient for the $\int \sin t$ function as

$$\frac{1}{R_2 C} = 1$$

Moreover, the gain for $2 \int y$ term is

$$\frac{1}{R_1 C} = 2.$$

**Example 10.15 Design a circuit that builds the dynamics of the following system.**

$$\ddot{y} + 2\dot{y} + y = u.$$

*Solution.* This circuit requires two integrators to form the state space equations as follows:

$$y = x_1$$
$$\dot{x}_1 = x_2$$
$$\dot{x}_2 + 2x_2 + x_1 = u$$

This results in

$$\dot{x}_2 = -x_1 - 2x_2 + u$$

Implementing these two equations through integrators is obtained as follows:

$$\dot{x}_1 = x_2 \rightarrow x_1 = \int x_2$$

(continued)

**Example 10.15** (continued)

$$\dot{x}_2 = -x_1 - 2x_2 + u \rightarrow x_2 = \int -x_1 - 2x_2 + u$$

The implemented circuit is shown in Fig. 10.25.

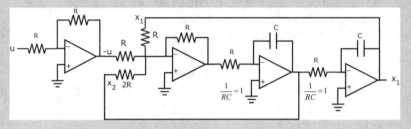

**Fig. 10.25**   Circuit realization of Example 10.15

## Negative Immittance Converter

The NIC has one input and one output port. The impedance inversion might occur at certain frequencies, but at least one frequency at which the impedance measured at the input port is negative of the impedance connected at the output port. The input and output ports are interchangeable, and the circuit is reciprocal. The NIC can be constructed from any two-port device with a voltage gain higher than 2. Chapter 12 discusses two-port networks.

## Negative Impedance

The circuit of Fig. 10.26 shows how an Opamp can be utilized in impedance-converting circuits.
The current $i$ equals

$$i = \frac{V_{in} - V_o}{Z}$$

The voltage of inverting port equals the $V_{in}$. Therefore, the current flows through resistors $R_1$ as $i_1$ and $i_2$ become

**Fig. 10.26** Opamp-based
negative impedance
converter

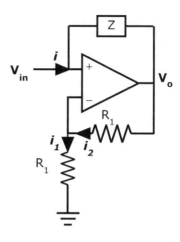

$$i_1 = \frac{V_{in}}{R_1}$$

$$i_2 = \frac{V_o - V_{in}}{R_1}$$

And

$$i_1 = i_2$$

Therefore,

$$V_o = 2V_{in}$$

Replacing this into the current equation results in

$$i = \frac{V_{in} - 2V_{in}}{Z}$$

$$\frac{V_{in}}{i} = -Z$$

The entire circuit shows the negative feedback impedance at the input. Therefore,

$$Z_{in} = -Z$$

**Fig. 10.27** Opamp-based
negative resistance
converter

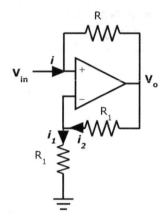

## Negative Resistance (Fig. 10.27)

Following the same approach as in the negative impedance converter, the input
resistance becomes

$$R_{in} = -R$$

## Negative Capacitance (Fig. 10.28)

Following the same approach as in the negative impedance converter, the input
impedance becomes

$$Z_{in} = -Z = -\frac{1}{j\omega C}$$

**Fig. 10.28** Negative
capacitance converter

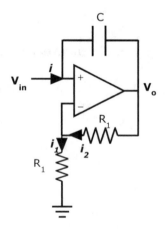

$$Z_{in} = \frac{j}{\omega C}$$

## Negative Inductance (Fig. 10.29)

The current $i$ equals

$$i = \frac{V_{in} - V_o}{R_1}$$

The voltage of inverting port equals the $V_{in}$. Therefore, the current flows through resistors $R_1$ as $i_1$ and $i_2$ becomes

$$i_1 = \frac{V_{in}}{R_1}$$

$$i_2 = \frac{V_o - V_{in}}{\frac{1}{j\omega C}} = j\omega C \ (V_o - V_{in})$$

And

$$i_1 = i_2$$

Therefore,

**Fig. 10.29** Negative capacitance converter

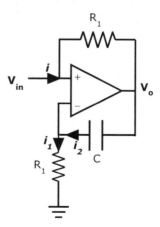

$$V_o = V_{in}\left(1 + \frac{1}{j\omega R_1 C}\right)$$

Replacing this into the current equation results in

$$i = \frac{V_{in} - V_{in}\left(1 + \frac{1}{j\omega R_1 C}\right)}{R_1}$$

$$R_1 i = V_{in}\frac{-1}{j\omega R_1 C}$$

$$\frac{V_{in}}{i} = -j\omega R_1^2 C$$

$$Z_{in} = -j\omega R_1^2 C$$

## Gyrator

A gyrator is an element with an input and an output port, known as a two-port device shown in Fig. 10.30. There is a specific relation between the input and output parameters.

The input current $i_1$ is a linear scale of the output voltage $v_2$, and the current output $i_2$ is linearly proportional to the inverted input voltage $v_1$. This means

$$i_1 = \alpha \, v_2$$

$$i_2 = -\alpha \, v_1$$

The scaling factor $\alpha$ is also known as gyration conductance. The impedance conversion of an ideal gyrator is obtained as follows:

$$Z_1 = \frac{v_1}{i_1}$$

Replacing $i_1$ and $v_1$ values results in

**Fig. 10.30**  Circuit of a
Gyrator with conductance $\alpha$

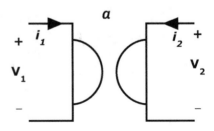

$$Z_1 = \frac{v_1}{i_1} = \frac{-\frac{i_2}{\alpha}}{\alpha v_2} = \frac{1}{\alpha^2} \frac{-i_2}{v_2}$$

Considering the output admittance $Y_2$ and the direction of the current that is out of the admittance yields

$$Y_2 = \frac{-i_2}{v_2}$$

The input impedance becomes

$$Z_1 = \frac{1}{\alpha^2} Y_2$$

**Example 10.16** Find the input impedance of a gyrator when a load resistance $R_L$ is connected to the output port, as shown in Fig. 10.31.

*Solution.* The output admittance is

$$Y_2 = \frac{1}{R_L}$$

Therefore,

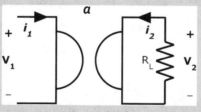

**Fig. 10.31** Circuit of Example 10.16 $Z_1 = \frac{1}{\alpha^2} \frac{1}{R_L}$

**Example 10.17** Find the input impedance of a gyrator when a capacitance $C$ is connected to the output port, as shown in Figs. 10.17 and 10.32.

*Solution.* The admittance of the capacitor is

(continued)

**Example 10.17** (continued)

$$Y_2 = j\omega C$$

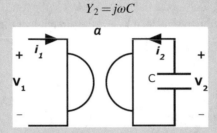

**Fig. 10.32** Circuit of Example 10.17

Replacing the gyrator equation results in

$$Z_1 = \frac{1}{\alpha^2} Y_2 = \frac{1}{\alpha^2} j\omega C = j\omega \frac{C}{\alpha^2}$$

This imitates an inductor impedance of

$$Z_1 = j\omega \frac{C}{\alpha^2} = j\omega L_{eq}$$

Therefore, at the input, the output capacitor imitates an inductor of value $\frac{C}{\alpha^2}$.

$$L_{eq} = \frac{C}{\alpha^2}$$

**Example 10.18 Find the input impedance of a gyrator when an inductor $L$ is connected to the output port, as shown in Fig. 10.33.**

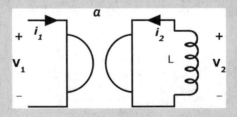

**Fig. 10.33** Circuit of Example 10.18

(continued)

**Example 10.18** (continued)

*Solution.* The output admittance of an inductor is

$$Y_2 = \frac{1}{j\omega L}$$

Replacing the gyrator equation results in

$$Z_1 = \frac{1}{\alpha^2} Y_2 = \frac{1}{\alpha^2} \frac{1}{j\omega L} = \frac{1}{j\omega L \alpha^2}$$

Therefore, at the input, the output inductor imitates a capacitor inductor of value $\alpha^2 L$.

$$Z_1 = \frac{1}{j\omega L \alpha^2} = \frac{1}{j\omega C_{eq}}$$

Therefore,

$$C_{eq} = \alpha^2 L$$

**Example 10.19 Find the input equivalent of a gyrator when a voltage source $E$ is connected to the output port, as shown in Fig. 10.34.**

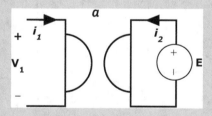

**Fig. 10.34** Circuit of Example 10.19

*Solution.* The input voltage and current relation is

$$i_1 = \alpha \ v_2$$

(continued)

**Example 10.19** (continued)

$$i_2 = -\alpha \ v_1$$

When a voltage source is connected to the output, the voltage $v_2 = E$ regardless of the current output $i_2$. Therefore,

$$i_1 = \alpha v_2 = \alpha \ E$$

The input voltage is as follows:

$$v_1 = \frac{-1}{\alpha} i_2$$

Since the current $i_2$ of a voltage source $E$ can ideally be any value, this equation imitates that the voltage in the input can take any value. However, the amount of current in the input is bounded by the value of the voltage source $E$. This fits into the operation of a current source in the input.

Therefore, the voltage source $E$ is converted to a current source of value $I = \alpha E$.

**Example 10.20 Find the input equivalent of a gyrator when a current source $I$ is connected to the output port, as shown in Fig. 10.35.**

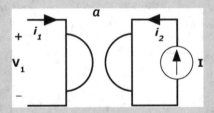

**Fig. 10.35** Circuit of Example 10.20

*Solution.* The input voltage and current relation is

$$i_1 = \alpha \ v_2$$

$$i_2 = -\alpha \ v_1$$

When a current source is connected to the output, the current $i_2 = I$ regardless of the output voltage $v_2$. Therefore,

(continued)

**Example 10.20** (continued)

$$v_1 = \frac{-1}{\alpha} I$$

The output voltage across the current source can take any value. Therefore, the input current can take any value as follows:

$$i_1 = \alpha v_2$$

Since the voltage $v_2$ of a current source $I$ can ideally be any value, this equation imitates that the current in the input can take any value. However, the voltage in the input is bounded by the value of the current source $I$. This fits into the operation of a voltage source in the input.

Therefore, the current source $I$ is converted to a voltage source of value $V = \frac{1}{\alpha} I$.

## Realization of a Gyrator in Circuits

A gyrator can be realized using operational amplifiers. The circuit creates a voltage-controlled current source (VCCS) when the input current is zero, and the input voltage controls the output current shown in Fig. 10.36.

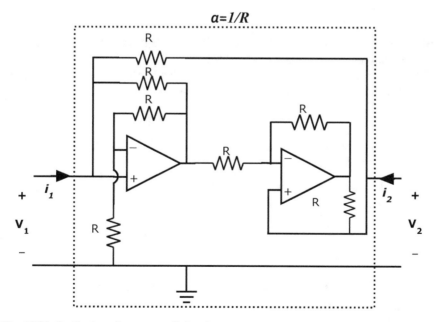

**Fig. 10.36** Realization of a gyrator utilizing Opamps

## Problems

10.1. Design an inverting amplifier to obtain a gain of 100. What is the output voltage if the input is $v_{in}(t) = 0.2 \sin t$ V. Discuss your input and feedback impedance choices.

10.2. An inverting amplifier has a gain of 500. Calculate and sketch the output voltage when the input is 0.2 V at 1 kHz.

10.3. An inverting amplifier has a gain of 500 and the source voltage of $\pm 15$ V. Calculate and sketch the output voltage when the input is 0.2 V at 1 kHz.

10.4. The frequency response of an inverting amplifier with a gain of 50 is shown. Find the output voltage at the given frequencies.

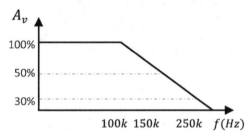

(a) $v_{in} = 5$ at $f = 10$ kHz
(b) $v_{in} = 5$ at $f = 50$ kHz
(c) $v_{in} = 5$ at $f = 100$ kHz
(d) $v_{in} = 5$ at $f = 150$ kHz
(e) $v_{in} = 5$ at $f = 250$ kHz

10.5. The frequency of a 100 mV signal to be amplified is 40 kHz. Find the slew rate required from an operational amplifier.

10.6. Slew rate of an operational amplifier is 2 V/ms. Determine whether its slew rate is suitable to amplify the following signals.

(a) 10 V, 5 Hz
(b) 10 V, 10 Hz
(c) 10 V, 50 Hz
(d) 10 V, 500 Hz

10.7. Find the slew rate required for each of the following signals.

(a) 1 V, 500 Hz
(b) 1 V, 1 kHz
(c) 1 V, 100 kHz
(d) 1 V, 10 MHz
(e) 5 V, 500 Hz
(f) 5 V, 10 kHz
(g) 5 V, 100 kHz

(h) 5 V, 10 MHz
(i) 15 V, 500 Hz
(j) 15 V, 1 kHz
(k) 15 V, 100 kHz
(l) 15 V, 10 MHz

10.8. Design a noninverting amplifier at a gain of 51. What is the output voltage if the input is $v_{in}(t) = 0.2 \sin t$ V.

10.9. Design a noninverting amplifier at a gain of 1001. What is the output voltage if the input is $v_{in}(t) = 0.2 \sin t$ V.

## Add and Subtract

10.10. Design a circuit to perform the following operations, $x_1$, $x_2$, $x_3$, inputs; $y$, output.

(a) $y = x_1 + 10x_2$
(b) $y = x_1 + 5x_2 - 6x_3$
(c) $y = 10(x_1 + x_2 - x_3)$
(d) $y = -10(x_1 + x_2 - x_3)$

## Differentiators

10.11. Design a differentiator circuit using inductors to have a gain of 30. What is the output of the amplifier if the input is $v_{in} = 0.5 \sin 500t$.

10.12. Design a differentiator circuit using a capacitor to have a gain of 100. What is the output of the amplifier if the input is $v_{in} = 0.05 \sin 500t$.

10.13. Using Opamps build the following signals, wherein $x$, $x_1$, $x_2$, input; $y$, output.

(a) $y = 20\frac{dx}{dt}$
(b) $y = -0.1\frac{dx_1}{dt} + \frac{dx_2}{dt}$
(c) $y = 150\frac{d^2x}{dt^2}$

## Integrators

10.14. Design an integrator circuit using inductors to have a gain of 30. What is the output of the amplifier if the input is $v_{in} = 0.5 \sin 500t$.

10.15. Design an integrator circuit using a capacitor to have a gain of 100. What is the output of the amplifier if the input is $v_{in} = 0.05 \sin 500t$.

10.16. Using Opamps build the following signals, wherein $x$, $x_1$, $x_2$, input; $y$, output.

(a) $y = 12 \int x dt$

(b) $y = 50 \int x_1 dt + 0.1 \int x_2 dt$

(c) $y = 800 \int \int x dt$

## Build Analog Computers

10.17. Build an operational amplifier circuit to implement the following systems ($x$, input; $y$, output):

(a) PID controller utilizes an error $e$ as the difference between the input $x$ and the output $y$ as $e = x - y$. A proportional, integral, and differential gain is applied to the error signal individually, and their effects are added to form the control signal. Build the controller using Opamp. Control $=$ $k_p e + k_I \int e + k_d \frac{d}{dt} e$.

(b) $\dot{y} + 3y = x$

(c) $\ddot{y} + \dot{y} + 3y = x$

(d) $\ddot{y} + 5\dot{y} + 6y = \int x dt$

(e) $\ddot{y} + 11\dot{y} + 30y = \dot{x} + 5x$

(f) $\ddot{y} + 11\dot{y} + 30y = \ddot{x} + \dot{x} + x$

10.18. Design an eight-bit digital to analog converter.

10.19. Find the output voltage in the following circuit.

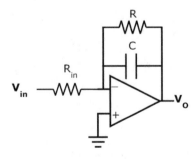

10.20. In the following circuit $R = 100\ \Omega$, $C = 50\ \mu F$. Find the output if the input signal is $v_{in} = 10 \sin 1000t$.

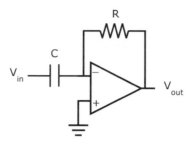

10.21. In Problem 10.20, if the power supply of the Opamp is limited to ±9 V, determine the output voltage. Sketch a rough estimate of the output voltage.

10.22. Determine the function at the output.

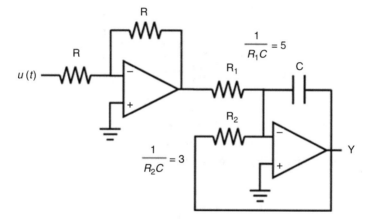

10.23. Find the output voltage in the following circuit.

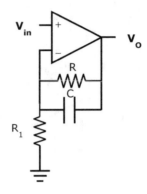

10.24. Find the output voltage in the following circuit.

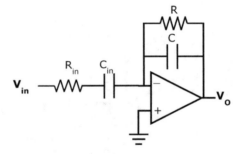

10.25. Find the output voltage in the following circuit.

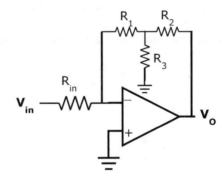

# Chapter 11
# Active Filters

## Introduction

Filters are frequency-selective circuits that deliberately allow or block a range of frequencies while passing from input to output. Passive filters were circuits that accomplish this task using passive circuit elements such as $R$, $L$, and $C$. The maximum gain of the output signal is 100% of the input signal when operated at the resonant or a frequency with orders of magnitude higher or lower than the cutoff frequency.

However, to amplify and filter the signals at the same time, an active element such as Opamp can be utilized. These circuits are called active filters. The gain of these filters can be adjusted, as explained in this chapter. Several devices can create active circuits: (1) a high-gain operational amplifier (at least 60 dB), (2) a low-gain voltage amplifier (20 dB or less), (3) a two-port device that can make the impedance connected on one terminal appear negative on the other terminal called negative immittance converter (NIC), and (4) a gyrator that converts capacitance to inductance and inductance to capacitance.

## Active Low-Pass Filter

As explained earlier, the gain of an inverting amplifier utilizing Opamp can be written as the negative ratio of the impedance at the feedback branch over the impedance at the input branch. Figure 11.1 is a reminder of how the operational amplifier gains when being used as an inverting amplifier can be obtained.

To design any filter, it should be noted that decreasing impedance at the feedback decreases the gain of the amplifier, and decreasing the impedance at the input line increases the gain of the amplifier. The impedance on these branches can be adjusted by a frequency that turns the circuit into an active filter.

A. Izadian, *Fundamentals of Modern Electric Circuit Analysis and Filter Synthesis*, https://doi.org/10.1007/978-3-031-21908-5_11

**Fig. 11.1** Inverting amplifier gain is directly proportional to the impedance at the feedback and inverse of the input impedance. The output voltage waveform is $180°$ out of phase from the input voltage

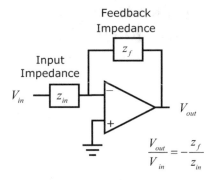

## *Active Low-Pass Filters Using Feedback Impedance*

To design a low-pass filter, if considered on the feedback line, the impedance must decrease as the frequency increases. In the meantime, it must be limited to a resistive circuit at DC to provide suitable amplification.

Consider a parallel RC circuit. As the frequency drops, the equivalent of an open loop capacitor and a parallel resistor is the resistance of the resistor. As the frequency increases, the capacitor becomes a short circuit and makes the equivalent circuit a short circuit. Once used on the feedback line, the change of impedance and its consequent amplifier gain change is aligned with what an LPF does. Figure 11.2 shows an active LPF circuit.

The transfer function of the filter is obtained by dividing the impedances of feedback over the input as follows:

$$\text{TF}(s) = -\frac{R_f \left\| \frac{1}{sC_f} \right.}{R_{\text{in}}}$$

$$\text{TF}(s) = -\frac{\frac{R_f \frac{1}{sC_f}}{R_f + \frac{1}{sC_f}}}{R_{\text{in}}} = -\frac{R_f}{R_{\text{in}}}\frac{1}{R_f C_f s + 1} = -\frac{R_f}{R_{\text{in}}}\frac{\frac{1}{R_f C_f}}{s + \frac{1}{R_f C_f}}$$

$$\text{TF}(s) = -K\frac{\omega_c}{s + \omega_c}$$

where $K = \frac{R_f}{R_{\text{in}}}$ is the gain of the filter and $\omega_c = \frac{1}{R_f C_f}$ is the cutoff frequency of the circuit.

**Fig. 11.2** Realizing an
active LPF using a parallel
*RC* circuit at the feedback.
The capacitance in parallel
to the resistor controls the
impedance of the feedback
at various frequencies.
Hence, the gain at various
frequencies can be
controlled

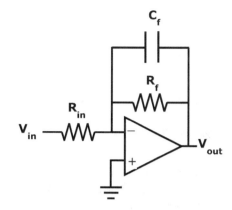

**Fig. 11.3** Realizing an
active LPF using a series *RL*
circuit at the input

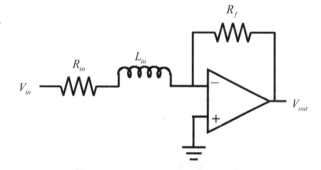

## *Active Low-Pass Filters Using Input Impedance*

The input impedance, as stated before, has a reverse effect on the gain of an inverted
amplifier. The input impedance should be low at low frequencies and increase at
higher frequencies to form a low-pass filter. This means the utilization of an
inductor. To prevent high-gain at low frequencies (the inductor becomes a short
circuit), a resistor needs to be connected in series with the inductor. Figure 11.3
shows the low-pass filter circuit realized with an *RL* series circuit in the input
impedance.

The transfer function of the filter can be obtained as follows:

$$\text{TF}(s) = -\frac{R_f}{R_{\text{in}} + sL_{\text{in}}}$$

$$\text{TF}(s) = -\frac{R_f}{R_{\text{in}}} \frac{1}{1 + s\frac{L_{\text{in}}}{R_{\text{in}}}} = -\frac{R_f}{R_{\text{in}}} \frac{\frac{R_{\text{in}}}{L_{\text{in}}}}{s + \frac{R_{\text{in}}}{L_{\text{in}}}}$$

$$TF(s) = -K\frac{\omega_c}{s + \omega_c}$$

where $K = \frac{R_f}{R_{in}}$ is the gain of the filter and $\omega_c = \frac{R_{in}}{L_{in}}$ is the cutoff frequency of the circuit.

## Active High-Pass Filters

In high-pass filters, the gain should increase as the frequency increases. That means if a feedback impedance is utilized, it should increase by frequency, and if an input impedance is utilized to realize the filter, the value of the input impedance should decrease as the frequency increases. At DC, the gain should be zero.

### Active High-Pass Filters Using Feedback Impedance

Inductors' impedance increases by frequency. Therefore, a parallel connection of an *RL* circuit at the feedback performs as a high-pass filter. Figure 11.4 shows the high-pass filter realized using an *RL* at the feedback. At low frequencies, the inductor becomes short, and the gain of the amplifier is negligible. At high frequencies, the inductor shows an extremely high impedance. The equivalent resistance in the feedback becomes the value of $R_f$, which limits the gain of the amplifier. This type of frequency response resembles a high-pass filter.

The transfer function of such a filter can be obtained as follows:

$$TF(s) = -\frac{R_f \| sL_f}{R_{in}}$$

**Fig. 11.4** Realizing an active HPF using a parallel *RL* circuit at the feedback

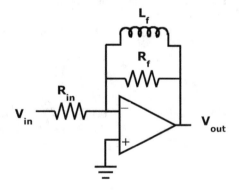

$$\text{TF}(s) = -\frac{\frac{R_f sL_f}{R_f+sL_f}}{R_{in}} = -\frac{R_f}{R_{in}}\frac{sL_f}{R_f+sL_f} = -\frac{R_f}{R_{in}}\frac{s}{s+\frac{R_f}{L_f}}$$

$$\text{TF}(s) = -K\frac{s}{s+\omega_c}$$

where $K = \frac{R_f}{R_{in}}$ is the gain of the filter and $\omega_c = \frac{R_f}{L_f}$ is the cutoff frequency of the circuit.

## Active High-Pass Filters Using Input Impedance

A series $RC$ circuit is utilized to realize an active high-pass filter by adjusting the input impedance of an inverting amplifier. The capacitor generates a high impedance at DC, which makes the amplifier gain very small. As the frequency increases, the input impedance decrease and causes an increase in the gain. The filter circuit of an active high-pass filter is shown in Fig. 11.5.

The transfer function of the circuit can be obtained as follows:

$$\text{TF}(s) = -\frac{R_f}{R_{in} + \frac{1}{sC_{in}}}$$

$$\text{TF}(s) = -\frac{R_f}{R_{in} + \frac{1}{sC_{in}}} = -\frac{R_f}{R_{in}}\frac{s}{s+\frac{1}{R_{in}C_{in}}} = -K\frac{s}{s+\omega_c}$$

where $K = \frac{R_f}{R_{in}}$ is the gain of the filter and $\omega_c = \frac{1}{R_{in}C_{in}}$ is the cutoff frequency of the circuit.

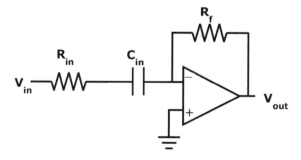

**Fig. 11.5** Realizing an active HPF using a series $RC$ circuit at the input

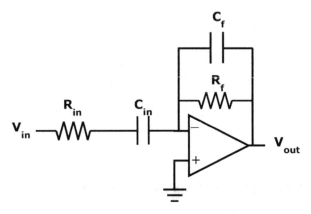

**Fig. 11.6** Realizing an active BPF using a parallel $RC$ at the feedback and a series $RC$ circuit at the input. This is a combination of LPF and HPF, where the LPF and HPF cutoff frequencies are selected to form a BPF

## Active Band-Pass Filters

Band-pass filters can be made by either combining the low- and high-pass filters or designing a new circuit approach. Figure 11.6 shows an active band-pass filter as a combination of both impedances on the feedback and in the input. Using $RC$ circuits requires a parallel branch at the feedback and a series branch at the input. The gain is small at DC because of the high-input impedance, and the gain also drops at high frequencies because of the small impedance at the feedback.

The circuit transfer function is obtained as follows:

$$\mathrm{TF}(s) = -\frac{R_f \left\| \frac{1}{sC_f}\right.}{R_{\mathrm{in}} + \frac{1}{sC_{\mathrm{in}}}}$$

$$\mathrm{TF}(s) = -\frac{\frac{R_f \frac{1}{sC_f}}{R_f + \frac{1}{sC_f}}}{\frac{R_{\mathrm{in}} C_{\mathrm{in}} s + 1}{sC_{\mathrm{in}}}} = -\frac{\frac{R_f \frac{1}{sC_f}}{\frac{R_f C_f s + 1}{sC_f}}}{\frac{R_{\mathrm{in}} C_{\mathrm{in}} s + 1}{sC_{\mathrm{in}}}} = -\frac{\frac{R_f \frac{1}{sC_f}}{\frac{R_f C_f s + 1}{sC_f}}}{\frac{R_{\mathrm{in}} C_{\mathrm{in}} s + 1}{sC_{\mathrm{in}}}} = -\frac{\frac{R_f}{R_f C_f s + 1}}{\frac{R_{\mathrm{in}} C_{\mathrm{in}} s + 1}{sC_{\mathrm{in}}}}$$

$$TF(s) = -\frac{R_f C_{in} s}{(R_f C_f s + 1)(R_{in} C_{in} s + 1)} = -\frac{R_f C_{in} s}{R_f C_f R_{in} C_{in}\left(s + \frac{1}{R_f C_f}\right)\left(s + \frac{1}{R_{in} C_{in}}\right)}$$

$$= -\frac{R_f C_{in} s}{R_f C_f R_{in} C_{in}\left(s + \frac{1}{R_f C_f}\right)\left(s + \frac{1}{R_{in} C_{in}}\right)}$$

$$= -\frac{1}{C_f R_{in}}\frac{s}{\left(s^2 + \left(\frac{1}{R_f C_f} + \frac{1}{R_{in} C_{in}}\right)s + \frac{1}{R_f C_f}\frac{1}{R_{in} C_{in}}\right)}$$

$$= -\frac{1}{C_f R_{in}\left(\frac{1}{R_f C_f} + \frac{1}{R_{in} C_{in}}\right)}\frac{\left(\frac{1}{R_f C_f} + \frac{1}{R_{in} C_{in}}\right)s}{\left(s^2 + \left(\frac{1}{R_f C_f} + \frac{1}{R_{in} C_{in}}\right)s + \frac{1}{R_f C_f}\frac{1}{R_{in} C_{in}}\right)}$$

$$= -\frac{1}{\left(\frac{R_{in}}{R_f} + \frac{C_f}{C_{in}}\right)}\frac{\left(\frac{1}{R_f C_f} + \frac{1}{R_{in} C_{in}}\right)s}{\left(s^2 + \left(\frac{1}{R_f C_f} + \frac{1}{R_{in} C_{in}}\right)s + \frac{1}{R_f C_f}\frac{1}{R_{in} C_{in}}\right)}$$

$$= -k\frac{BWs}{(s^2 + BWs + \omega_0^2)}$$

where $k = \frac{1}{\left(\frac{R_{in}}{R_f} + \frac{C_f}{C_{in}}\right)}$ bandwidth $BW = \left(\frac{1}{R_f C_f} + \frac{1}{R_{in} C_{in}}\right)$ and the natural frequency are $\omega_0^2 = \frac{1}{R_f C_f}\frac{1}{R_{in} C_{in}}$.

Normally, to form a band-pass filter, the cutoff frequency of the high pass is lower than the cutoff frequency of the low pass. This creates a band of frequencies that pass to the output. This circuit has no resonant frequency; therefore, the *unamplified maximum* value of the transfer function is less than 100%. However, the operational amplifier can boost the maximum.

## Active Band-Pass Filter Using a Combination of Low- and High-Pass Filters

A cascade combination of low-pass and high-pass filters can also perform as a band-pass filter as long as the low-pass and high-pass cutoff frequencies meet certain criteria. The filter is shown in Fig. 11.7.

The following procedures can be utilized to design and implement active band-pass filters (Fig. 11.8):

1. Design an active low-pass filter with unity gain and cutoff frequency of $\omega_{CLP}$.
2. Design an active high-pass filter with unity gain and cutoff frequency of $\omega_{CHP}$.

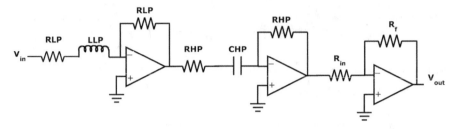

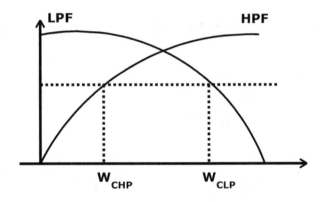

**Fig. 11.7** Cascade connection of two active filters with LPF and HPF configurations and a final gain stage. Overall, the circuit operates as a BPF. The cutoff frequency of LPF is larger than the cutoff frequency of HPF

**Fig. 11.8** In a BPF, the cutoff frequency of LPF can be considered larger than HPFs

To create a band-pass filter, the cutoff frequency of the high-pass filter should be lower than the cutoff frequency of the low-pass filter, or

$$\omega_{CHP} < \omega_{CLP}$$

3. The last stage is to apply gain to the filtered signal and amplify the signals in the desired band of frequencies.

Unity gain low-pass and high-pass filters are desired because they do not amplify the signal filtered by the next stage. Once the signal is filtered, a suitable gain is applied to amplify it to the desired level.

## Transfer Function of a Band-Pass Filter

The circuit transfer function is obtained by a series connection of three blocks, as shown in Fig. 11.9.

**Fig. 11.9** Block diagram equivalent of the BPF, which was a combination of the transfer function of the LPF and HPF, and the gain stage

The transfer function of the overall system is obtained as follows:

$$\text{TF}(s) = -\frac{\frac{R_{LP}}{L_{LP}}}{s + \frac{R_{LP}}{L_{LP}}} \times -\frac{s}{s + \frac{1}{R_{HP}C_{HP}}} \times -\frac{R_f}{R_{in}}$$

$$\text{TF}(s) = -k\left(\frac{\omega_{CLP}}{s + \omega_{CLP}} \times \frac{s}{s + \omega_{CHP}}\right)$$

The gain of this filter is set by the last stage and equals $\frac{R_f}{R_{in}}$, and the cutoff frequencies $\omega_{CLP}$ and $\omega_{CHP}$ can be obtained from individual filters.

## Active Band-Reject Filters

To obtain a band-reject filter, the low-pass and high-pass filters should change from series to parallel, and their summative effort should be amplified in the third stage. A band-reject filter circuit formed by adding the output of a low-pass filter to a high-pass filter is shown in Fig. 11.10.

The following procedures can be utilized to design and implement active band-reject filters (Fig. 11.11):

1. Design an active low-pass filter with unity gain and cutoff frequency of $\omega_{CLP}$.
2. Design an active high-pass filter with unity gain and cutoff frequency of $\omega_{CHP}$.

To create a band-reject filter, the cutoff frequency of the high-pass filter should be higher than the cutoff frequency of the low-pass filter, or

$$\omega_{CHP} > \omega_{CLP}$$

3. The last stage is to apply gain to the filtered signal and amplify the signals in the desired band of frequencies.

The filter transfer function can be obtained as follows (Fig. 11.12):

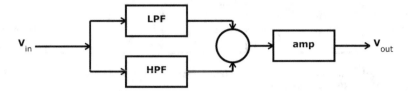

**Fig. 11.10** Realizing an active BRF using a combination of LPF and HPF

**Fig. 11.11** Realizing a BPF
by a combination of LPF
and HPF. The cutoff
frequency of the LPF is
lower than the HPF's.
$\omega_{CHP} > \omega_{CLP}$

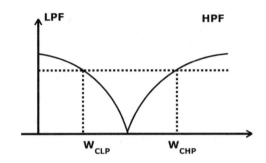

**Fig. 11.12** Realization of a
BRF by summation of an
LPF and HPF

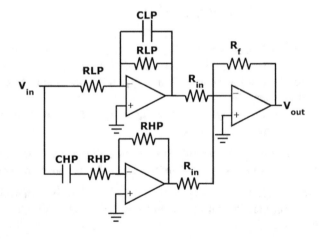

$$TF(s) = -\left(\frac{\frac{1}{R_{LP}C_{LP}}}{s + \frac{1}{R_{LP}C_{LP}}} + \frac{s}{s + \frac{1}{R_{HP}C_{HP}}}\right) \times -\frac{R_f}{R_{in}}$$

$$TF(s) = k\left(\frac{\omega_{CLP}}{s + \omega_{CLP}} + \frac{s}{s + \omega_{CHP}}\right)$$

## Multiple Feedback (MFB) Opamp Circuits

This circuit receives two sets of feedback from the output voltage. Figure 11.13 shows an MFB circuit in which nodes 1 and 2 receive feedback through resistors $R_3$ and $R_5$, respectively.

The circuit analysis shows that the voltage of node 2 is zero because it is virtually connected to the noninverting port of the Opamp. Therefore, the current $I_{R_5}$ becomes

$$I_{R_5} = \frac{0 - V_o}{R_5}$$

It is required to find the voltage of node 1. This results in the currents of resistors. KCL in node 1 shows: (It is assumed that the input voltage $V_{in}$ sources the current to the circuit through $R_1$. Therefore,

$$-\frac{V_{in} - V_1}{R_1} + \frac{V_1 - V_o}{R_3} + \frac{V_1}{R_2} + \frac{V_1 - V_2}{R_4} = 0$$

$$V_2 = 0$$

$$V_1 \left( \frac{1}{R_1} + \frac{1}{R_2} + \frac{1}{R_3} + \frac{1}{R_4} \right) = \frac{V_{in}}{R_1} + \frac{V_o}{R_3}$$

Since the input impedance of the Opamp is infinite,

$$I_{R_5} = I_{R_4}$$

$$-\frac{V_o}{R_5} = \frac{V_1}{R_4}$$

Or

$$V_1 = -\frac{R_4}{R_5} V_o$$

**Fig. 11.13** Multiple feedback circuits

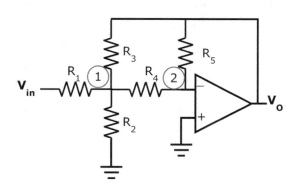

Replacing the KCL equation results in

$$-\frac{R_4}{R_5}V_o\left(\frac{1}{R_1}+\frac{1}{R_2}+\frac{1}{R_3}+\frac{1}{R_4}\right)=\frac{V_{in}}{R_1}+\frac{V_o}{R_3}$$

$$-\left(\frac{R_4}{R_5}\left(\frac{1}{R_1}+\frac{1}{R_2}+\frac{1}{R_3}+\frac{1}{R_4}\right)+\frac{1}{R_3}\right)V_o=\frac{V_{in}}{R_1}$$

$$V_0=-\frac{V_{in}}{R_1\left(\frac{R_4}{R_5}\left(\frac{1}{R_1}+\frac{1}{R_2}+\frac{1}{R_3}+\frac{1}{R_4}\right)+\frac{1}{R_3}\right)}$$

Considering

$$\frac{1}{R_1}=G_1,\frac{1}{R_2}=G_2,\frac{1}{R_3}=G_3,\frac{1}{R_4}=G_4,\frac{1}{R_5}=G_5$$

$$\frac{V_0}{V_{in}}=\frac{-G_1G_4}{G_5(G_1+G_2+G_3+G_4)+G_3G_4}$$

## Creating a Low-Pass Filter

The circuit can be converted to a low-pass filter with the following replacements:

- The conductance of $G_2$ is replaced by a capacitor of admittance $j\omega C_2$.
- The conductance of $G_5$ is replaced by a capacitor of admittance $j\omega C_5$.

Considering these replacements, the circuit of an LPF is shown in Fig. 11.14. The transfer function in steady state sinusoidal analysis becomes

**Fig. 11.14** Low-pass filter using multiple feedback operational amplifier

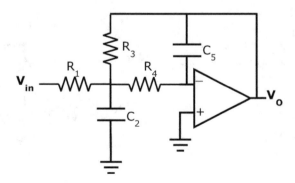

$$\frac{V_0}{V_{in}} = \frac{-G_1G_4}{j\omega C_5(G_1 + j\omega C_2 + G_3 + G_4) + G_3G_4}$$

Simplification results in

$$\frac{V_0}{V_{in}} = \frac{-G_1G_4}{(G_3G_4 - \omega^2 C_2C_5) + j\omega C_5(G_1 + G_3 + G_4)}$$

At low frequencies,

$$\omega \to 0, \quad \frac{V_0}{V_{in}} = \frac{-G_1G_4}{(G_3G_4 - 0^2 C_2C_5) + j0C_5(G_1 + G_3 + G_4)} = -\frac{G_1}{G_3}$$

The DC gain of the amplifier is

$$\text{gain} = \frac{G_1}{G_3}$$

At high frequencies

$$\omega \to \infty, \quad \frac{V_0}{V_{in}} = \frac{-G_1G_4}{(G_3G_4 - \infty^2 C_2C_5) + j\infty C_5(G_1 + G_3 + G_4)} = -\frac{G_1G_4}{\infty} = 0$$

This shows the behavior of an LPF. Considering

$$C_2 = C_5 = C \& G_1 = G_3 = G_4 = G$$

$$\omega_c = \frac{1}{RC}$$

The transfer function of the filter in the Laplace domain can be expressed by considering $s = j\omega$ as follows:

$$\frac{V_0}{V_{in}} = \frac{-G_1G_4}{s^2 C_2C_5 + sC_5(G_1 + G_3 + G_4) + G_3G_4}$$

$$\frac{V_0}{V_{in}} = \frac{-\frac{G_1G_4}{C_2C_5}}{s^2 + s\frac{(G_1+G_3+G_4)}{C_2} + \frac{G_3G_4}{C_2C_5}}$$

## Creating a High-Pass Filter

To build an HPF, the following elements need to be replaced.

- The conductance of $G_1$ is replaced by a capacitor of admittance $j\omega C_1$.
- The conductance of $G_3$ is replaced by a capacitor of admittance $j\omega C_3$.
- The conductance of $G_4$ is replaced by a capacitor of admittance $j\omega C_4$.

The circuit configuration is shown in Fig. 11.15.
Considering the circuit element replacements, the ratio of voltages becomes

$$\frac{V_0}{V_{in}} = \frac{\omega^2 C_1 C_4}{G_5(j\omega C_1 + G_2 + j\omega C_3 + j\omega C_4) - \omega^2 C_3 C_4}$$

$$\frac{V_0}{V_{in}} = \frac{\omega^2 C_1 C_4}{(G_2 - \omega^2 C_3 C_4) + j\omega G_5(C_1 + C_3 + C_4)}$$

At low frequency,

$$\omega \to 0, \frac{V_0}{V_{in}} = \frac{0^2 C_1 C_4}{(G_2 - 0^2 C_3 C_4) + j0G_5(C_1 + C_3 + C_4)} = 0$$

At high frequency,

$$\omega \to \infty, \frac{V_0}{V_{in}} = \frac{-\omega^2 C_1 C_4}{\sqrt{(G_2 - \omega^2 C_3 C_4)^2 + (\omega G_5(C_1 + C_3 + C_4))^2}} = \frac{-\omega^2 C_1 C_4}{-\omega^2 C_3 C_4} = \frac{C_1}{C_3}$$

The zero DC gain, low gains at low frequencies, and high-frequency gain correspond with HPF characteristics. Considering

$$C_1 = C_3 = C_4 = C \& G_2 = G_5 = G$$

The cutoff frequency is

**Fig. 11.15** High-pass filter
using multiple feedback
operational amplifiers

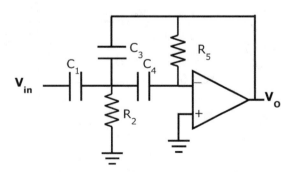

$$\omega_c = \frac{1}{RC}$$

The transfer function of the HPF is

$$\frac{V_0}{V_{in}} = \frac{-s^2 C_1 C_4}{s^2 C_3 C_4 + s G_5 (C_1 + C_3 + C_4) + G_2 G_5}$$

$$\frac{V_0}{V_{in}} = \frac{-s^2 \frac{C_1 C_4}{C_3 C_4}}{s^2 + s \frac{G_5 (C_1 + C_3 + C_4)}{C_3 C_4} + \frac{G_2 G_5}{C_3 C_4}}$$

$$\frac{V_0}{V_{in}} = \frac{-s^2 \frac{C_1}{C_3}}{s^2 + s \frac{G_5 (C_1 + C_3 + C_4)}{C_3 C_4} + \frac{G_2 G_5}{C_3 C_4}}$$

## Creating a Band-Pass Filter

The following replacements in the MFB circuit make it a BPF.

- The conductance of $G_3$ is replaced by a capacitor of admittance $j\omega C_3$.
- The conductance of $G_4$ is replaced by a capacitor of admittance $j\omega C_4$.

The circuit configuration is shown in Fig. 11.16.
The circuit transfer function becomes

$$\frac{V_0}{V_{in}} = \frac{-j\omega G_1 C_4}{G_5 (G_1 + G_2 + j\omega C_3 + j\omega C_4) - \omega^2 C_3 C_4}$$

$$\frac{V_0}{V_{in}} = \frac{-j\omega G_1 C_4}{-\omega^2 C_3 C_4 + G_5 (G_1 + G_2) + j\omega G_5 (C_3 + C_4)}$$

At low frequencies

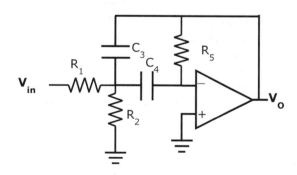

**Fig. 11.16** Band-pass filter using multiple feedback operational amplifier

$$\omega \to 0, \frac{V_0}{V_{in}} = \frac{-j0G_1C_4}{-0^2C_3C_4 + G_5(G_1 + G_2) + j0G_5(C_3 + C_4)} = 0$$

At high frequencies

$$\omega \to \infty, \frac{V_0}{V_{in}} = \frac{-j\omega G_1 C_4}{-\omega^2 C_3 C_4 + G_5(G_1 + G_2) + j\omega G_5(C_3 + C_4)} = \frac{1}{\omega} = 0$$

Somewhere between these high and low frequencies, the amplitude increases and becomes non-zero. This resembles the behavior of a BPF.

The transfer function of the BPF becomes

$$\frac{V_0}{V_{in}} = \frac{-sG_1C_4}{s^2C_3C_4 + sG_5(C_3 + C_4) + G_5(G_1 + G_2)}$$

$$\frac{V_0}{V_{in}} = \frac{-s\frac{G_1}{C_3}}{s^2 + s\frac{G_5(C_3+C_4)}{C_3C_4} + \frac{G_5(G_1+G_2)}{C_3C_4}}$$

The bandwidth of the filter is

$$BW = \frac{G_5(C_3 + C_4)}{C_3C_4}$$

The filter gain is

$$Gain = \frac{G_1C_4}{G_5(C_3 + C_4)}$$

## Problems

11.1. Design an inverting active low-pass filter to amplify the frequencies below 1 kHz with a gain of $K = 15$.

11.2. Design an inverting active high-pass filter to amplify the frequencies above 500 Hz with a gain of $K = 10$.

11.3. Design an active band-pass filter to amplify the frequencies between 500 and 1000 Hz with a gain of $K = 20$.

11.4. Design an active band-reject filter to eliminate the frequencies between 500 and 1000 Hz with a gain of $K = 20$.

11.5. Determine the type of filter in the following circuit.

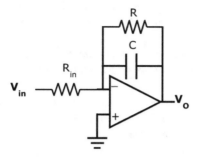

11.6. Determine if the following circuit can be a filter?

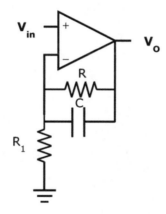

11.7. Determine the type of filter in the following circuit.

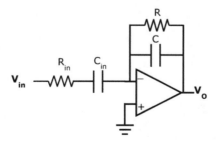

11.8. Design MFB second-order active low-pass filter to amplify the frequencies below 1 kHz with a gain of $K = 15$. Sketch the circuit and determine the component values.

11.9. Design MFB second-order active high-pass filter to amplify the frequencies above 500 Hz with a gain of $K = 10$. Sketch the circuit and determine the component values.

11.10. Design MFB second-order active band-pass filter to amplify the frequencies between 500 and 1000 Hz with a gain of $K = 20$. Sketch the circuit and determine the component values.

11.11. Design a second-order active band-reject filter to eliminate the frequencies between 500 and 1000 Hz with a gain of $K = 20$. Sketch the circuit and determine the component values.

# Chapter 12
# Two-Port Networks

## Introduction

A network combines one or several electric circuits performing a specific action together. The network might have several inputs that receive excitations and outputs to show the results. A port is a set of two terminals that allows a source to connect and excite the network or connect a measurement device and record the response. For instance, a voltage source connected to a port of a network may cause currents to flow through the network and voltage drops to appear across the elements. To measure any of these currents or voltages, terminals may be extended to demonstrate some measurement locations, forming an output port. This forms a two-port network. Similarly, two-port networks can be easily expanded to multiple-port networks, each showing a parameter in the circuit. Figure 12.1 shows a circuit and the process of considering it as a two-port network.

The benefit of showing a circuit as a two-port network is the capability of representing the network in mathematical forms. Similar to transfer functions, the mathematical terms are defined for a specified input port to the desired output port. The mathematical expressions present characteristics such as impedance, admittance, transmission, or a combination of these terms in hybrid forms. Each of these characteristics is unique and is defined regardless of the type of the input and the type of the output. It means that only ports matter and the type of source or element connected to the port is irrelevant to the mathematical expression of the network. Once the network is presented in terms of impedance, admittance, or others, the network can be used in mathematical forms, and the response to several excitations can be obtained. Networks can also be connected in series or parallel; depending on their characteristics, their mathematical terms can be combined to represent a larger network. However, that network is also presented as a two-port network. Figure 12.2 shows a network matrix $N$ and the set of input and output ports. In this network, the set of voltages at the input and output ports are $V_1$ and $V_2$, and

**Fig. 12.1** A circuit showing an input terminal connecting to a source V1 and an output port, the measured voltage across the capacitor

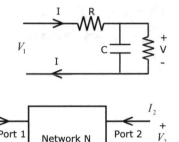

**Fig. 12.2** Terminals of a two-port network system $N$

the currents at those ports are $I_1$ and $I_2$. These voltages and currents might be AC or DC.

In a two-port network, the voltage and current in the input port will cause a voltage or current in the output port. The network presentations have to fit in a 2 × 2 matrix. Depending on the grouping of the parameters, several presentations can be obtained.

For instance,

$\begin{bmatrix} V_1 \\ V_2 \end{bmatrix} = N \begin{bmatrix} I_1 \\ I_2 \end{bmatrix}$ where $N$ shows the impedance of the network and hereafter is shown by $Z$

$$\begin{bmatrix} V_1 \\ V_2 \end{bmatrix} = Z_{2 \times 2} \begin{bmatrix} I_1 \\ I_2 \end{bmatrix}$$

$\begin{bmatrix} I_1 \\ I_2 \end{bmatrix} = N \begin{bmatrix} V_1 \\ V_2 \end{bmatrix}$, where $N$ shows the admittance of the network and hereafter is shown by $Y$

$$\begin{bmatrix} I_1 \\ I_2 \end{bmatrix} = Y_{2 \times 2} \begin{bmatrix} V_1 \\ V_2 \end{bmatrix}$$

$\begin{bmatrix} V_1 \\ I_1 \end{bmatrix} = N \begin{bmatrix} V_2 \\ I_2 \end{bmatrix}$, where $N$ shows a transmission matrix and hereafter is shown by $T$

$$\begin{bmatrix} V_1 \\ I_1 \end{bmatrix} = T_{2 \times 2} \begin{bmatrix} V_2 \\ I_2 \end{bmatrix}$$

This chapter discusses these mathematical representations and the process of obtaining them.

## Impedance Matrix of a Two-Port Network

In this network representation, the entire circuit is expressed by a $Z_{2 \times 2}$ matrix that defines transfer functions as the ratio of port voltages to the network's port currents. The matrix elements show how the voltage at the network ports varies by the current variations through the ports. It is considered that the direction of currents is entering the ports (Fig. 12.3).

The equations that represent an impedance network are as follows:

$$\begin{bmatrix} V_1 \\ V_2 \end{bmatrix} = Z_{2 \times 2} \begin{bmatrix} I_1 \\ I_2 \end{bmatrix}$$

$$\begin{cases} V_1 = Z_{11}I_1 + Z_{12}I_2 \\ V_2 = Z_{21}I_1 + Z_{22}I_2 \end{cases}$$

$$\begin{bmatrix} V_1 \\ V_2 \end{bmatrix} = \begin{bmatrix} Z_{11} & Z_{12} \\ Z_{21} & Z_{22} \end{bmatrix} \begin{bmatrix} I_1 \\ I_2 \end{bmatrix}$$

The impedance matrix has four elements, $Z_{11}$, $Z_{12}$, $Z_{21}$, and $Z_{22}$:

- $Z_{11}$ shows how the voltage of port 1 is related to the current of port 1.
- $Z_{12}$ shows how the voltage of port 1 is related to the current of port 2.
- $Z_{21}$ shows how the voltage of port 2 is related to the current of port 1.
- $Z_{22}$ shows how the voltage of port 2 is related to the current of port 2.

To measure each impedance matrix element, only the voltage and current related to the element will have values, and the other parameters are considered zero. For instance,

$$Z_{11} = \frac{V_1}{I_1}\Big|_{I_2 = 0}$$

**Note 12.1** This shows that to measure or calculate $Z_{11}$, only $V_1$ and $I_1$ are required while imposing $I_2 = 0$. This requires removing or disconnecting the source at the second port to prevent current $I_2$ from flowing.

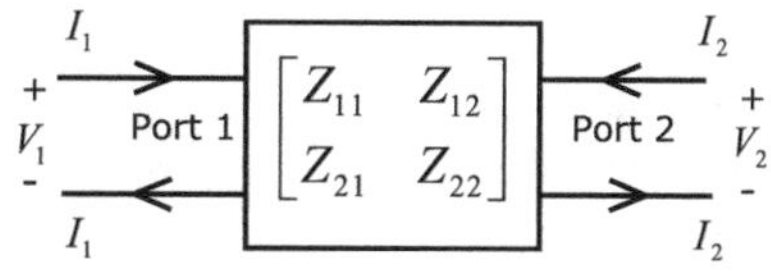

**Fig. 12.3** System network is shown as impedance

Similarly,

$$Z_{12} = \frac{V_1}{I_2}\Big|_{I_1=0}$$

$$Z_{21} = \frac{V_2}{I_1}\Big|_{I_2=0}$$

$$Z_{22} = \frac{V_2}{I_2}\Big|_{I_1=0}.$$

**Note 12.2**  In networks *without dependent sources*, the matrix elements $Z_{12}$ and $Z_{21}$ will become similar, i.e.,

$$Z_{12} = Z_{21}$$

These networks are called *reciprocal*.

## The Equivalent of an Impedance Network

Most of the time, a two-port network can contain multiple loops and nodes. A matrix representation may also not be suitable when cascade networks exist or when part of the original network is missing. Therefore, the two-port network can be simplified to a *T*-equivalent network.

## Reciprocal Networks

### T Model

As mentioned earlier, an impedance network might be reciprocal if it does not contain dependent sources. A two-loop network that shares a common element is presented in a reciprocal network. The element shared in both loops is either $Z_{12}$ or $Z_{21}$, as $Z_{12} = Z_{21}$. Other elements in the first loop are $Z_{11} - Z_{12}$ and in the second loop, $Z_{22} - Z_{12}$. This forms a *T* network, as shown in Fig. 12.4.

**Fig. 12.4**  *T* equivalent for an impedance network

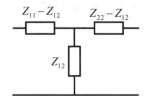

## Nonreciprocal Networks

At least one dependent source makes the $Z_{12}$ and $Z_{21}$ different in a nonreciprocal network. There are two approaches to modeling the impedance matrix (1) having two separate loops or (2) having loops that share one element.

## Separate Loop Model

In this model (shown in Fig. 12.5), the first loop current $I_1$ flows through $Z_{11}$. The loop contains a current-dependent voltage source with the value of $Z_{12}I_2$. Therefore, the KVL becomes

$$V_1 = Z_{11}I_1 + Z_{12}I_2$$

In the second loop, the current $I_2$ flows through impedance $Z_{22}$. A current-dependent voltage source with the value of $Z_{21}I_1$ is also added to form a KVL as

$$V_2 = Z_{21}I_1 + Z_{22}I_2$$

## Element-Sharing Loops

When two loops share one element, the impedance matrix must be written such that a positive element exists in each of the four components, as described below.
  Consider the original impedance matrix as follows:

$$Z = \begin{bmatrix} Z_{11} & Z_{12} \\ Z_{21} & Z_{22} \end{bmatrix}$$

Consider a shared element $Z_m$ exists in all four components of $Z_{11}, Z_{12}, Z_{21}$, and $Z_{22}$. Therefore, the impedance matrix becomes

**Fig. 12.5** Impedance network equivalent in separate loops. The dependent voltage sources show the dependency of the generated voltages

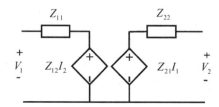

**Fig. 12.6** Element-sharing
impedance equivalent loops

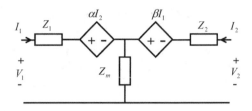

$$Z = \begin{bmatrix} Z_1 + Z_m & Z_m + \alpha \\ Z_m + \beta & Z_2 + Z_m \end{bmatrix}$$

The equivalent element in loop 1 of the $T$ model becomes impedance $Z_1$ and a current-dependent voltage source with the value of $\alpha I_2$. The shared element between the loops becomes $Z_m$, and the second loop becomes the impedance $Z_2$ and a current-dependent voltage source with the value of $\beta I_1$. The equivalent circuit is shown in Fig. 12.6.

---

**Example 12.1 Find the impedance matrix of the circuit shown in Fig. 12.7.**

*Solution.* The two-port network shown in Fig. 12.1 forms a $T$ network with a shared element. There are multiple methods to obtain the impedance matrix, some of which are explained in this example.

**Method 1**

Classical Approach

In this method, the definition equations such as $Z_{11} = \frac{V_1}{I_1}|_{I_2=0}$ are used. To obtain $Z_{11}$, port 2 is considered without a source; therefore, $I_2 = 0$. That makes the circuit, as shown in Fig. 12.8.

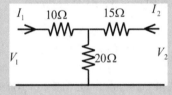

**Fig. 12.7** Circuit of Example 12.1

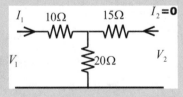

**Fig. 12.8** Circuit when the output current is forced to zero

(continued)

**Example 12.1** (continued)

Since $I_2 = 0$, a KVL in this loop results in

$$V_1 = 10I_1 + 20I_1 = 30I_1$$

The ratio $\frac{V_1}{I_1}$ is obtained as follows:

$$Z_{11} = \frac{V_1}{I_1} = 30$$

In the same circuit, the voltage generated at port 2 can be calculated as

$$V_2 = 20I_1$$

The impedance element $Z_{21}$ can be calculated as follows:

$$Z_{21} = \frac{V_2}{I_1}\Big|_{I_2=0} = \frac{20I_1}{I_1} = 20$$

Now considering the current of port 1 as zero, $I_1 = 0$, the circuit is shown in Fig. 12.9.

In this case, $Z_{22}$ and $Z_{12}$ can be calculated as follows:

$$Z_{22} = \frac{V_2}{I_2}\Big|_{I_1=0}$$

KVL in loop 2, when $I_1 = 0$, results in

$$V_2 = 15I_2 + 20I_2 = 35I_2$$

The ratio results in

$$Z_{22} = \frac{V_2}{I_2}\Big|_{I_1=0} = \frac{35I_2}{I_2} = 35 \ \Omega$$

**Fig. 12.9** Circuit when the input current is forced to zero

(continued)

**Example 12.1** (continued)

When $I_1 = 0$, the voltage in port 1 is calculated as

$$V_1 = 20I_2$$

The element $Z_{12}$ can be calculated as follows:

$$Z_{12} = \frac{V_1}{I_2}\Big|_{I_1=0} = \frac{20I_2}{I_2} = 20 \ \Omega$$

Since there is no dependent source in the circuit, the impedance matrix is reciprocal, meaning that

$$Z_{12} = Z_{21}$$

Moreover, by calculations, it was demonstrated that

$$Z_{12} = Z_{21} = 20 \ \Omega.$$

The impedance matrix becomes

$$Z = \begin{bmatrix} 30 & 20 \\ 20 & 35 \end{bmatrix}$$

**Method 2**
Using $T$ Matrix Analysis
As Fig. 12.7 shows, a $T$ network exists. Compared with the $T$ network, $Z_{11}$ can be obtained by adding all impedances existing in loop 1. It means the following:
In loop 1:

$$Z_{11} = 10 + 20 = 30 \ \Omega$$

In loop 2:

$$Z_{22} = 15 + 20 = 35 \ \Omega$$

The shared element between two loops is $Z_{12} = Z_{21}$. Therefore,

(continued)

**Example 12.1** (continued)

$$Z_{12} = Z_{21} = 20 \ \Omega$$

The impedance matrix becomes

$$Z = \begin{bmatrix} 30 & 20 \\ 20 & 35 \end{bmatrix}$$

## Alternative Approach in Impedance Matrix

Consider the circuit shown in Fig. 12.10.
The impedance matrix becomes

$$Z = \begin{bmatrix} Z_a + Z_c & Z_c \\ Z_c & Z_b + Z_c \end{bmatrix}$$

**Example 12.2** Find the impedance matrix of the circuit shown in Fig. 12.11.
*Solution*

$$Z = \begin{bmatrix} 20 + 40 & 40 \\ 40 & 30 + 40 \end{bmatrix} = \begin{bmatrix} 60 & 40 \\ 40 & 70 \end{bmatrix}$$

**Fig. 12.10** A $T$ network

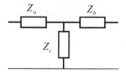

**Fig. 12.11** Circuit of Example 12.2

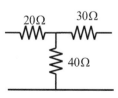

**Example 12.3** Find the impedance matrix of the circuit shown in Fig. 12.12.

*Solution*

$$Z = \begin{bmatrix} 20 + 2s & 2s \\ 2s & 30 + 2s \end{bmatrix}$$

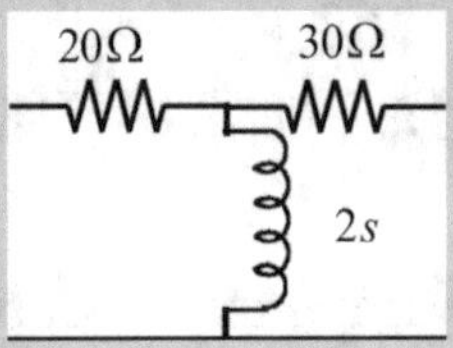

**Fig. 12.12** Circuit of Example 12.3

**Example 12.4** Find the impedance matrix of the circuit shown in Fig. 12.13.

*Solution*

$$Z = \begin{bmatrix} 16 + 12 & 12 \\ 12 & 0 + 12 \end{bmatrix} = \begin{bmatrix} 28 & 12 \\ 12 & 12 \end{bmatrix}$$

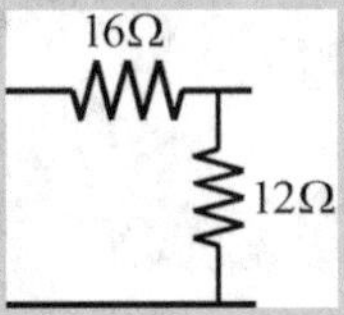

**Fig. 12.13** Circuit of Example 12.4

**Example 12.5** Find the impedance matrix of the circuit shown in Fig. 12.14.

*Solution*

$$Z = \begin{bmatrix} 0+15 & 15 \\ 15 & 0+15 \end{bmatrix} = \begin{bmatrix} 15 & 15 \\ 15 & 15 \end{bmatrix}$$

**Fig. 12.14** Circuit of Example 12.5

**Example 12.6** The output of an impedance matrix (Fig. 12.13) feeds a 10 Ω resistor. If the input port voltage is $V_1 = 100 \angle 0 \ V$, find the currents $I_1$ and $I_2$ and the output voltage (Figs. 12.14 and 12.15).

*Solution.* The impedance matrix shows the relations of the input and output port voltages and currents as

$$\begin{bmatrix} V_1 \\ V_2 \end{bmatrix} = \begin{bmatrix} Z_{11} & Z_{12} \\ Z_{21} & Z_{22} \end{bmatrix} \begin{bmatrix} I_1 \\ I_2 \end{bmatrix}$$

Therefore,

$$\begin{bmatrix} V_1 \\ V_2 \end{bmatrix} = \begin{bmatrix} 10 & j10 \\ j15 & 20 \end{bmatrix} \begin{bmatrix} I_1 \\ I_2 \end{bmatrix}$$

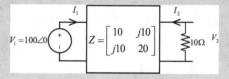

**Fig. 12.15** Circuit of Example 12.6

(continued)

**Example 12.6** (continued)

The output port voltage $V_2$ is a function of the load resistance and current as follows:

$$V_2 = -10I_2$$

Replacing $V_1 = 100 \angle 0$ and $V_2 = -10I_2$, the network equations become

$$\begin{bmatrix} 100 \\ 0 \end{bmatrix} = \begin{bmatrix} 10 & j10 \\ j15 & 20 \end{bmatrix} \begin{bmatrix} I_1 \\ I_2 \end{bmatrix}$$

Bringing the $-10I_2$ back to the right side, the equations read as

$$\begin{bmatrix} 100 \\ 0 \end{bmatrix} = \begin{bmatrix} 10 & j10 \\ j15 & 20+10 \end{bmatrix} \begin{bmatrix} I_1 \\ I_2 \end{bmatrix}$$

The currents can be found as follows:

$$\begin{bmatrix} 100 \\ 0 \end{bmatrix} = \begin{bmatrix} 10 & j10 \\ j15 & 20+10 \end{bmatrix}^{-1} \begin{bmatrix} I_1 \\ I_2 \end{bmatrix}$$

$$\begin{bmatrix} I_1 \\ I_2 \end{bmatrix} = \frac{1}{10 \times 30 - j15 \times j10} \begin{bmatrix} 30 & -j10 \\ -j15 & 10 \end{bmatrix} \begin{bmatrix} 100 \\ 0 \end{bmatrix}$$

$$\begin{bmatrix} I_1 \\ I_2 \end{bmatrix} = \frac{1}{450} \begin{bmatrix} 30 & -j10 \\ -j15 & 10 \end{bmatrix} \begin{bmatrix} 100 \\ 0 \end{bmatrix} = \begin{bmatrix} \dfrac{3000}{450} \\ \dfrac{-j15 \times 100}{450} \end{bmatrix} = \begin{bmatrix} 6.67 \\ -j3.33 \end{bmatrix} A$$

$$\begin{bmatrix} I_1 \\ I_2 \end{bmatrix} = \begin{bmatrix} 6.67 \\ -j3.33 \end{bmatrix} A$$

*Another approach* to finding currents is through Cramer's method as follows:

$$I_1 = \frac{\begin{vmatrix} 100 & j10 \\ 0 & 20 \end{vmatrix}}{\begin{vmatrix} 10 & j10 \\ j15 & 20 \end{vmatrix}} = \frac{3000}{450} = 6.67 A \quad I_2 = \frac{\begin{vmatrix} 10 & 100 \\ j15 & 0 \end{vmatrix}}{\begin{vmatrix} 10 & j10 \\ j15 & 20 \end{vmatrix}} = \frac{-j1500}{450} = -j3.33 A$$

# Finding Impedance Matrix in Multi-loop Networks

If the network contains any configuration except the $T$ form, it may form more than two loops. However, two ports still represent the entire network, i.e., the impedance matrix still has a $2 \times 2$ dimension.

# Current and Voltage Considerations

- The voltage and current in the input port are $V_1$ and $I_1$, with the current entering the port.
- The voltage and current in the output port are $V_2$ and $I_2$, with the current entering the port.
- The direction of current in other loops is arbitrary.
- If the impedance of input and output ports is not placed in the first and second equations, the rows and columns can be exchanged to shift the desired equations to their designated places. It is recommended to have the input equation in the first row and the output equation in the second row.

# Finding the Matrix Dimension and Its Elements

- Consider a square matrix $Z_{n \times n}$ with its dimension matching the number of loops, i.e., for a three-loop system, $n = 3$, $Z_{3 \times 3}$ matrices are obtained.
- The element $Z_{ii}$ on the diagonal is obtained by summating all impedances in loop $i$.
- The off-diagonal elements $Z_{ij}$ are the shared elements between loop $i$ and loop $j$.

  - If the current direction of these loops is similar through the shared element, $a + Z_{ij}$ is obtained.
  - If the current direction of these loops does not match through the shared element, $a - Z_{ij}$ is obtained.

# General Form of KVL Equations

Following the abovementioned rules, a general impedance matrix equation is obtained with zero constant matrix elements except for the $V_1$ and $V_2$. For a network with $n$ loops, the general form is

$$
\begin{bmatrix} Z_{11} & \cdots & Z_{1n} \\ \vdots & \ddots & \vdots \\ Z_{n1} & \cdots & Z_{nn} \end{bmatrix} \begin{bmatrix} I_1 \\ \vdots \\ I_n \end{bmatrix} = \begin{bmatrix} V_1 \\ V_2 \\ 0 \end{bmatrix}
$$

A $2 \times 2$ matrix can be selected/partitioned from this general model that contains equations that have values other than zero. The system can be expressed as follows:

$$\begin{bmatrix} A_{2 \times 2} & M_{2 \times (n-2)} \\ N_{(n-2) \times 2} & D_{(n-2) \times (n-2)} \end{bmatrix} \begin{bmatrix} I_1 \\ I_2 \\ \vdots \\ I_n \end{bmatrix} = \begin{bmatrix} V_1 \\ V_2 \\ 0_{(n-2) \times 1} \end{bmatrix}$$

## Matrix Size Reduction

The size of the split general equation can be reduced to the non-zero element constant matrix size. In this case, $V_1$ and $V_2$, i.e., the size is 2. The impedance matrix of a two-port representation of the multi-loop is as follows:

$$Z_{2 \times 2} = A_{2 \times 2} - M_{2 \times (n-2)} D^{-1}_{(n-2) \times (n-2)} N_{(n-2) \times 2}$$

Removing the matrix dimensions results in

$$Z = A - MD^{-1}N$$

**Example 12.7** Find the impedance matrix of the circuit shown in Fig. 12.16.
*Solution.* There are multiple approaches to obtaining the impedance matrix of this network. In this example, the impedance matrix is obtained by two methods. Later in this chapter, another method is introduced to simplify the solution.

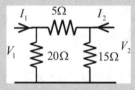

**Fig. 12.16** Circuit of Example 12.7

(continued)

**Example 12.7** (continued)

**Method 1**

Using the definitions,

$$Z_{11} = \frac{V_1}{I_1}\Big|_{I_2=0} \Rightarrow V_1 = I_1\left(20\,\middle\|\,(5+15)\right) = 10I_1 \rightarrow Z_{11} = \frac{V_1}{I_1} = \frac{10I_1}{I_1} = 10$$

$$Z_{21} = \frac{V_2}{I_1}\Big|_{I_2=0} \Rightarrow V_2 = \frac{15}{5+15}V_1;\,(V_1 = 10I_1) \rightarrow Z_{21} = \frac{V_2}{I_1}\Big| = \frac{\frac{15}{5+15}10I_1}{I_1}$$

$$= 7.5$$

$$Z_{22} = \frac{V_2}{I_2}\Big|_{I_1=0} \Rightarrow V_2 = I_2\left(15\,\middle\|\,(20+5)\right) = 9.375I_2 \rightarrow Z_{22} = \frac{V_2}{I_2}$$

$$= \frac{9.375I_2}{I_2} = 9.375$$

$$Z_{12} = \frac{V_1}{I_2}\Big|_{I_1=0} \Rightarrow V_1 = \frac{20}{5+20}V_2;\,(V_2 = 9.375I_2) \rightarrow Z_{12} = \frac{V_1}{I_2}$$

$$= \frac{\frac{20}{5+20}(9.375I_2)}{I_2} = 7.5$$

$$Z = \begin{bmatrix} 10 & 7.5 \\ 7.5 & 9.375 \end{bmatrix}$$

**Method 2**

Using a three-loop system and matrix reduction. In the circuit of Fig. 12.16, consider the current circulation in loop 1 clockwise, in loop 2 counterclockwise, and loop 3 the same as loop 1.

Therefore, the elements of a general KVL by considering the assumed direction of currents are obtained as follows:

$$\begin{bmatrix} 20 & 0 & -20 \\ 0 & 15 & 15 \\ -20 & 15 & 5+20+15 \end{bmatrix}\begin{bmatrix} I_1 \\ I_2 \\ I_3 \end{bmatrix} = \begin{bmatrix} V_1 \\ V_2 \\ 0 \end{bmatrix}$$

Since the third-row equation has a zero voltage source on the right-hand side of the equation, the third equation can be eliminated, and the remaining system becomes $2 \times 2$ matrices, representing the impedance matrix of a two-port network.

(continued)

**Example 12.7** (continued)

The size reduction suggests the following formulation, wherein $n = 3$:

$$A_{2\times 2} = \begin{bmatrix} 20 & 0 \\ 0 & 15 \end{bmatrix}$$

$$M_{2\times(n-2)} = M_{2\times 1} = \begin{bmatrix} -20 \\ 15 \end{bmatrix}$$

$$N_{(n-2)\times 2} = N_{1\times 2} = [-20 \ 15]$$

$$D_{(n-2)\times(n-2)} = D_{1\times 1} = 40$$

$$Z = A - MD^{-1}N = \begin{bmatrix} 20 & 0 \\ 0 & 15 \end{bmatrix} - \begin{bmatrix} -20 \\ 15 \end{bmatrix} 40^{-1}[-20 \ 15]$$

$$Z = \begin{bmatrix} 10 & 7.5 \\ 7.5 & 9.375 \end{bmatrix}$$

**Example 12.8 Consider the circuit of Fig. 12.17, and find the impedance matrix.**

*Solution.* Considering the total impedance in loops 1 and 2 and the shared element, the impedance can be obtained as follows:

The impedance of loop 1: $Z_{11} = \frac{V_1}{I_1}\big|_{I_2=0} \Rightarrow Z_{11} = sL_1 + R$.

The impedance of the shared element considering that the current directions are similar:

$$Z_{21} = \frac{V_2}{I_1}\big|_{I_2=0} \Rightarrow Z_{21} = R$$

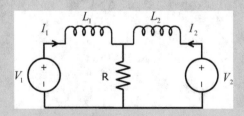

**Fig. 12.17**  Circuit of Example 12.8

(continued)

**Example 12.8** (continued)

The impedance of loop 2: $Z_{22} = \frac{V_2}{I_2}\big|_{I_1=0} \Rightarrow Z_{22} = sL_2 + R$.

The impedance of the shared element considering that the current directions are similar:

$$Z_{12} = \frac{V_1}{I_2}\big|_{I_1=0} \Rightarrow Z_{12} = R$$

Therefore,

$$Z = \begin{bmatrix} 10 & 7.5 \\ 7.5 & 9.375 \end{bmatrix}$$

**Example 12.9 In the circuit of Fig. 12.18, find the impedance matrix.**
*Solution.* Considering the direction of currents as determined in the figure (the direction of ports 1 and 2 is fixed, and the direction of current in loop 3 is arbitrary), the elements of a general KVL are obtained as follows:

$$\begin{bmatrix} R_1 + sL & sL & -R_1 \\ sL & R_2 + sL & R_2 \\ -R_1 & R_2 & R_1 + R_2 + R_3 \end{bmatrix} \begin{bmatrix} I_1 \\ I_2 \\ I_3 \end{bmatrix} = \begin{bmatrix} V_1 \\ V_2 \\ 0 \end{bmatrix}$$

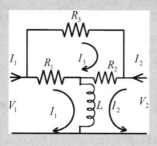

**Fig. 12.18** Circuit of Example 12.9

(continued)

**Example 12.9** (continued)

The elements of the third row, shown in red, are negative because the currents that pass the shared element between loops 1 and 3 are in opposite directions. Since the third-row equation has a zero voltage source on the right-hand side, the third equation can be eliminated, and the remaining system becomes a $2 \times 2$ matrix, representing the impedance matrix of a two-port network.

The size reduction suggests the following formulation, wherein $n = 3$:

$$A_{2 \times 2} = \begin{bmatrix} R_1 + sL & sL \\ sL & R_2 + sL \end{bmatrix}$$

$$M_{2 \times (n-2)} = M_{2 \times 1} = \begin{bmatrix} -R_1 \\ R_2 \end{bmatrix}$$

$$N_{(n-2) \times 2} = N_{1 \times 2} = \begin{bmatrix} -R_1 & R_2 \end{bmatrix}$$

$$D_{(n-2) \times (n-2)} = D_{1 \times 1} = R_1 + R_2 + R_3$$

$$Z = A - MD^{-1}N = \begin{bmatrix} R_1 + sL & sL \\ sL & R_2 + sL \end{bmatrix}$$

$$- \begin{bmatrix} -R_1 \\ R_2 \end{bmatrix} (R_1 + R_2 + R_3)^{-1} \begin{bmatrix} -R_1 & R_2 \end{bmatrix}$$

$$Z = \begin{bmatrix} R_1 + sL - \dfrac{R_1^2}{R_1 + R_2 + R_3} & sL + \dfrac{R_1 R_2}{R_1 + R_2 + R_3} \\ sL + \dfrac{R_1 R_2}{R_1 + R_2 + R_3} & R_2 + sL - \dfrac{R_2^2}{R_1 + R_2 + R_3} \end{bmatrix}$$

## Impedance Matrix Existence

The circuit and the two-port network must be such that the transfer functions of $Z_{11} = \frac{V_1}{I_1}\big|_{I_2 = 0}$, $Z_{12} = \frac{V_1}{I_2}\big|_{I_1 = 0}$, $Z_{21} = \frac{V_2}{I_1}\big|_{I_2 = 0}$, and $Z_{22} = \frac{V_2}{I_2}\big|_{I_1 = 0}$ exist. Transfer functions show the dynamics of a system defined to directly explain the ratio of the desired output over a desired input parameter. If the desired input does not influence the output, the transfer functions may not exist.

**Example 12.10 Consider an ideal transformer with transformation ratio $n$, shown in Fig. 12.19, and find the current and voltage ratios.**
*Solution.* Considering the transformer as a two-port network, the output voltage $V_2$ is a function of its input voltage $V_1$ as follows:

$$V_2 = nV_1$$

The output current $I_2$ is also a function of its input current $I_1$ as

$$I_2 = \frac{1}{n}I_1$$

As these equations show, the input voltage of an ideal transformer is not a function of the input or output current. Therefore, $Z_{11}$ and $Z_{12}$ do not exist. Similarly, the output voltage is not a function of the input and output currents. Therefore, $Z_{21}$ and $Z_{22}$ do not exist. A different form of transformation can be defined for a transformer, which is introduced later in this chapter.

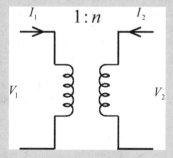

**Fig. 12.19** Ideal transformer

## Admittance Matrix of a Two-Port Network

Admittance matrix shows the transfer functions existing in a two-port network defined as ratios of the currents over the voltages seen from each of the ports with respect to itself or the other ports. In this network representation, the entire circuit is expressed by a $Y_{2\times2}$ matrix that defines transfer functions as the ratio of port currents to the network's port voltages. The matrix elements show how the current at the network ports varies by voltage variations. Current sources are utilized at the ports to control the number of input currents, and the voltages are measured (Fig. 12.20).

**Fig. 12.20** Admittance
matrix

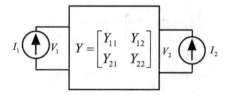

**Fig. 12.20** Admittance
matrix

The equations that represent an admittance network are as follows:

$$\begin{bmatrix} I_1 \\ I_2 \end{bmatrix} = Y_{2\times2}\begin{bmatrix} V_1 \\ V_2 \end{bmatrix}$$

$$\begin{cases} I_1 = Y_{11}V_1 + Y_{12}V_2 \\ I_2 = Y_{21}V_1 + Y_{22}V_2 \end{cases}$$

$$\begin{bmatrix} I_1 \\ I_2 \end{bmatrix} = \begin{bmatrix} Y_{11} & Y_{12} \\ Y_{21} & Y_{22} \end{bmatrix}\begin{bmatrix} V_1 \\ V_2 \end{bmatrix}$$

The admittance matrix has four elements of $Y_{11}$, $Y_{12}$, $Y_{21}$, and $Y_{22}$:

- $Y_{11}$ shows how the current of port 1 is related to the voltage of port 1.
- $Y_{12}$ shows how the current of port 1 is related to the voltage of port 2.
- $Y_{21}$ shows how the current of port 2 is related to the voltage of port 1.
- $Y_{22}$ shows how the current of port 2 is related to the voltage of port 2.

To measure each admittance matrix element, only the voltage and current related to the element should be considered, and the other parameters are considered zero. For instance,

$$Y_{11} = \frac{I_1}{V_1}\Big|_{V_2=0}$$

**Note 12.3** This shows that to measure or calculate $Y_{11}$, only $I_1$ and $V_1$ are required while imposing $V_2 = 0$. This requires that the terminals of the second port be short-circuited.

Similarly,

$$Y_{12} = \frac{I_1}{V_2}\Big|_{V_1=0}$$

$$Y_{21} = \frac{I_2}{V_1}\Big|_{V_2=0}$$

$$Y_{22} = \frac{I_2}{V_2}\Big|_{V_1=0}$$

**Note 12.4** In networks that do not have *dependent sources*, the matrix elements $Y_{12}$ and $Y_{21}$ become similar, i.e.,

$$Y_{12} = Y_{21}$$

*These networks are called reciprocal.*

## The Equivalent of Admittance Network

Often, a two-port network can contain multiple loops and nodes. A matrix representation may also not be readily obtained when parallel networks exist or when part of the originally known network is missing. Therefore, the admittance matrix of a two-port network can be simplified to a $\Pi$-equivalent network.

## Reciprocal Network

### *$\Pi$ Model*

As mentioned earlier, an admittance network might be reciprocal if it does not contain dependent sources. A reciprocal network is a two-node network that shares a common element, as presented in Fig. 12.21. The shared element is either $-Y_{12}$ or $-Y_{21}$, as $Y_{12} = Y_{21}$. The not-shared element in the first node is $Y_{11} + Y_{12}$ and in the second node, $Y_{22} + Y_{12}$. This forms a $\Pi$-equivalent network.

### *Nonreciprocal Network*

At least one dependent source makes $Y_{12}$ and $Y_{21}$ different in a nonreciprocal network. The equivalent model presented in this chapter uses two voltage-dependent current sources at each node and a shared element to model a nonreciprocal admittance matrix in a $\Pi$ network.

**Fig. 12.21** $\Pi$-equivalent of admittance network

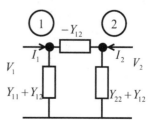

**Fig. 12.22** An element-
sharing node of an
admittance network

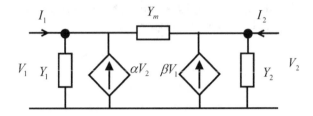

## *Element-Sharing Nodes*

The equivalent circuit of an element-sharing two-port network is shown in
Fig. 12.22.

Consider the original admittance matrix as

$$Y = \begin{bmatrix} Y_{11} & Y_{12} \\ Y_{21} & Y_{22} \end{bmatrix}$$

Consider a shared element $Y_m$ exists in all four components of $Y_{11}$, $Y_{12}$, $Y_{21}$, and
$Y_{22}$. Therefore, the admittance matrix can be written as

$$Y = \begin{bmatrix} Y_1 + Y_m & -\alpha - Y_m \\ -\beta - Y_m & Y_2 + Y_m \end{bmatrix}$$

Parameters $\alpha$ and $\beta$ show the existence of dependent sources in the $\Pi$-equivalent
network and are used to model the nonreciprocal networks. The equivalent element
in node 1 of the $\Pi$ model becomes admittance $Y_1$ and a voltage-dependent current
source with the value of $\alpha V_2$. The shared element between the nodes becomes $Y_m$,
and the second node has the admittance $Y_2$ and a voltage-dependent current source
with the value of $\beta V_1$.

---

**Example 12.11 In the circuit of Fig. 12.23, find the $Y$ matrix.**
*Solution.* The two-port network shown in Fig. 12.23 forms a $\Pi$ network with a
shared element. There are multiple methods to obtain the admittance matrix,
some of which are explained in this example.
  **Method 1**
  Classical Approach
  In this method, the definition equations such as $Y_{11} = \frac{I_1}{V_1}\big|_{V_2=0}$ are used. To
obtain $Y_{11}$, port 2 is considered a short circuit; therefore, $V_2 = 0$, and the 8 $\Omega$
resistor is shorted out. That makes the circuit, as shown in Fig. 12.24.

---

(continued)

**Example 12.11** (continued)

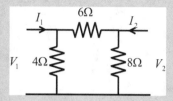

**Fig. 12.23** The circuit in Example 12.11

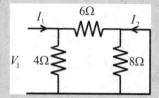

**Fig. 12.24** Short circuit

Since $V_2 = 0$, a KVL in node 1 results in

$$V_1 = I_1 \left(4 \| 6\right) = 2.4 I_1$$

The ratio $\frac{I_1}{V_1}$ is obtained as follows:

$$Y_{11} = \frac{I_1}{V_1} = \frac{I_1}{2.4 I_1} = 0.417 \ \Omega^{-1}$$

In the same circuit, the current of $I_2$ is opposite to the current of the 6 $\Omega$ resistor. A current division finds the current as

$$I_2 = - \frac{4}{4 + 6} I_1 = -0.4 I_1$$

The admittance element $Y_{21}$ can be calculated as follows:

$$Y_{21} = \frac{I_2}{V_1} \big|_{V_2 = 0} = \frac{-0.4 I_1}{V_1} = -0.4 \ \frac{I_1}{V_1} = -0.4 \times 0.417 = -0.166$$

(continued)

**Example 12.11** (continued)

Now considering the node voltage of port 1 as zero, $V_1 = 0$, the short circuit in the input node is shown in Fig. 12.25.

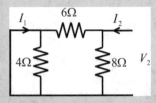

**Fig. 12.25** When $V_1 = 0$ is imposed, a short circuit in node 1 is created

In this case, $Y_{22}$ and $Y_{12}$ can be calculated as follows:

$$Y_{22} = \frac{I_2}{V_2}\Big|_{V_1 = 0}$$

KVL in loop 2, when $V_1 = 0$, shortens out the 4 $\Omega$ resistor. Therefore,

$$V_2 = I_2\left(6\|8\right) = 3.42 I_2$$

The ratio $\frac{I_2}{V_2}$ is obtained as

$$Y_{22} = \frac{I_2}{V_2}\Big|_{V_1 = 0} = \frac{I_2}{3.42 I_2} = 0.292$$

When $V_1 = 0$, the current in port 1 is similar to the opposite of the current that flows through the 6 $\Omega$ resistor, as follows:

$$I_1 = -\frac{8}{6 + 8} I_2 = -0.571 I_2$$

The element $Y_{12}$ can be calculated as follows:

$$Y_{12} = \frac{I_1}{V_2}\Big|_{V_1 = 0} = \frac{-0.571 I_2}{V_2} = -0.571 \frac{I_2}{V_2} = -0.571 \times 0.292 = -0.166$$

Since there is no dependent source in the circuit, the impedance matrix is reciprocal, meaning that

(continued)

**Example 12.11** (continued)

$$Y_{12} = Y_{21}$$

Moreover, by calculations, it was demonstrated that

$$Y_{12} = Y_{21} = -0.166$$

The admittance matrix becomes

$$Y = \begin{bmatrix} 0.417 & -0.166 \\ -0.166 & 0.292 \end{bmatrix}$$

**Method 2**
Using Π Matrix Analysis
As Fig. 12.21 shows, a Π network exists. Comparing side by side with the Π network explained in Fig. 12.19, the $Y_{11}$ can be obtained by adding all admittances connected to node 1. It means the following:
In node 1:

$$Y_{11} = \frac{1}{4} + \frac{1}{6} = 0.416$$

In node 2:

$$Y_{22} = \frac{1}{6} + \frac{1}{8} = 0.292$$

The negative of shared admittance between two nodes is $Y_{12} = Y_{21}$. Therefore,

$$Y_{12} = Y_{21} = -\frac{1}{6} = -0.166$$

The admittance matrix becomes

$$Y = \begin{bmatrix} 0.417 & -0.166 \\ -0.166 & 0.292 \end{bmatrix}$$

**Fig. 12.26** A Π-equivalent
circuit is used in the
admittance matrix

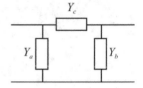

## Alternative Approach in Admittance Matrix

Consider the circuit shown in Figs. 12.8 and 12.26.
    The admittance matrix becomes

$$Y = \begin{bmatrix} Y_a + Y_c & -Y_c \\ -Y_c & Y_b + Y_c \end{bmatrix}$$

**Example 12.12  Find the admittance matrix of the circuit shown in Fig. 12.27.**
*Solution.* The element values in the Π network are provided in impedance values. Therefore, in their summation, their admittance should be considered as follows:

$$Y = \begin{bmatrix} \dfrac{1}{4} + \dfrac{1}{2s} & -\dfrac{1}{2s} \\ -\dfrac{1}{2s} & \dfrac{1}{8} + \dfrac{1}{2s} \end{bmatrix}$$

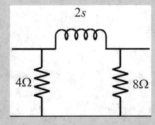

**Fig. 12.27** Circuit of Example 12.12

**Example 12.13 Find the admittance matrix of the circuit shown in Fig. 12.28.**

*Solution*

$$Y = \begin{bmatrix} \dfrac{1}{2s} + \dfrac{1}{4} & -\dfrac{\frac{1}{4}}{s} \\[2mm] \dfrac{\frac{1}{4}}{s} \\[2mm] -\dfrac{\frac{1}{4}}{s} & 5 + \dfrac{\frac{1}{4}}{s} \end{bmatrix} = \begin{bmatrix} \dfrac{1}{2s} + \dfrac{s}{4} & -\dfrac{s}{4} \\[2mm] -\dfrac{s}{4} & 5 + \dfrac{s}{4} \end{bmatrix}$$

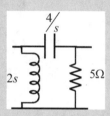

**Fig. 12.28** Circuit of Example 12.13

**Example 12.14 Find the admittance matrix of the circuit shown in Fig. 12.29.**

*Solution*

$$Y = \begin{bmatrix} 16 + 12 & -12 \\ -12 & 0 + 12 \end{bmatrix} = \begin{bmatrix} 28 & -12 \\ -12 & 12 \end{bmatrix}$$

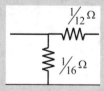

**Fig. 12.29** Circuit of Example 12.14

**Example 12.15 Find the admittance matrix of the circuit shown in Fig. 12.30.**

*Solution*

$$Y = \begin{bmatrix} 0+15 & -15 \\ -15 & 0+15 \end{bmatrix} = \begin{bmatrix} 15 & -15 \\ -15 & 15 \end{bmatrix}$$

**Fig. 12.30** Circuit of Example 12.15

**Example 12.16 The output of an admittance matrix (Fig. 12.31) feeds a 0.1 $\Omega^{-1}$ conductance. If the input port current source is $I_1 = 50 \angle 0\,A$, find the currents $V_1$ and $V_2$ and the output current.**

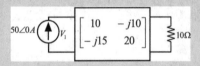

**Fig. 12.31** Network of Example 12.16

*Solution.* The admittance matrix shows the relations of the input and output port voltage and currents as

$$\begin{bmatrix} I_1 \\ I_2 \end{bmatrix} = \begin{bmatrix} Y_{11} & Y_{12} \\ Y_{21} & Y_{22} \end{bmatrix} \begin{bmatrix} V_1 \\ V_2 \end{bmatrix}$$

Therefore,

$$\begin{bmatrix} I_1 \\ I_2 \end{bmatrix} = \begin{bmatrix} 10 & -j10 \\ -j15 & 20 \end{bmatrix} \begin{bmatrix} V_1 \\ V_2 \end{bmatrix}$$

(continued)

**Example 12.16** (continued)

The output port voltage $V_2$ is a function of the load resistance and current as follows:

$$I_2 = -\frac{1}{0.1} V_2$$

Replacing $I_1 = 50 \angle 0$ and $I_2 = -10V_2$, the network equations become

$$\begin{bmatrix} 50 \\ -10V_2 \end{bmatrix} = \begin{bmatrix} 10 & -j10 \\ -j15 & 20 \end{bmatrix} \begin{bmatrix} V_1 \\ V_2 \end{bmatrix}$$

Bringing the $-\frac{1}{10} V_2$ back to the right side, the equations read as

$$\begin{bmatrix} 50 \\ V_2 \end{bmatrix} = \begin{bmatrix} 10 & -j10 \\ -j15 & 20+10 \end{bmatrix} \begin{bmatrix} V_1 \\ V_2 \end{bmatrix}$$

The currents can be found as follows:

$$\begin{bmatrix} I_1 \\ I_2 \end{bmatrix} = \begin{bmatrix} 10 & -j10 \\ -j15 & 30 \end{bmatrix}^{-1} \begin{bmatrix} 50 \\ 0 \end{bmatrix}$$

$$\begin{bmatrix} I_1 \\ I_2 \end{bmatrix} = \frac{1}{10 \times 30 - j15 \times j10} \begin{bmatrix} 30 & +j10 \\ +j15 & 10 \end{bmatrix} \begin{bmatrix} 50 \\ 0 \end{bmatrix}$$

$$\begin{bmatrix} I_1 \\ I_2 \end{bmatrix} = \frac{1}{450} \begin{bmatrix} 30 & +j10 \\ +j15 & 10 \end{bmatrix} \begin{bmatrix} 50 \\ 0 \end{bmatrix} = \begin{bmatrix} \dfrac{1500}{450} \\ \dfrac{-j15 \times 50}{450} \end{bmatrix} = \begin{bmatrix} 3.33 \\ +j1.66 \end{bmatrix} A$$

$$\begin{bmatrix} I_1 \\ I_2 \end{bmatrix} = \begin{bmatrix} 3.33 \\ +j1.66 \end{bmatrix} A$$

*Another approach* to finding the currents is through Cramer's method.

$$I_1 = \frac{\begin{vmatrix} 50 & -j10 \\ 0 & 30 \end{vmatrix}}{\begin{vmatrix} 10 & -j10 \\ -j15 & 30 \end{vmatrix}} = \frac{1500}{450} = 3.33 \ A \quad I_2 = \frac{\begin{vmatrix} 10 & 50 \\ -j15 & 0 \end{vmatrix}}{\begin{vmatrix} 10 & -j10 \\ -j15 & 30 \end{vmatrix}} = \frac{j750}{450} =$$

$$+j1.66 \ A$$

## Finding Admittance Matrix in Multi-node Networks

If the network contains any configuration other than the standard $\Pi$ form, it may form more than two nodes. However, two ports are still utilized to represent the entire network, i.e., the admittance matrix still has a $2 \times 2$ dimension. A procedure to find the admittance matrix can be expressed as follows:

## Current and Voltage Considerations

- The voltage and current in the input node are $V_1$ and $I_1$, with the current entering the port.
- The voltage and current in the output node are $V_2$ and $I_2$, with the current entering the port.
- The order of naming the voltage of other nodes is arbitrary.
- If the admittance of input and output nodes is not placed in the first and second equations, the rows and columns can be exchanged to shift the desired equations to their designated places. It is recommended to have the input node equation in the first row and the output node equation in the second row.

## Finding the Matrix Dimension and Its Elements

- Consider a square matrix $Y_{n \times n}$ with its dimension matching the number of nodes, i.e., for a three-node system $n = 3$, a $Y_{3 \times 3}$ matrices is obtained.
- The element $Y_{ii}$ on the diagonal is obtained by summating all admittances connected to node $i$.
- The off-diagonal elements $Y_{ij}$ are the negative of shared elements between node $i$ and node $j$.

## General Form of KCL Equations

Following the abovementioned rules, a general admittance matrix equation is obtained with constant matrix elements being zero except for $I_1$ and $I_2$. For a network with $n$ nodes, the general form is

$$\begin{bmatrix} Y_{11} & \cdots & Y_{1n} \\ \vdots & \ddots & \vdots \\ Y_{n1} & \cdots & Y_{nn} \end{bmatrix} \begin{bmatrix} V_1 \\ \vdots \\ V_n \end{bmatrix} = \begin{bmatrix} I_1 \\ I_2 \\ 0 \end{bmatrix}$$

A $2 \times 2$ matrix can be selected from this general model that contains equations that have values other than zero. The system can be expressed as follows:

$$\begin{bmatrix} A_{2 \times 2} & M_{2 \times (n-2)} \\ N_{(n-2) \times 2} & D_{(n-2) \times (n-2)} \end{bmatrix} \begin{bmatrix} V_1 \\ V_2 \\ \vdots \\ V_n \end{bmatrix} = \begin{bmatrix} I_1 \\ I_2 \\ 0_{(n-2) \times 1} \end{bmatrix}$$

## Matrix Size Reduction

The size of the split general equation can be reduced to the non-zero element constant matrix size. In this case, $I_1$ and $I_2$, i.e., the size is 2.

The impedance matrix of a two-port representation of the multi-loop is as follows:

$$Y_{2 \times 2} = A_{2 \times 2} - M_{2 \times (n-2)} D^{-1}_{(n-2) \times (n-2)} N_{(n-2) \times 2}$$

Removing the matrix dimensions results in

$$Z = A - MD^{-1}N$$

**Example 12.17 Find the admittance matrix of the circuit shown in Fig. 12.32.**

*Solution.* There are multiple approaches to obtaining the admittance matrix of this network. In this example, the admittance matrix is obtained by two methods. Later in this chapter, another method is introduced to simplify the solution.

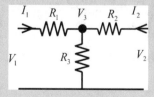

**Fig. 12.32** Circuit of Example 12.17

(continued)

**Example 12.17** (continued)

**Method 1**

Using the definitions,

$$Y_{11} = \frac{I_1}{V_1}\Big|_{V_2=0} \Rightarrow I_1 = \frac{V_1}{R_1 + \left(R_2 \| R_3\right)} \rightarrow Y_{11} = \frac{I_1}{V_1} = \frac{1}{R_1 + \left(R_2 \| R_3\right)}$$

$$= \frac{R_2 + R_3}{R_1 R_2 + R_2 R_3 + R_1 R_3}$$

$$Y_{21} = \frac{I_2}{V_1}\Big|_{V_2=0} \Rightarrow I_2 = -\frac{R_3}{R_2 + R_3} I_1; \left(I_1 = \frac{R_2 + R_3}{R_1 R_2 + R_2 R_3 + R_1 R_3} V_1\right) \rightarrow Y_{21}$$

$$= \frac{-\dfrac{R_3}{R_2 + R_3}\left(\dfrac{R_2 + R_3}{R_1 R_2 + R_2 R_3 + R_1 R_3} V_1\right)}{V_1}$$

$$= -\frac{R_3}{R_2 + R_3}\left(\frac{R_2 + R_3}{R_1 R_2 + R_2 R_3 + R_1 R_3}\right)$$

$$Y_{21} = -\frac{R_3}{R_1 R_2 + R_2 R_3 + R_1 R_3}$$

$$Y_{22} = \frac{I_2}{V_2}\Big|_{V_1=0} \Rightarrow I_2 = \frac{V_2}{R_2 + \left(R_1 \| R_3\right)} \rightarrow Y_{22} = \frac{I_2}{V_2} = \frac{R_1 + R_3}{R_1 R_2 + R_2 R_3 + R_1 R_3}$$

$$Y_{12} = \frac{I_1}{V_2}\Big|_{V_1=0} \Rightarrow I_1 = -\frac{R_3}{R_1 + R_3} I_2; \left(I_2 = \frac{R_1 + R_3}{R_1 R_2 + R_2 R_3 + R_1 R_3} V_2\right) \rightarrow Y_{12}$$

$$= \frac{I_1}{V_2} = -\frac{R_3}{R_1 + R_3}\left(\frac{R_1 + R_3}{R_1 R_2 + R_2 R_3 + R_1 R_3}\right) = -\frac{R_3}{R_1 R_2 + R_2 R_3 + R_1 R_3}$$

$$Y = \begin{bmatrix} \dfrac{R_2 + R_3}{R_1 R_2 + R_2 R_3 + R_1 R_3} & -\dfrac{R_3}{R_1 R_2 + R_2 R_3 + R_1 R_3} \\[4mm] -\dfrac{R_3}{R_1 R_2 + R_2 R_3 + R_1 R_3} & \dfrac{R_1 + R_3}{R_1 R_2 + R_2 R_3 + R_1 R_3} \end{bmatrix}$$

**Method 2**

Using a three-loop system and matrix reduction

In the circuit of Fig. 12.32, consider the node voltages; the elements of a general KCL are obtained as follows:

(continued)

**Example 12.17** (continued)

$$
\begin{bmatrix}
\dfrac{1}{R_1} & 0 & -\dfrac{1}{R_1} \\
0 & \dfrac{1}{R_2} & -\dfrac{1}{R_2} \\
-\dfrac{1}{R_1} & -\dfrac{1}{R_2} & \dfrac{1}{R_1}+\dfrac{1}{R_2}+\dfrac{1}{R_3}
\end{bmatrix}
\begin{bmatrix} V_1 \\ V_2 \\ V_3 \end{bmatrix}
=
\begin{bmatrix} I_1 \\ I_2 \\ 0 \end{bmatrix}
$$

Since the third-row equation has a zero-value current source on the right-hand side, the third equation can be eliminated, and the remaining system becomes a $2 \times 2$ matrices, representing the admittance matrix of a two-port network.

The size reduction suggests the following formulation, wherein $n = 3$:

$$
A_{2\times2} =
\begin{bmatrix}
\dfrac{1}{R_1} & 0 \\
0 & \dfrac{1}{R_2}
\end{bmatrix}
$$

$$
M_{2\times(n-2)} = M_{2\times1} =
\begin{bmatrix}
-\dfrac{1}{R_1} \\
-\dfrac{1}{R_2}
\end{bmatrix}
$$

$$
N_{(n-2)\times2} = N_{1\times2} =
\begin{bmatrix}
-\dfrac{1}{R_1} & -\dfrac{1}{R_2}
\end{bmatrix}
$$

$$
D_{(n-2)\times(n-2)} = D_{1\times1} = \dfrac{1}{R_1}+\dfrac{1}{R_2}+\dfrac{1}{R_3}
$$

$$
Y = A - MD^{-1}N =
\begin{bmatrix}
\dfrac{1}{R_1} & 0 \\
0 & \dfrac{1}{R_2}
\end{bmatrix}
$$

$$
-
\begin{bmatrix}
-\dfrac{1}{R_1} \\
-\dfrac{1}{R_2}
\end{bmatrix}
\left(\dfrac{1}{R_1}+\dfrac{1}{R_2}+\dfrac{1}{R_3}\right)^{-1}
\begin{bmatrix}
-\dfrac{1}{R_1} & -\dfrac{1}{R_2}
\end{bmatrix}
$$

$$
Y =
\begin{bmatrix}
\dfrac{R_2+R_3}{R_1R_2+R_2R_3+R_1R_3} & -\dfrac{R_3}{R_1R_2+R_2R_3+R_1R_3} \\
-\dfrac{R_3}{R_1R_2+R_2R_3+R_1R_3} & \dfrac{R_1+R_3}{R_1R_2+R_2R_3+R_1R_3}
\end{bmatrix}
$$

## Admittance to Impedance Conversion

KVL and KCL equations can be converted to each other if their systems are not singular. Accordingly, the admittance and impedance matrices can be converted to each other as follows:

$$\text{If } (\det(Y) \neq 0) \Rightarrow Z = Y^{-1}$$

and

$$\text{If } (\det(Z) \neq 0) \Rightarrow Y = Z^{-1}$$

**Example 12.18 Consider a network with an impedance matrix** $Z = \begin{bmatrix} 10 & -j10 \\ -j15 & 20 \end{bmatrix}$. **Find the admittance matrix of the same network.**

*Solution*

$$Y = Z^{-1} = \begin{bmatrix} 10 & -j10 \\ -j15 & 20 \end{bmatrix}^{-1}$$

$$= \frac{1}{10 \times 20 - (-j15) \times (-j10)} \begin{bmatrix} 20 & j10 \\ j15 & 10 \end{bmatrix} = \begin{bmatrix} 0.0574 & j0.0286 \\ j0.0429 & 0.0286 \end{bmatrix}$$

## Admittance Matrix Existence

The circuit and the two-port network must be such that the transfer functions of $Y_{11}$, $Y_{12}$, $Y_{21}$, and $Y_{22}$ exist. Transfer functions show the dynamics of a system defined to directly explain the dependency of the desired output to the desired input. If the desired input does not influence the output, the transfer functions may not exist.

**Example 12.19 Consider a network with an impedance matrix** $Z = \begin{bmatrix} 10 & 10 \\ 10 & 10 \end{bmatrix}$. **Find the admittance matrix of the same network.**

*Solution.* $Y = Z^{-1} = \begin{bmatrix} 10 & 10 \\ 10 & 10 \end{bmatrix}^{-1}$. However, $\det(Z) = 0$. Therefore, this circuit does not have admittance representation.

**Example 12.20** Consider a network with an admittance matrix $Y =$
$\begin{bmatrix} 10 & -1 \\ -1 & 5 \end{bmatrix}$. **Find the impedance matrix of the same network.**
*Solution*

$$Z = Y^{-1} = \begin{bmatrix} 10 & -1 \\ -1 & 5 \end{bmatrix}^{-1} = \begin{bmatrix} 0.102 & 0.0204 \\ 0.0204 & 0.2041 \end{bmatrix}$$

**Example 12.21** Consider a network with an impedance matrix $Z =$
$\begin{bmatrix} 0.1 + j0.3 & -j0.2 \\ j0.4 & 0.5 + j0.1 \end{bmatrix}$. **Find the admittance matrix of the same network.**
*Solution*

$$Y = Z^{-1} = \begin{bmatrix} 0.1 + j0.3 & -j0.2 \\ j0.4 & 0.5 + j0.1 \end{bmatrix}^{-1}$$
$$= \begin{bmatrix} -0.47 - j2.94 & 1.09 - j0.411 \\ -2.19 + j0.82 & 1.43 - j1.16 \end{bmatrix}.$$

**Note 12.5** A $\Pi$ network shown in Fig. 12.33 has a $Y$ matrix obtained as follows:

$$Y = \begin{bmatrix} \sum Y_{1j} & -Y_{12} \\ -Y_{21} & \sum Y_{2j} \end{bmatrix} = \begin{bmatrix} Y_a + Y_b & -Y_a \\ -Y_a & Y_a + Y_c \end{bmatrix}$$

**Note 12.6** Adding an impedance element $z$ ($\Omega$) connected between the input port and the output port, as shown in Fig. 12.34; a new admittance matrix can be found as follows:

**Fig. 12.33** A $\Pi$ network $Y_a$, $Y_b$, and $Y_c$

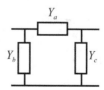

**Fig. 12.34** An admittance
network bypassed by an
impedance Z

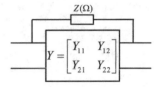

The bypass impedance has an admittance matrix parallel to the given network
admittance. Therefore, the new admittance matrix becomes

$$Y_{new} = \begin{bmatrix} Y_{11} & Y_{12} \\ Y_{21} & Y_{11} \end{bmatrix} + \begin{bmatrix} \dfrac{1}{Z} & 0 \\ 0 & \dfrac{1}{Z} \end{bmatrix} = \begin{bmatrix} Y_{11} + \dfrac{1}{Z} & Y_{12} - \dfrac{1}{Z} \\ Y_{21} - \dfrac{1}{Z} & Y_{11} + \dfrac{1}{Z} \end{bmatrix}$$

**Example 12.22** **The admittance matrix of a two-port network is given**
**as** $Y = \begin{bmatrix} 0.5 & -0.5 \\ -0.5 & 0.5 \end{bmatrix}$. **The network is augmented by an inductor**
**of** $L = 2$ **mH bypassing the input port to the output port, as shown**
**in Fig. 12.35. Find the new admittance matrix.**

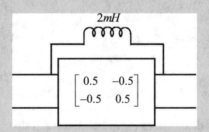

**Fig. 12.35**  Circuit of Example 12.22

*Solution.* The new admittance matrix becomes

$$Y_{new} = \begin{bmatrix} Y_{11} & Y_{12} \\ Y_{21} & Y_{11} \end{bmatrix} + \begin{bmatrix} \dfrac{1}{Z} & 0 \\ 0 & \dfrac{1}{Z} \end{bmatrix} = \begin{bmatrix} 0.5 & -0.5 \\ -0.5 & 0.5 \end{bmatrix} + \begin{bmatrix} \dfrac{1}{sL} & -\dfrac{1}{sL} \\ \dfrac{1}{sL} & \dfrac{1}{sL} \end{bmatrix}$$

$$= \begin{bmatrix} 0.5 + \dfrac{500}{s} & -0.5 - \dfrac{500}{s} \\ -0.5 - \dfrac{500}{s} & 0.5 + \dfrac{500}{s} \end{bmatrix}$$

**Example 12.23** The impedance matrix of a two-port network is $Z = \begin{bmatrix} 1 & j0.2 \\ j0.5 & 1+j0.5 \end{bmatrix}$. This network is augmented by a bypass resistor of $R = 2\ \Omega$ from the input to the output, as shown in Fig. 12.34. Find the new impedance matrix (Fig. 12.36).

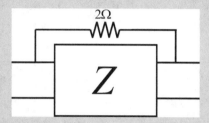

**Fig. 12.36** Circuit of Example 12.23

*Solution.* Since the augmented element is a bypass from input to output, there is a need to find the admittance matrix, as the added element can be easily integrated into the admittance matrix.

$$Y = Z^{-1} = \begin{bmatrix} 1 & j0.2 \\ j0.5 & 1+j0.5 \end{bmatrix}^{-1} = \begin{bmatrix} 0.92 + j0.03 & -0.06 - j0.15 \\ -0.17 - j0.37 & 0.75 - j0.34 \end{bmatrix}$$

Considering the effect of the added resistor in the $Y$ matrix,

$$Y_{\text{new}} = \begin{bmatrix} Y_{11} & Y_{12} \\ Y_{21} & Y_{11} \end{bmatrix} + \begin{bmatrix} \dfrac{1}{z} & 0 \\ 0 & \dfrac{1}{z} \end{bmatrix}$$

$$= \begin{bmatrix} 0.92 + j0.03 & -0.06 - j0.15 \\ -0.17 - j0.37 & 0.75 - j0.34 \end{bmatrix} + \begin{bmatrix} \dfrac{1}{2} & -\dfrac{1}{2} \\ -\dfrac{1}{2} & \dfrac{1}{2} \end{bmatrix}$$

$$= \begin{bmatrix} 1.42 + j0.03 & -0.56 - j0.15 \\ -0.67 - j0.37 & 1.25 - j0.34 \end{bmatrix}$$

To find the new impedance matrix,

(continued)

**Example 12.23** (continued)

$$Z_{\text{new}} = \begin{bmatrix} 1.42 + j0.03 & -0.06 - j0.15 \\ -0.17 - j0.37 & 1.25 - j0.34 \end{bmatrix}^{-1}$$

$$= \begin{bmatrix} 0.68 - j0.001 & 0.01 + j0.08 \\ 0.03 + j0.21 & 0.71 + j0.21 \end{bmatrix}$$

**Example 12.24 Find the $Y$ for the circuit of Fig. 12.37.**
*Solution.* Consider the elements of $L_1$, $L_2$, and $R_1$ forming a $T$ connection. Therefore, the impedance matrix can be written as

$$Z_T = \begin{bmatrix} sL_1 + R_1 & R_1 \\ R_1 & sL_2 + R_1 \end{bmatrix}$$

Now the element $R_2$ is considered a bypass from the input to the output. Writing the $Y$ matrix, this element can be integrated as follows:

$$Y = \frac{1}{s((L_1 + L_2)R_1 + L_1L_2s)} \begin{bmatrix} sL_2 + R_1 & -R_1 \\ -R_1 & sL_1 + R_1 \end{bmatrix}$$

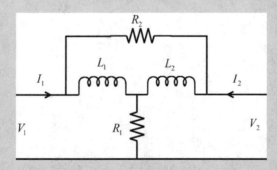

**Fig. 12.37** Circuit of Example 12.24

(continued)

**Example 12.24** (continued)

Adding the bypass element, $R_2$ results in

$$Y = \frac{1}{s((L_1 + L_2)R_1 + L_1 L_2 s)} \begin{bmatrix} sL_2 + R_1 & -R_1 \\ -R_1 & sL_1 + R_1 \end{bmatrix} + \begin{bmatrix} \dfrac{1}{R_2} & -\dfrac{1}{R_2} \\ -\dfrac{1}{R_2} & \dfrac{1}{R_2} \end{bmatrix}$$

$$Y = \begin{bmatrix} \dfrac{sL_2 + R_1}{s((L_1 + L_2)R_1 + L_1 L_2 s)} + \dfrac{1}{R_2} & \dfrac{-R_1}{s((L_1 + L_2)R_1 + L_1 L_2 s)} - \dfrac{1}{R_2} \\ \dfrac{-R_1}{s((L_1 + L_2)R_1 + L_1 L_2 s)} - \dfrac{1}{R_2} & \dfrac{sL_1 + R_1}{s((L_1 + L_2)R_1 + L_1 L_2 s)} + \dfrac{1}{R_2} \end{bmatrix}$$

**Note 12.7** Adding a series element $Z$ ($\Omega$) to the input port of a two-port network adds the same impedance to the $z_{11}$ of the impedance matrix, as shown in Fig. 12.38.

$$\mathbf{Z_{new}} = \begin{bmatrix} z_{11} + Z & z_{12} \\ z_{21} & z_{22} \end{bmatrix}$$

**Note 12.8** Adding a series element $Z$ ($\Omega$) to the output port of a two-port network adds the same impedance to the $z_{22}$ of the impedance matrix, as shown in Fig. 12.39.

$$\mathbf{Z_{new}} = \begin{bmatrix} z_{11} & z_{12} \\ z_{21} & z_{22} + Z \end{bmatrix}$$

**Note 12.9** Adding a parallel element $Z$ ($\Omega$) to the input port of a two-port network adds the same admittance $\frac{1}{Z}$ to the $Y_{11}$ of the admittance matrix, as shown in Fig. 12.40.

**Fig. 12.38** An impedance $Z$ is added to the input of an impedance network

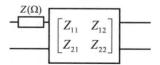

**Fig. 12.39** An impedance $Z$ added to the output of an impedance network

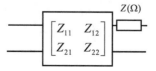

**Fig. 12.40** Impedance $Z$ is
added parallel to the
admittance network

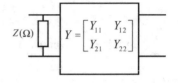

**Fig. 12.41** Impedance $Z$ is
added in parallel to the
output of an admittance
network

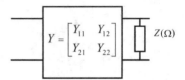

$$Y_{new} = \begin{bmatrix} Y_{11} + \dfrac{1}{Z} & Y_{12} \\ Y_{21} & Y_{22} \end{bmatrix}$$

**Note 12.10**   Adding a parallel element $Z\,(\Omega)$ to the output port of a two-port network adds the same admittance $\frac{1}{Z}$ to the $Y_{22}$ of the admittance matrix, as shown in Fig. 12.41.

$$Y_{new} = \begin{bmatrix} Y_{11} & Y_{12} \\ Y_{21} & Y_{22} + \dfrac{1}{Z} \end{bmatrix}$$

**Example 12.25   The admittance matrix of a circuit is given as $Y = \begin{bmatrix} 2s+4 & -1 \\ -1 & s+2 \end{bmatrix}$. Sketch the circuit.**

*Solution*

• The admittance matrix can be presented as a $\Pi$ network.

  The elements of the network can be found as follows:

• Since the admittance matrix is reciprocal, the shared element is $Y_{12} = Y_{21} = 1\ \Omega^{-1}$.

• Separating the shared element, the rest of the system reads

$$Y = \begin{bmatrix} 2s+3+1 & -1 \\ -1 & s+1+1 \end{bmatrix}$$

(continued)

**Example 12.25** (continued)

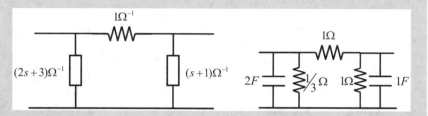

**Fig. 12.42** The circuit found for the admittance matrix of Example 12.25

- Form $2s + 3$ in the input node to the ground.
  - Admittance $2s + 3$ is a parallel of a 2 F capacitor and a $\frac{1}{3}$ Ω resistor.
- Form $s + 1$ in the output node to the ground.
  - Admittance $s + 1$ parallels a 1 F capacitor and a 1 Ω resistor.

  The circuit is shown in Fig. 12.42.

**Example 12.26 The impedance of a two-port network is given as** $Z = \begin{bmatrix} s + 10 & 15 \\ 25 & \dfrac{1}{s} + 10 \end{bmatrix}$. **Sketch the circuit.**

*Solution*

- The impedance matrix is better coordinated with a $T$ network.

  The elements on the $T$ network can be found as follows:

- Since the network is not reciprocal, meaning $Z_{12} \neq Z_{21}$, an arbitrary-shared element must be found in all impedance matrix elements. One option for the shared element of the $T$ network can be the 10 Ω resistor as follows:

$$Z = \begin{bmatrix} s + 10 & 5 + 10 \\ 15 + 10 & \dfrac{1}{s} + 10 \end{bmatrix}$$

(continued)

**Example 12.26** (continued)

- Besides the element $s$ (taken from $Z_{11}$ element), there is a dependent voltage source in loop 1, with the value of $5I_2$ (taken from $Z_{12}$ element).
- Besides the element $\frac{1}{s}$ (taken from $Z_{22}$ element), there is a dependent voltage source in loop 2, with the value of $15I_1$ (taken from $Z_{21}$ element).

    The circuit is shown in Fig. 12.43.

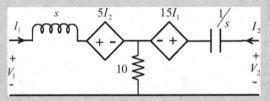

**Fig. 12.43** The circuit found for the impedance matrix of Example 12.26

**Example 12.27 The impedance of a two-port network is given as** $Z = \begin{bmatrix} s+1 & -2 \\ 3 & \frac{1}{s}+1 \end{bmatrix}$**. Sketch the circuit.**

*Solution*

- The impedance matrix is better coordinated with a $T$ network.

    The elements on the $T$ network can be found as follows:

- Since the network is not reciprocal, meaning $Z_{12} \neq Z_{21}$, an arbitrary-shared element must be found in all impedance matrix elements. One option for the shared element of the $T$ network can be the $1\Omega$ resistor as follows:

$$Z = \begin{bmatrix} s+1 & -3+1 \\ 2+1 & \frac{1}{s}+1 \end{bmatrix}$$

- Besides the element $s$ (taken from $Z_{11}$ element), there is a dependent voltage source in loop 1, with the value of $-3I_2$ (taken from $Z_{12}$ element).
- Besides the element $\frac{1}{s}$ (taken from $Z_{22}$ element), there is a dependent voltage source in loop 2, with the value of $2I_1$ (taken from $Z_{21}$ element).

(continued)

**Example 12.27** (continued)

The circuit is shown in Fig. 12.44.

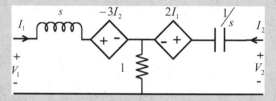

**Fig. 12.44** The circuit was found for the impedance matrix of Example 12.27

**Example 12.28** The impedance of a two-port network is given as $Z = \begin{bmatrix} s+5 & 1 \\ 4 & s+\dfrac{1}{s}+2 \end{bmatrix}$. **Sketch the circuit.**

*Solution*

- The impedance matrix is better coordinated with a $T$ network.

  The elements on the $T$ network can be found as follows:

- Since the network is not reciprocal, meaning $Z_{12} \neq Z_{21}$, an arbitrary-shared element must be found in all impedance matrix elements. One option for the shared element of the $T$ network can be the $2\Omega$ resistor as follows:

$$Z = \begin{bmatrix} s+3+2 & -1+2 \\ 2+2 & s+\dfrac{1}{s}+2 \end{bmatrix}$$

- Besides the element $s + 3$ (taken from $Z_{11}$ element), there is a dependent voltage source in loop 1, with the value of $-I_2$ (taken from $Z_{12}$ element).
- Besides the element $s + \frac{1}{s}$ (taken from $Z_{22}$ element), there is a dependent voltage source in loop 2, with the value of $2I_1$ (taken from $Z_{21}$ element).
  The circuit is shown in Fig. 12.45.

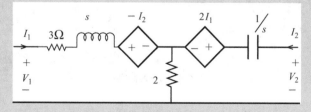

**Fig. 12.45** The circuit found for the impedance matrix of Example 12.28

## Nonreciprocal Admittance Matrix

**Example 12.29** The admittance matrix of a two-port network is given as $Y = \begin{bmatrix} 10 & -15 \\ -2 & 7 \end{bmatrix}$. **Sketch the circuit.**

*Solution.* Compare the admittance to what was introduced earlier:

$$Y = \begin{bmatrix} Y_1 + Y_m & -\alpha - Y_m \\ -\beta - Y_m & Y_2 + Y_m \end{bmatrix}$$

- Since the elements of $Y_{12} \neq Y_{21}$, the system is nonreciprocal. The equivalent $\pi$ circuit contains dependent sources.
- Consider a value for the shared element $Y_m$, for instance, $Y_m = +1$.
- The admittance matrix becomes.

$$Y = \begin{bmatrix} 9+1 & -14-1 \\ -1-1 & 6+1 \end{bmatrix}$$

- From element $Y_{11}$, there is a shared admittance $1\,\Omega^{-1}$ connection between nodes 1 and 2, and there is $Y_1 = 9\,\Omega^{-1}$ admittance from node 1 to the common node.
- From element $Y_{12}$, there is the same shared element $1\,\Omega^{-1}$ and the value of the voltage-controlled current source that feeds node 1 by a value of $\alpha = 14V_2$.
- From element $Y_{21}$, there is the same shared element $1\,\Omega^{-1}$ and the value of the voltage-controlled current source that feeds node 1 by a value of $\beta = +1V_1$.
- From element $Y_{22}$, there is the shared admittance $1\,\Omega^{-1}$ connection between nodes 2 and 1, and there is $Y_2 = 6\,\Omega^{-1}$ admittance from node 2 to the common node.

The circuit is shown in Fig. 12.46.

**Fig. 12.46** A circuit was found for the admittance matrix of Example 12.29

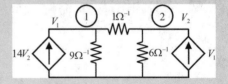

**Example 12.30** The admittance matrix of a circuit is given as $Y = \begin{bmatrix} 2s+4 & -2 \\ -3 & s+2 \end{bmatrix}$. **Sketch the circuit.**

*Solution*

- Since the elements of $Y_{12} \neq Y_{21}$, the system is nonreciprocal. The equivalent $\pi$ circuit contains dependent sources.
- Consider a value for the shared element $Y_m$, for instance, $Y_m = +1$
- The admittance matrix becomes.

$$Y = \begin{bmatrix} 2s+3+1 & -1-1 \\ -2-1 & s+1+1 \end{bmatrix}$$

- From element $Y_{11}$, there is a shared admittance $1\ \Omega^{-1}$ connection between nodes 1 and 2, and there is $Y_1 = 2s + 3\ \Omega^{-1}$ admittance from node 1 to the common node. This is a parallel of a $2F$ capacitor and conductance of $3\ \Omega^{-1}$.
- From element $Y_{12}$, there is the same shared element $1\ \Omega^{-1}$ and the value of the voltage-controlled current source that feeds node 1 by a value of $\alpha = +1V_2$.
- From element $Y_{21}$, there is the same shared element $1\ \Omega^{-1}$ and the value of the voltage-controlled current source that feeds node 1 by a value of $\beta = +2V_1$.
- From element $Y_{22}$, there is the shared admittance $1\ \Omega^{-1}$ connection between nodes 2 and 1, and there is $Y_2 = s + 1\ \Omega^{-1}$ admittance from node 2 to the common node. That is a parallel of $1F$ capacitor and conductance of $1\ \Omega^{-1}$ (Fig. 12.47).

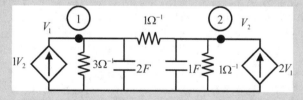

**Fig. 12.47** A circuit was found for the admittance matrix of Example 12.30

**Example 12.31 Find the admittance matrix of the circuit shown in Fig. 12.48.**

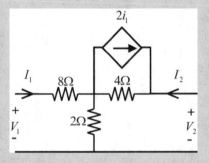

**Fig. 12.48** Circuit of Example 12.31

*Solution.* The circuit has a *T* network and a current source parallel to the 4 Ω resistor. A Norton to Thevenin conversion results in a circuit shown in Fig. 12.49.

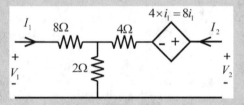

**Fig. 12.49** Source conversion in the circuit of Fig. 12.48

The circuit has a nonreciprocal $Z$ matrix. The shared element is a 2 Ω resistor, and in the first loop, there is an 8 Ω resistor and no current-dependent voltage source. Therefore, $Z_{11}$ is 2 + 8. However, since there is a 4 Ω resistor in the second loop, $Z_{22}$ becomes 2 + 4. The value of the $8i_1$ voltage source adds to the $Z_{21}$ element, and it becomes 2 + 8. Therefore, the impedance matrix is

$$Z = \begin{bmatrix} 2+8 & 2 \\ 2+8 & 2+4 \end{bmatrix} = \begin{bmatrix} 10 & 2 \\ 10 & 6 \end{bmatrix}$$

The admittance matrix becomes

$$Y = Z^{-1} = \begin{bmatrix} 10 & 2 \\ 10 & 6 \end{bmatrix}^{-1}$$

(continued)

**Example 12.31** (continued)

$$= \frac{1}{10 \times 6 - 10 \times 2} \begin{bmatrix} 6 & -2 \\ -10 & 10 \end{bmatrix} = \begin{bmatrix} \dfrac{6}{40} & -\dfrac{2}{40} \\ -\dfrac{10}{40} & \dfrac{10}{40} \end{bmatrix}$$

$$Y = \begin{bmatrix} 0.15 & -0.05 \\ -0.25 & 0.25 \end{bmatrix}$$

## Hybrid Parameters

This system representation uses a mix of impedance, voltage gain, current gain, and admittance. The system equations are as follows:

$$H = \begin{bmatrix} h_{11} & h_{12} \\ h_{21} & h_{22} \end{bmatrix}$$

$$\begin{bmatrix} V_1 \\ I_2 \end{bmatrix} = \begin{bmatrix} h_{11} & h_{12} \\ h_{21} & h_{22} \end{bmatrix} \begin{bmatrix} I_1 \\ V_2 \end{bmatrix}$$

Expansion of these equations demonstrates their units as follows:

$$V_1 = h_{11}I_1 + h_{12}V_2$$
$$I_2 = h_{21}I_1 + h_{22}V_2$$

As a result,

$$h_{11} = \frac{V_1}{I_1}\Big|_{V_2 = 0}$$

$$h_{12} = \frac{V_1}{V_2}\Big|_{I_1 = 0}$$

$$h_{21} = \frac{I_2}{I_1}\Big|_{V_2 = 0}$$

$$h_{22} = \frac{I_2}{V_2}\Big|_{I_1 = 0}$$

If the $Z^{-1}$ and $Y^{-1}$ matrices exist,

$$h_{11} = \frac{\det(Z)}{z_{22}}, h_{12} = \frac{z_{12}}{z_{22}}$$

$$h_{21} = -\frac{z_{21}}{z_{22}}, h_{22} = \frac{1}{z_{22}}$$

## Inverse Hybrid Parameters

This system representation uses a mix of impedance, voltage gain, current gain, and admittance. The system equations are as follows:

$$G = \begin{bmatrix} g_{11} & g_{12} \\ g_{21} & g_{22} \end{bmatrix}$$

$$\begin{bmatrix} I_1 \\ V_2 \end{bmatrix} = \begin{bmatrix} g_{11} & g_{12} \\ g_{21} & g_{22} \end{bmatrix} \begin{bmatrix} V_1 \\ I_2 \end{bmatrix}$$

Expansion of these equations demonstrates their units as follows:

$$I_1 = g_{11}V_1 + g_{12}I_2$$
$$V_2 = g_{21}V_1 + g_{22}I_2$$

As a result,

$$g_{11} = \frac{I_1}{V_1}\Big|_{I_2=0}$$

$$g_{12} = \frac{I_1}{I_2}\Big|_{V_1=0}$$

$$g_{21} = \frac{V_2}{V_1}\Big|_{I_2=0}$$

$$g_{22} = \frac{V_2}{I_2}\Big|_{V_1=0}$$

This shows the following:
- Element $g_{11}$ is an admittance equal to $y_{11}$.
- Element $g_{12}$ is a current gain.
- Element $g_{21}$ is a voltage gain.
- Element $g_{22}$ is an impedance equal to $z_{22}$.

## Transmission Matrix Parameters

The transmission matrix has some interesting characteristics: Large systems can be split into smaller sections. Once the transmission matrix of these sections is identified, they can be connected back together to define the entire system.

Transmission matrix $T$ relates the voltage and current at a system's entry (or sending end) to its parameters at the receiving end. Accordingly, the inverse of the transmission matrix relates the parameters at the receiving end to the parameters at the sending end. Figure 12.50 shows the connection of the system and its transmission parameters.

The equations for a system presented by its transmission matrix are as follows:

$$\begin{bmatrix} V_1 \\ I_1 \end{bmatrix} = \begin{bmatrix} A & B \\ C & D \end{bmatrix} \begin{bmatrix} V_2 \\ -I_2 \end{bmatrix}$$

The expanded system is obtained as

$$V_1 = AV_2 - BI_2$$

$$I_1 = CV_2 - DI_2$$

The components of the transition matrix can be found by either opening or shorting the receiving end and solving for the relations of the voltages and currents from the sending end to the receiving end.

$$A = \frac{V_1}{V_2}\Big|_{I_2=0}$$

$$B = -\frac{V_1}{I_2}\Big|_{V_2=0}$$

$$C = \frac{I_1}{V_2}\Big|_{I_2=0}$$

$$D = -\frac{I_1}{I_2}\Big|_{V_2=0}$$

**Note 12.11** The transmission matrix in *a reciprocal network* has a unique characteristic: its determinant is always 1. This means

$$AD - BC = 1$$

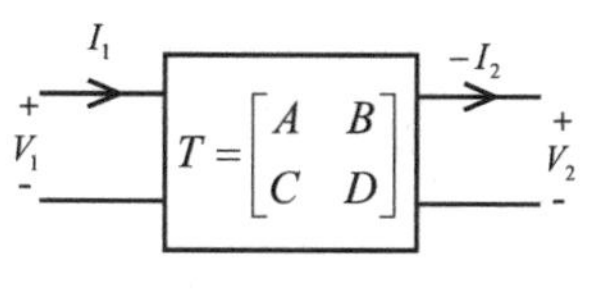

**Fig. 12.50** Transmission matrix

## Presenting the Transmission Matrix Parameters in Terms of Impedance and Admittance Matrices

Parameters of the transmission matrix can be presented using the admittance and impedance matrices, provided that both $Z$ and $Y$ matrices exist. Otherwise, the transmission matrix cannot be defined and does not exist. The transmission matrix elements must be converted to the $Z$ and $Y$ parameters to obtain the needed impedance and admittance parameters.

**Note 12.12** In redefining the transmission matrix parameters in terms of impedance or admittance, the condition that the transmission matrix parameters defined must be strictly enforced. For instance, $A$ is defined where the current $I_2 = 0$. Therefore, its expansion can be found as follows:

$$A = \frac{V_1}{V_2}\Big|_{I_2=0} = \frac{V_1}{I_1}\Big|_{I_2=0}\frac{I_1}{V_2}\Big|_{I_2=0}$$

$$A = \frac{z_{11}}{z_{21}} \quad \text{or} \quad A = -\frac{y_{22}}{y_{21}}$$

Accordingly,

$$B = -\frac{V_1}{I_2}\Big|_{V_2=0} = -\frac{V_1}{I_1}\Big|_{V_2=0}\frac{I_1}{I_2}\Big|_{V_2=0}$$

$$B = \frac{1}{y_{11}}\frac{z_{22}}{z_{21}}$$

$$C = \frac{I_1}{V_2}\Big|_{I_2=0}$$

$$C = \frac{1}{z_{21}} \quad \text{or} \quad C = \frac{1}{z_{11}}\frac{-y_{22}}{y_{21}}$$

$$D = -\frac{I_1}{I_2}\Big|_{V_2=0}$$

$$D = \frac{z_{22}}{z_{21}} \quad \text{or} \quad D = \frac{y_{11}}{y_{21}}$$

**Note 12.13** This approach is extremely useful when there are dependent sources in the circuit.

**Example 12.32 Find the transmission matrix of the circuit shown in Fig. 12.51.**

*Solution.* The circuit forms a $T$ matrix with a dependent source. The impedance matrix can be easily obtained as follows:

The $T$ structure without the dependent source results in

$$Z = \begin{bmatrix} 15 + 10 & 15 \\ 15 & 15 + 0 \end{bmatrix} = \begin{bmatrix} 25 & 15 \\ 15 & 15 \end{bmatrix}$$

The effect of the dependent source is on the off-diagonal element in the second loop as follows:

$$Z = \begin{bmatrix} 25 & 15 \\ 15 - 5 & 15 \end{bmatrix} = \begin{bmatrix} 25 & 15 \\ 10 & 15 \end{bmatrix}$$

The admittance matrix is obtained by

$$Y = Z^{-1}$$

$$Y = \begin{bmatrix} 0.0677 & -0.0677 \\ -0.0444 & 0.1111 \end{bmatrix}$$

The transmission matrix parameters become

$$A = \frac{z_{11}}{z_{21}} = \frac{25}{15} = 1.667$$

$$B = \frac{1}{y_{11}} \frac{z_{22}}{z_{21}} = \frac{1}{0.0677} \frac{15}{10} = 22.156$$

$$C = \frac{1}{z_{21}} = \frac{1}{10} = 0.1$$

$$D = \frac{z_{22}}{z_{21}} = \frac{15}{10} = 1.5$$

**Fig. 12.51** Circuit of Example 12.32

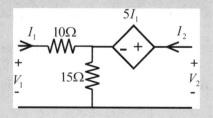

**Fig. 12.52** Impedance in parallel

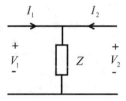

## Parallel Connection of an Element

Consider a circuit as shown in Fig. 12.52.

The transmission matrix of the network when the parallel element has an impedance of $z$ ($\Omega$) can be obtained as follows:

$$T = \begin{bmatrix} 1 & 0 \\ \dfrac{1}{z} & 1 \end{bmatrix}$$

**Example 12.33 Find the transmission matrix of the circuit shown in Fig. 12.53.**

*Solution.* The capacitor $C$ (F) has the impedance of $z = \frac{1}{sC}$ in Laplace. Therefore, the transmission impedance is

$$T = \begin{bmatrix} 1 & 0 \\ \dfrac{1}{\dfrac{1}{sC}} & 1 \end{bmatrix} = \begin{bmatrix} 1 & 0 \\ sC & 1 \end{bmatrix}$$

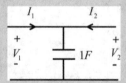

**Fig. 12.53** Circuit of Example 12.33

## Series Connection of an Element

Consider a circuit as shown in Fig. 12.54.

**Fig. 12.54** Impedance in series

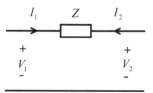

The transmission matrix of the network when the series element has an impedance of $z$ ($\Omega$) can be obtained as follows:

$$T = \begin{bmatrix} 1 & z \\ 0 & 1 \end{bmatrix}$$

**Example 12.34 Find the transmission matrix of the circuit in Fig. 12.55.**
*Solution.* The inductor $L$ ($H$) has the impedance of $z = sL$ in Laplace. Therefore, the transmission impedance is

$$T = \begin{bmatrix} 1 & z \\ 0 & 1 \end{bmatrix} = \begin{bmatrix} 1 & sL \\ 0 & 1 \end{bmatrix}$$

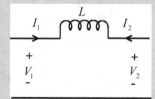

**Fig. 12.55** Circuit of Example 12.34

# Transmission Matrix of Cascade Systems

Consider the cascade connection of several systems with transmission matrices of $T_1, T_2, T_3, \ldots T_n$, as shown in Fig. 12.56.

**Fig. 12.56** Cascade connection of transmission matrices

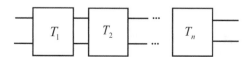

The transmission matrix of the entire system is obtained as the product of the transmission of the subsystems in cascade connection. Therefore,

$$T = \prod_{i=1}^{n} T_i$$

**Example 12.35 Find the transmission of the circuit shown in Fig. 12.57.**
*Solution.* The circuit can be broken into several parallel and series elements connected in a cascade. In particular,

- $R_1$ is a series element with a transmission matrix $\begin{bmatrix} 1 & R_1 \\ 0 & 1 \end{bmatrix}$.

- $L$ is a parallel element with a transmission matrix $\begin{bmatrix} 1 & 0 \\ \frac{1}{sL} & 1 \end{bmatrix}$.

- $C$ is a series element with a transmission matrix $\begin{bmatrix} 1 & \frac{1}{sC} \\ 0 & 1 \end{bmatrix}$.

- $R_2$ is a parallel element with a transmission matrix $\begin{bmatrix} 1 & 0 \\ \frac{1}{R_2} & 1 \end{bmatrix}$.

Therefore, the transition matrix of the network is the product of the transmission matrices as follows:

$$T = \begin{bmatrix} 1 & R_1 \\ 0 & 1 \end{bmatrix} \begin{bmatrix} 1 & 0 \\ \frac{1}{sL} & 1 \end{bmatrix} \begin{bmatrix} 1 & \frac{1}{sC} \\ 0 & 1 \end{bmatrix} \begin{bmatrix} 1 & 0 \\ \frac{1}{R_2} & 1 \end{bmatrix}$$

$$T = \begin{bmatrix} \dfrac{(LC(R_1 + R_2))s^2 + (L + R_1 R_2 C)s + R_1}{R_2 LCs^2} & \dfrac{R_1 CLs^2 + Ls + R_1}{LCs^2} \\ \dfrac{R_1 LC + Ls + R_1}{R_2 LCs^2} & \dfrac{LCs^2 + 1}{LCs^2} \end{bmatrix}$$

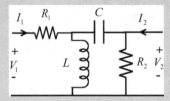

**Fig. 12.57** Circuit of Example 12.35

**Example 12.36 Find the input impedance of the circuit shown in Fig. 12.58.**

*Solution.* The transmission matrix is used to find the circuit's input impedance.

$$z_{in} = \frac{V_1}{I_1} = \frac{AV_2 - BI_2}{CV_2 - DI_2}$$

Considering the relation of the voltage and current at the output or receiving end as

$$V_2 = -R_L I_2$$

The replacement in the input impedance results in

$$z_{in} = \frac{A(-R_L I_2) - BI_2}{C(-R_L I_2) - DI_2}$$

Simplifying the expression yields

$$z_{in} = \frac{AR_L + B}{CR_L + D}$$

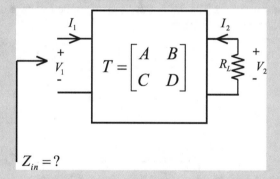

**Fig. 12.58** Circuit of Example 12.36

## Finding Thevenin Equivalent Circuit from Transmission Matrix

Consider the circuit shown in Fig. 12.59.

With known input or sending end values of $V_1$ and $I_1$, the circuit equations are obtained as

$$V_1 = AV_2 - BI_2$$
$$I_1 = CV_2 - DI_2$$

The load must be disconnected to obtain the Thevenin voltage $V_{th}$, which imposes the current $I_2 = 0$. This results in

$$V_1 = AV_2$$

Therefore,

$$V_{th} = V_2 = \frac{V_1}{A}$$

To obtain the Thevenin impedance, all independent sources in the circuit that includes the input voltage must be zero. Applying a voltage at the output indicates how much current is drawn. The ratio of the known applied voltage over the current determines the Thevenin impedance.

$$z_{th} = \frac{(V_2 = 1)}{I_2}\Big|_{V_1 = 0}$$

$$V_1 = AV_2 - BI_2 \rightarrow 0 = A - BI_2$$

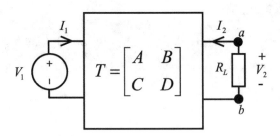

**Fig. 12.59** A transmission matrix and its load connection

**Fig. 12.60** Thevenin
equivalent of the
transmission matrix

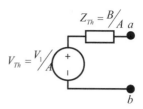

$$z_{\text{th}} = \frac{1}{I_2} = \frac{B}{A}$$

Therefore, the Thevenin equivalent circuit is shown in Fig. 12.60.

**Example 12.37** A system has a transmission matrix of $T = \begin{bmatrix} 2 & 1+j \\ 1 & 1+0.5j \end{bmatrix}$.
**Find the Thevenin equivalent when the input voltage is 50 V. At what load
impedance the maximum power is transferred to the load?.**
*Solution.* The Thevenin impedance and voltage are obtained as follows:

$$V_{\text{th}} = \frac{V_1}{A} = \frac{50}{2}$$

$$z_{\text{th}} = \frac{B}{A} = \frac{1+j}{2}\,\Omega$$

The load impedance must be the complex conjugate of the Thevenin
impedance to transfer the maximum power. Therefore,

$$z_L = \frac{1-j}{2}\,\Omega$$

## Problems

12.1. Find the impedance parameters of the following circuit.

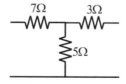

**12.2.** Find the impedance parameters of the following circuit.

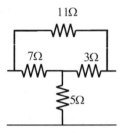

**12.3.** Find the impedance parameters of the following circuit.

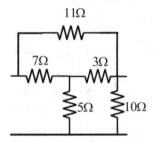

**12.4.** Find the impedance parameters of the following circuit.

**12.5.** Find the impedance parameters of the following circuit.

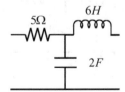

**12.6.** Find the impedance parameters of the following circuit.

12.7. Find a two-port network that results in the following Z parameters.

$$Z = \begin{bmatrix} \dfrac{3s+1}{s} & \dfrac{1}{s} \\ \dfrac{1}{s} & \dfrac{5s^2+1}{s} \end{bmatrix}$$

$$Z = \begin{bmatrix} 5s+4 & 4 \\ 4 & 3s+4 \end{bmatrix}$$

12.8. Find admittance parameters in the following circuits.

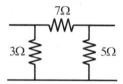

12.9. Find admittance parameters in the following circuits.

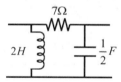

12.10. Find admittance parameters in the following circuits.

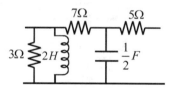

12.11. Find a two-port network that results in the following Y parameters.

$$Y = \begin{bmatrix} \dfrac{2s+1}{s} & -2 \\ -2 & s+2 \end{bmatrix}$$

$$Y = \begin{bmatrix} \dfrac{s^2+1}{s} & -s \\ -s & \dfrac{s^2+1}{s} \end{bmatrix}$$

12.12.  Find $Y$ and $Z$.

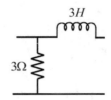

12.13.  Find $Y$ and $Z$.

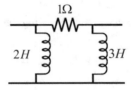

12.14.  Find $Y$ and $Z$.

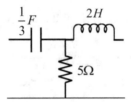

12.15.  Find $Y$ and $Z$.

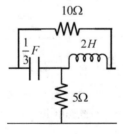

12.16.  Find $Y$ and $Z$.

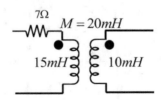

12.17. Find a two-port network that has the following impedance matrix.

$$Z = \begin{bmatrix} 2s + 1 & 5s + 4 \\ 5s + 1 & s + \dfrac{1}{s} \end{bmatrix}$$

12.18. Find a two-port network that has the following admittance matrix.

$$Y = \begin{bmatrix} \dfrac{s+1}{s} & s \\ -2 & 2s + 7 \end{bmatrix}$$

12.19. Find the transmission matrix of the following two-port network.

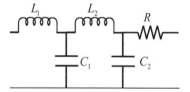

12.20. Find the impedance and admittance matrix of the circuit in the previous problem when $L_1 = 1$ H, $L_2 = 3$ H, $C_1 = \frac{1}{2}$ F, $C_2 = \frac{1}{5}$ F, and $R = 5$ $\Omega$.

# References

Electric Circuits Theory, BY: Charles Desoer, Ernest Kuh
Fundamentals of Electric Circuits, BY: Charles Alexander and Matthew Sadiku
Electric Circuits, BY: James W. Nilsson and Susan Riedel
Electric Circuits, BY: William Hayt

# Index

## A

Active filters, 591–606
  band-pass (*see* Band pass filter (BPF))
  BRF, 599, 600
  cutoff frequency, 591, 597, 598
  definition, 591
  high-pass (*see* High pass filter (HPF))
  low-pass (*see* Low-pass filter (LPF))
  MFB (*see* Multiple feedback opamp
    circuits (MFB))
Active high-pass filter, 594, 595, 597, 599
Active low-pass filter, 591–593, 597, 599, 606
Active power, 272–275, 280, 309
Admittance, 241–243, 258, 269, 301–303
Admittance matrix, 628
  circuit, 649, 651
  classical approach, 630–633
  definitions, 640
  dependent sources, 629
  elements, 628, 648
  equations, 628
  impedance conversion, 642
  impedance values, 634
  input and output port voltage and
    currents, 636
  L=2mH, 644
  multi-node networks
    current and voltage considerations, 638
    KCL equations, 638
    matrix dimension and elements, 638
    matrix size reduction, 639
  non-reciprocal, 652, 653
    element-sharing nodes, 630

reciprocal
    Π model, 629
  resistor, 645
  short circuit, 631, 632
  source conversion, 654
  T connection, 646
  3-loop system and matrix reduction, 640
  T network, 649–651
  transfer functions, 627, 642
  two-port network, 652
  2 ($\Omega$), 645
  Z ($\Omega$), 643, 647, 648
  Π-equivalent circuit, 634
  Π matrix analysis, 633
  Π network, 643
Admittance network, 628, 629, 644, 648
Apparent power, 271–273, 275–277, 282, 309
Average, 148–150, 167

## B

Band pass filter (BPF)
  bandwidth, 512
  frequency response, 512, 513
  LC parallel, 517–520
  LC series, 513–517
  LPF and HPF, 597, 598
  magnitude and phase variation, 513
  quality factor, 516, 519
  RC circuits, 596
  transfer function, 598, 599
  unamplified maximum value, 597
Band reject filter (BRF), 599, 600

Printed in the United States
by Baker & Taylor Publisher Services